CAR

SFL
QC981.8.G56 L47 2005
Leroux, Marcel
Global warming : myth or
reality : the erring ways o
climatology

1/20105

Global Warming – Myth or Reality?

The Erring Ways of Climatology

Marcel Leroux

Global Warming – Myth or Reality?

The Erring Ways of Climatology

Professor Marcel Leroux
Directeur du Laboratoire de Climatologie
Risques Naturels, Environnement
Université Jean Moulin
CNRS-UMR 5600
18 rue Chevreul
69362 Lyon Cedex 07
France
leroux@univ_lyon3.fr
marsleroux@wanadoo.fr

The satellite picture of 25 January, 2005, shows three MPHs, directly arrived from the Arctic and agglutinated over the northern Atlantic and Europe. These MPHs are responsible for a very severe cold wave (and new low-temperature records), frost and snowfall reaching North Africa into the Sahara. The iced cars and trees in Geneva (CH), along the Leman lakeside correspond to this northern winds synoptic situation. Temperature is controlled by the weather dynamics, not by hypothetical equations ...

SPRINGER–PRAXIS BOOKS IN ENVIRONMENTAL SCIENCES
SUBJECT *ADVISORY EDITOR*: John Mason B.Sc., M.Sc., Ph.D.

ISBN 3-540-23909-X Springer-Verlag Berlin Heidelberg New York

Springer is part of Springer-Science + Business Media (springeronline.com)

Bibliographic information published by Die Deutsche Bibliothek

Die Deutsche Bibliothek lists this publication in the Deutsche Nationalbibliografie; detailed bibliographic data are available from the Internet at http://dnb.ddb.de

Library of Congress Control Number: 2005923727

Apart from any fair dealing for the purposes of research or private study, or criticism or review, as permitted under the Copyright, Designs and Patents Act 1988, this publication may only be reproduced, stored or transmitted, in any form or by any means, with the prior permission in writing of the publishers, or in the case of reprographic reproduction in accordance with the terms of licences issued by the Copyright Licensing Agency. Enquiries concerning reproduction outside those terms should be sent to the publishers.

© Praxis Publishing Ltd, Chichester, UK, 2005
Printed in Germany

The use of general descriptive names, registered names, trademarks, etc. in this publication does not imply, even in the absence of a specific statement, that such names are exempt from the relevant protective laws and regulations and therefore free for general use.

Cover design: Jim Wilkie
Translator: Bob Mizon, 38 the Vineries, Colehill, Wimborne, Dorset, UK
Project management: Originator Publishing Services, Gt Yarmouth, Norfolk, UK

Printed on acid-free paper

Contents

Preface

I do not claim to be able to reveal all the sources of our errors concerning this limitless subject; when we think that we have found the truth in one place, it escapes us in a thousand others. However, I hope that, by exploring the principal areas of the mind, I may be able to observe essential differences, and dispel a good number of those imaginary contradictions which ignorance fosters.

Luc de Clapiers, Marquis de Vauvenargues,
Introduction à la Connaissance de l'Esprit Humain, Paris, 1746.

The preface is, by definition, not the whole of the text. It can, however, throw light upon it, and give an insight into the author's mind and motivation, and the reasons why he has undertaken to write that text. Why has he thought it necessary right now to intervene in the 'greenhouse effect' or 'global warming' debate?

In the 'climate change' issue of the journal *Energy and Environment* (Vol. 14, 2003), editor Sonya Boehmer-Christiansen referred in her editorial to my article, also entitled *Global Warming: myth or reality?*, as the product of an 'angry man'. It is just that.

Clive Horwood, of Praxis Publishing, commented on seeing the article: '*It is an excellent topic for a book*'. I was tempted ... an article is too short to allow one to really demonstrate everything that one would like to say. I hesitated, for there are already many books on 'climate change', not that host of opuscules which content themselves with the servile rehashing of pages from Intergovernmental Panel on Climate Change (IPCC) reports, but real works of deep and original reflection on aspects of this vast subject. Excellent books have already appeared in French, for example those of Roqueplo (1993), Lenoir (1992, 2001), and Kohler (2002). In English, there are even more: works by Idso (1989), Singer (1990), Michaels (1992), Lomborg (2001), Essex and McKitrick (2002), Kondratyev (2003), Kondratyev *et al.* (2003), and many others ...

I finally decided to go ahead for two main reasons. First, to comment upon the sad state into which climatology has drifted during the last 20 years or so, since its entering into politics, and second to show that climatology is also itself to blame for this drift, for complacently accepting it, and even encouraging it with rash and irresponsible adventures.

Recent happenings in the field of climatology give cause for complaint, as do the approaches of some of its practitioners, especially those who, lacking any real qualification, claim to belong to the climatological community, but give it an erroneous image. It is galling to see the media 'hype' which ensues every time a meeting of the IPCC is announced, every time an extra drop of rain falls here, or fails to fall there, or every time a door slams because the wind is blowing a bit more strongly than is 'normal'. How irksome it is to hear the simplistic slogans, and sometimes barefaced lies, churned out yet again; to have to put up with the *Diktat* of an 'official line' and the parroted pronouncements of the 'climatically correct', numbing all reflection. It becomes ever more difficult to stomach the kind of well-intentioned naïvety or foolishness which, through the medium of tearful reportage, tugs at our heartstrings with tales of doomed polar bears, or islanders waiting for the water to lap around their ankles ...

Hardly a week goes by without some new 'scoop' of this nature filling our screens and the pages of our newspapers. 'Global warming', caused by the 'greenhouse effect', is our fault, just like everything else, and the message/slogan/misinformation becomes ever more simplistic, ever cruder! It could not be simpler: if the rain falls or drought strikes; if the wind blows a gale or there is none at all; whether it's heat or hard frost; it's all 'because of the greenhouse effect', and we are to blame! An easy argument – but stupid! The Fourth Report of the IPCC might just as well decree the suppression of all climatology textbooks, and replace them in our schools and universities with its press communiqués!

Is climatology itself without fault, given all the current 'climatomania'? What risks (especially to its credibility) is it running by letting itself be taken over, distorted, tarnished, and manipulated by political bodies whose cause is not that of climatology? Most importantly, has climatology the right, is it in a position, to lord it over the world of today and tomorrow? Is it everywhere irreproachable, worthy of trust? Has it reached the zenith of its maturity, built on solid foundations? What is its basic scientific consensus, or is there one? What profit, or loss (more surely!), will it make from all the 'hype' occurring in its name, or in fact without it, since all is already decided, coming from a direction which is neither scientific nor open to scrutiny?

Another aim of this book is to denounce the '*erring ways of climatology*', where necessary and without fear or favour, in the knowledge that everyone is responsible for their writings, and has consequently to uphold them. The fact that one disaster movie follows another is made possible by scientists (some self-styled) who provide the scenarios upon which they are based. The Hollywood treatment is merely a digest, a striking miniature version, of the 'scientific' literature, which sometimes leaves us wondering whether it is meant to be science, or science fiction. This book also tries to defend climatology against its own failings, or, as some might say, its smug self-satisfaction.

For whom has this book been written? Certainly not for those delighting in superficial certainties, nor those so deeply involved in, or compromised by, the IPCC's scenario, that the greenhouse effect now pays their bills! They will undoubtedly not take kindly to this book, but they should remember that arguments based on science must be – honestly – countered by similarly scientific arguments. Also, this book will be of interest to those who would like to hear something other than what the 'censors' allow, and want objective information, and/or those who are tired of the ever greater liberties taken in this field by newspapers, magazines, some scientific journals, and broadcasters. For 20 years now, the media have been repeating the same stock phrases, clichés, and nonsense, inflating the story to caricature (as does the IPCC) as time goes on. They know of no other way of fanning our curiosity, as they have never been able to furnish the slightest tangible, truly visible, proof of what they are claiming.

The subject is presented here in as deliberately simple a form as possible, as best we can (and it is not always easy), in order to make it accessible to the many. This book does not deal in rash suppositions, but only in concrete facts based on direct observation, with simple proofs: satellite images, synoptic charts, and curves from hard data. It will not be easy reading, even so, since it is not possible to take simplification to extremes, and climatology is a complex discipline, always requiring a basic knowledge and a holistic vision of phenomena.

I was going to omit certain facts, but the passionate nature of the debate suggests that they be mentioned. Doubly a doctor, from University and from the state, in Climatology, I am a member of the Société Météorologique de France and of the American Meteorological Society. As a Professor of Climatology, my employer is the French Republic, which has adopted the official religion of 'climate change', to which I do not adhere. I am not beholden to any 'slush fund', and my Laboratoire de Climatologie, Risques, Environnement (LCRE), in spite of its links with the Centre National de la Recherche Scientifique (CNRS), has never received any funding from this state institution, certainly by reason of heresy. I am neither a militant nor an armchair 'eco-warrior', but I live in the countryside, near the little village of Vauvenargues, near Aix-en-Provence, on the 'Grand Site Sainte Victoire' (immortalised by the painter Paul Cézanne), a listed and protected area of mountains and wild forests. I grow vegetables in my (small) 'organic' kitchen garden. I am naturally inclined to question things, and I am basically a Cartesian, living by René Descartes' primary precept of 'never assuming anything to be true which I did not know evidently to be such' (*Discours de la Méthode*, 1637).

A task of this nature, which has its risks (not least that of upsetting people), needs the encouragement and support of others. I have received these, and am very grateful to those who have offered them. I apologise to any whom I have inadvertently omitted.

I thank: J. Comby, my colleague at the LCRE, whose collaboration has been most valuable, as has that of members of the laboratory, L. Barthélémy, and research *attachés* A. Pommier, A. Favre, and also C. Gérard for the abundant documentation. All the doctoral students have offered, and still offer, their scientific contributions to further this project, particularly: E. Barbier, J. Reynaud. O. Dione, J.-B. Ndong, O. Haroun Suleiman, Z. Nouaceur, P. Sagna, L. Amraoui, A. Aubert, D. Soto, I. Kane, and many others. I also wish very sincerely to thank B. Mizon for his excellent translation, and for supplying extracts from the British press.

I owe much to the following for their support, and for exchanges of ideas: the late H. Faure, and C. Toupet, Y. Lenoir, J. Bethemont, J. Demangeot, Y. Lageat, J.P. Vigneau, A. Godard, E. Grenier, P. Kholer, G. Rivière-Wekstein, C. Vernin, G. Kukla, S.F. Singer, K.Y. Kondratyev, S. Boehmer-Christiansen, R. Lindzen, H. Jelbring, T. Bryant, G.D. Sharp, A. Gershunov, A. Gil-Olcina, H. Thieme, T. Hämeranta, and many others. I have found many of the less 'conventional' websites very useful, especially: *Still waiting for greenhouse* by J. Daly (sadly no longer with us); *Techcentralstation*, by W. Soon, *Center for the Study of Carbon Dioxide and Global Change* by C.D. Idso, and *Mytosyfrauds, FAEC* by E. Ferreira ... I have received regular information bulletins, many from The Science & Environmental Policy Project (SEPP): *The week that was (TWTW)*, distributed by S.F. Singer, and from *CCNet (Cambridge-Conference*) by B. Peiser. The online permanent discussion forum, moderated by T. Hämeranta, of the *Climatesceptics* think tank has graciously allowed me to subscribe to its rapidly growing list of members. These bulletins and permanent contacts have meant that I am kept informed of any new articles, and of the latest responses from the climatologists.

I have also realised to what degree the scientific community – or that part of it not unquestioningly obedient to the IPCC – cares, and actively wishes to see climatology regain its currently much faded dignity. This perspective of a climatology free from any political thrall and solely devoted (as it must be) to scientific concerns, has considerably encouraged me in my preparation of this book.

M.L.
Vauvenargues, 10 January 2005

Figures

Abbreviations

AA	Anticyclonic Agglutination
ACIA	Arctic Climate Impact Assessment
AFP	Agence France Presse
AGU	American Geophysical Union
AMS	American Meteorological Society
AO	Arctic Oscillation
CCO	Contemporary Climate Optimum
CERFACS	Centre Européen de Recherche et de Formation en Calcul Scientifique
CFC	chlorofluorocarbon
CGCP	Canadian Global Change Programme
CNRM	Centre National de Recherche Météorologique
CNRS	Centre National de la Recherche Scientifique
COP	Conference of the Parties
CRII-Rad	Independent Regional Commission for Information on Radioactivity
CRU	Climatic Research Unit
DVI	dust veil index
EDF	Electricité de France
EEA	European Environmental Agency
ENSO	El Niño–Southern Oscillation
ESA	European Space Agency
FAO	Food and Agriculture Organisation
FGGE	First Global GARP Experiment
GARP	Global Atmospheric Research Programme
GCM	General Circulation Model
GCR	galactic cosmic ray
GDF	Gaz de France
GHCN	Global Historical Climate Network
GLOSS	Global Sea Level Observing System

HC	Hadley Centre
HCO	Holocene Climatic Optimum
ICSU	International Council for Scientific Union
IDNDR	International Decade for National Disaster Reduction
IGY	International Geophysical Year
IME	inclined meteorological equator
IMO	International Meteorological Organisation
IPCC	Intergovernmental Panel on Climate Change
IVI	ice volcano index
LFO	Low-Frequency Oscillation
LGM	Last Glacial Maximum
LIA	Little Ice Age
LMD	Laboratoire de Météorologie Dynamique
MAB	Man and the Biosphere (Programme)
ME	meteorological equator
MF	Météo-France
MPH	Mobile Polar High
MSU	microwave sounding unit
MWP	Medieval Warm Period
NAD	North Atlantic Dynamic
NAM	Northern Annular Mode
NAO	North Atlantic Oscillation
NGO	non-governmental organisation
NPi	North Pacific index
NPO	North Pacific Oscillation
NTF	National Tidal Facility
OTP	Ozone Trend Panel (NASA)
PARCA	Programme for Arctic Regional Climate Assessment
PDO	Pacific Decadal Oscillation
PNA	Pacific North American
RCO	Recent Climatic Optimum
SAT	surface air temperature
SCS	Soil Conservation Service
SLP	sea level pressure
SMF	Societe Météorologique de France
SO	Southern Oscillation
SST	sea surface temperature
TCR	terrestrial counter-radiation
TD	Trade Discontinuity
TEJ	Tropical Easterly Jet
THP	Tropical High Pressure
TI	Trade Inversion
TWTW	The Week That Was
UNEP	United Nations Environmental Programme
UNESCO	United Nations, Educational, Scientific, Cultural Organisation

UNFCCC	United Nations Framework Convention on Climate Change
UNO	United Nations Organisation
VEI	volcanic explosivity index
VME	vertical meteorological equator
WAIS	West Antarctic Ice Sheet
WCC	World Climate Conference
WCED	World Commission on Environment and Development
WCRP	World Climate Research Programme
WHO	World Health Organization
WMO	World Meteorological Organization

1

Introduction

That's why I'm talking to you. You are one of the rare people who can separate your observation from your preconception. You see what is, where most people see what they expect.

John Steinbeck. *East of Eden*, 1952.

'Climate change' (or the 'greenhouse effect') is very much THE subject of the moment for scientists, the media, politicians and the general public. Among its many more or less transitory 'fashionable' aspects have been: the threat of a new Ice Age; the nuclear winter; acid rain; the Sahel drought; the forests (and especially the Amazon) as the 'lung' of the planet; the hole in the ozone layer, which seems to be closing up again; *El Niño*, ruler of the world ... Some of these subjects are still very much 'alive', and share the spotlight with the greenhouse effect itself. The greenhouse effect, however, is the undisputed star of the show, while we await the possible appearance, sooner or later, of some new fad or passing craze, as we tire of the old, or change our (manifold) interests...

The idea of temperature rising as CO_2 levels increase is an old one, and seems logical enough in most minds. We can understand it and experience it (e.g., at a business meeting or social event in a closed and poorly ventilated space). The engineer Charles Vernin has calculated that 100 people gathered together will produce about 26 litres of CO_2 in one hour, at the rate of 16 respirations per minute (i.e., a total of 2,600 litres); if the space they occupy measures 200 m^2 and is 4 m high, giving a volume of 800 m^3, the concentration of CO_2 at the end of the hour will be of the order of 0.003 25, or 3,250 ppm, in other words about 10 times greater than the 'normal' concentration of CO_2 in the atmosphere. Even if the temperature has risen slightly during the gathering (and not only because of the CO_2), it will not have been necessary to clear the room, since the critical level would be more than 20,000 ppm.

So we do not have to resort to complicated models to tell us that an increase in CO_2 brings about, theoretically, an increase in temperature: a simple rule of thumb, a 'back-of-the-envelope' calculation, will suffice. However, this hypothesis has never been demonstrated as far as climate is concerned, and remains in the realm of the virtual. To hold increasing CO_2 levels responsible for all phenomena, floods, storms, droughts, heatwaves and even freezing spells, without limit or distinction, and especially without reflection, is seriously to undervalue the dynamics of weather and climate, to reduce them to bar-room chatter. 'It must be true' – especially if we proclaim, in heavily scientific tones, that it's all because of the 'greenhouse effect'! There are limits which should not be traversed, even by those trying do outdo each other: the scientists and non-scientists of the Intergovernmental Panel on Climate Change (IPCC); the press and other media; ecologists; politicians; those in government; non-governmental organisations; lobby groups; clubs; associations (too numerous to remember)... whose preoccupations are further and further removed from real climatology.

1.1 HOT AIR...

The idea of 'climate change' has become widely synonymous with that of 'global warming'. This is, paradoxically, construed in a very negative way as inevitably catastrophic, even in countries like Canada, in the grip of prolonged and intense cold spells, of extreme severity. Such countries ought to welcome the idea of a few extra degrees. In Russia, though, the prospect of some warming is welcomed by some, as a recent IPCC meeting in Moscow showed; it could only be of benefit to lands where winters are long and bitter.

Day after day, the same mantra – that '*the Earth is warming up*' – is churned out in all its forms. As 'the ice melts' and 'sea level rises' the Apocalypse looms ever nearer! Without realising it, or perhaps without wishing to, the average citizen is bamboozled, lobotomised, lulled into mindless acceptance. People dread those few extra degrees, and are, even with a time lag, terrorised by the '0.8-degree rise' of the last 140 years. In the winter, though, and especially when the electricity bill drops through the door, we dream of the summer to come ... People even indulge themselves with ultraviolet treatments, and 'bright light therapy', often on the advice of their doctors, and, if they can afford it, off they go to some distant isle where the weather is warm. They rush off in summer to beaches by the Mediterranean or in the Tropics, soaking up every last drop of sunshine. And when they think about retirement, people in western Europe dream of settling down in the South of France, or in Spain, Italy, Greece, or perhaps further south. Americans aim for California, Florida, Arizona ... The risks they run! Sheer recklessness! This conscious or unconscious fear of warming, which is constantly fuelled, is especially incomprehensible to this climatologist, who has lived for 40 years in Africa, 22 of them in tropical Africa! Is shivering with cold really such a marvellous thing? Does warmth really have to bring catastrophe?

1.2 CONFUSION REIGNS...

There is much confusion about this subject. It encompasses all sorts of things:

- *Pollution and climate.* Climate has become an excuse and a bugbear. Its future behaviour seems set in stone, and anybody casting doubts on the predicted warming is thought to be tolerant of pollution, or 'branded as mad, bad, or in the pay of the oil industry' (Singer, 2001) – as if anybody would be in favour of foul and unbreathable city air! Of course, it is absolutely obvious that anyone 'profiting' – in the widest sense, directly or indirectly – from this climatic 'manna' (source of an enormous amount of business) is above suspicion!
- *Good intentions and (declared and undeclared) interests.* '*The planet is in danger, and must be saved.*' In the words of Bjorn Lomborg, when he listed the fears of the ecologists: '*We are defiling our Earth, the fertile topsoil is disappearing, we are paving over nature, destroying the wilderness, decimating the biosphere, and will end up killing ourselves in the process*' (Lomborg, 2001, p. 4). However, at the same time we are talking about 'permits to pollute', the infamous negotiable emission quotas. There is also a sentimentalist approach, playing upon guilt feelings, since humankind is held responsible for all ills – a self-accusation which was already being heard 30 years ago as the Sahel drought began to worsen. However, one does not have to look far to find corners being fought, and the ambiguous attitudes of those who would defend their own more or less personal interests, interests which are often only thinly disguised. There is much talk of doing good, with generous intent, but is it always tinged with altruism?
- *Prognostications and realities, theories within models and real mechanisms, hypotheses about the climate of the future and the evolution of current weather.* The further into the future the predictions reach, the more gratuitous they tend to be: 2100 seems a fairly safe date, since one doesn't run much risk of being found wanting! Contemporary observations emphasise signs of the coming catastrophe, and there is a careful *triage* of information, with no emphasis on cold, attributed (without any exact reason) to (little understood) natural variability, while heat is 'talked up', of course, since it can only confirm the predictions of the models.
- *Sensationalism and serious science.* The quest for a science 'scoop' interferes more and more with well-founded fact. The simulations of the models perpetuate this, and politicians and the media add to the confusion; some of the scientists themselves do not help, by coming out with hasty and conjectural statements which are, unfortunately, often unauthorised.

1.3 *BIS REPETITA...PLACENT?*

This 'global warming' scenario has something *déjà vu* about it: haven't we already lived through something like it before? Weren't we once taken in by something brazenly similar? Ashamed as we are to admit believing in them for a moment, we can think back to a succession of 'scares': the Ice Age coming back, indisputably; our forests will soon be killed by acid rain; western industrial activities being held to blame for the Sahel drought, and (already) amends must be made! Do you recall the 'ozone panic' and its repercussions, with Concorde and the Space Shuttle implicated in the 'aerosol wars', and the subsequent 'hype'? The imminent threat of cancers and cataracts, whipped up by news items such as this: '*It is said that, near Punta Arenas, at the edge of the Antarctic ozone hole, shepherds, their sheep, and also rabbits, are going blind*'! (*L'Express*, February 1992, echoing the *New York Times*, July 1991, and *Newsweek*, December 1991).

What do we say now about the deluge of campaigns to 'save the ozone layer'? The ozone fad is now behind us: why do we see the same methods used to promote the new bugbears, often with the same names at the helm? Do we recall the apocalyptic predictions of James Hansen in November 1987, made during a particularly cold spell, according to which, '*the global warming predicted in the next 20 years will make the Earth warmer than it has been in the past 100,000 years*'? So, mark well, Hell opens its gates in 2007! So in three years' time, we'll be living in another interglacial period like that of the Eemien. How marvellous! We can say or write anything we like, without commitment, because the main thing is not to inform, but to impress!

Humankind is constantly in the dock, and the accusing finger is pointed, as we pollute too much, cherish our cars (or even covet cars we haven't got), drive too fast, have a house to ourselves, live intensively, and produce too much rubbish ... perhaps we even breathe too deeply! We are always made to feel guilty by more virtuous souls ... Of course people take from nature, and sometimes too readily, and they do not always appreciate it or respect it; it is not always the one who exploits, destroys, and pollutes, who complains, repairs, and, what is more, pays the bill. It is not complicated – humans are responsible for everything, always, commit all manner of sins, and are capable of anything, even of dictating the weather, a power once reserved for the gods! And now we must pay! Does this remind you of something? The heat to come, held up as our 'punishment': it sounds like some vision of Hell. Whatever it might imply, it's just the same old stew, reheated!

1.4 THE GOOD, THE BAD ...

Citizens fall into two categories. The majority are the 'good': often sincere, and occasionally militant, or just trusting, or (most often) following like sheep; others, the 'bad' minority, finally weary of the catastrophist pronouncements, have lost faith in them, or prefer some other road. Non-believers in the greenhouse scenario are in the position of those long ago who doubted the existence of God ... fortunately for them, the Inquisition is no longer with us!

There is a similar split within the climatological community. Most climatologists go along with the IPCC view, in good faith, though there are those who have some self-interest (declared or otherwise). These latter are the most ardent apologists, defenders of the faith, and travelling preachers who sometimes show symptoms of chronic psittacism. There are others, less numerous, who will not go along with 'what everybody knows', and they are branded heretics, troublemakers, obstructionists, and sceptics. They are considered suspect, and may even have sold their souls to some occult power! Rochas (2002), president of the editorial committee of the journal *La Météorologie*, took Lenoir (2001) to task for maintaining '*an opposing stance*' to what he (revealingly) felt was '*the prevailing ideology*'. Ideology ... an echo of the good old days of Lysenkoism, when 'officially approved science' ranked with 'the religion of the state'! Moreover, because of the blurring of the distinction between climate and pollution, those who dare to think differently are seen as enemies of nature, the 'baddies', as opposed to the true and single-minded followers of the climatically correct path! They can only be discredited and rejected, or, as the following 'right-thinking' critique puts it: '*Ranged against the honesty, modesty and prudence of researchers who weigh every word ... are the peremptory declarations of a few isolated characters, blown out of all proportion in press articles ... What is the weight of these opinions compared with the almost total consensus of researchers all over the world? Certainly not great ...*' (Fellous, 2003, pp. 18–19). There is no need to ask which camp the good author sits in!

The term 'consensus' as used above, a term often employed by the IPCC, implies only that the 'good' keepers of the true faith are in the majority, meaning that mere weight of numbers (not necessarily synonymous with better quality) may control and dismiss discordant voices from publications and papers. The debate is either swept under the carpet, or soon loses the required rationality, since the idea of disinterested and objective science, *stricto sensu*, has long since been absent. Opponents are even labelled 'charlatans', without any scientific discussion! This burdensome situation is particularly inherent in France. Like other freethinkers, I have tested it out, being content not to enter the parrots' cage, where the *leitmotiv* is the servile repeated chorus of the IPCC catechism. In 2000, the President of the Centre National de la Recherche Scientifique (CNRS), a chemist and specialist on the subject of ozone, but introduced for the occasion as a leading climatologist, proclaimed: '*The artificial creation of scientific controversies certainly does not help to clarify an already very complex debate. It can only contribute to fuel sterile discussions about scientific uncertainty, and thereby weaken the image of science*'. So, science without debate, without 'sterile' discussions, is the ideal! If only we could go back to burning books in the public square, as they did in *Fahrenheit 451*! This process of the elimination of opponents, which is general at all climatic conferences, has been denounced, notably at an IPCC meeting at the Moscow Academy of Sciences in July 2004. Some British scientists, great proponents of the official doctrine, committed 'intellectual terrorism' by excluding 'climato-sceptics' from the proceedings, even though they were internationally recognised: modeller R. Lindzen, entomologist P. Reiter, oceanographer N.-A. Mörner, and meteorologist R. Khandehar. One of the principal advisors to the Russian government, A. Illarionov, called it 'totalitarian ideology'! And is not the idea of censorship unacceptable in so-called democratic regimes?

1.5 POLLUTION AND THE 'PRECAUTIONARY PRINCIPLE'

Let us make one thing clear from the start: there should be no confusion of pollution and climate. Pollution is a serious and urgent enough subject by itself, and would require dealing with separately, by appropriate specialists. The fight against pollution would certainly be more effective if it were not turned into some hypothetical climatic consequence. So, in this book, we deal not with pollution, but only with climatology.

The greenhouse effect scenario, which blames pollution on the activities of humans and holds them responsible for all catastrophes, may even be beneficial if it helps to limit the said pollution, and in the same spirit, make everyone more respectful of the environment. A worthy outcome, much to be applauded. If we can breathe more easily in our cities, if ailments such as asthma, in particular, were a thing of the past, and nature were better protected, and so on, then there is no harm to anyone. Quite the reverse ...

However, there is no justification in trying to achieve such an end by making a hostage of climatology, by using it as a sounding board for all sorts of pronouncements until it often comes to accept them itself. Is it a justification for distorting reality, and for linking everything to 'global warming': drought, flooding, storms, heatwaves ... simply to stir up the emotions aroused by catastrophes? We know the moral of the tale of the boy who cried wolf ... so much so that, in the end, nobody believed him. Similarly, climatology runs the grave risk of sacrificing its future credibility, already a fragile thing. And more so because the message is a virtual one: a climatic catastrophe is predicted, but the prediction is only hypothetical, and no-one can readily demonstrate any concrete consequences. The search for 'proof at any price' has become a veritable quest for the Holy Grail with the IPCC, which seizes upon the slightest glimmering of 'evidence'. Is it the lack of tangible and indisputable proof which has lengthened the prediction period as far as 2100? Precautions have to be taken, assuming that the predictions are accurate – 'no risk' research. This is what was described in the journal *L'Ecologiste* ('Appeal against climatic destabilisation') in 2000 (Vol. 1, No. 2): '*all the measures listed above are necessary, whether or not the climate is in danger* ...' So we have to adopt so-called precautionary measures, even if they are useless and we do not know whether the danger is real. However, we might meanwhile be aiming at the wrong target, and wasting time and energy better spent elsewhere. Is this possibility considered when the indispensable precautions are drawn up? Is the sacred 'precautionary principle' gradually becoming an element in the pollution of the spirit?

1.6 IS 'CLIMATOLOGY' STILL WITH US?

The debate also involves (and this is the reason for its appeal) one of the oldest of myths, that of 'understanding' the weather; it is indeed our most frequent topic of conversation, and we all have our own views about it. Interpretations often border on the magical, with the mystery of the weather being as always a matter of faith, and bar-room discussions, 'prognostications' and 'weather lore' all reflect the permanent confusion between climate and the evolution of the weather.

Just as much mystery surrounds the 'models', and the less we understand their mechanisms, the more mysterious they are. The fabulous batteries of computers seem (in all reports) like sacred beasts from one of humanity's most ancient dreams, that of a 'weather-making machine'. Imagine that we might one day be able to affect weather directly, in the spirit of anti-hail guns, cloud-seeding, and Project Stormfury, through which it was hoped that we might control tropical cyclones.

However, definitely the most important factor in this debate and something which makes it even less credible is that *climate change* is a climatological subject which is dealt with, as an annexe to that of pollution for which it is held responsible, largely by non-climatologists. Not too long ago, if you were a meteorologist, your task was the prediction of the weather, and not the study of climate. Within weather agencies, the climatologists busied themselves with the filing of observational data, and the notion of means drawn from data was what more or less defined 'climatology'. Climatology as we now understand it (i.e., the analysis of the interface between Earth and atmosphere), was the territory of geophysicists, as palaeoclimatology was that of the geologists and geomorphologists. Nowadays, the number of climatologists, real or self-styled, is increasing at dizzying speed. All they usually need to do is to parrot faithfully what the IPCC has said in its press handouts. The art of mimicry has, it seems, become qualification enough: the carbon copy words, dogmatically reproduced, are all that is required.

The contributions from the various areas of specialisation which have widened the discipline of climatology represent an important asset in themselves, but it must not be forgotten that it is, first and foremost, climatology. A glaciological chemist, for example, is not really in a position to give opinions on the frequency of tropical cyclones, and a telecommunications engineer is unlikely to give an expert opinion on the likelihood of flooding. A data-processing expert does not automatically possess insights into meteorological phenomena. Climatology practised 'on the side', so to speak, is unfortunately not prohibited! So the discipline has gradually become distanced from the treatment of real facts, especially under the growing influence of modelling. One example will suffice to illustrate this distancing. As part of a conference on Mobile Polar Highs (MPHs), which will be mentioned at length in this book, a scheduled topic was an in-depth discussion based on synoptic charts and satellite imaging. A modeller from Météo-France associated with the Laboratoire de Météorologie Dynamique (LMD), a priori against the MPH concept, agreed to take part. He announced that a synoptic meteorologist would accompany him, because, as he himself stated, '*I'm not used to reading charts*'! How can one be a meteorologist, yet be unfamiliar with the basic tools of observation (and not know how to draw up a synoptic chart)? Can one estimate the validity of a concept if one makes no observations of real phenomena? The consequences of this compartmentalisation of ideas, and of the influence of modelling, if it is disconnected from reality, are considerable for our perception of climatic phenomena.

1.7 THE CONFUSIONS OF MANNERS

At the same time, the media, which have a priori no reason to hold back, and which more often than not have no way of distinguishing truth from falsehood (nor, perhaps, do they feel the need to do it), rehash the IPCC's handouts; they add a dash of sensationalism and doom, concentrating, as usual, on bad news and disaster. Year upon year, from conference after conference, the same slogans, the same propaganda, and the same 'exceptional' images emerge! Again we see the same articles and pictures, the same opinions, the same clichés about 'polar bears in peril' and the 'Maldives syndrome'. Just change the date as (ever more frequent) circumstances dictate ... It gets quite wearing when one hears and sees the relentless fables, which, by mere force of repetition, acquire the semblance of truth!

The media are always on the lookout for the slightest 'scoop', and well orchestrated leaks mean that they can announce some forthcoming article which will show, for example, how 'the melting ice sheet gushes forth' (*La Tribune de Genève*, 6–7 November 2004)! The magic phrase for the media may well be 'scientifically proven', whether they are describing skin cream, hair conditioner, or a climatic phenomenon; for a few scientists mindful of stardom, it is 'as seen on TV'! For the power of a microphone or a camera is extraordinary, since the interviewee is rendered omniscient on the spot! The main aim, then, is getting that scoop ... something of which researchers are aware, and they, and the press officers of the scientific organisations to which they belong, will ensure publicity.

A recent item in the journal *La Météorologie* (No. 45, May 2004) was a very good example of the way in which aspects are mixed together. An article by C. Cassou on 'Climate Change and the North Atlantic Oscillation (NAO)' began with a reference to *Time* magazine's '*Wild weather: Global warming is bringing deluge and thunderstorms in Europe'*, and went on to say that, '*the high positive NAO values observed during the last twenty years can be largely attributed to human activities*' (Cassou, 2004). This was followed by an insert entitled '*The Author's View*', which stated that, '*our "fevered" climate is in line with the simulations of future climate scenarios integrating the impact of human activities*'. There follow the usual ecological clichés and slogans: a reference to Gaïa, '*we are appropriating the Earth from future generations*', and '*the sacred character of our understanding of and respect for the planet*', etc ...

So, there is a reference to a press item confirming the greenhouse effect idea, rather than a reference to original scientific articles, putting the press on a par with science; a page from the 'hymn book' of the IPCC about the cause of the greenhouse effect, and an anthology of fine thoughts from the 'Little Green Book'. It is a good thing that this journal states on page 1 that its '*editorial line is strictly scientific*' – because the reference to human activities is based merely upon the fact that the greenhouse gas curve is on the rise, as is the curve for the North Atlantic Oscillation; but so are the curves for the price of petrol, and bread, and crime rates in France! So all rising curves can be 'correlated' with the greenhouse effect! It is fortunate that *La Météorologie* includes the article under 'theoretical meteorology'!

In the same vein, we might often ask ourselves if the simulations created by models really apply to climatic phenomena, or are better suited to electronic games or Hollywood productions. Science has been rapidly edged out of the debate since 1992, mainly because of the direct involvement of governments, especially as a result of the United Nations Framework Convention on Climate Change (UNFCCC). It has become primarily a 'political affair'. In France, within such an authority as the CNRS, only the 'state religion' is allowed, and imposed, having become a 'national priority'. 'Global warming' has become a political programme, and once this is established, just follow the official line! Those who are known as 'policymakers' may not have the necessary qualifications in climatology, or, lacking the time or the inclination, cannot acquire them; perhaps they do not wish to appear conspicuous; they prefer to go along with ready-made recipes, dished out by authority. The politics of the greenhouse effect, now an official 'umbrella', has developed into an extraordinary exercise in the collective abrogation of responsibility!

The politicisation of what is still called science is apparent within the IPCC, which might appear to some as a kind of 'International Prime Council of Climatology', a Supreme Soviet on Climate, dispenser of 'universal truth'! To the extent that, according to glaciological chemist Jean Jouzel, '*working for the IPCC or being published in its reports is deemed an honour*' (Chevassus-au-Louis, 2003). And, as we all know, the IPCC is principally an intergovernmental body (i.e., essentially political and certainly not a scientific one).

1.8 THE REAL STATE OF CLIMATOLOGY

The widespread lack of a 'climatological' qualification can lead to a blind faith in some idealised science of meteorology. It is generally not admitted (knowingly or otherwise) that meteorology has been in a veritable conceptual *impasse* for more than fifty years. This explains the almost naive confidence, and the almost unquestioning and uncritical belief (not a healthy thing!) of self-styled climatologists, politicians, the unqualified media, and the uninformed public in the quality of models and their predictions. Meteorology, the science of the weather and of its prediction, has always been, both in the past and even more so in the present, the target of criticism and even jibes, the standard of its forecasting leaving much to be desired. In the words of one humorist among many, '*Weather forecasting is a science which tells us about the weather we should have had*!' (P. Bouvard). We can't really know what the weather will be like more than two or three days ahead, as many forecasters will confirm. But now, all this has been erased in a trice! Now, it is unhesitatingly claimed, we can predict weather and climate (which is the sum of weather) as far ahead as the year 2100, from our viewpoint a century earlier! Astrology or science? This is in fact all rather gratuitous, since who among us will be around to confirm our predictions in 2100? This seems laughable, and we might well smile if only short-term predictions about the hazards of the weather served to alleviate suffering, for there are already too many victims; and if there were not so many vested interests involved, in the fields of science, politics, and the economy.

This certainty about predictions is not normal in science. It is usually agreed that, in spite of what is learned, there remain many imperfections, contradictions, and unknowns; there is much to be done before we hold the keys to weather and climate. Such 'certainty' approaches arrogance, if it signals that meteorology–climatology has assumed (or affected) the mantle of absolute self-belief! In reality, things are different. For example, researchers have been investigating the causes of the Sahel drought for 30 years now, and, in spite of the deluge of studies and publications, they have still not arrived at any unanimous, formal response (except, of course, the recent and miraculous greenhouse effect!). Similarly, we are still in the dark about the causes of storms 'Lothar' and 'Martin' which caused tragedy in western Europe in late December, 1999 ... but we can still predict (without knowing the causes) that storms like these will increase in number! The floods in southern France and in central Europe in August 2002 were blamed upon the greenhouse effect. The heatwave and drought which affected western Europe in August 2003, and whose origin is unexplained, were also said to be undoubtedly linked (just like the floods and storms the year before) to the greenhouse effect! If a cold snap comes along, or it snows in low-lying areas, then the greenhouse effect will be the culprit; quite curious really, with cold being the result of 'warming', but let it go ... The success of the scenario may be appreciated when it is claimed that the thermohaline circulation in the North Atlantic could be the cause of glaciation(!). The arcane route by which the greenhouse effect might lead to cold weather is hidden from us, but there are those who manage this sleight-of-hand, and now even the cold is a 'greenhouse consequence'! The more paradoxical the models' 'explanations', the more likely they are to be welcomed – especially by the media!

Seen from the outside, this 'triumphant' meteorology presents a façade of solid science. The impression is bolstered by the magic of information technology and the myth-like attraction of digital modelling, marvels conjured by ever more powerful computers, which, like the Smurfs' 'weather-controlling machine', excite much interest, but are extremely fallible nevertheless. However, in reality, the real face of meteorology is lined with inexactitudes, shortcomings, paradoxes, and errors. There are considerable differences of opinion on every aspect, and the concepts used are often based upon 'physical miracles' (mysterious postulates). Generally, the discipline of meteorology has reached a conceptual dead end. Is climatology, with all its flaws, really in a position to make such confident statements about the future of the climate? It is now dealing not with just a few days hence, or even with the weather next weekend; the so-called 'predictions' now reach decades and even centuries into the future ...

1.9 THE APPROACH OF THIS BOOK

It is therefore necessary to look deeply and rigorously, and in a focused way, at climatology, without complacency or concessions. The mechanisms of weather and climate must be analysed, in order to determine whether or not meteorology–

climatology is really capable of reliable predictions concerning the future of the climate. Any likely errors, shortcomings, or blind alleys should also be noted.

Part One (Chapters 2–8) presents the subject, the players, decisions, and the principle basis. First of all, we should look at the various elements of the 'greenhouse effect' debate. How did it originate and become such a talking point? What is its scientific basis? And how has it 'escaped' from the arena of science to become just about everybody's business?

In Chapters 2–4, we shall examine the subject, the players, and the decisions made.

Chapter 2 will look into:

- The history of the scientific concept, reviewing the slow evolution of the notion of climate change, and its association with greenhouse gases.
- The interaction between science and politics, and the definition of ecology (the science of nature) and of the ecological movement, with its militant, politicised philosophy censuring the effects of human activity on the environment.
- The way in which the defence of the environment has gradually taken on an international dimension, and has been formulated through various world conferences and events: from the Stockholm Conference of 1972, through the World Climate Conference of 1979 in Geneva, and the Villach conferences (the decisive one being that of 1985).
- The birth of the 'greenhouse panic' in 1988, and finally the Climate Change Convention and the Rio Earth Summit in 1992.

Chapter 3 sums up the work done so far, under the banner of the UN, by the IPCC, where science and politics meet. Its three main reports, of 1990, 1995, and 2001, whose recommendations are contained essentially in the *Summary for Policy-makers*, have determined the world's political stance on climate. The First Report already contains the core ideas of what is known as 'global warming', but its tone is moderate and makes no mention of human responsibility for it. The Second Report contributes nothing new scientifically, but now the human race is held responsible for climatic warming. In spite of the way in which this theory was imposed, and all the subsequent controversy, the idea was never refuted. The Third Report increased the value of the predicted rise in temperature, and clinched the argument with the 'hockey stick' diagram, stating that temperatures in recent times are greater than they have been for a thousand years. Moreover, the spectrum of the consequences of the greenhouse effect was considerably broadened, to the extent that it included every meteorological phenomenon.

Chapter 4 seeks to establish the contributions of the various players – scientists, ecologists, politicians, and the media – involved in what now passes for climatology. Various unanswered questions are revealed and this book will attempt some answers. There will be an analysis of how the media have been primed to compose their interpretation of the greenhouse effect, and of the interplay between scientists and the media, not forgetting the way in which science has gradually given way to politics. The role of vested interests (of scientists, environmentalists, politicians, economists, and the media) which have created a 'people's climatology', leads inexorably to the question: where is climatology in all this?

Chapters 5–8 are devoted to basic principles, as regards the greenhouse effect, the causes of climatic changes, methods of prediction, and concepts (especially that of the general circulation of the atmosphere).

Chapter 5 examines the reality of the greenhouse effect. Assuming first of all that there is a greenhouse effect (as does the IPCC), we look at the basic principles of radiation, the respective contributions of the so-called greenhouse gases, and the uncertainties surrounding the part played by clouds and aerosols. Then we discuss major criticisms: the fact that the role of greenhouse gases is minimal compared with that of water vapour, the main cause of the effect; the fact that the greenhouse effect does not warm up the atmosphere, which, without the contribution of perceptible and latent heat, would cool; and the fact that the greenhouse effect is the cause of thermal gradients (i.e., of general circulation). The upshot of all this is that the greenhouse effect is not as definite as might be claimed, and that the scenario based upon it is essentially a water effect, which ought to be taken into account.

Chapter 6 is devoted to an analysis of the causes of climatic variations. Since the IPCC considers that the greenhouse effect is the main cause of climate change, it will be necessary to examine the relevant factors. Variations in solar radiation depending on orbital factors, involving overlapping cycles modifying the Earth's orbit, its eccentricity, the inclination of the polar axis and the precession of the equinoxes, explain climatic changes on the palaeoclimatic scale (as described in Chapter 9). Thanks to data from satellites, variations in the Sun's activity, long since recognised through sunspot studies, now offer promising new perspectives, but their direct influence on climate, and especially temperature, requires further investigation. Volcanism also affects radiation, causing aperiodic and temporary variations, characterised by abrupt drops in temperature transmitted by general circulation.

Chapter 7 deals with climate models, upon which the whole of the IPCC scenario, and all the climate prediction hypotheses, are based. This chapter does not claim to be an exhaustive analysis of these models, since only a modeller would be able to perform this. There is a widespread and unjustified faith in the predictions of the models, though not all the faithful know quite why they believe. So we look at the 'architecture' of the models, and the limitations imposed by the manner of their construction: these limitations arise from the initial approach to phenomena at 'base cell' level, severely restricting any holistic vision. Then we examine the way in which models envisage the hypothesis of the future evolution of global or polar temperatures, rainfall, and disturbances. The hold that models have over climatology, meteorology, and modes of thinking, is considerable. This goes a long way toward explaining the decline of interest in conceptualisation of the dynamics of phenomena.

Chapter 8 looks at general circulation in the troposphere. Frequent reference to general circulation highlights it as the vehicle for variations in the climate. However, climatic models especially do not take this into account, and there exists therefore no coherent, general schema within which each element can be allotted its proper place in the chain of processes. A historical look at concepts of general circulation underlines their shortcomings, and shows that we have not answered all the questions. Then, the concept of the MPH is introduced. This concept explains the dynamic of weather, the division of air circulation into units within the lower layers, and general circulation in the troposphere. Seasonal variations and aerological compartmentalisation and stratification are evidence of the rigorous organisation of meteorological phenomena, within a definite context applicable to all scales of intensity, space, and time.

Part Two (Chapters 9–14) stress the primary importance of real observation. The greenhouse effect scenario is essentially a theoretical one, and so it is vital to test it using facts drawn from observation of the real world.

Chapter 9 analyses the supposed relationship between palaeoclimatology and the greenhouse effect. This relationship is constantly evoked in the discussion of global warming as a long-term phenomenon, and as a result, palaeoclimatology has become a solid prop for partisans of the greenhouse effect. First of all, we ask the important question of the true representativeness of CO_2 levels as measured using ice cores. It is upon these results that the supposed relationship is based, a fact of which we should be wary. The MPH concept, as applied to palaeoclimatic variations, establishes two modes of general circulation: a rapid mode, corresponding to a strong polar thermal deficit, and a slow mode, corresponding to an attenuated deficit. For the purposes of comparison, we examine the dynamic of past climates as described by climate models. Then, the MPH concept is applied, in order to determine the dynamical conditions involved in glaciation, both in the Antarctic and in northern regions (not necessarily only in the Arctic), and the conditions involved in deglaciation. It would seem that the 'greenhouse effect' scenario is not a long-term phenomenon, and that it has never controlled climate change in the past.

Chapter 10 examines the evolution of temperature, the climatic parameter which, according to the scenario, determines the evolution of the climate. The reconstitution of the climate of the last thousand years using the 'hockey stick' diagram has led to a lively debate. The IPCC refutes previously held views of climatologists, especially those concerning the existence of warm periods in the past. The 'stick' is finally seen for what it is – junk. The standard or 'official' curve for the evolution of mean global temperature is brandished by the IPCC as formal proof of the warming of our planet during the period 1860–2000. In fact, the co-variation of CO_2 and temperature suggests a possible relationship for only the last few decades (the Recent Climatic Optimum, or RCO). This recent rise may also be connected with the urban greenhouse effect, but it remains unconfirmed by satellite and radiosonde data. The evolution of temperature is not global, but regional, and diverse, with some regions warming up, while others cool. We look at high latitudes, because they are of considerable importance in driving general circulation. Most of the Antarctic is getting colder, while in the Arctic there is both cooling and warming; the organisation of thermal fields invites a closer examination of aerological dynamics. In the final analysis, with reference to thermal phenomena, nothing remains of the role of the greenhouse effect: it has not been demonstrated.

Chapter 11 discusses weather. The models would have us believe in an intensification of the hydrological cycle, based on the schematic equation 'T/R', which links temperature, evaporation, and rain. Since rain is essentially a product of disturbed weather, we revisit schools of thought and meteorological concepts, to emphasise further the part played by the dynamic of weather. MPHs are proposed as a factor in extratropical weather, and a discussion of the North Atlantic FASTEX experiment of 1997 reveals profound disagreements in this area. A brief look at tropical cyclones points out ambiguities in our view of the tropical dynamic. Recalling the conditions required for pluviogenesis, we examine the prediction that rainfall will increase, analysing, for example, typical extreme rainfall in southern France. Then, we look at the dynamic of the weather across the Mediterranean basin. We discuss drought (which we are told will increase) through examples such as that of the heatwave and dry conditions experienced in western Europe in the summer of 2003, winter drought and lack of snow in mountainous areas, the 'dust bowl' saga in America, and the Great Sahel Drought. These discussions will show that the greenhouse effect does not control the weather, and therefore does not control rainfall, or the lack of it, since the models take no account of the dynamic of weather.

Chapters 12 and 13 aim to place the dynamic of weather and climate in the context of particular aerological spaces, with the North Atlantic and the North Pacific as detailed examples. The dynamic of circulation in the North Atlantic space (Chapter 12) is revealed through study of the North Atlantic Oscillation (NAO), based on the mean pressure field, and by the North Atlantic Dynamic (NAD), based on the real dynamic. The indices of variation represented by the NAO and NAD show positive phases, characterised by intense meridional exchanges, while negative phases exhibit diminished intensity. Since the 1970s, the phase has been positive. Recent climatic evolution in the North Atlantic space is examined, for the Arctic, for North America east of the Rockies, for the Gulf of Mexico/Central America/Caribbean area, for the central and south-eastern Atlantic, and for the north-east Atlantic, which is at present a focus of attention. The chapter concludes that the greenhouse effect does not control the evolution of the climate, but rather that it is wholly determined by cooling in the western Arctic. There is no 'mean Atlantic climate', and the 1970s represent a veritable climatic watershed. In that conclusion, the 'hijacking' of bad or extreme weather by the IPCC is deplored: factors working in the opposite way to the greenhouse effect lie at the origin of such weather.

Chapter 13 examines the dynamic of circulation within the North Pacific space. The variations in its intensity are described by the North Pacific Oscillation (NPO) or the Pacific Decadal Oscillation (PDO), which show alternating positive and negative phases. The focus is on the eastern Pacific, and its evolution is studied via indices of anticyclonic and depressionary activity. Details involve the following parameters: temperature, pressure, precipitation, and disturbances, in the north-eastern Pacific, the central and eastern Pacific, and the North Pacific off Central America. As in the North Atlantic, meridional exchanges are intensified and the greenhouse effect definitely does not control climatic evolution. The second part of the chapter deals with the distinction between El Niño as a climatic 'star' and the real

El Niño. We ask whether 'superstar', 'World Master' El Niño really rules the world, and we analyse the real or traditional El Niño of the equatorial eastern Pacific, as well as the equatorial El Niño associated with the Southern Oscillation (ENSO). The mechanisms at the origin of El Niño are examined, with slow evolution linked to a southward shift of the meteorological equator (ME), and rapid evolutions probably owing more to volcanism, with general circulation as the distributing factor. In conclusion, El Niño gives an insight into recent happenings in climatology, which has drifted away from the facts: a particular example of this being the zonal 'Walker Cell', derived from a meteorology which has lost track of reality. The current perception of El Niño is evidence of a decisive step towards the hijacking of climatology by models, with the inevitable and obvious sliding from the ENSO phenomenon to the greenhouse effect.

A conclusion to the two chapters on aerological matters stresses that there is no longer an 'Atlantic' climate, or a 'Pacific' climate, nor even (and there are four other units of air circulation) any 'global' climate, the very notion being an aberration. Chapters 12 and 13 offer a fundamental reply to the official curve of global temperature: it is an artefact. So there is no proof of the supposed warming associated with the greenhouse effect. However, this curve from the IPCC may well be of use as an index of the intensity of meridional exchanges. What is necessary is carefully to distinguish between the apparent and the real evolution of climate. The apparent evolution is derived from means and models, while the real evolution can be seen in the various climatic behaviours marshalled by the aerological dynamic, with cooling being propagated by MPHs, and warming by cyclonic advections provoked by MPHs.

Chapter 14 examines marine variations, including those of the thermohaline circulation. The 'dangers' of rising sea levels are constantly brandished by the IPCC, which considers that they are the result of the thermal expansion of warmed water, and the melting of ice. Are sea levels really rising? There is no certain answer; perhaps the sea has risen by about fifteen centimetres over the last hundred years, but there is no proof of any recent acceleration in the trend. The meteorological factor, in the form of wind and atmospheric pressure, plays its part, which causes us to reconsider the rise of 2.5 mm per year claimed as a result of the Topex–Poseidon project. In fact, this rise is limited to the equatorial Pacific and to areas of low pressure along the ME; there is certainly no general global rise. The role of the Antarctic, Greenland, and the world's glaciers is also a possibility – but the sea is not rising, and there is no immediate danger of coastal areas being submerged. Thermohaline circulation in the North Atlantic is said to be causing regional cooling. This unlikely scenario is analysed, with reference to the contributions of surface water density, circulation in the Arctic, and the circulation of air and water in the North Atlantic. The conclusion reached is that the climate cannot 'topple'. The way in which the sea is dealt with says much about the approach of the IPCC to the greenhouse effect: in the absence of proof, a slogan is coined, playing on our fears, and so-called scientists are accomplices to this process of mystification.

The General Conclusion is indisputable: the 'greenhouse effect' or 'global warming' scenario is a myth. Climatology has better things to do rather than waste its time and resources, and ultimately lose its credibility, in this way. It does not exist to provide ideas for the plots of disaster movies. Its priority should be to minimise real weather risks, now, and not to gaze into the future trying to imagine improbable dangers to come.

Part One

The subject, the players, and the principle basis

2

History of the notion of global warming

I wish to gauge the most important characteristics of our state of the world – the fundamentals. And these should be assessed not on myths but on the best available facts.

B. Lomborg. *The Skeptical Environmentalist*, 2001.

In every clime, and since the dawn of humankind, our most frequent topic of conversation has been the weather. All over the world, trying to predict, or rather guess what the weather has in store for us is undoubtedly our major preoccupation. Even at this basic level, we cannot always agree: while the farmer might wish for rain, the would-be holidaymaker would prefer heat and a dry spell. *Knowing* or at least trying to know if the weather is going to be warmer, colder, drier, or wetter, and whether it will bring us calm or storms: this debate holds elements of the magical, the subjective, and the emotional.

Humans have always modified their environment by their very presence and through their activities, using its resources and developing it, or 'humanising' it. Only recently have we recognised that this might involve degradation or even destruction of the environment and, as a corollary, its reparation and preservation.

It was inevitable that these concepts of predicting the weather and protecting the natural world should meet ...

2.1 EVOLUTION OF THE SCIENTIFIC VIEW

The initial hypothesis put forward by Fourier in 1827 was that the atmosphere gains heat as if it were beneath a pane of glass, a process now known as the greenhouse effect. Molecules of gas intercept radiation from the Earth's surface and some of this energy is re-radiated towards the surface, making the ground warmer than if there were no atmosphere above it. The gas principally responsible for this phenomenon is water vapour. Apparently, Fourier was also the first to suggest that human activity could modify the climate.

In 1861, Tyndall carried out experiments on the radiative properties of atmospheric gases, and their capacity for absorption. He found that, as well as water vapour, other gases, for example methane and carbon dioxide, could also act as a barrier to infrared radiation re-emitted by the Earth's surface. In 1884, Langley added to our understanding of this phenomenon by analysing the effects of selective absorption of gases on surface temperature levels and this work led to further advances by Arrhenius. It should be remembered that in 1873, standardised meteorological observations began, using methods which would allow measurements to be compared throughout the world.

More than a century ago, in 1896, Arrhenius declared that human activity continuously causes more and more CO_2 to be introduced into the atmosphere, and that this might cause climatic change. His study *The Influence of Carbonic Acid in the Air Upon the Temperature of the Ground* set out to show that carbon dioxide deficits could explain the onset of ice ages. He put forward five scenarios, involving CO_2 levels lower ($\times 0.67$) and higher ($\times 1.5$, $\times 2.0$, $\times 2.5$, and $\times 3$) than in his own time, when the amount was of the order of 300 ppmv. His calculations led him to conclude that glacial periods were the result of reduced CO_2 levels. It should be noted that Arrhenius welcomed the prospect of warmer times, having feared that a new ice age might be coming.

In 1903, he suggested that most of the CO_2 produced by the combustion of fossil fuels could be absorbed into the oceans. He also hinted that there was '*a modification in CO_2 concentration in relation to that of water vapour*' (Lenoir, 2001). He calculated how CO_2 intercepts radiation and announced that, if its level doubled, the average temperature of the Earth's surface would rise by 5–6°C, though his timescale for this was at least several centuries. He also stated that, '*the effect will be greater in winter than in summer ... and over the continents rather than the oceans ... it will be at a maximum in the polar regions*'. It is of great interest that these 100 year old predictions have a very modern look about them, especially considering the proposed amplitude of mean temperature increase and the 'warming' in higher latitudes!

For half a century after Arrhenius, most scientists thought his theory implausible, and it was treated with scepticism. For example, Lotka (1924) reiterated that the oceans contain 50 times as much CO_2 as the atmosphere, and therefore, '*the sea acts as a vast equaliser*', absorbing 95% of all CO_2 introduced into the atmosphere, so that fluctuations '*are ironed out and moderated*'. It was widely believed that nature was capable of automatic restoration of the equilibrium, as evidenced by millions of years of Earth's history. The optimistic Lotka spoke confidently of a world of 'practically imperishable' resources, and looked forward to 'universal prosperity'. In 1942, Blair reflected the general opinion of the decade when writing: 'we can say with confidence that climate is not influenced by the activities of man except locally and transiently'. Such confidence in the future and in man and his technology would last until the 1950s.

One major objection to the greenhouse theory, taken up by Weart in 1997, is

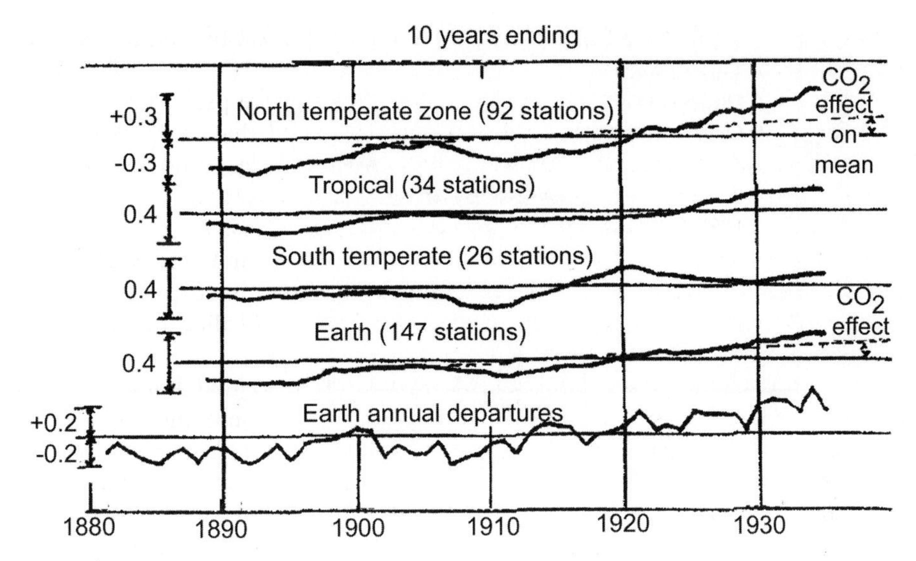

Figure 1. Global rise in average temperature associated with the 'CO_2 effect'. Average temperature changes over 10-year periods (in °C) from statistical studies of the late 1930s. From Callendar (1938).

linked to the pre-eminent role of water vapour, which is much more abundant than CO_2. The American Meteorological Society's *Compendium of Meteorology* stressed in 1951 that the theory that carbon dioxide would alter the climate '*was never widely accepted, and was abandoned when it was found that all the long-wave radiation (that might be) absorbed by CO_2 is already absorbed by water vapor*' (in Weart, 1997).

Callendar was one of the few who disagreed. He moved on from Arrhenius' theory, which concentrated essentially on glacial epochs, to a discussion of anthropic origins of climatic variability. In 1938 he estimated that, since 1890, about 150 million tonnes of CO_2 had been introduced into the atmosphere (corresponding to a 10% increase) and that 75% of this additional CO_2 remained. This increase was, to him, the reason for an increase in temperature of 1°C during that period. His later work (1949, 1958) examined human intervention in climate forcing, which might explain rising temperatures during the first 40 years of the 20th century (cf. Figure 1): what Callendar called the 'CO_2 effect'.

Finally, by the end of the 1950s, scientists were treating the greenhouse effect as a serious possibility, or even a real threat. This revision in thinking was largely due to progress in infrared spectroscopy, especially that undertaken by Plass (1955), who showed that additional CO_2 in the atmosphere leads to greater interception of infrared radiation. He estimated that, if CO_2 levels doubled, there would be a global rise in temperature of 3.6°C, and that if levels were halved, there would be a fall of 3.8°C.

Studies by Revelle and Suess (1957) showed that '*the average lifetime of a CO_2 molecule in the atmosphere before it is dissolved into the sea is of the order of 10 years*',

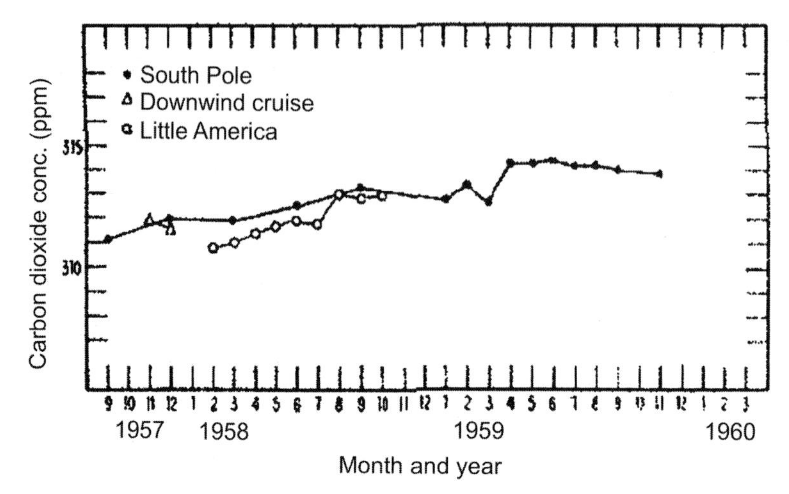

Figure 2. Level of CO_2 as first reported by Keeling (1960) based on two years' measurements in Antarctica.
From Weart (1997).

and they firmly concluded that '*most of the CO_2 released by artificial fuel combustion since the beginning of the Industrial Revolution must have been absorbed by the oceans*'. This optimistic view became, however, less prevalent as it was realised that absorption is less rapid than previously supposed, and that, '*80% of the CO_2 added to the atmosphere will stay there*' (Weart, 1997): it seemed that the oceans were not absorbing all the carbon dioxide emitted by industrial processes.

Now, in the post-war period (and in the shadow of the mushroom cloud), a more pessimistic vision began to appear, reflected in a shift in public perception. Now, '*factory smoke was becoming less an emblem of prosperity than of dangerous pollution*', and the notion of climatic change arising from human activity began to seem more plausible. But tangible proof of increasing levels of atmospheric CO_2 was still lacking. In 1957–1958, the United Nations sponsored the International Geophysical Year (IGY), involving international programmes of long-term research. As part of the IGY, Keeling embarked upon a series of direct CO_2 measurements, and the resultant curve showed a slight but continuous rise over two years (Figure 2). This curve became, as the years went on, '*an icon of the greenhouse effect*'.

So by the 1960s the notion of 'global warming' as a result of human activity seemed established, and began to appear, albeit with strong reservations, as a definite threat, another attack on nature to challenge the ecologists.

2.2 ECOLOGY AND ENVIRONMENTALISM

The term 'ecology' (German *Ökologie*) is specifically defined by the biologist Haeckel (1866) as the study of the relationship between living creatures and their surroundings: a definition which puts ecology among the natural sciences. It is the science of the environment, whether natural or modified by man (Martin-Ferrari, 1992). An 'ecosystem' is a '*system of interacting living organisms together with their physical environment ... from very small spatial scales to, ultimately, the entire Earth*' (cf. IPCC, 2001, glossary).

Environmentalism is a political and philosophical movement, which examines human actions detrimental to the environment. The noosphere (from Greek *noos*, intelligence) encompassing the whole of humanity and its technosphere, transforms the biosphere and thereby the habitability of the Earth (Grinevald, 1993). The fundamental principles of environmentalism are, according to Kingsnorth (2001): rejection of the growth economy, biocentrism, decentralisation, and biological, geographical, political, and cultural diversity, attachment to the land, and systematic mistrust of technology. This is not merely a cult of nature: according to Lenoir (2001), 'ecological militancy', a branch of environmentalism, is '*a highly necessary and useful activity in its persistent opposition to the degradation of the biosphere*'. On the political front, then, environmentalism becomes an ideology preaching control from above over human activity: '*the environmentalist movement must be radical, since its aim is to regain the very basics of our modern world, and to fight for a better one*' (Kingsnorth, 2001). The work of environmentalists is therefore seen as '*the alternative to the end of the world*', its aim being to 'save the imperilled planet' from human aggression, especially but not exclusively against irremediable damage from pollution. Such goals can only be praiseworthy, because it is obviously impossible not to love nature!

Since the beginning of the 18th century, interest in the environment has followed two distinct paths:

- firstly, the 'naturalist' path, mainly preoccupied with the protection of nature and its resources; and
- secondly, the 'humanist' path, concerned chiefly with protecting humanity against the excesses of its technology (Deléage, 1993).

2.2.1 The protectionist awareness

The first, protectionist thread evolved during the 19th century. As early as 1846, von Humboldt considered the Earth to be '*one great whole animated by the breath of life*', anticipating the notion of a living planet. In 1858, concern for the preservation of nature was illustrated, for example, by what has been called the first expression of environmentalism in France: painters of the Barbizon school, supported by writers such as Victor Hugo and George Sand, called for the protection of areas of the forest of Fontainebleau.

A vigorous preservationism of wild nature flowered in the USA: the first great National Park, at Yellowstone, was founded in 1872, at the same time as various associations dedicated to protecting nature. The British established national parks in their colonies, especially in Africa. Holland and Germany did the same, and in 1909 Sweden created Europe's first nature reserves. In 1948, the International Union for the Protection of Nature was founded in Switzerland. It is now known as the World Conservation Union.

2.2.2 The Dust Bowl of the Great Plains

The second, 'humanist' thread developed out of the over-exploitation of the enormous expanses of the American Great Plains, and the dire imbalance which resulted from it. The Plains experienced intense soil erosion: one of the greatest ecological disasters ever known in the USA. Intensive stock and crop methods combined with severe drought denuded and weakened the topsoil, and winds raised dust storms ('dusters') of ever greater frequency. The dusters would appear after unusually dry, warm spells. Dust would be carried up to altitudes of one or two kilometres, on the leading edges ('fronts') of cold polar air masses (Mobile Polar Highs – MPHs), with greatest frequency in winter. In 1933, weather stations in the western USA recorded 179 dusters, one of which deposited 12 million tonnes of dust on Chicago (Flaherty *et al.*, 1985). After the most severe dusters, sunlight was unable to pierce the dense veil of dust for several days.

The storms swept across the American plains, and clouds of fine dust fell as far away as the East Coast. Famously, when a dust storm fell upon Washington D.C., Congress took the decision to create the Soil Conservation Service (SCS). Among the measures introduced by the SCS to protect the soil was the planting of 220,000 trees to act as windbreaks: agricultural methods were reviewed in order to protect the land. The region worst hit by wind and drought, the 'Dust Bowl', encompassed 40 million hectares of Kansas, Oklahoma, Colorado, New Mexico, and Texas. Impoverished farmers and labourers were forced to leave the Dust Bowl and migrate westwards. John Steinbeck, in *The Grapes of Wrath*, described the misery and wanderings of these destitute 'Okies' in California.

Wind erosion of the land continued into 1940, when the return of normal rainfall levels caused some to neglect the existing conservation measures. Drought returned, however, in the 1950s, this time for a shorter interval but even more severely. Then the rains, and prosperity, returned. In the 1970s, a period when frequent cold air masses (MPHs) descended across the American continent, drought and more dust were seen. One of the most severe dusters, tracked by satellite, arose over New Mexico and Colorado on 23 February 1977. It spread across 650,000 square kilometres, following a path towards the Atlantic, which it reached three days later. Even in the 1980s severe erosion continued to affect some parts of the Great Plains, and dusters arising from denuded soils still posed a constant threat during dry spells. These episodes of intense wind erosion played an important part in the rise of the American ecological movement, and still resonate when 'global warming' is discussed. Memories of the Dust Bowl go a long way towards explaining American sensitivity on the subjects of heat and drought, and were a major factor in the triggering of the 'greenhouse panic' in 1988. Moreover, aspects of the Dust Bowl later reappeared in the case of the Sahel drought: over-exploitation of the soil; overgrazing; reforestation; and more recently, links with sea temperatures ... in spite of the fact that, there, meteorological conditions are very different.

The Second World War and its aftermath were a great spur to the environmentalist movement. The nuclear arms race could only increase the threat of world ecological catastrophe, as Martin pointed out in 1955 in his work *Has H-Hour Struck for the World?*, a precursor of the 1980s 'nuclear winter' scenario. As Grinevald (1993) points out again: '*The ecological conscience of the modern world was born in the shadow of Hiroshima*'.

The first world organisations dedicated to the protection of nature were founded shortly after the war. In 1948 the founding conference of the International Union for the Protection of Nature took place. Successive meetings and conferences led to the decision in 1968 to hold the Stockholm Conference of 1972. Again in 1948, it was decided to replace the International Meteorological Organisation (IMO) with a World Meteorological Organisation (WMO), a specialist agency of the United Nations with effect from 1951. The WMO has therefore existed in its present configuration for only about 50 years. It is also worth mentioning that the first orbiting weather satellite, *Tiros*, was not launched until 1960.

Alongside these national and international bodies, numerous non-governmental organisations and groups of private citizens have arisen, with a view to countering the blind destruction of the natural world and averting the risk of a nuclear war which will spare nobody. The 'environmental/ecological revolution' which characterised the 1970s was expressed in the proliferation of demonstrations, both against nuclear tests and for the defence of the environment.

2.3 THE 1970s

2.3.1 Ecological pessimism

This decade saw a transition from positivism, an optimistic outlook with faith in technical progress, to an ecological pessimism in the face of the degradations caused by human activity. This pessimism came more and more to the fore, as Lenoir stressed in 2001: a new generation of ecologists, generally better informed, roundly criticised the consumer society and clamoured for a better one, nearer to nature. A confrontational theme was also developing, based on a quasi-religious view of nature, extending even to a kind of anti-progress sectarianism and the active militancy of 'political environmentalism'. This undercurrent came to the surface most obviously in the Hippie movement, calling for 'peace, love, and flower power', in the call to return to the land, and in the upheavals which occurred in universities and workplaces in 1968.

The 1970s saw the emergence of the debate on the ozone layer, a subject which soon became a major cause of contention: catastrophist scenarios based on the effects of ultraviolet radiation suggested dramatic consequences for human health. The 'ozone saga' had several aspects, the first, in 1970, involving supersonic flights (a reaction against Concorde?). Then came the Space Shuttle debate, superseded by that involving chlorofluorocarbons (CFCs), which was all the rage in 1975 and 1976: the so-called 'aerosol wars'. In 1977 the United Nations Environment Programme

(UNEP) organised a co-ordinating committee on the question of the ozone layer, with a view to monitoring its evolution and ensuring its protection. Rowland and Molina (1975) were instrumental in convincing the USA, producer of more than half the world's CFCs, to enforce a national ban on their use as propellants in aerosol cans, though this had the effect of stimulating their production elsewhere! In 1982, the British Antarctic Survey announced its discovery of the 'ozone hole' (though the observed diminution had already been noted during the IGY, and was again observed in 1983 and 1984). The publication of these observations in *Nature* in 1985 led to measures being taken internationally.

In 1968, an Italian industrialist, Aurelio Peccei, had founded the Club of Rome, and in March 1972 this 'think tank' issued its report 'Limits to Growth', sometimes known as the Meadows Report after one of its editors. The report stressed the impact of economic activity on the biosphere, defining a world strategy for the reconciliation of development and environment. It thus proposed the first 'world model', and its diagnosis was summed up by Grinevald (1993): '*the pursuit of world economic growth is not socially and environmentally sustainable*'.

The international 'Friends of the Earth' association began in North America in 1969, and spread rapidly throughout Europe. Militant environmentalists would now come onto the political stage, like 'die Grünen' in Germany and 'les Verts' in France in 1984. 'Earth Day' was founded in the USA in 1970, and Greenpeace made its appearance in Vancouver in 1971. Originally an anti-nuclear forum, Greenpeace now involves itself in the disarmament debate and promotes the protection of the natural world. In 1971, under the aegis of UNESCO, the Man and the Biosphere (MAB) programme was launched, to encourage research into reconciling the management of Earth's resources, protection of natural environments, and human activities. In 1972, it was decided to go ahead with UNEP.

2.3.2 The 1972 Stockholm Conference

The Stockholm Conference on the Human Environment, proposed by the UN in 1968, took place in 1972. The report by Ward and Dubos (1972) (*Only One Earth*) defined a diagnosis of crisis:

> The two worlds of Man, the Biosphere he has inherited and the Technosphere he has created, are in disequilibrium and virtually in conflict, and Man finds himself in the middle of that conflict.

The conference put forward a plan of action against pollution and for vigilance in the protection of nature, the watchwords being 'sustainable development'. This official conference was mirrored by a 'parallel forum' involving thousands of young people from the same campuses which saw the student revolt of 1968, alongside representatives from environmental protection organisations. The forum coined the first great motto of the ecology movement: '*We have only one Earth*'. Among other eco-slogans there appeared 'the lung of the planet', referring to the Amazon rain forest, under threat from logging: '*excessive deforestation can reduce* *the rate of natural elimination of CO_2 in the atmosphere via the foliage of trees*' (Ward and Dubos, 1972).

The 1970s also saw intense interest in climatic questions at the United Nations Organisation (UNO). A notable outcome was the setting up of the Global Atmosphere Research Programme (GARP) by the WMO in 1974. GARP organised its 'First Global GARP Experiment' in 1979, christened FGGE (or 'Figgy'). Meanwhile, the geostationary *Meteosat* had been launched in 1977 by the European Space Agency (ESA).

2.3.3 The Sahel drought

Interest in matters climatic was also stimulated in this decade by the Sahel drought, which struck the peoples of the southern Sahara from the Atlantic across to Ethiopia. This drought intensified through 1972 and the years following, with particularly dramatic effects. It was attributed to mainly external causes, but (just as the 'Dust Bowl' had been the fault of the American farmers) lack of rainfall was partly due, it was claimed, to the activities of the people of the Sahel: they were held responsible for their own drought (cf. Chapter 11). At first marginally, and then with increasing emphasis among other explanations, the declining rainfall figures were ascribed to the industrial pollution of the developed nations: though the question remains as to how this occurs. So those nations were 'duty bound' to pay compensation for the damage caused to Africa – an idea found in the 1992 Climate Change Convention. The summer droughts of 1976 led to much reflection in France and western Europe too.

The so-called Gaia hypothesis, a geophysiological, 'Mother Earth' theory more religious than scientific, also flourished during the 1970s. In 1979 Lovelock wrote *Gaia: A New Look at Life on Earth*. According to Gaia, the Earth system, a living planet, is viewed as an organism rather than as a machine, its globe covered with '*a gigantic, aqueous, mountainous, hairy, and living organism*', in short an enormous self-organising system.

2.3.4 Cooling or warming?

The 1970s was also a time of discussion focused on the vulnerability of the biosphere, a fragility evoked in the 'nuclear winter' theory, at the heart of the preoccupation of the time with the arms race. But in all these decades there remained doubts about the future of the climate: were we heading for cooling – or warming?

For example, Schneider (Boulder, CO) confidently announced in the early 1970s that a new ice age was coming, and that humanity would perish. According to Schneider, because of the saturation of the principal absorbing band of carbon dioxide, the rise in temperature linked to CO_2 would be no greater than 0.1 K over the next 30 years (1971, in *Science*, Vol. 173, pp. 138–141). However, industrial dust would so increase the opacity of the atmosphere that global temperature would fall by 3.5 K, enough 'to trigger an ice age'!

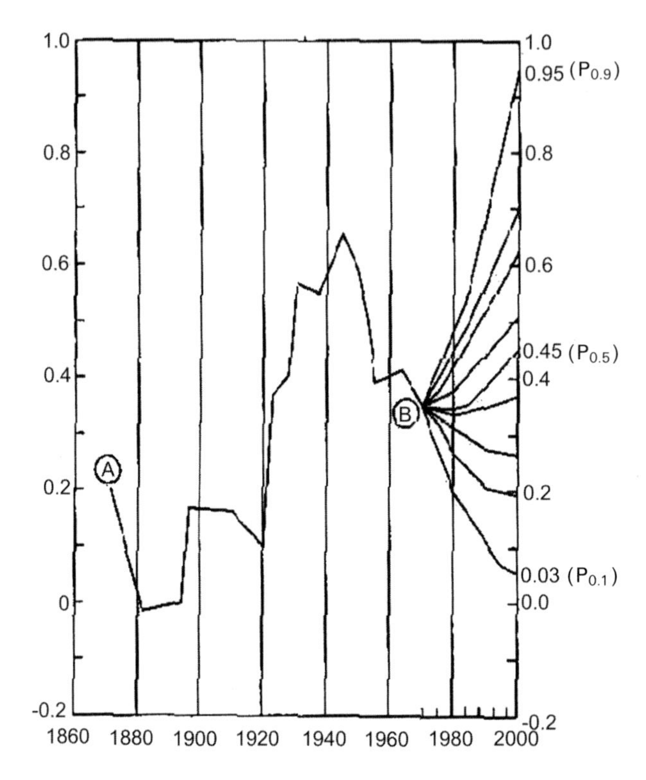

Figure 3. Evolution of average temperature (°C) in the northern hemisphere and temperature predictions until the year 2000. Between A (1870) and B (1970) represents an observed change, after Murray and Mitchell (NOAA). After B, until the year 2000 represents the summary of experts' opinions.
From *Climate Change to the Year 2000*, US Ministry of Defence (1977).

Among the many commentaries of the period, that of Kukla *et al.* (1977) illustrated well the prevailing mood: '*the oscillatory cooling observed in the past 30 years in the northern hemisphere has not yet reversed*'. Michaels (1992) observed that 'the popular vision' of the time was of an impending 'Little Ice Age' ... and magazines carried disturbing images: the Empire State Building toppled by the onward march of a new ice sheet!

This indecision was clearly in evidence in the probabilities expressed in temperature change scenarios for the northern hemisphere from 1970 to 2000. Figure 3 shows these predictions. In the late 1970s, a study involving international specialists, of views on the likely evolution of temperatures before 2000, produced the following results:

- The great majority (80%) of experts consulted thought that, by 2000, temperatures would change between latitudes 0° and 80°N, from values (B, Figure 3) for 1965–1969, by amounts ranging from −0.3°C to +0.6°C (between $P_{0.1}$ and $P_{0.9}$).

- The mean ($P_{0.5}$), suggests by 2000 an increase of +0.45°C compared with the mean value of 1880–1884 (reference 0 of the temperature curve) (i.e., +0.15°C compared to the mean value of 1965–1969 (B)).

So uncertainty was rife, and the main finding of the study was, according to the specialists, '*to demonstrate our lack of knowledge*' about the real future of the climate. And with this in mind, two climate conferences were held.

2.4 1979 – FIRST WORLD CLIMATE CONFERENCE, GENEVA, SWITZERLAND

The World Climate Conference (WCC) which took place in Geneva in 1979 was not 'a specialist meteorological conference', but was called by the WMO to examine the relationship between on the one hand climate, and on the other, human activities and the environment. Delegates came from far and wide, with representatives from the International Council for Science (ICSU, founded in 1931), the FAO (food and agriculture), UNESCO (ecology), the World Health Organization (WHO) (health) and UNEP (environment). This conference was envisaged as a coming together of experts in their own right, and *not as representatives of their governments*: it is important to stress this point, because this means that the scientists involved had a personal input and were not defending some political stance. Sadly, not necessarily the case nowadays with members of the Intergovernmental Panel on Climate Change (IPCC) ...

The WCC set out to assess current climatic knowledge and reach a better understanding of the ways in which climatic variability affects our activities and the environment. Justification for the conference lay in the growing concerns about the repercussions of natural climatic variations on 'world food production, resources and energy requirements, water availability, land management and other social aspects'. Added to this: '*disquieting signs that suggest that humankind, through its own activities, runs the risk of provoking significant climatic changes*'. Here was a new and global vision, the climate considered as 'a communal resource which humanity should protect'.

Delegates were not unanimous about the levels of possible damage to the atmosphere due to human intervention, nor did they agree about the level of urgency when recommending ways in which to counter it. CO_2 was certainly on the increase, it seemed, but the complex cycle of this gas and its consequences for the atmosphere were not well enough understood, and a realistic forecast of future climatic evolution was therefore not possible. Some scientists feared a potentially dangerous warming, and advocated preventive measures (whence the 'precautionary principle'), while more realistic colleagues held such measures to be premature and over-prescriptive.

The final statement called for the elaboration of a common global strategy with a view to better understanding climate and relating more rationally to it. The tendency for a slow cooling observed in certain areas of the northern hemisphere for some decades was thought analogous to similar phenomena in the past, '*and thus whether it will continue or not is unknown*' (WMO, 1979). It could not be ruled out that modern man was unwittingly modifying the climate on a local scale, and to a lesser extent regionally. As a result of the burning of fossil fuels, deforestation, and changing land-use techniques, the carbon dioxide content of the atmosphere had increased by about 15% in the last hundred years, and was continuing to do so at the rate of about 0.4% a year. It was thought plausible that increasing concentrations of this gas in the atmosphere could contribute to progressive warming in its lower layers, especially at high latitudes. This process threatened changes in temperature and rainfall distribution, and in other meteorological parameters, but the fact remained that a detailed explanation of how these things might come to pass was lacking.

Even the authors of the various models thought them too rough to be able to make forecasts for the near future, or for decades or a century to come. One of the essential aims of the future World Climate Programme, organised in collaboration with other international, intergovernmental, and non-governmental bodies, was specifically to improve the models (cf. press release, WMO No. 348, Geneva, 1979).

The WCC thought that the likelihood of cooling in the near future, on the scale of a century, was small. But it could not deny a slight warming due essentially to the greenhouse effect of gases introduced by humankind. Opinions were very mixed on whether preventive and restrictive measures needed to be taken forthwith, or whether it was better to await the results of current research. Neither was there agreement on staging another conference without delay, to discuss measures to be taken world wide.

Given these circumstances, it was still not formally accepted that humanity could be a factor in climate, and such was the conclusion of the first Villach conference in 1980.

2.5 1980 – THE FIRST VILLACH CONFERENCE (AUSTRIA)

The first international gathering of experts on the role of CO_2 in climatic variation took place in November 1980 in Villach, under the aegis of UNEP, the WMO, and the ICSU (cf. WCP, 1981). Discussions revolved largely around a scenario developed by Rotty and Marland (1980). Their projections of CO_2 concentrations by 2025 held that 40–55% of total emissions would remain in the atmosphere, leading to a probable level of 450 ppmv. Calculation of the corresponding temperature rise gave figures from 1.5°C to 3.5°C for a twofold increase in CO_2 (Bolin *et al.*, 1986).

The report of the World Climate Programme (WCP, 1981) then concluded that:

- Because of the lack of certainty, the development of a programme for controlling atmospheric CO_2 levels and prevention of impacts upon society would be premature.
- Priority should be given to the establishment of a solid scientific basis for viewing the problem, always bearing in mind that '*agreement (was needed) on some basic issues*' (Bolin, 1986).
- Emissions, however, due to deforestation and changing land use were deemed

insufficient to provoke climatic change. But fossil fuel reserves were considered sufficient to have an effect on the environment if their exploitation continued at current rates. In the knowledge that the two economic crises of 1973 and 1979 had slowed fuel consumption, a real sense of urgency was still not apparent.

For some, there was still not enough evidence to justify modification of the use of combustibles; others claimed that immediate action was necessary. This disagreement is still with us.

In 1980 the WMO and the ICSU started the World Climate Research Programme (WCRP). Its aims were to understand the physical system and processes of climate, and to determine our capacity to be able to predict climatic evolution. The Moscow WCRP Conference of 1982 held that: '*science can only state the facts (to the extent that they are known) and determine the probability that the climate change has been identified and attributed to some given cause*' (WMO/WCP, 1983). A large number of scientific programmes followed. There was a research programme on climatic forecasting. The TOGA project studied interactions between ocean and atmosphere in the tropics. Project WOCE examined characteristics and changes of global circulation in the oceans, with a view to the elaboration of coupled ocean/atmosphere models. GEWEX probed the hydrological cycle and energy budgets in ocean and atmosphere. In the 1990s came others, such as ACSYS, studying the Arctic system, and CLIVAR, which analysed climatic variability and predictability. From now on, research into the climate would be on an international scale, with global perspectives.

At the time of the first Villach Conference, another threat loomed larger than that of CO_2. '*Leaving aside the consequences for humanity, a world-wide nuclear conflict would degrade the natural environment and might entail global climatic changes*': the 'nuclear winter' hypothesis (Golitsyn and McCracken, 1987), born of the 1980s at various research centres, centres which will be mentioned again, in a different guise, modelling future climates ... This preoccupation would lead the US Academy of Sciences to adopt, in 1983, a resolution on nuclear war and arms control.

As early as the 1970s, the first warnings had been sounded on the subject of acid rain in Europe and on the eastern side of North America, highlighting damage caused to lakes and rivers. Forests withered in Germany, and then in Switzerland in 1982, and in the Vosges in 1983. Sick forests were mainly attributed to atmospheric pollution, especially that caused by sulphur emissions from factories. The consequences of these concerns are well known: 'clean' cars with their catalytic converters, lead-free petrol, and even a supplementary motorway tax in Switzerland (to save the forest), which is still in force.

In the same spirit, international accords on the environment were signed and acted upon. A good example is the Vienna Convention of 1985, its aims being to safeguard human health and the environment from the negative effects of 'ozone layer' changes. In 1987, the Convention was followed by the Montreal Protocol, whose signatories undertook a step-by-step reduction, based on 1986 values, with a view to ending the use and production of CFCs.

2.6 1985 – CLIMATE CHANGE CONFERENCE AND THE POLITICS OF DEVELOPMENT, VILLACH, AUSTRIA

The WMO, UNEP, and ICSU organised the second Villach Conference 'to assess the evolution and the role of increased carbon dioxide and other radiatively active constituents of the atmosphere (*collectively known as greenhouse gases and aerosols*) on climate changes and associated impacts' (WMO, 1986). The conference also firmly supported the idea of a link between greenhouse gases and climate change. Major themes were the increase in average temperatures and the rise in sea levels.

2.6.1 The Villach 'climatic alert'

Among the main conclusions, we may note:

- Quantities of certain trace gases (mainly CO_2, N_2O, CH_4, O_3, and CFCs) are increasing in the troposphere.
- The role of greenhouse gases other than CO_2 is already as important as that of CO_2 in climatic modification.
- The most advanced general circulation models show a rise in mean surface temperature of between 1.5°C and 4.5°C for a twofold increase in CO_2 (or equivalence) concentration. A twofold increase is considered possible over decades, and may well be reached by 2030.
- Other factors (aerosols, solar energy, vegetation) may influence the climate, but the greenhouse effect will be the most important cause of climate change in the next century.
- Changes on the regional scale have not yet been modelled with certainty, but it can already be said that warming will be greatest in high latitudes rather than in the tropics.
- Global warming of between 1.5°C and 4.5°C will lead to a rise in sea levels of 20 to 140 cm. Significant melting is not expected in the western Antarctic during the next century.
- The estimated rise in mean global temperature of between 0.3°C and 0.7°C during the last 100 years is compatible with that expected from an increase in greenhouse gases, but it cannot be definitely attributed to that factor alone;
- There is little doubt that future climate change as predicted by the models will not have profound effects on a global scale on ecosystems, agriculture, water resources, and sea ice.
- Great uncertainty remains in global and regional predictions of rainfall and temperatures, and the responses of ecosystems are not well understood. But, '*some warming of climate now appears inevitable*'.
- A clear difference is established between the results from models and future scenarios: '*At present, and certainly for a number of years to come, climate change studies by means of models provide possible scenarios, not predictions of future climate*'. It must be stressed that the precision 'not predictions' is underlined in the original WMO text (1986).

Such was the statement of beliefs arising from the Villach Conference, which may be seen as a 'climatic alert'. Its conclusions differed from those reached before 1985, especially in that warming due to the greenhouse effect was now considered an established fact (though how it worked was not demonstrated). It said nothing about man's direct responsibility for this, admitting that factors other than the greenhouse effect might be in play.

This statement of beliefs would be constantly reiterated and modified; initially played down in 1990 by the IPCC's first report, it then became 'dramatised', notably in the second report (on human responsibility) and the third (which estimated that current temperatures are the highest for 1,000 years), and which added all sorts of (fictitious) contributory factors.

2.6.2 The International Geosphere–Biosphere Programme

This programme (IGBP), also known as 'Global Change', brought into the 'Earth system' not only the atmosphere and oceanic hydrosphere but also the continental and marine biosphere, the land and the life forms which these milieux support. New communities of scientists (pedologists, ecologists, biologists, chemists) stepped into climate research, as biogeochemical cycles were integrated into the Earth–atmosphere–ocean–biosphere system. Examples of the projects which resulted are: examination of past global changes (PAGES), global modelling (GAIM), global atmospheric chemistry (IGAC), biospheric aspects of the hydrological cycle (BAHC), the flux of substances in the ocean (JGOFS), land and ocean interactions in the coastal zone (LOICZ), changes in land use and cover (LUCC), and the response of terrestrial ecosystems to global change (GCTE).

Two events occurred in 1986 to concentrate minds:

- In January 1986 the Space Shuttle *Challenger* exploded shortly after launching, a deeply traumatic event for America, NASA, and the world (cf. Section 2.7).
- On 26 April 1986, at 1 a.m., reactor no. 4 of the Chernobyl nuclear installation blew up, sending out a radioactive cloud which rapidly spread across western Europe. The passage of a contaminated cloud was not announced in France until 10 May. The reaction of ecologists to the French authorities' obvious lack of concern, and negligence, was the creation of the Independent Regional Commission for Information on Radioactivity (CRII-Rad), dedicated to independent measurements. The Chernobyl event, following on from disasters at Flixborough (UK) in 1974, Seveso (Italy) in 1976, Three Mile Island (USA) in 1979, and Bhopal (India) in 1984, could only add to concerns about the failings of technology and strengthen the influence of a fast-growing ecologist movement.

The conferences of Villach (1987) and Bellagio (1987) defined 'measures to face up to climate change' (WCP, 1988). There was a 90% chance that temperatures would rise between 0.06°C and 0.08°C per decade, and that sea levels would change by between

−1 cm and +24 cm per decade. And new ground was covered: climate change might now lead to extreme meteorological events such as droughts, floods, and storms.

The Brundtland Report (from the name of the president) was issued by the World Commission on Environment and Development (WCED) in 1987. Entitled *Our Common Future*, this report identified the most important environmental problems overshadowing the development of many countries, especially in the southern hemisphere: rising populations, excessive demands on grazed and cultivated land, deforestation, destruction of species, and modification of the atmosphere, leading to imbalance in the world's climate. The Brundtland Report called for the immediate adoption of decisive measures to ensure resources guaranteeing 'durable progress' and 'the survival of humanity'. 1987 was also declared European Environment Year.

2.7 1988 – BIRTH OF THE 'GREENHOUSE PANIC'

In January 1988, freezing cold air masses (MPHs) moved down across Canada and the USA east of the Rockies, while California remained dry. There was nothing unusual about this situation, and, in line with theory then current in North America (cf. Chapter 11), it was associated with conditions at altitude ('*that condition will prevail as long as there is a jet stream*'). Spring brought little rain to the Corn Belt, and during the next three months records were broken for drought conditions in the region from Nebraska to Ohio. At the same time, the eastern USA was experiencing normal or even abundant rain. The spring ploughing and sowing period raised disturbing dust clouds from Omaha to Cincinnati, and plants began to wither as the dry spell moved into June. Soil was denuded, became even warmer, and very fine particles were carried off (cf. Michaels, 1992); all very reminiscent of the well-remembered 'Dust Bowl' scenario of the Great Plains (Section 2.2.2).

Given these circumstances, Hansen could state before the US Congress on 23 June 1988 that there was '*a high degree of confidence*' that the situation was linked to an enhanced (anthropic) greenhouse effect. He had already announced in November 1987 that world temperatures would be at their highest for the last 100,000 years by 2007, to little effect. This time it struck home, and became the 'lead story'!

Hansen immediately encountered considerable (and amply justified) peer opposition (notably from Kerr in *Science*), but the scientists' protestations were not as newsworthy as '*the ubiquitous footage of dead chickens and tractors lost in the dust*' (Michaels, 1992). It must be said that Hansen's presentation left much to be desired scientifically (cf. Figure 4). Its 'statistical incongruity', uniquely appealing to receptive (or even over-receptive) minds, was swallowed whole and without discussion by members of Congress, and by ecologists and journalists. As Schneider put it: '*journalists loved it, environmentalists were ecstatic*'!

How did this 'greenhouse panic', based on such poor 'scientific' argument, become so readily accepted? Without delay, Congress demanded in its Climate Change Act that the administration should propose, in 1989, measures for limiting greenhouse gases. It is not difficult to pinpoint the many reasons for the ease with

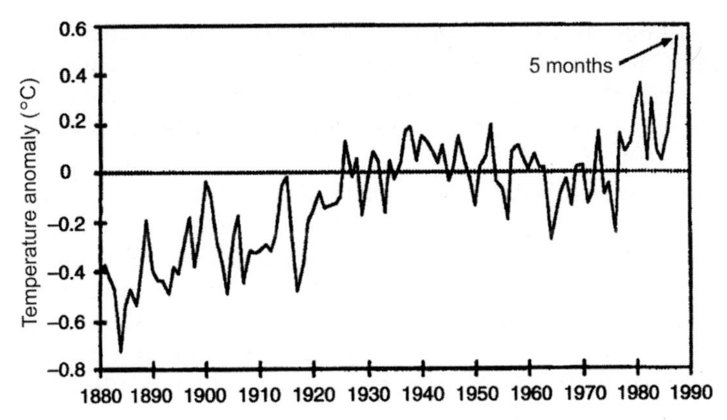

Figure 4. Temperature curve presented by Hansen (NASA) to the US Congress, 23 June 1988. This temperature curve mixed, according to Michaels, 'apples and oranges', because it compares a 1988 five-month value to annual averages.
From Michaels (1992, p. 17).

which the 'message' was accepted. These include an international pessimism such as that manifested at the second Villach Conference and associated workshops; recent events such as Chernobyl and the radioactive cloud it sent across western Europe; the resultant environmentalist hold; and the Great Plains drought and dramatic reminders of the Dust Bowl. Another reason, as Lenoir (2001) thought, is that Hansen himself, as director of NASA, was using the occasion, after the *Challenger* trauma, to restore the image of his Agency and head off the threat of a drastic reduction in funding (a ploy often used in other places too, especially by laboratories and weather services).

Hansen also in this way contributed to a belief in the coming climatic catastrophe, the 'ecocide' predicted by eco-militants. He helped to create the popular vision of 'gloom and doom', of an Armageddon which was a reworking of the nuclear war threat. The repercussions of this were a determining factor in the decision to set up the IPCC.

In that same year (November 1988), the IPCC was created by the WMO and UNEP to carry on the work started at Villach in 1985, 1986, and 1987 within the official framework of the United Nations Organisation. The first IPCC report would appear two years later. Also in 1988, the Toronto Conference called for a 20% reduction in CO_2 emissions from industrialised nations by 2005.

In 1989, scientists such as Seitz, Bendetsen, Jastrow, and Nierenberg brought out their report *Scientific Perspectives on the Greenhouse Problem* (George Marshall Institute). They considered that, given the many uncertainties, the best policy would be to put much greater efforts into research before introducing drastic reduction measures (cf. Roqueplo, 1993). In spite of support from prestigious names like Newell and Namias, and Lindzen, who expressed doubts on the subjects of convection, non-radiative transfers, and water vapour, the catastrophist 'global warming panic' scenario was inexorably up and running.

So the summer 1988 droughts in America and Europe conveniently averted any reservations about the attribution of responsibility and doubts about imperfect measurements and models. The idea quickly became a 'runaway train', while no fundamental discovery came forth to turn the debate round with some scientifically valid 'proof'. It was considerably amplified when the Rio 'Earth Summit' was announced in 1989.

In 1990 the second WCC took place in Geneva. The recommendations of the first IPCC report were adopted. This report showed, moreover, that not all scientists were Hansens: it significantly put the brake on the trend of the 1980s (Chapter 3).

2.8 1992 – THE UNFCCC AND RIO 'EARTH SUMMIT'

2.8.1 The United Nations Framework Convention on Climate Change

In 1990, nearly 20 years after Stockholm, the UN general assembly created an intergovernmental negotiating committee charged with the elaboration of a Framework Convention on Climate Change. The project was adopted on 8 May 1992 at the UN headquarters in New York, and the convention put up for signature in June 1992 at the Rio Earth Summit. Representatives of 154 countries signed, and it came into force on 21 March 1994.

The Convention on Climate Change introduced 'universal citizenship', the principle of equal access for all to the enjoyment of the environment. One preamble in particular echoed the Vienna Convention (1985) and the Montreal Protocol (1987) on the protection of the ozone layer, and previous UN resolutions on the environment and development (1989), on the protection of world climate (1988 to 1991), rising sea levels (1989), and desertification (1989). Particular emphasis was given to:

- the worrying and '*harmful effects*' of climate change;
- '*increased average warming*' resulting from a heightened greenhouse effect;
- the fact that '*climate change forecasts harbour a great number of uncertainties*'; and
- the signatories '*resolving to preserve the climatic system for present and future generations*'.

The convention particularly stipulated:

- Within its definitions, that '*climate change*' should mean '*changes in the climate directly or indirectly attributable to some human activity*', changes over and above the natural variations of climate. This slanted definition is simplistic as a notion of climate change.
- That the ultimate goal is to '*stabilise atmospheric greenhouse gas concentrations at a level obviating any dangerous anthropic damage to the climatic system*' (cf. UNFCCC, Article 2 – extract).
- That '*the absence of total scientific certainty should not be used as a pretext to defer the adoption of measures ...*', in order to '*predict, prevent or attenuate the causes of climate change and limit harmful effects*' (cf. UNFCCC, Article 3). Thus was born the sacred principle of precaution ...

The scientific aspects discussed above form but a small part of the convention, which includes financial and technological dispositions and the obligations, especially of developed countries, to '*cover in total the agreed costs incurred by developing countries*' (in UNFCCC, Article 4.3). 'Developing countries' in this context are, especially: small island states; those with coastal regions near sea level, or with arid or semi-arid zones; those with forest cover; those subject to natural disasters, floods, drought, or desertification; those with mountain systems at risk; those without coastlines, or reliant on the exploitation of fossil combustibles ... So nobody was left out, and Article 4.3 was nothing if not particularly federal in nature, and a sure provider of votes! Also evident was the concept that the greenhouse effect was now being seen as responsible for (almost) all meteorological phenomena ...

The economic aspects of the convention divert our attention from the scientific preoccupations proper to this book, and make the underlying science seem incidental; the 'intergovernmental' character of the IPCC comes into its own here especially with procedures for the adoption of the *Summary for Policymakers* (cf. Chapter 3). Moreover, Article 6 of UNFCCC, on education and training, obliges participants to sensitise the public, at a national level, to climate change and its effects. States signatory to the convention are thus bound to adopt the concept of 'global warming' at the highest institutional level, to impose it as incontrovertible dogma (i.e., a sort of 'state religion' impervious to debate).

The Conference of the Parties (COP), the 'supreme organ' of the Climate Change Convention, furthers and regularly reviews the business of the Convention. Since it came into force in 1994, annual sessions have taken place. COP3 sat in Kyoto in 1997: the Kyoto Protocol is a legally binding agreement stating that industrialised nations will reduce total emissions of six greenhouse gases by 5.2% between 2008 and 2012. The Protocol involves three mechanisms: the first refers to clean development, the second to an exchange system on 'emission rights' (the much-vaunted negotiable emission licences or 'permits to pollute'), and the third describes application criteria. Meeting after meeting has ended without agreement and the Kyoto Protocol is still not in force, as there is still an insufficient number of signatory nations.

2.8.2 June 1992 – The 'Earth Summit', Rio de Janeiro

This UN conference on Environment and Development had the slogan 'The world in our hands'. As was the case 20 years earlier in Stockholm, there was a parallel 'global forum' of local organisations, a forum for industries looking to the progressive environmental involvement of large groups, and representatives of 760 non-governmental organisations (NGOs). This coming together of NGOs on the subjects of the environment and development had been preceded in 1991, in Paris, by a meeting of more than a thousand NGOs, whose manifesto was entitled '*Ya Wananchi*' (in Swahili, 'children of the Earth').

Agenda 21 considered the NGOs as firm partners in the execution of its resolutions. The forum held that the Earth is the common property of us all, and established a lyrical 'Earth Charter', the flavour of which is given by this extract from the preamble: '*We the people, plants and animals, rains and oceans, airs of the forests and tides of the oceans, honour the Earth, dwelling place of all living beings. Let us cherish the beauty and diversity of life ...*' From this conference came the proposition for the signature of the UNFCCC, which came into force two years later.

Principles for sustainable management of the planet are gathered together in Agenda 21 under four headings: social and economic dimensions; conservation and management of resources for development; support of the role of principal groups; and means of execution. Agenda 21 brought together the whole range of measures necessary for the promotion of sustainable development, with some 2,500 recommendations. It may therefore be seen as a catalogue of good intentions. Section II Chapter 9 in Agenda 21, on the protection of the atmosphere, reminds us that '*our atmosphere is under increasing pressure from greenhouse gases, which threaten the Earth with a modification of the climate*', and refers to existing international agreements in this field. Industrialised nations commit themselves, without putting a figure to it, to limiting greenhouse gas emissions, foreshadowing the future Kyoto Protocol.

It is worth noting that, as the Rio Earth Summit began, the so-called 'Heidelberg Appeal' appeared in *Le Monde* (3 June 1992). Signed by 400 scientists, 59 of them Nobel prize winners, this appeal states: '*We are ... worried ... at the emergence of an irrational ideology which is opposed to scientific and industrial progress and impedes economic and social development ... we herewith demand that this stock-taking (of a scientific ecology), monitoring and preservation (of natural resources) be founded on scientific criteria and not on irrational preconceptions*'.

In 1995, the second IPCC evaluative report appeared, stating that '*cumulative observations suggest a discernible human influence on the global climate*'.

In 2001, IPCC's Group I résumé, adopted in Shanghai in January, claimed that '*observations to date present a picture of a world in the grip of warming and other changes in the climatic system*'.

In 2002, 10 years after Rio, the Global Summit on Sustainable Development met in Johannesburg. This second Earth Summit was inconclusive, but Russia, China, and Canada made promises on the ratification of the Kyoto Protocol.

In 2003, in Moscow, Russia refused to ratify the Kyoto Protocol in the short term, principally because of '*uncertainties of global climate change observations and simulation modelling*' (Kondratyev, 2003).

2.9 SUMMARY

The notion of 'global warming', which teaches that humans are responsible for climate change, has been forming for more than a century and a half, at first very slowly, and then, since the 1985 second Villach Conference, very rapidly. Since 1985 to the present day, and more especially since 1988, the 'certainty' that man is an essential factor in climate change, indeed the principal factor, seems established. The expected global warming is bound to bring in its wake the modification of various elements of the climate, and meteorological parameters will be increasingly modified.

This assurance emerges in the conclusions of the IPCC, in its *Summary for Policymakers*: a result of the previously mentioned *démarches*, a blending of scientific and ecological processes driven by international politics.

3

Conclusions of the IPCC (Working Group I)

... the role and the responsibility of scientists are important. They must draw public attention to environmental problems, insisting on what is truly known, what we think we know, and what we do not know ...

I. Rasool. *Systeme Terre*, Flammarion, 1993.

The conclusions of the Intergovernmental Panel on Climate Change (IPCC) were published three times, in three successive reports, in 1990, 1995, and 2001. These reports were generally very much alike, differing only in the amplitude and nature of the envisaged climatic changes, and the range of possible modifications, which became progressively wider.

Weather and climate

The IPCC foresees changes in the weather as well as in the climate. We therefore should specify the IPCC definitions (IPCC, 2001, glossary) for these two different but complementary notions.

Weather

Weather, for the IPCC, is '*the fluctuating state of the atmosphere around us, characterised by temperature, wind, precipitation, clouds and other weather elements*'. Weather is the result of weather systems, rapidly growing and fading as high- and low-pressure systems in middle latitudes, with their frontal zones and associated showers and tropical cyclones.

Our ability to predict the weather is limited. The average convective systems can be described only for a few hours into the future. On the synoptic scale, cyclones are predictable over a few days, perhaps a week. The behaviour of individual weather systems cannot be predicted more than a week ahead. Also, on a smaller scale which is not simulated in the models, that of storms or tornadoes, forecasting is simply impossible.

Climate

Climate, for the IPCC, '*refers to the "average weather" in terms of the mean and its variability over a certain time-span and a certain area*'. Our traditional appreciation of weather and climate is based on mean values of variables: maximum and minimum temperatures, surface winds, precipitation in all its forms, humidity, mist and cloud types, solar radiation, etc. Climate is determined by atmospheric circulation and its interactions with ocean currents on a large scale, and with continental characteristics such as relief, albedo, vegetation, and land humidity, among other factors.

Climate varies from place to place, as a function of latitude, distance from the sea, vegetation, and the presence or absence of mountains or some other geographical factor. Climate also varies over time: seasonally, annually, by decades, or through even longer periods such as glacial eras. Statistically significant variations from the mean state of the climate or its variability, lasting for decades or longer, are considered to be 'climate change' (cf. IPCC, 2001, glossary).

It should be pointed out that, for the IPCC, the notion of climate change involves '*any change in climate over time, whether due to natural variability, or as a result of human activity*' (IPCC, 2001). This definition is more compatible with reasoned climatology than that of the Climate Change Convention, which attributes climate change, in too restrictive and slanted a way, 'directly or indirectly to a human activity' (UNFCCC, art.1, defn. 2. p. 8, 1992).

Climatic variations provoked by external causes may be partially predictable, especially on the very largest (continental and global) scales. For the IPCC, the Earth's climate depends as a whole mainly on factors influencing radiative balance (e.g., atmospheric composition, solar radiation, and volcanic eruptions).

The IPCC stresses that '*to understand the climate of our planet Earth and its variations, and to understand and possibly predict the changes of the climate brought about by human activities, one cannot ignore any of these many factors and components that determine the climate*'. It will therefore be useful, as we proceed, to examine different affirmations and to verify if all the 'factors and components that determine the climate' have been effectively examined, and if any of them may have been neglected or even ignored, whether involuntarily or voluntarily.

Our account of the IPCC's conclusions will be based upon each of the reports of the *Summary for Policymakers* of Working Group I, the group charged with the scientific aspect. This summary, aimed at decision-makers, has the apparent and main 'benefit' of having been read, re-read, and discussed, and then approved line by line and word by word by IPCC 'experts'. According to the current procedure the first text is submitted to the assembly of the IPCC, comprising representatives of governments (i.e., 'experts' who may have been allotted for essentially political reasons, and may not therefore have the required scientific background). Then the text is more or less worked over. It therefore represents the sum of the discussions (which may not always have been scientific in their preoccupation), and the most official pronouncement, of this agency: it sets out the essential points for outside consumption. This distillation is then very widely distributed by the IPCC press service, becoming the document which non-specialists in climatic matters are likely to read, and it will very often be the only document seen by politicians and the media.

3.1 THE FIRST REPORT OF THE IPCC, 1990

CLIMATE CHANGE: THE IPCC SCIENTIFIC ASSESSMENT

Edited by J. T. Houghton, G. J. Jenkins and J. J. Ephraums

(Cambridge University Press, published for the IPCC/WMO/UNEP)

The Executive Summary of the *Summary for Policymakers* of the IPCC's Working Group I stresses essential points, which we may summarise as follows:

- **Certainties**: a natural greenhouse effect exists, making the Earth warmer than it otherwise would be, whilst emissions due to human activities substantially increase the concentration of greenhouse gases (carbon dioxide, methane, CFCs, and nitrous oxide). This increase must add to the greenhouse effect, bringing in its wake additional warming of the Earth's surface. As a consequence, the amount of water vapour, the principal greenhouse gas, must also increase.
- **Reliability of calculations**: certain gases are more active than others, carbon dioxide being responsible for more than half the enhanced greenhouse effect. Emissions of long-lived gases must be immediately reduced.
- **Predictions based on the results of models** (simple models of the 'box-diffusion' type):
  - according to scenario A, mean global temperature will rise by 0.3°C per decade over the next century (±0.2° to 0.5°C), i.e., by approximately +1°C by about 2025, and +3°C before the end of the 21st century. Scenario B predicts 0.2°C/decade, scenario C just over 0.1°C/decade and scenario D about 0.1°C/decade;
  - land masses will warm more quickly than oceans, and high northern latitudes will be warmer than the global average in winter;
  - regional climatic variations may differ from the global average, though the reliability of this statement based upon models is not firm;
  - sea levels will rise by about 6 cm/decade (with a possible uncertainty suggesting between 3 to 10 cm) as a result of thermal expansion of the oceans and the melting of ice sheets. The mean predicted rise is by +20 cm around 2030, and +65 cm by the end of the next century.
- **Uncertainties**: there are numerous uncertainties regarding the timescale, amplitude and geographical spread of climate change. They stem from incomplete knowledge of:
  - sources and sinks of greenhouse gases when predicting future concentrations;

- cloud cover, a strong influence upon the degree of change;
- the oceans, which influence the timescale and regional spread of changes;
- the behaviour of polar ice, in predicting future sea levels.

- **The diagnosis**:
  - globally, mean surface air temperatures have risen by between 0.3° to 0.6°C over the last 100 years, the warmest years having occurred during the 1980s;
  - in this same period, the mean increase in global sea level has been between 10 and 20 cm, in an irregular, non-uniform way;
  - the amount of warming has been roughly that predicted by climatic models. However, this warming is also of the same amplitude as might occur given natural climatic variability, without external influences, and therefore this may be largely responsible. Unequivocal identification, through observations, of the enhanced greenhouse effect **'is not likely for a decade or more**' (IPCC, 1990, p. xii);
  - there is no evidence that the climate has become more variable over recent decades.
- **To improve predictive capacity there must be**:
  - a better understanding of the diverse processes involved in climate, and more especially those associated with clouds, oceans and the carbon cycle;
  - improved systematic observation of variables associated with climate on a global scale, and determination of those variations which have taken place in the past;
  - improved, reliable models of the Earth's climatic system.

The executive *résumé* of this first IPCC report already holds within it all the core ideas of what we normally understand by the term 'global warming'. The report in fact contains the basics, especially that of radiative forcing, the Sun being '*the driving energy for weather and climate*'. The greater part (in fact, more than half) of the document dwells upon the greenhouse effect, with the natural greenhouse effect considered to be one of the most important factors, and water vapour always the greatest contributor. Measurements of polar ice cores show that in the last 160,000 years temperature has closely mirrored changes in levels of carbon dioxide and methane, though is has not been possible to distinguish between cause and effect. Past climatic evolution is not deemed, in this report, to illustrate the greenhouse effect. The various greenhouse gases, their future evolution and their potential to cause warming are all considered, as are the evolutionary scenarios of temperatures as a function of an enhanced greenhouse effect before 2030.

Winter precipitation should increase by between 5% and 10% over land in high- and mid-latitudes (35–55°N), though regional estimates are somewhat less reliable. There is no clear evidence that the variability of the weather will change in the future. As for large-scale weather *régimes*, it is not known if, or indeed how, they might change. What will be the future evolution of perturbations? Tropical storms (typhoons and hurricanes) ought – theoretically – to be on the increase, as they form over waters with temperatures above 26°C (we shall return to this in Chapter 11): but the models offer no information on this subject. Storms of middle latitudes '*are driven by the equator-to-pole temperature contrast*'. As this contrast will lessen in a warmer world, these storms should also decrease in force or change their trajectories, leading to a possible reduction in the variability of their paths in winter: a hypothesis examined below (cf. Chapters 7 and 11).

The report asks the fundamental question: '*Has man already begun to change the global climate?*'

This essential question introduces the enhanced or anthropic greenhouse effect. Warming is borne out by observation: 0.3–0.6°C between 1861 and 1989, urbanisation being considered as responsible for less than 0.05°C. This warming fits the predictions of the models, and is 'confirmed' by retreating glaciers and a rise in sea levels of between 1 and 2 mm annually. But **it is not possible to attribute this warming to man any more than to natural variability**, given that '*we do not know what the detailed signal (of human influence) looks like because we have limited confidence in our predictions of climate change patterns*' (IPCC, 1990, p. xxix). There is no clearer statement of the fact that, in 1990, it was impossible to attribute any climate change to humans.

The rise in sea levels, calculated using simple models, should be about 6 cm per decade (±3 cm to 10 cm) over the next century. This change would not be uniform: thermal expansion, variations in ocean circulation, and surface pressure would vary regionally with warming, '*but in an as yet unknown way*' (IPCC, 1990, p. xxx).

Ecosystems should be affected by climate change, since photosynthesis and respiration depend upon climatic factors and carbon dioxide concentrations. Current models are not capable of allowing for modifications of these parameters on the localised scales involved.

At the end of the report, we are reminded that scientific uncertainties are many, and international research is invited to 'improve our capability to observe, model and understand the global climatic system' (IPCC, 1990, p. xxxii), a programme which can only be for the good:

- *observation* is judged to be 'of vital importance' to understand natural variability, parameterise key processes for models and verify their simulations;
- *modelling* climate change requires the development of global models linking atmosphere, land, sea and ice, integrating the most realistic formulations of the processes and interactions involved; and
- *understanding* of the climatic system will be promoted by analysis of observations and the results of simulations by models.

The main thrust will derive from the link between ocean and atmosphere, regional predictions and the representation of small-scale processes. But of course, given the complexity of the problems, we can hardly expect 'rapid results'.

In summary

The First Report of the IPCC (1990) recounts basic knowledge, especially on the subject of the greenhouse effect, to which is devoted the longest development.

It recognises its priority: everything hinges on the rise in the greenhouse gas content of the atmosphere, and the supposed increase in temperature which will result.

It is not without a few approximations or statements of the obvious: '*as mean temperatures increase, episodes of higher temperature will probably become more frequent in the future, and cold spells less frequent*' (IPCC, 1990, p. xii), which in the final analysis seems highly likely!

However, **no 'culprit' is formally identified**, and the claimed warming, less than that predicted by the initial Villach Conference (1985), might as easily be caused 'naturally' as anthropically. We cannot expect such an unequivocal response for the time being. The rise in temperature is associated with a rise in sea levels through the 21st century. On the other hand, evolution in precipitation is hardly discussed (for mid-latitudes), and no noticeable change in weather intensity is envisaged. Perturbations should also remain unaffected, be they tropical cyclones or temperate depressions. Neither is any 'oscillation' mentioned, not even El Niño.

Moreover, the report stresses unambiguously the uncertainties, the limits of and gaps in our knowledge, and the necessity of further observation and research to ensure a better understanding of the climatic system. **It can therefore be seen as an honest if informal overview of the state of meteorological science at the end of the 1980s**: what was known and not known; the faith in models and also the perceived limits of their reliability, their inability to respond satisfactorily to certain questions and the impossibility of their answering others. This was undoubtedly, from a scientific standpoint, the most objective IPCC report.

This report, then, contained little to 'spread alarm and despondency', with the possible exception of the supposed warming, and rising sea levels. Compared with previous pronouncements such as the conclusions from Villach on rising temperatures, it was almost reassuring. It was a moderate exposé, even though it insisted on the existence of the enhanced greenhouse effect. So there was little here that differed from Arrhenius' and Callendar's hypotheses of 1896 and 1938. Models were still unable to give firm answers.

This 1990 report will be used as a reference with which to compare IPCC's later publications.

3.2 SUPPLEMENTARY REPORT TO THE FIRST SCIENTIFIC REPORT OF THE IPCC

CLIMATE CHANGE 1992

Edited by J.T. Houghton, B.A Callender and S.K. Varney

(Cambridge University Press, published for the IPCC/ WMO/ UNEP)

This intermediate report reaches essentially the same conclusions as that of 1990: '*The results of scientific research since 1990 have not altered our fundamental understanding of the science of the greenhouse effect, any more than they confirm or inva-*

lidate the main conclusions of the First Scientific Report' (IPCC, 1992, p. 5). What *was* confirmed:

- the increase in greenhouse gas concentrations, as a result of human activity;
- evidence from climatic models that global temperatures will only rise by between 1.5°C and 4.5°C if carbon dioxide levels double;
- the increase stated by the models is of the same amplitude as natural variability, to which rising temperatures can be largely attributed;
- there remain many uncertainties, especially in the area of the detection of the enhanced greenhouse effect: **a definitive result may not be expected for 10 years or more**.

There are also some significant results, summarily presented below:

- cooling due to aerosols (especially sulphates from volcanoes or industrial processes) has been able partly to compensate for warming in the Northern Hemisphere;
- climatic models have continued to improve, but the part played by other factors, especially that of clouds, has not yet been resolved;
- average warming in northern mid-latitudes is characterised by a more pronounced rise in night-time temperatures, the rise in daytime temperatures being more moderate. This warming of the Northern Hemisphere has not been uniform: there have been seasonal and geographical variations, and the north-west Atlantic is notable for its lack of warming.

This 1992 supplement did not depart from the fundamental conclusions of the 1990 report. It remarked upon a slow improvement in the models, noting that any increase in their reliability, especially on the regional scale, would need new simulations; there was no strong evidence pointing to climatic variability or an increase in stormy weather. The supplement also draws attention to 'our incomplete knowledge of the climate' (IPCC, 1992, p. 15).

3.3 THE SECOND REPORT OF THE IPCC, 1995

CLIMATE CHANGE: THE IPCC SCIENTIFIC ASSESSMENT

Edited by J. T. Houghton, L. G. Meira-Filho, J. Bruce, Hoesung Lee, B. A. Callender, E. Haites, N. Harris and K. Maskell

(Cambridge University Press, published for the IPCC/WMO/UNEP)

The *Summary for Policymakers* begins by stating that our knowledge on climate change has 'considerably' improved since 1990, and new data and analyses are to hand. The main points underlined by this report are:

- **Greenhouse gases have continued to increase**:
  This increase, which has been continuing since pre-industrial times, has

produced a radiative forcing which is gradually warming the planet and 'producing other climatic changes'. The tone is soon set: the principle of global warming is reaffirmed, whilst additional consequences are envisaged for other aspects of the climate.

If CO_2 emissions were maintained at 1994 levels, their concentration would reach 500 ppmv by the end of the 21st century (i.e., about twice the 280 ppmv concentration of the pre-industrial era). According to the models, the CO_2 content of the atmosphere could be stabilised 40 years on, at around 450 ppmv, only if emissions of human origin went back to 1990 levels (or 140 years on, at 650 ppmv).

- **Aerosols of human origin are responsible for a negative radiative forcing**: Microscopic particles suspended in the air (aerosols) originate from the burning of fossil fuels, from the biomass and from other (mainly terrigenous) sources. They have caused a direct mean negative forcing of $-0.5\,W/m^2$ globally. Their concentration above certain regions means that their role locally may outweigh positive forcing due to the greenhouse effect.
- **Climate has evolved over the last century,** as observation has shown:
  - Globally, temperatures have risen by between 0.3°C and 0.6°C, with recent years being the warmest. Night-time temperatures have generally increased more than daytime temperatures. On the regional scale, changes have been apparent: although there has been greater warming over the land at mid-latitudes in winter and spring, other regions, for example the North Atlantic, are seen to be cooling. These are important observations, for this is the first time that the 'warming' scenario is not considered to be uniformly global!
  - Also noted is an increase in precipitation over land at high latitudes in the Northern Hemisphere, especially in winter;
  - Sea levels have risen by between 10 and 25 cm in the course of a century. This rise is largely attributable to warming;
  - There are still insufficient data to determine whether climatic variability or extreme meteorological conditions are happening;
  - The long, warm period of the *El Niño*-Southern Oscillation (ENSO) between 1990 and 1995 was unusual compared with the previous 120 years. It is worth noting here that this is the first time that *El Niño* appears in the IPCC scenario. The IPCC considers that this phenomenon '*is a cause of drought and floods in many regions*' (cf. Chapter 13).
- **'The balance of evidence suggests a discernible human influence on global climate'** According to the IPCC, there has been progress in assessing the additional greenhouse effect of sulphurous aerosols, which it says have led to a more realistic evaluation of radiative forcing due to human activity. 'Important' results have been able to verify this 'attribution', linking cause and effect, and making mankind responsible for climatic evolution.
  - '*Mean global air temperature is as high in the 20th century as it has been in any other period between 1400 AD and the present*', referring to the long period following the so-called Medieval Warm Period, and which included the Little

Ice Age. Data from before 1400 AD are thought too fragmentary for any reliable evaluation of mean global temperatures (an idea which it may be useful to bear in mind when reading the third report);
- '*The tendency towards warming is probably not of uniquely natural origin, but our capacity to measure the influence of man on the global climate remains limited, since the expected signal is still difficult to distinguish from the background noise of natural variability, and there is uncertainty on various important factors*'. Whilst man's 'responsibility' seems well established, as is suggested by the title above, this extract from the report mentions 'natural variability', the faintness of the 'signal', and various uncertainties ... It looks like a discreet 'denial', a sort of disguised U-turn, and again reveals the absence of any consensus on this crucial subject. But the die is cast, the reservations have vanished, and the shock phrase is now writ large, proclaiming the all-important fact that **mankind has been found well and truly guilty!** And so it would now be, across all the media.

- **'Climate is expected to continue to change in the future'** (a mere truism, since climate has always been changing!):
  - According to the mean scenario (based on the most probable figures), the predicted rise in mean global temperature will be around 2°C between 1990 and 2100 (i.e., one-third lower than the 1990 value);
  - According to the lowest estimate scenario, the value for future warming will be about 1°C, the highest estimates giving a value of about 3.5°C;
  - '*Regional fluctuations in temperature could be noticeably different from the global mean*', according to the IPCC (1995, p. 23), but these regional values are not precisely determined. Also it might be thought from this that the mean value is known, in advance of the regional values, which allow us precisely to calculate the mean! North polar latitudes will perhaps see a much greater increase in winter;
  - A rise in mean sea levels of about 50 cm by 2100 (the scenarios vary between 15 and 95 cm) is predicted, a result of the warming of the oceans and the melting of polar ice caps and glaciers on mountains. This will be 25% less than was predicted in 1990;
  - A more intense hydrological cycle is predicted globally, with increased precipitation at high latitudes in winter. There is less certainty on the regional scale, since '*increasing temperatures will bring about a reinforcement of the hydrological cycle, increasing the risk of droughts and/or floods in certain places and a possibility of a reduced amplitude of these phenomena in other places*' (we are therefore unlikely to get it wrong, and the scientific precision of this prediction may be appreciated by all!). Moreover, several models foresee a growth in extreme rainfall events;
  - There is still insufficient knowledge for any prediction of the number or location of severe storms, especially tropical cyclones;
  - Higher rainfall values are expected, according to the models, in the summer monsoon area of Asia – but, if aerosols are taken into account, then there will be a diminution in monsoon rains ...

In summary

The Second Report of the IPCC (1995) does not add much to its 1990 report. The First Report laid considerable stress upon the greenhouse effect, while the Second Report dwelt particularly upon aerosols, which would partly counterbalance the predicted warming. So the amplitude of the thermal change was narrowed, and this report even seems to play down the 'danger', with less rapid and smaller rises in temperatures, and also in sea levels.

New, still tentative 'overtures' appeared occasionally, notably on the subjects of the water cycle and rainfall, with perhaps an increase in torrential rains.

But doubts remained: about monsoons (considered only in their Asian context), temperate perturbations, cyclones, regional weather behaviour, and natural variability ... The *Summary for Policymakers* concluded that many uncertainties persisted, pointing out that '*numerous factors limit our capacity to predict and detect coming climatic changes*'. (IPCC, 1995).

This Second Report was therefore generally 'reassuring'. So it became necessary, doubtless to forestall the defection of those with an interest in pursuing the IPCC process, to 'back-pedal'. This was the reason why 'human influence on the global climate' again surfaced ...

The controversy over 'human responsibility'

This concept provoked an immediate outcry against its perpetrator, Santer, who put it into the final version of the report after it had been approved by the members of the IPCC. The 'responsibility' in question came from a report, then unpublished and unreviewed, the main author of which was Santer. Its results were published in *Nature* only in 1996 (after the IPCC report) and caused immediate and violent criticism from climatologists. Chief among the critics were Michaels and Knappenberger (*Nature*, December, 1996), demonstrating the inanity of the proposed attribution. However, there is a world of difference between the impacts of a report widely distributed throughout the media, and an argument meant for the eyes of scientists who read *Nature*. Santer's desired effect was largely achieved, since the point in the IPCC report most widely covered by the media was precisely this accusation levelled against humanity.

To gain an insight into this controversy we need look no further than the exchanges in the *Bulletin* of the American Meteorological Society (AMS), criticising or supporting the chief editor of this *Summary for Policymakers* who made direct changes to chapter 8 of the Report:

- Seitz (1996) pointed out that the original version contained, in particular, the following sentences: '*None of the studies cited above has shown clear evidence that we can attribute the observed (climate) changes to the specific cause of increases in greenhouse gases*'; '*No study has positively attributed all or part (of the climate change observed to date) to anthropogenic causes*'. These sentences were omitted in the final document.

- Bolin (president of the IPCC) *et al.* (1996) considered that the changes made to the revised version were aimed only at '*producing the best possible and most clearly explained assessment of the science*'. The 'best possible', agreed, but in what spirit and to what end; and is it really science that is being safeguarded here?
- Avery *et al.* (1996) lent their support to Santer, editor of the *Summary*, in the name of the Executive Committee of the American Meteorological Society.
- Kondratyev (1997), on the other hand, greatly deplored IPCC's lack of scientific ethics in 1995.
- Singer *et al.* (1997) denounced the way in which climate science was being used and stressed that '*the real issue then is the political misuse of the IPCC report and of climate science rather than the ongoing debate about procedure*'.

Nevertheless, and as is well known, this controversy was to no avail, since the formula has passed without amendment into the public domain, no retraction ever having been effected by the IPCC in 1996 and in spite of the censure, it survived intact in the IPCC's 2001 Report! So, the **Second IPCC Report represents a definite change from its first**, and from 1995 onwards it has been possible to question the real motives of this institution.

3.4 THE THIRD REPORT OF THE IPCC, 2001

CLIMATE CHANGE: THE IPCC SCIENTIFIC ASSESSMENT

Edited by J. T. Houghton, Y. Ding, D. J. Griggs, M. Noguer, P. J. van der Linden, X. Dai, K. Maskell and C. A. Johnson

(Cambridge University Press, published for the IPCC/ WMO/ UNEP)

The Third Report was approved by IPCC members in Shanghai in January 2001. The *credo* of the IPCC is proclaimed in the first lines of the report:

'An increasing body of observations gives a collective picture of a warming world and other changes in the climate system'.

The reference to an increasing body of observations attests that we are not dealing here with hypotheses: the claim is rather that we can prove that the planet is warming up generally. The 'global warming' scenario has now therefore **moved on from being a hypothesis, and has become an observed fact, planet-wide**.

The projected increase in temperature of 1995 is now greater: '*the globally averaged surface temperature is projected to increase by 1.4 to 5.8°C over the period 1990 to 2100*'. The higher figure of 5.8°C has been added after later revision by experts, a procedure already seen in 1995. Moreover, several other climate changes are remarked upon, and we shall restrict ourselves to specifying what they are, and how they operate.

The IPCC also remarks on a '*better understanding of climate change*'. Is this a completely new element, a 'revolution' even in the concepts explaining this 'better understanding' during the five years since the previous report?

The essence of the Second Report (1995), which concluded that '*the balance of evidence suggests a discernible human influence on global climate*', is reaffirmed. Or even reinforced: '*There is new and stronger evidence that most of the warming observed over the last 50 years is attributable to human activities*'. Most of the observed warming is attributed to the increase in the concentration of greenhouse gases (the concentration of CO_2 having risen by 31% since 1750, to a level of 370 ppmv in 2000). The IPCC considers it most improbable that warming is caused by the inherent variability of the climate, or is purely natural in origin, which suggests that the factors normally evoked (solar irradiance and stratospheric aerosols from volcanic explosive eruptions) have contributed only weakly to radiative forcing during the last 100 years.

The first strong argument is found in the rise in temperature during the 20th century of $0.6°C \pm 0.2°C$, which is 0.15°C higher than the 1995 estimate as a result of higher temperatures during the period 1995–2000. These figures take account of different adjustments, including the effect of urban 'heat islands'.

The main argument for warming is based on two periods: 1910–1945 and 1976–2000. It is thought very probable (i.e., with a certainty of over 90%) that the 1990s were the warmest decade since 1961, 1998 having been the warmest year.

Average minimum night-time temperatures rose by 0.2°C per decade between 1950 and 1993, but maximum daytime temperatures rose during the same period by only 0.1°C. Frost-free periods were observed to be getting longer in many areas.

Temperatures have risen in the lowest 8 km of the atmosphere in the last four decades, as has happened at the surface, by +0.15°C per decade (±0.05°C). Measurements taken by weather balloons and, from 1979 onwards, by satellites too, show a rise of 0.05°C per decade (±0.1°C) (i.e., three times less than observations taken at lower levels). This difference is judged to be statistically significant.

Snow cover has diminished by about 10% since 1960, and mountain glaciers in non-polar regions have retreated considerably during the 20th century. In the northern hemisphere, sea ice cover in spring and summer has shrunk by 10–15% since 1950, and ice has become less thick in places depending to the season.

The second strong argument is found in new analyses of indirect data for the northern hemisphere: '*observations of tree rings, corals and ice cores, and evidence from historical sources*'. It would appear that the 1990s were the warmest decade, while '*the rapidity and duration of warming in the 20th century have been much greater than in any of the nine preceding centuries*'. These statements have as their basis the so-called 'hockey stick curve' (Mann *et al.*, 1999), according to which temperatures fell slowly through nine centuries (1000–1900), only to show a rapid increase in the twentieth.

Daly (1997) calls this curve, which, it is claimed, shows that temperatures are higher than they have been in 1,000 years, the 'infamous hockey stick'. Its adoption by the IPCC constitutes a veritable scientific *coup*, reversing the flow of climatic history and creating an immediate hue-and-cry. We shall revisit it in Chapter 10.

Average sea levels have risen and the quantity of heat in the oceans has increased. The mean increase in global sea level has been between 10 and 20 cm during the 20th century. The rise in sea surface temperature between 1950 and 1993 is considered to be about half of the average rise in land temperature ... Given the high probability that warming has contributed significantly to the 'thermal expansion of surface water', to what extra volume would this correspond, considering that the overall heat in the oceans has increased since the 1950s?

It is predicted that sea levels should rise by between 0.09 and 0.88 m in the period 1990–2100, mainly because of thermal expansion and the melting of the glaciers. Since 1950 a probable diminution of 40% has been noted in the Arctic at the end of summer/beginning of autumn. A modification of thermohaline circulation is similarly conceivable ...

Other important aspects of the climate have also undergone changes:

- It is very probable that precipitation has increased by between 0.5% and 1% per decade in high- and mid-latitudes in the north, and by only between 0.2% and 0.3% per decade in the tropics between 10°N and 10°S.
- In the northern hemisphere, at temperate and polar latitudes, the frequency of severe rainfall events has increased by 2–4%, in association with '*changes in atmospheric moisture, thunderstorm activity and large-scale storm activity*'.
- The subtropical latitudes of the northern hemisphere, between 10°N and 30°N, have experienced less rain (down by about 0.3% per decade) during the 20th century.
- On the scale of a century, there has been only a slight extension of the regions prone to severe drought or severe humidity. In certain areas, for example in Asia and Africa, the frequency and intensity of droughts have increased during recent decades.
- Warm El Niño–Southern Oscillation (ENSO) episodes have been more frequent and more intense, and have lasted longer since the mid-1970s. The IPCC states that ENSO '*consistently affects regional variations of precipitation and temperature over much of the tropics, sub-tropics and some mid-latitude areas*'. It is unlikely that there will be any great modifications over the next 100 years, but even without changes '*in El Niño amplitude, global warming is likely to lead to greater extremes of drying and heavy rainfall, and increase the risk of drought and floods that occur with El Niño events in many different regions*' (IPCC, 2001, p. 16)!
- No significant trend is suggested concerning modifications in the intensity and frequency of tropical and extra-tropical storms during the 20th century.
- As for extreme weather events, it is thought likely that there will be more instances of intense rainfall, that land masses will be drier in summer with increased drought risks, and that tropical cyclones will harbour stronger

maximum winds and deliver more rain, of both normal and extreme volumes. Note that these predictions are somewhat out of line with that in the preceding paragraph. Also, the IPCC points out that small-scale phenomena such as tornados, thunderstorms, hail, and lightning are not simulated within the models.

- It is probable that warming '*will cause an increase of Asian summer monsoon precipitation variability. Changes in monsoon mean duration and strength depend on the details of the emission scenario*'. However, the reliability of such a prediction is limited by the way the seasonal evolution of the monsoon is simulated by the models ... Is the above reference to an 'increase in variability' in the monsoon meant to suggest that we know, at last, how to explain the irregularity of the monsoon? And of its rains? And of the perturbations within the monsoon?

In summary

The Third Report of the IPCC (2001) repeats the same *credo* of global warming, and extends the reach of the rise in temperature from 1.4° to 5.8°C. It reaffirms and reinforces the controversial formula of the Second Report (1995) regarding the human contribution to climate change. It also drops in the 'hockey-stick shock', according to which the 20th century has known the greatest upheaval in temperatures for a millennium.

It also embraces new aspects of the climate: temperature and sea level are now joined by precipitation, extremes of weather, ice, ocean circulation, perturbations, monsoons, El Niño ...

Not content with predicting climatic evolution until 2100, it drives far into the future, since 'anthropogenic climate change will persist for many centuries', and even for 'thousands of years'!

This Third Report retains a highly scientific look, but lacks reservations and seems overloaded with 'certainties'. With its peremptory pronouncements, its embracing of all aspects of the climate, and its adoption of the 'hockey stick' diagram which despises the patient contributions of climate research, it might be said to have become an instrument of eco-militancy, or a propaganda bulletin for an unconditional greenhouse effect scenario.

3.5 THE IPCC REPORTS: A SYNTHESIS

An analysis of the three successive IPCC reports (1990, 1995, 2001) reveals some points in common, and some differences. Utilisation of the *Summaries for Policymakers*, as worked over by approving IPCC 'experts', points to a deal of intervention from outside the scientific community *stricto sensu*, mostly from ecologists, politicians, and the media.

The points in common are:

- The IPCC reports retain the spirit of the second Villach Conference (1985), in that they discuss: (a) the anthropogenic or enhanced greenhouse effect, and the human contribution; (b) the notion of global warming, associated with an enhanced greenhouse effect; and (c) rising sea levels, mainly as a result of the expansion of the water.

- Analyses are based upon numerical climate models and their predictions, and although these are wide-ranging, the accent is on maximal effects.
- Uncertainties are still mentioned, not only as regards to lack of knowledge of climatic mechanisms but also to their representation (parameterisation) in the models. Such reservations are too often afterwards forgotten, glossed over or sidelined.
- The reports present observed facts, supposed to confirm the predictions made by the models. So, for example, warming is mentioned, but cooling hardly at all – does it never get cold? So **'climate change' becomes synonymous with 'warming'!** What is more, everyday, common phenomena, familiar to all climatologists, are well highlighted and might seem to be something new to the uninitiated ...
- The content sometimes labours the obvious: in discussing rising temperatures, the conclusion on the subject of temperature minima and frosts more or less states that if it's getting warmer, it will be less cold!
- 'Climate change', which necessarily means 'global warming', **is always presented as a harmful eventuality**, very much in the spirit of the definition within the Climate Change Convention of 1992.

The differences are:

- At first, carbon dioxide was considered to be more or less the only culprit, then other greenhouse gases entered the frame, now that their natures and evolution were better understood. But the telling role of water vapour, the principal greenhouse gas, remained a poor relation, or was even (deliberately?) ignored.
- The presumed rise in temperature was at first played down in comparison with Villach in 1990 (by 1–3°C), and in 1995, in the face of the importance ascribed to aerosols (about 2°C, one-third less than in 1990). It took ten years (1990–2001) for the IPCC to bring it to the 1985 figure, a prediction of a mean margin of 1.5–4.5°C, though a rise of 6°C was already on the cards for 2100. The extension of the margin to 5.8°C in 2001 was not though a new idea, not least because it matches Arrhenius' estimate of a century earlier!
- In the last report, estimates are given in percentages per decade, whilst probabilities are expressed in percentages according to the following scale: virtually certain = >99% confidence; very probable = 90–99%; probable = 66–90%; medium probability = 33–66%; improbable = 10–33%; very improbable = 1–10%; exceptionally improbable = <1%. This preciseness, in an area so full of uncertainties, seems unwarranted, if not derisory.
- Global warming was at first attributed to natural variability in 1990: to classic factors like the Sun and volcanism ... No firm response was expected in the short term, let it be remembered. Then, **in 1995, human beings definitely entered the equation** and the 'anthropic signal', once so difficult to discern, was now much in evidence: lightning progress!

- After the long period known as the 'Little Ice Age', the 20th century seemed to be the warmest since 1400. Or so the 1995 report had it, and all climatologists were aware of this ... However, data for the period before 1400 were thought too fragmentary to allow any reliable temperature estimates.
- This did not prevent the 2001 report from looking back to the year 1000, and putting forward the much-vaunted 'hockey stick', which cheerfully did away with the Medieval Optimum vouched for by so many previous climatic studies.
- As each report appeared, the field under investigation widened and, as well as temperature and sea levels, the IPCC embraced precipitation, ice, extremes of weather, tropical and temperate perturbations, monsoons, El Niño ... the IPCC rake went far and wide, drawing in everything around! So climatology became more and more simplified ...
- The IPCC now gave the impression that all was known, the causes of everything were understood, to the extent that it was now possible to explain any meteorological variation.
- Above all, the IPCC gave the impression that the greenhouse effect causes everything! There is nothing left to discover, since the one factor determines all. Nothing was safe from global warming, and by definition, from man ... we were not far now, according to this argument, from seeing man as the principal factor in climate, responsible for warming (and sometimes for the cold too!), in fact for all climatic disasters. The power we wield!
- **Reliance on models continues to increase**. This may signify that the ability of models to predict the climate has become greater, especially given the improved capacity of the equipment; it may also mean that more confidence is being expressed in their predictions. It could also mean that the modellers now have ultimate confidence in themselves, or that their influence within the IPCC is now such that nobody is applying the brake to their peremptory pronouncements. Or it may even be, and how welcome, that brilliant discoveries have been made, erasing all the inherent deficiencies of modelling (Chapter 7). And obviously, the uncertainties will at the same time fade away.
- Predictions made by models, which at first looked ahead towards 2025 (in 1990), were extended to 2050, then soon afterwards to 2100, and in the end well beyond that. Even to the point of looking several centuries, or millennia, into the future ... are we still talking about the climate, or are we now trespassing on Nostradamus' territory?

IN SUMMARY

Differences exist between the basic reports of Group I of the IPCC (mostly their scientific bases) and the *Summaries for Policymakers*, which deal with non-scientific aspects. The principal editor of the reports, Houghton, reminds us that '*the science must never be compromised*', but it seems that this is not always easy to achieve, for he adds '*but scientists need to be patient with politicians*' (Masood, 1997). How far might patience stretch in this field?

Now these *Summaries for Policymakers* are official documents, and are very widely distributed and quoted: they carry the message of 'climate change'. If they contain exaggerations or errors, the IPCC scientists ought to be making sure that such imperfections do not eventually appear. Backing them, even if the final text has not been approved by the 'experts', gives these documents a scientific standing which will not be questioned ... and they become the climatic 'bible'. Hence the above analysis.

The IPCC has issued three successive reports (1990, 1995, 2001). These reports further the work of the second Villach Conference of 1985, where an '*indisputable body of knowledge*' was affirmed, as Lenoir (2001) points out. Since then the IPCC has played down Villach (1990), not adding much that was basically new, but enlarging the range of possible outcomes. By announcing a 5–6°C temperature rise in 2001, the IPCC rehabilitated the 1985 prediction, but it was really the same as that of Arrhenius in 1903, and he used no models.

Where the IPCC broke new ground was in its two successive *coups* of 1995 (making humans responsible for climate change) and 2001 (with the 'hockey stick' confirming that temperatures had never been higher in at least 1,000 years). These 'revelations' are not really born of the science of climate, but smack of militancy. Moreover, as presumed consequences multiplied, the IPCC progressively made the greenhouse effect, and in a manner of speaking, humans, responsible for everything. The greenhouse effect becomes the universal panacea, explaining all.

So climate science has gradually given ground, and climatology has become so much simpler and very 'accessible' as it is reduced to a few slogans, or even to just one *leitmotiv*. One question remains to be asked, though: is the third letter of the abbreviation IPCC really justified?

4

Science, media, politics...

This public perception of science, organised by the media with the active or passive complicity of scientists, no longer allows even an educated public truly to distinguish between real and bogus discoveries, nor between real and bogus science. It heads doggedly onwards in the same dangerous direction, blurring the distinction between science and magic.

François Lurçat. *De la Science à l'Ignorance*, Editions du Rocher, Paris, 2003.

What has become of the Intergovernmental Panel on Climate Change (IPCC) reports, and more especially the *Summaries for Policymakers*, the most widely distributed of the documents and usually the only ones to be read? These have been very widely published in various forms by the IPCC press service, as agency releases, information packs and publications, on Internet sites (www.ipcc.ch) and at press conferences. The IPCC relies very largely upon the media to ensure maximum publicity for its reports, and even issues an advance press pack for meetings and conferences. In this way, on the very day of the event, the headlines will appear, and sometimes even the main conclusions of forthcoming discussions! With the great proliferation of meetings and conferences on the subject of climate, and the associated media coverage, we have reached the stage where it is '*difficult today to pick up a newspaper or view television without encountering a story or an interview on the subject of global climate warming, now familiarly embraced in the vernacular under the rubric of the greenhouse effect*' (White, 1989). This 'bludgeoning' by the media is nothing new, having occurred in other guises before all the interest in 'climate change'.

4.1 PREVIOUS MEDIA EVENTS

Without dwelling upon the Dust Bowl saga of 1930, which received considerable attention in media reports the like of which are still with us, we can still point, unfortunately, to a great number of dramatic events offering much scope for wide *reportage*.

4.1.1 The Sahel drought

In the 1970s, and particularly in 1972 and 1973, the severe lack of rainfall in the southern Sahara, commonly known as the 'Sahel drought', brought together all the ingredients for a media 'catastrophe': sun-scorched landscapes, encroaching sand, desertification, famine, mass migration, dead animals. That there was undoubtedly real distress goes without saying: and there would be still more in 1982–1984, but its treatment in the media was already apocalyptic in tone. Images of desolation abounded: fields of withered vegetation (a normal sight in savannah areas in the dry season, when the pictures were taken); cracks across the dry ground (a frequent occurrence in dried-out clay soils, especially if essentially composed of quickly contracting montmorillonite); and that dead zebu by the laterite or rocky path (a precursor image, perhaps, of that block of ice, crashing and crashing, again and again, into the sea). Added to these were images from the 'charity bizness' when charitable events were held on behalf of the Sahel, and when there was for example a famous 'concert for Ethiopia'.

In 1976, when drought struck France, the appropriate formulae were already in place and we soon learned that '*southern France is threatened with desertification*', below a photo of a field of withered maize (echoing those Sahelian sorghum fields). Images of poultry which had died of heatstroke in battery farms, recalling the 'dead chickens' of the Dust Bowl era, now reappeared and continued to do so, especially in the dog days of summer 2003, when the chickens were joined by turkeys ...

In 1981, *Der Spiegel*, having already pictured Berlin locked in ice, an abiding image of terror, introduced the idea of acid rain, its cover declaring that 'THE FOREST IS DYING'. Soon, all over Europe, we were hearing of factory fumes and devastated forests in every newspaper and on every television screen. And another forest, this time the rain forest of the Amazon, went up in flames in all the media. The idea of the forests as '*the lung of our planet*' soon developed, and much compassion was expended on behalf of the slashed and burned forest and its Indian inhabitants: the media shed tears as only they know how.

4.1.2 The ozone affair

The ozone question is one which deserves to be mentioned: not for its own sake, since, until it is proved otherwise, no one has yet shown that the concentrations of stratospheric ozone has been a direct factor in climate change. Locally, this is not the case with the ozone in lower atmospheric layers. We mention it because of its similarity to other facets of the catastrophist scenario.

The 'nuclear summer' scenario was in the news in the early 1970s, since one of the main results of a nuclear conflict would be the destruction of the ozone layer and the subsequent irradiation of the Earth by deadly ultraviolet rays, annihilating all life. Refuted in 1973, this scenario gave way in 1983, with little astonishment shown by the media, to the 'nuclear winter' scenario (cf. Maduro and Schauerhammer, 1992). This was immediately mirrored in the climatic field by images in an eager and mindless press of Europe's cities buried in ice instead of being invaded by sand dunes. In the field of so-called 'information', it's 'here today and gone tomorrow'.

The ozone and global warming scenarios have something in common: the fact that there is *nothing* (spectacular) to see, since a direct relationship between (supposed) causes and effects is not immediately obvious. Everything stems from a laboratory hypothesis, or indeed some co-variation, and no physical, causal link is demonstrated. What can be done about the absence of images of an ozone catastrophe? Firstly, there are those much-vaunted, false-colour 'ozone hole' images, with garish hues, all the better to bring out the diminution. This type of misleading image would be replaced, when climate was discussed, by thermal printouts from the models, in bright red for the most part. Then there were the impressive headlines, a memorable one being 'the spectre of cancer ...', certain to hit home. On the other hand, what journal was it that reported the measurements over the Antarctic made during 1957–1958, the International Geophysical Year (IGY), measurements which showed that such a seasonal diminution was already happening? Chlorine levels were very much dictated by the volcano Mount Erebus. Whatever became of all those catastrophist scenarios, for example the one in 1975 which predicted an 18% diminution in the ozone layer by 1990?

So the media approach to global warming follows tried and trusted 'recipes', hurriedly disseminated as they are rapidly reproduced. They are provided ready-made, or the old discs and CDs may be recycled, more and more frequently as meetings and conferences about 'the climate' succeed each other.

4.2 THE MEDIA PRESENTATION OF 'GLOBAL WARMING'

The 'mediatisation' of the idea of the climate in the press and on the television, here analysed through extracts from the French media, creates headlines and images of an 'eloquence' designed to impress the reader.

There follows a selection of such headlines, some of them rather jumping to conclusions, picked at random from the French press.

'*The Earth is hotting up, and mankind has given it fever*' (*Le Monde*, June 2000).

'*In twenty years from now, Copenhagen will have a climate like that of Paris, and Paris will be as hot as Madrid*' (!) This prediction was made in 1984, so it should by now have come true, in 2004! But there seems to have been some delay: *Paris-Match* (issue 2847) stated in December 2003 that '*climatologists are sure of it: in less than 50 years, vines will be growing in Lille, palm trees will thrive in Lyon, and there will be new outbreaks of (in the original French) "le paludisme et la malaria"*' – an interesting pairing, since *le paludisme* is the French for malaria, originally an Italian word evoking the 'bad air' of marshland thought to be associated with the disease.

'*A severe heatwave in 2050 or thereabouts will disrupt landscapes all over the planet*'.

'*All the studies agree: in the next 100 years, our planet will suffer a drastic heatwave – it is feared that deserts will spread into the temperate zones, and (since catastrophes never come singly) tropical typhoons and hurricanes will move into the temperate zones*' (*Libération*, August 1988). Since when have cyclones not reached temperate zones?

'*Can Man escape the climatic catastrophe?*' (*VSD*, 1994).

'*4° warmer? There would be near-Siberian winters in Europe, while the rest of the world sweats buckets*'. Here we have cooling caused by warming! Did journalists invent this particular scoop?

'*Consensus reigns within this group of climate experts called upon by the United Nations ... from now on, there will not be two extreme and opposing camps when it comes to our distant and largely imperceptible future. There is certainty, near enough to touch ...*' (*Le Monde*, 26 November 1997). Consensus? 'Experts'? Certainty? Are we being a little too hasty here?

'*The paradise islands of the Pacific are sinking into the ocean*' (*Le Progrès*, 7 November 1999). Of course, the name of any island which has disappeared beneath the Pacific is not supplied!

'*Scientists are convinced that, for all these imbalances, there is only one culprit: El Niño*' (*Le Monde*, 4 October 1997). The 'universal answer' has been found!

'*Why there will be more storms*' (*Science et Avenir*, November 2000). After the storms of December 1999, whose causes are still unexplained, the causes of future storms will be revealed!

'*Storms heralding evil weather, cars spewing out tomorrow's climate, and rising waters threatening a new Atlantis*' (*GEO*, 2000).

Whether written or spoken, in newspapers, in magazines, and on screens, the words are accompanied by similarly evocative images: images not always totally relevant to the subject, but chosen for their shock value, purely and simply. Most prized of these images is usually the satellite view (*Meteosat*, 'full Earth', of course), since the subject is the climate. The photo, however, is of no use at all, since it has nothing to do with the subject! Then there is the inevitable iceberg (a reference to the *Titanic* will suggest catastrophe), and **the** famous ice block crashing into the sea (always the same block, much reproduced, encapsulating the coming 'inexorable' melt). Or the polar bear in distress, sometimes seen floating on its back in the sea! The photo showing dried-up clay soil has long featured in the album, next to the coconut palm bending before the wind, fronds whipping around in the gusts to show the violence of the cyclone. The tornado is also inescapable, even if a photo of a tornado in the USA doesn't have much in common with strong winds and storms in France ... Other contenders are the plumes of volcanic eruptions, waves crashing onto sea walls, a tangle of boats in a harbour at low tide, and pictures of floods, with mention of those in Bangladesh ... and the inevitable, ritual computer screen, symbolising triumphant modern science and the omnipotent *model*!

In most cases it is not the primary concern of the media to inform, but to make the most out of the event, impressing readers and viewers and thereby selling papers and boosting audience ratings. Thus the '*perverse taste of the public for thrills and apocalyptic forecasts*' (Lenoir, 1992) is satisfied. So it is thought best to add fuel to the fire, preaching climate change as a synonym of warming and catastrophe, as these sensationalist statements show:

> '*Rarely has the planet known such a succession of climatic disasters ... The climate has gone mad. Why?*'
>
> '*Questions about an unsettled climate ... Droughts, tornadoes, floods: great climatic chaos ... a planet in climatic turmoil*' (*Le Monde*, August 2002).
>
> '*We must save the Earth ... Nature in crisis ... Ten years to save the world ... 5000 days to save the Earth ... These computers, which will save the planet ... The Earth is going crazy ... Operation Survival ... The greenhouse effect and a climate at war ... Heal the fevered Earth ... Our Earth in danger ... The end of the road for humanity ...*'!

Kohler, himself a journalist but with a scientific background and therefore in a position to offer an informed view, wrote in 2003: '*Each meteorological event which is just a little excessive (a severe drought, a flood, a lack of snow for winter sports, a storm, or a violent cyclone) is presented not as a piece of news but as another argument for the theory that man is causing climatic warming*'. So the media show not the slightest doubt, and any uncertainty or reservations are either kept out or very rarely expressed in the small print ... Attempts at correction sent to editors by thinking readers stand little chance of being considered for publication, as they would overshadow or destroy the magic of the sensational! The subtle mix of observed facts, even if they have little to do with the subject, and computer predictions, which may not be allied to those facts, gives the public the impression that the facts are well established, part of an 'undisputed' reality. The reality is all the more undisputed since the news-sheets have no interest in contributing to scientific discussion, but deal with the facts in isolation from their context. Thus the debate about climate variation is ignored: it would take up too many lines or too much airtime. A proper, in-depth debate would not be of interest to the world at large.

The way the facts are presented and illustrated in the media is all the more 'undisputed' since they originate from the scientists themselves, which guarantees their acceptance.

4.3 THE INFLUENCE OF SCIENTISTS UPON THE MEDIA

As Kohler underlined in 2003, it is worth looking at how the press deals with scientific matters. Certain events such as volcanic eruptions, earthquakes, and eclipses are newsworthy in their own right, and these are usually treated objectively and the facts duly reported. Other events are 'managed'. For example, a press release may be sent to news agencies by a researcher or an organisation (e.g., NASA, WMO, IPCC, CNRS) in order to publicise some personal research or a laboratory project seeking funds. Such releases are usually short and to the point, making them easy to process, and as such they are often published 'as seen' by journalists on a deadline.

Because of the '*tendency for the various media to copy each other*' (Kohler, 2003), communiqués are widely disseminated, and media which might not have published them feel bound to catch up on them later on. So some rather commonplace piece of news, after many repetitions and overstatements in the press, will appear to the public to be a real 'event'. It is worth noting the sad practice of mentioning in the communiqués or press reports only those things which are, or are supposed to be, grist to the mill of global warming: heatwaves, droughts, storms, floods. The accumulation of such reports gives the impression that the weather is 'in trouble', but if considered in their wider geographical context, such happenings are normal for the places where they occur. However, ignorance of climatic matters is so widespread that only a few will realise this. As a result of this selective information, the only winters reported are those without snow on the mountains, the only glaciers we see are those which are retreating (the advancing ones seem not to exist!), and it is never cold any more, anywhere. Falling temperatures, even if dramatic, slip through the information net! But just let the thermometer rise a little, and the media are full of it! This can only reinforce the myth of warming, so constantly thrust upon us. Warm periods are seen as catastrophes, a common enough but paradoxical vision since everybody spends the long winters dreaming of summer sunshine and heat! The persuaded public asks for more of the same, and is interested only in the official pronouncements, suspecting those who might not be 'climatically correct'!

4.3.1 Use of media

Scientists are in the habit of using the media to disseminate their messages more rapidly. Maduro and Schauerhammer (1992) mentioned the treatment of the ozone question by Watson in 1988 and 1991. In 1988, Watson, a chemist and president of NASA's Ozone Trend Panel (OTP), announced at a press conference, before papers were published, a diminution in ozone by 2% to 3% between 1969 and 1986 (i.e., by 0.2% annually over 17 years). This was proof, he said, of the destruction of the ozone layer by CFCs. Taking up the question again in 1991, as vice president of the United Nations Environmental Programme (UNEP) ozone committee, Watson stated that new (unpublished) results showed a diminution of 8%, implying 275,000 extra skin cancer cases each year! In 1992, this became 300,000 more cancers and 1.6 million more cataracts. Singer commented scathingly in the *Washington Times* of 20 November 1991 that '*environmental policy seems once again to be driven by press release rather than by proven scientific data*'.

This did not prevent Watson, being a real 'professional' in communication, from repeating the performance when he was president of the IPCC (and still a chemist). In 2001, he inserted into the final communiqué a prediction of an average temperature rise of 5.8°C, stating that '*it would give an extra impetus to governments to find ways of respecting their commitments to reduce greenhouse gas emissions*' (see Grenier, 2002). It will be recalled that the press repeated this value as a major prediction of the Third IPCC Report!

This utilisation of the media is very frequent, and accepted by scientists themselves: Schneider, for example, was an ardent defender of the theory of a return to glaciation, then later became an equally fervent supporter of warming. Having said that '*scientists were ethically bound by the scientific method*' (which apparently, is not of the first order of importance), he also said: '*We need to get some broad-based support, to capture the public's imagination ... getting loads of media coverage. So we have to offer up scary scenarios, make simplified, dramatic statements ...*' (in Maduro and Schauerhammer, 1992).

4.3.2 'Scientifically settled...'

Journalists draw reassurance from scientists and their work, since their own training is of course inadequate to equip them with the necessary critical base. Are there many journalists with a qualification in climatology? So they content themselves with trusting in the publications, and more especially those of the IPCC. Their trust is reinforced if they believe that, as *Le Monde* put it on 24 November 2000, '*the IPCC expresses the views of the international community of climatologists*'. Are they deliberately ignoring the fact that the IPCC is by no means merely a collection of 'climate experts'? There are many stock phrases beneath which the media shelter: 'The climatologists agree ...'; 'all the studies confirm ...'; '*scientists are convinced ...*'; '*a consensus has been reached ...*'. They may represent only a fragile umbrella, but what matters is to keep the responsibility with somebody else! It must be said that the IPCC reports are a 'gift' to journalists, who could not invent such things. The IPCC Group II reports are especially welcomed, since they discuss the possible impacts of the supposed warming (i.e., threatening simulations '*imagining maximum damage ... and alarming everybody*' (Lenoir, 2001)). It takes little further effort to add a few 'shock horrors' and appropriate images: to picture, for example, Paris locked in ice, Paris invaded by sand dunes, Paris under water with the Eiffel Tower projecting from the flood ... Ideas which one might think must be made up, since they are so exceptional, stem directly from the work of the IPCC, lending its 'scientific' endorsement. Scientists themselves over-contribute in this area. For example, Le Treut and Jancovici (2001), having noted that the '*possibility of polar ice melting is almost nil at this time*', also pointed out that '*partial melting in the Antarctic would have phenomenal and irreversible consequences*', and that rising waters 'also threaten the existence of certain islands'. In other words, nothing to fear for the next 100 years, but we ought nevertheless to prepare for the worst at any time! How should we respond to the apocalyptic predictions of the same authors: '*how can we stay healthy if there is no longer enough food, if toxic products abound or if intense stress levels increase the consumption of drugs and alcohol*?' If ... if ...! Do journalists really need to add to this terrifying and ultimately grotesque gossip?

Thanks to its 'scientific' *cachet* and its endorsement by 'scientists', all this has become certain fact. Global warming as a result of human activities is now a reality in all the media. So now almost anything goes, bearing the label of 'scientific coverage', like this explanation of the 1988–1989 drought in France: '*Through the interplay of ocean waters, girdling the Earth at our latitude, the meteorological situation in France is linked to that in California. This "teleconnection", as specialists know it, is itself under the influence of the equatorial zone. Given a shortfall or excess of energy in this very active belt, the shifting boundary between high and low pressure areas moves northwards or southwards. This is why la douce France is occasionally visited by rough weather...*' (*Le Point* 1990, No. 906). What a dazzling exposition! And the dry spell of summer 2003 was to be cast in the same mould! Whether or not it is intelligible is neither here nor there, as long as the 'specialists' umbrella is open and – most importantly – the magical character of the climate is invoked!

4.4 THE INFLUENCE OF THE MEDIA ON SCIENTISTS

The media needs the work of scientists, and their 'authorised' endorsements. A strange metamorphosis sometimes overtakes 'scientists' when they are guest speakers or during interviews. Necessarily introduced as a 'specialist' on the subject in hand, the scientist may then be declared a 'climatologist'; the media creates its own complaisant 'climatologist'. Remember that the number of 'climatologists' – or those who are so labelled or bestow the title upon themselves – has grown amazingly quickly over the last decade. One eloquent example is that of Jancovici, who said on a public broadcast that he was a 'climatological engineer'! Here is a seemingly new specialism, though he has never done any analysis in the field of climate science; he can, however, produce chapter and verse from the IPCC on demand!

At the microphone, very few of us are able to admit our ignorance on the subject in question, or say that it is beyond our competence. Take for example the following reply to a question about the causes of the floods along the River Somme in northern France: '*There is no reason to say that the flooding in Picardy is due to the greenhouse effect ... On the other hand, we can say that these heavy rains are exactly the kind of event which the models tell us will become more frequent as the century proceeds*'. To put it in fewer words: 'we can say but we can't say'! Petit, who made that statement, was (oddly) a member of the French delegation to the IPCC Working Group I(!), and was labelled a 'climatologist', although he has no diploma or experience in that field. He continued: '*To those who, like me, are convinced that the warming of our planet is a reality, this is the kind of thing we must expect. Be aware though, this does not mean that there will be floods every year. No we'll see some dry years too*' (Petit, *Le Monde*, 19 April 2001). It is astonishing that such a futile declaration can be made based upon 'conviction' (in what area are we speaking from conviction?), predicting anything and nothing. Equally astonishing is the fact that a serious newspaper sees fit to publish such a meaningless thing!

What is more, certain scientists may be 'used', actively or passively, by the media to rubber stamp the prevailing theory, with certain (partial) journalists encouraging the 'right' replies. The journalists have a list of 'on-side' specialists, to guarantee the impact of what is published or broadcast. Some of these specialists have become unassailable as media referees, and they may be brought out to comment even on some scientific news item marginal to their specialism. They are happy to do this, since they need the press to further their 'brand image', generally with some new popular science book in mind (Kohler, 2003). The upshot of this is that it is always the same scientists (or pseudo-scientists) who make those sensational statements, using the fame of the institution to which they belong and often acting under instructions in order to secure funding. Sometimes the 'enlightened' media will offer the 'zest' of some lively debate, but they are careful not to invite some sharp adversary who will put their star witnesses in the shade and (especially) risk bringing out the inanity of their views ...

In France both the National Assembly and the Senate have declared the battle against the greenhouse effect to be a 'national priority' (cf. UNFCCC, 1992, chapter 2). The 'servants' of the state, and, in their name, both audio-visual media and laboratories, feel bound to propagate the official dogma, just like a certain press agency in the East in its heyday; echoing the triumph of Lysenkoism, they mould public opinion in favour of the official theses. They organise broadcasts, sometimes (falsely) labelling them 'debates', but the guest 'scientists' are hand-picked to obviate any unwanted opposition. And so year after year they recite the old refrain, as is expected of them, to their own advantage. Reports and documents are selected in the same spirit. Some journalists try occasionally to respect differing opinions, bringing in contradictory views, but the magazine or TV editors will put them back on the 'right track', in the well-worn furrow. For example, *Arte*, a national channel, did a *reportage* at our own climate laboratory in Lyon, LCRE, for five hours in April 2002. However, a few weeks later, this sceptical account of global warming was withdrawn at the editing stage before it ever appeared on screen. In the same vein, *France 2*, a public TV channel, featured a programme called *Climaction* under the sponsorship of a state-run company, to 'mobilise the public against climate change' and spark a 'citizens' response'. After the usual pictures of storms, floods and melting glaciers, a so-called 'climatology expert' (obviously not a climatologist but paid to act the part) came out with the kind of shallow stuff heard in bar-room debates. Among other naïve statements was advice against using cars, even for a weekend trip to the beach or the mountains, and against long-distance holidays, using up jet fuel ... Apparently he didn't think of advising against breathing, with all the extra CO_2 it produces, or against the inopportune venting of methane. Also, he didn't even say whether he himself acts in support of such odd views. This deliberate patronising of the public on a national TV channel was supported by Electricité de France/Gaz de France (EDF/GDF, national power companies) who pride themselves on their production of (atomic) energy without the emission of CO_2 ...

4.4.1 The scientific media

Within the media there are the self-styled scientific magazines, always on the lookout for articles on global warming and its consequences. The most obvious banalities can be disseminated under this heading, as long as they contain the current 'buzz-words'. For example, *La Recherche* carried an article in 1991 on the future of the Camargue (i.e., the Rhône delta), entitled *The Camargue in the 21st Century*. Its author was not a climatologist, nor a hydrologist, nor an oceanographer, and had never shown any previous interest in sea levels, or in the Camargue. The reasoning was simple: temperature rises, sea level follows suit, Camargue drowns! Anyone could have written this, just by following the map contours: one more metre, two, three, and

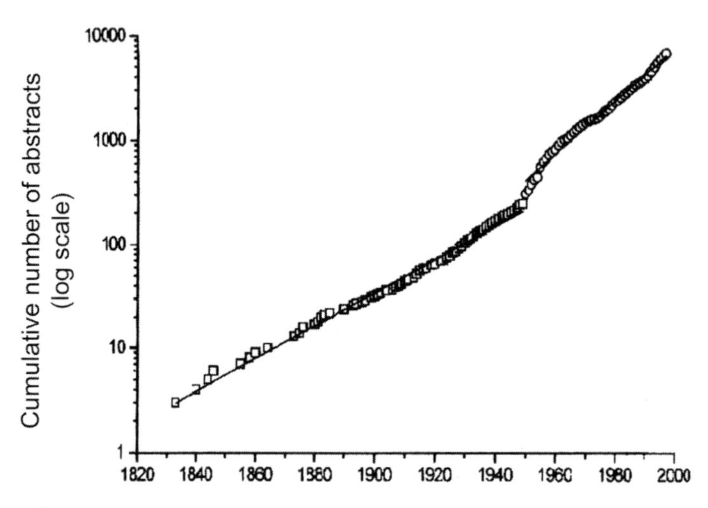

Figure 5. The growth of the scientific literature on climate change.
After Stanhill (1999).

the delta is progressively inundated, by simple cartography! Spectacular it may be, and it appeals to the emotions, since the Camargue, with its annual gypsy pilgrimage and its landscapes of horses, bulls and birds is a cherished heritage area ... but scientifically, the article has no value. However, it gets published. Such a simplistic procedure is apparent in many articles, notably on Bangladesh after the disaster of 1970, or it could be Holland, or the Nile delta, or the lagoon of Venice ... and low-lying islands and coastal areas, with IPCC graphics illustrating the 'Maldives syndrome'!

This editorial opportunism also causes the problem of quality in so-called peer review, which is supposed to guarantee the maintenance of standards in publications. Sometimes, the 'right' reviewers are called upon, a co-option system well known to the publishers, and to the IPCC. As a consequence of this, we all know nowadays that it is very difficult, if not impossible, to have work published in some publications (even those at the top of the range) if one expresses unconventional ideas (i.e., ideas not in line with 'climatical correctness'). It may be that publishers want to avert any risk of displeasing their readers, who expect their dose of catastrophe according to the 'gospel', or perhaps the publishers themselves are biased and eco-militant.

This selectivity has had a further effect. It has led to an inflated number of publications on the subject of climate change, to the detriment of other fundamental subjects, as shown in Figure 5.

So this aspect of climate seems to be (almost) the only one that matters. It is as if all other climatological problems have been solved, and interest in basic research is on the wane. This could give the impression that the scientific analysis of climatic evolution is cut and dried, and that there is nothing left to discover or demonstrate: all is (apparently) already known, and totally understood ... Which is, alas, very far from the truth!

4.5 TOWARDS A 'PEOPLE'S CLIMATOLOGY'

The interplay between scientists and the media has made climatology *the* subject of the moment *par excellence*, and 'climatomania' is now with us. What 'climatology' is really meant, though?

Climate change is dealt with, alongside pollution for which it is made the excuse, largely by non-climatologists whose qualifications extend only to repeating the IPCC *Summary for Policymakers*. Specialists and qualified people are much in evidence, however, but their fields of expertise are data processing, modelling, mathematics, statistics, agronomy, glaciology, chemistry, history, geology ... but not climatology and its dynamics. Climatological knowledge is thus generally limited, as the IPCC implicitly recognised when it stated that '*the aptitude of scientists to verify the predictions of models is often limited by incomplete knowledge of true climatic conditions*' (UNEP–WMO, 2002). 'Explanations' are therefore very simplified (or simplistic) as a rule, and they are presented in a very schematic way. They cannot reflect, or may only partially reflect, a complex scientific reality. Such superficial and elementary 'knowledge' stems in the first place from '*the inevitable simplifications built into the construction of models*' (Le Treut, 1997). This aspect will be analysed in Chapter 7, which deals with models. But the simpler (or more simplistic) the message is, the nearer it is to an easily absorbed 'slogan', and the greater is the likelihood that it will be taken up and disseminated, especially by the media, ecologists, unqualified self-styled climatologists, and an unsuspecting public. The system of 'immediate consumption' does away with the need for reflection or overcomplicated and lengthy (and therefore boring) explanations.

The kind of 'climatology' created by this Pavlovian process, depending on tenuous relationships involving two or at most three parameters, may be summed up thus: temperature, rain, weather, wind, or in a word, climate ... it all hangs on the greenhouse effect. It is getting warmer, or will do so; and things are going, or will go, from bad to worse; and this will happen everywhere across the world. With this kind of material, it is a fact that anyone can be a climatologist without the slightest intellectual effort.

So the 'people's climatology' is staggeringly mediocre. Let us consider an example chosen at random from among so many. On 2 January 2004, a documentary from Gédéon Programme on the national *France 5* channel, focusing mainly on rising sea levels, brought together work from the USA and Canada, supported by scientists from reputable American universities and laboratories. Two locations were featured: Shismaref, Alaska, on the Bering Strait, and Tuvalu, a nation of nine atolls athwart the 10°S line of latitude in the Pacific. The well-known argument was rehearsed: rising temperatures (illustrated, inevitably, by blocks of ice crashing into the sea) will lead to a rise in sea levels and an increase in the severity of storms (two closely associated aspects, for some unknown reason). The two low-lying islands in question are therefore threatened with imminent disappearance ... obviously, the programme could not leave out the theme of '*innocent islanders suffering the consequences of overconsumption by industrialised nations*' (the basis of Maldives syndrome). 'Serving scientists' showed eroded beaches, underlining the threat of submersion – although the actual submersion cannot be shown, since mean water level is not indicated! Honest viewers cannot help being moved by the plight of the islanders, and they are convinced that the danger is imminent by such an array of 'real science', striking images and positive, moralistic comments.

However, this so-called 'demonstration' is, climatologically speaking, extremely silly.

- First, by its choice of sites thousands of kilometres apart: do they really experience the same aerological and hydrological conditions?
- Is the choice of sites supposed to indicate that the rise in sea levels is global, and associated only with the presumed rise in temperature?
- Could a real rise in sea levels, of perhaps a few centimetres (but was it measured?), account for the erosion of beaches – might not storms be a factor here?
- Is there a properly established physical link between warming on the one hand, and a rise in sea levels and more severe storms on the other? In exactly what circumstances did this erosion of the sand occur?
- If storms have indeed increased in violence, is this because of the presumed warming? And would Shismarev and Tuvalu experience the same kind of perturbation?
- On Shismarev, as elsewhere in the extra-tropical zone, storms should be less severe, given that a warm scenario lessens the violence of weather interfaces (i.e., a warming episode ought to cause exactly the opposite effect to that predicted by the 'scientists' interviewed, who seem to think, however, that '*the hotter it gets, the more severe the weather will be*'!) On the contrary, if temperatures were really rising ... the weather would become increasingly milder (cf. Chapter 11). So what does explain, or what would explain (cf. Chapter 13), the erosion of the beaches, if storms were becoming less violent?
- On Tuvalu, which is normally not visited by tropical cyclones, why are these 'unusual' perturbations appearing, bringing increasingly violent weather to an ocean paradise?
- Also, are the atolls of the Pacific (and elsewhere) actually sinking beneath the waves? What is it that can raise, either temporarily or permanently, the level of the ocean around Tuvalu (cf. Chapter 14)?
- What could explain the rising temperature in Alaska? Floods of warm air from the south? Deeper depressions? An increase in the violence of winds? If all this is going on, what dynamic underlies such behaviour in the weather (cf. Chapter 13)?
- Are modifications occurring in the weather dynamics of the equatorial Pacific (cf. Chapter 14), and more especially is there a displacement in the lows defining the meteorological equator (ME), with a lowering of pressure which might bring about a rise in the sea level?

All these aspects and many more were certainly not analysed in this attractive-looking, impressive documentary: that would make for terribly heavy viewing, and would anybody watch that kind of 'documentary' right through? Nothing sensational to be seen! It is so much easier and quicker to go straight for the shaky 'CO_2 argument', and much more interesting if, in the bargain, you can bring tears to the eyes of simple folk as those islanders face their sad fate!

4.6 CONVERGING INTERESTS

This situation in climatology is a creation of converging interests (cf. Kohler, 2001): four principal sets of players, scientists, ecologists, politicians, and journalists, all share convergent views for diverse reasons. The distinction between these four groups, formerly separate but now moving in the same direction, is becoming more and more blurred as they mesh together ever more closely.

4.6.1 The scientists

This is the oldest of the groups, since the question of the greenhouse effect was originally an academic one, confined for a long time within the microcosm of universities. Scientists, generally moderate in their approach, analysed the facts (theoretically) in a neutral and objective way, more often than not. No consensus was ever reached among them on this subject, which sparked many debates, as is proper within science; the balance of arguments would swing first one way, then the other, with no definitive agreement. But certain scientists got somewhat carried away, as was the case in 1985 at the second Villach Conference, or in 1988, before the US Congress, or again in 2001, with the appearance of the temperature curve nicknamed the 'hockey stick' (cf. Chapter 10). Other scientists were quick to denounce publicly the so-called consensus, and they criticised in writing these aberrations of their discipline, while several opposing discussion groups were formed. Example of these were The Science and Environmental Policy Project (www.sepp.org) and the 'Climate Sceptics' (www.yahoo.com/group/climatesceptics). It was a good thing that this spirit of debate was present, in spite of (sometimes official) obstruction, though it was not as frequent and as widespread as it might have been. As Grenier (1992) pointed out: '*It is only when the freedom of debate is called into question that science disappears*'.

In the main, the scientific arguments were moderate and measured, because they took account of real advances as well as uncertainties, and did not in general deal with catastrophism. Evidence of this comes also from a comparison between the last IPCC report (already written in 2001) and the Position Statement of the American Geophysical Union (AGU) on climate change, established 'after a long and careful process' (Lanzerotti, 1999). '*As the premier scientific organization that incorporates all of the disciplines engaged in research to understand the climate system*', the AGU considered it its duty to make clear statements about our scientific understanding of the climatic system and the likelihood of it changing. To review some of its principal points:

- *'Although greenhouse gas concentrations and their climatic influences are projected to increase, the detailed response of the system is uncertain'.*
- *'The increase in global mean surface temperatures over the past 150 years appears to be unusual in the context of the last few centuries, but it is not clearly outside the range of climate variability of the last few thousand years'.*
- *'There are significant scientific uncertainties, for example, in predictions of local effects of climate change, occurrence of extreme weather events, effects of aerosols, changes in clouds, shifts in the intensity and distribution of precipitation and change in oceanic circulation'* (AGU, 1999).

However, this sense of moderation has had its recent 'mad moments', especially after the known facts had been worked into the context of political decisions as 'scientific expertise' (Roqueplo, 1993). This expertise '*transformed inconclusive scientific research into political argument*', since researchers 'let themselves be part of a process of political decision-making and therefore feel obliged to come up with an answer, while in the majority of cases science can give no answer' (Roqueplo, 1993). It therefore becomes difficult for a scientist to reason dispassionately, if funding and publication are often contingent upon the acceptance of conditions which depart from the strictly scientific; and then there are personal considerations of strategy and vanity ... The media attention given to the IPCC has even led to a supposition that '*working for the IPCC or being quoted in one of its reports is seen as an honour*' (Chevassus-au-Louis, 2003). So this very politicised body has become a 'super-jury' in matters of climate science! The position of the independent scientist, as far as independence is achievable, is becoming increasingly difficult, as the system marginalises anyone who refuses to sit in the parrot's cage.

The fragile nature of dispassionate science is demonstrated by the recent AGU (EOS) *Statement on Human Impacts on Climate* (December 2003), which, four years on, now takes the position that:

- *'Human activities are increasingly altering the Earth's climate'.*
- *'It is virtually certain that increasing atmospheric concentrations of carbon dioxide and other greenhouse gases will cause global surface climate to be warmer'.*
- *'"Scientists" understanding of the fundamental processes responsible for global climate change has greatly improved during the last decade'.*

4.6.2 The ecologists and environmentalists

This is a more recent group, emerging as an ecological one, concerned with the environment and having little to do with the climate. Humans were certainly put in the dock: their technology could destroy their surroundings, which it was necessary to preserve. Large-scale pollution, disasters with dramatic consequences, fears for a future overshadowed by the arms race: all these accelerated the evolution of the ecological movement, and the scenario of nuclear war forged a particular link with the climate. The protectors of nature leaned inevitably towards politics, in an effort to ensure preventive measures. Eco-militants thereafter strained every sinew, seizing upon whatever might be faintly favourable to their cause and labelling it a certainty, whether or not it was based on real science: they cited '*the quite extraordinary unanimity of the hundreds of specialists grouped within the IPCC*' (Goldsmith, 2000), referring to an illusory consensus to add weight to their propositions. The idea of human responsibility now carried upon its shoulders the increasingly heavy burden of the ecological conscience. Rightly or wrongly – what did it matter, since the actual question no longer stood in the realm of science?

There was no point now in expressing reservations, and gratuitous affirmations served only to 'upvalue' the environmental movement. As evidence, here are a few extracts from the *Appeal Against Climatic Destabilisation* published in the journal *L'Ecologiste* (Vol. 1, No. 2) in 2000:

- '*Proofs of human impact upon the Earth's climate are now irrefutable*'.
- '*Our climate will become more and more unstable, marked by extreme conditions uncharacteristic of the seasons*'.
- '*Our health and food resources will be dramatically affected by increasing droughts, heatwaves and the spread of insects and other creatures bringing diseases . . .*'. The IPCC predicts: '*millions of deaths across the planet, and millions of eco-refugees*' . . .

Now it was possible to indulge in such verbal flights as '*What good will it do us to have a Mediterranean climate on Spitsbergen if we have nothing to eat?*' (Lutzenberger, 2000), and '. . . *fifty years from now . . . we will have to face up to climatic disturbances which may well prove fatal to the future of our species*' (Bunyard, 2000)!

Such arguments, which may or may not greatly exaggerate the IPCC's predictions, are, in the final analysis, no more misleading than any other political message. And if some real contribution is made to improving the state of our planet, then why not! It may even be worth doing . . . but this takes us away from the object of this book, and from the discipline of climatology.

4.6.3 The politicians and economists

With this group, things are considerably more complex, and markedly less clear. Perhaps they will become clearer in a few years from now, as was the case with the ozone crisis, which nobody now believes was an impartial debate. Its greenhouse aspects will still linger, without mentioning possible interventions by all kinds of lobbies, or a political 'take-over' by certain economic interests: '*climate change science is now Big Science involving Big Money*' (Stanhill, 1999).

The idea of climate change has progressively moved onto the political plane, at the highest level, as an inevitable result of the internationalisation which has been seen since the Second World War, with the creation of various bodies such as the United Nations Organisation (UNO), the International Council for Science (ICSU), the World Meteorological Organisation (WMO), and UNEP. International conferences and worldwide scientific experiments have played their part. Lenoir (2001, p. 188, Figure 27) shows in two diagrams the amazing tangle of international organisations and programmes involved in questions of climate, what he calls the 'climatocracy'. Within a network such as this, the freedom of thought of government-sponsored 'experts' is necessarily limited, and through incentives and constraints the actual contribution of science risks being sidelined and used only as a label of 'good faith'. Lomborg (2001, p. 322) shows how this is especially true within the IPCC: '*Many scientists in the IPCC are undoubtedly professional, academically committed and clearheaded, but the IPCC works in a minefield of policy, and it has to take political responsibility for its seemingly scientific decisions, if they cause obvious biases in reporting*'.

The creation of the IPCC and its placement beneath the UNO umbrella give extraordinary scope for the control of information, or disinformation, about climate change. The nature of the debate has over time been modified: '*The subject of climate change has long ago left the realm of pure science and entered the political arena*' (Lanzerotti, 1999). Politics strongly influences the way things are done within the IPCC. Let us remember that the 'I' of IPCC does not stand for 'International', neither does it stand for 'Institute', since the IPCC is not concerned with scientific research, as is too often supposed. It stands for 'Intergovernmental', meaning that its members are directly nominated by their respective governments. Although some are certainly scientists, who may be more or less up-to-date with the discipline of climatology, most of them are politicians with a governmental brief. The office of the IPCC comprises thirty members elected by representatives of member states, and it is this office which chooses authors and reviewers, through co-option. The argument that reports come from 'hundreds' or even thousands of 'experts' should therefore be taken with a very large pinch of salt, since the small dominant group imposes its views upon a majority which has no climatological qualifications. The review process of the reports, and especially that of the *Summary for Policymakers*, also solicits comment from government representatives and even representatives of non-governmental organisations of the ever-changing ecological scene. Negotiation of the final text, a procedure which Trenberth says has become '*absurd and out of control*', has made the *Summary for Policymakers* '"*a piece of propaganda*" written by representatives of governments which all support the Kyoto Protocol' (according to Linzden, in Chevassus-au-Louis, 2003).

The governments which commit themselves to the defence of the environment, thereby easing their consciences and improving the brand-image of politics, have moreover signed up to the UN Framework Convention on Climate Change (UNFCCC, 1992), which was indeed the one which adopted the Kyoto Protocol in 1997. The IPCC and the UNFCCC, legally two distinct entities, often put forward the same representatives, and the confusion between science and politics is thereby furthered. The UNFCCC moreover obliges signatory states to raise the status of the concept of global warming to that of a 'state religion'. In France, where this concept has been declared of 'national priority', and everyone knows the 'dangers at the door', 'servants of the state' can have no option but to do their duty and serve the state: so there is no place for heretics; we must sing from the same hymn-sheet! It can therefore be said that, since 1992, there has been no proper scientific debate on this subject. Similarly, the designation of government delegates to the IPCC allows no ambiguity, and the adoption of its reports and resolutions therefore harbours no surprises!

4.6.4 The media

We have already discussed the media at some length. They love alarmist news items, and even when they refrain from popularisation, they still succumb inevitably to the lure of the sensational: even to the extent of the American weekly, *Time* magazine, declaring planet Earth to be its 'Man of the Year'! The contributions of scientists are used in dramatic and spectacular ways, and events are sometimes created out of nothing, or next to nothing. The media also benefit from the often unexpected help of pseudo-scientists and frenzied ecologists. They also broadcast so-called (unfounded) facts, which they have taken neither the time nor the precaution to verify, since tomorrow's edition will supersede today's, and no retraction is ever published! The greenhouse effect scenario offers journalists an almost inexhaustible range of catastrophist subjects. Since they copy each other in their search for a good story, the scenario is revived and fed to the public over and over again until it seems 'politically correct', although it is most often scientifically incorrect. The result of such media bludgeoning is that everything not conforming to the scenario appears incongruous!

The echo chamber that is the media changes its subjects and its views habitually at the slightest veering of the wind: it is ready to embrace one folly after another.

For example, *France 2*, a public TV channel, rebroadcast in its *Contre-courant* programme a BBC *Horizon* documentary *The Big Chill*, by Laverty, on 25 June 2004. Oceanographers Turell, Halley, Keigurn, and Broecker, and meteorologists Joyce and Woods took part in this fiction. According to the so-called 'conveyor belt' theory (Chapter 14), 'when salt water cools, it becomes denser and sinks straight down to the bottom of the ocean. It then moves southwards, to where the famous Gulf Stream forms' (in *Le Figaro*, 25 June 2004). But the addition of freshwater from melting ice and from rivers decreases the salinity and therefore the density of surface waters, slowing the descent of the water, and blocking the flow! Then the anguished question is posed: '*if the conveyor belt were to be blocked as it was 115,000 years ago, when the last glaciation began ...*' (in *Le Figaro*), what would happen? A wave of cold would sweep across Europe, and a new glaciation would begin in possibly only 20 years' time! So, according to *Le Figaro*, '*with heavy use of time-lapse film, virtual and archive imagery and realistic tableaux, with shifting glaciers, floods, blizzards, downed pylons and cities cut off from the world and plunged into darkness*', the Thames would freeze over, the Channel would become an icefield, tornadoes and cyclones would rage, and the Amazon rain forest would disappear ... (the whole media arsenal, in fact). Fiction would do its utmost to '*fill emotion-seeking viewers with dread*'. This is a proven technique, that of the door-to-door salesman: don't give them a moment for reflection, or the thinking person might soon see right through you! This fictive documentary is, on a scientific level, nothing but a tissue of nonsense.

In passing they attribute the Würm glaciation to the 'conveyor belt' effect, though if they have freezing seas to the south of England they do not know much about actual Ice Age conditions (cf. Chapter 9). They are ignoring the meteorological aspects governing the weather, which is by no means subordinated to the Gulf Stream (cf. Chapters 12 and 14). They are not evaluating the real differences in sea water densities, nor are they asking themselves where (warm?) surface water goes if it does not sink (cf. Chapter 14). This blinkered approach is just like that of a costermonger who persists in his single alarmist cry: '*the disappearance of the Gulf Stream will bring cold*'. For a film director, this production would be counted a success, since its aim is to hold the attention of the audience, which usually wants nothing more than to be entertained . . . Let them hear this rubbish: it doesn't really matter, because the next documentary will promise them palm trees growing along a 'Mediterranean' coast in England, or even the fires of Hell consuming the Earth. Responsibility for this rests solely with those scientists who approve of such wild imaginings, and who sometimes appear in these make-believe programmes.

In these productions, the sensation-seeking media create images based on the sometimes hastily conceived scenarios of scientists, resulting in a striking (but entertaining) demonstration of the inanity, or stupidity, of these scenarios. This ought to serve as a lesson to the scientists . . . but unfortunately it does not, in spite of the extreme depictions of these, and many other, untruths! The 'thermohaline circulation' hypothesis is very much in vogue at present (cf. Chapter 14); even in *La Recherche*, supposed to be a scientific journal, the possibility was aired in March 2004 of rapid cooling caused by a lessening of salinity, and again the question was asked: '*Can the climate collapse*?' (Bard, 2004)!

A recent book, *The Coming Global Superstorm* (Bell and Strieber, 2000), upon which the 2004 film *The Day After Tomorrow* was based, is an excellent example of what has become of climatology for public consumption. Michaels (2004) wrote about *The Coming Global Superstorm*: '*a book so bad, so inaccurate, so outrageous . . .*', stressing '*the absurdity of its views on global warming and climate change*'. It's all there: heat, cold, ice, hurricanes, victims, all taken to extremes with the 'wow' factor ever present, as today's tastes warrant! Still, why worry – it's only fiction, impressive and entertaining in spite of the mega-catastrophes!

So any really profound discussion of this subject will not come from the media. That is not their role, anyway. Scientists and science-based journals would do well to remember this. 'Sharp' press articles, sensational ideas, striking images, climatic horrors on paper and on screen, all these are symptomatic of the state of the media mind, and as such they have their uses. They show us that **climatology has finally become nothing but entertainment**, in which some people no longer believe; and soon nobody will believe in it!

4.7 SO WHERE IS CLIMATOLOGY NOW?

For more than 20 years now, since the 1980s, the various groups listed above have become conjoined and tangled. After a very long gestation, this important scientific subject reached a rapid *dénouement*, especially in 1985 with its new and frankly unexpected 'certainties'. This is really astonishing, and there is no shortage of questions which need asking:

- Has observation really provided irrefutable proofs?
- Is this certainty the fruit of a science of meteorology which has achieved all it can?
- Have models really made such considerable progress?
- Has there been a 'revolution' in ideas?
- Is this merely the result of scientists, ecologists and politicians trying to outdo each other?
- Have scientists allowed themselves to be led along, or have they actively contributed to the success of the greenhouse scenario?
- Did the sudden *dénouement* of 1985 offer a definite, unequivocal answer, especially on the subject of the physical link between the greenhouse effect and temperature?
- Has there been any definitive progress in the last 15 years?
- Why is there still no consensus among climatologists?

Is there a good answer to any of these questions, and to many others?

Leaving aside all the descriptions, analyses, and hypotheses, and allowing for the long-drawn-out business of the IPCC reports, there is only one point to be made in the end. It is a crucial point. **Everything is based upon one single hypothesis**. All the arguments, the predictions, and the supposed consequences count for nothing unless this basic hypothesis, debated for more than 100 years, is based on reality. **This has still to be proved**.

The hypothesis is this: the anthropic greenhouse effect *has* increased global temperatures, and *will go on* increasing them, and as a consequence will modify elements of the climate.

- The greenhouse effect 'has' increased global temperatures (according to observations):
  - the IPCC claims that the greenhouse effect has controlled climate on the palaeoclimatic scale;
  - the IPCC claims that 'global warming' has been observed, on the basis of the standard thermal curve for the period 1860–2003, which provides 'irrefutable' proof; and
  - the IPCC also claims that, according to the so-called 'hockey stick' curve, temperatures are higher than they have been for a thousand years or even more ...
- The greenhouse effect 'will go on' increasing global temperatures (according to the climatic models):
  - all numerical models predict a rise in mean global temperatures, and a variable rise as a function of latitude; and

– the models also predict that all elements of the climate will change, especially precipitation and disturbances...

It is therefore necessary to be really sure that this core hypothesis is well founded and has been verified, in order to be able to tackle the many unanswered questions. For **it must once again be underlined that there is NO evidence, nothing that can be directly stated**.

There is no tangible, indisputable proof of the above hypothesis, except for images of pollution in towns covered with a greyish halo, which is a very real proof but only a local one. Conversely, on the global scale, one can only brandish the threat of some imminent or distant peril on the basis of hypothetical 'links' which may be mere co-variations without any proven physical connection. Is this the reason for the mad rush of research during the last decade, for multiple proofs presented as definitive although they are perhaps just as illusory as each other, and for the hypothetical relationships between increasing numbers of parameters (sea level, glaciers, droughts, storms, floods, El Niño)? Inventions or realities?

The time has come to bring the supposed 'climatic warming', with its political, economic, and media overtones, **back into the field of climatology**! The whole community of meteorologists and climatologists, even those who might be gaining (especially financially) by drawing attention to themselves, is involved in this debate. Even though their work is 'exploited' by other interests, are the scientists themselves blameless? Are they not sometimes a little too ready to put at risk, purposely or otherwise, the credibility of climatology, and by extension their own credibility?

5

Greenhouse effect – water effect

Can we put it down to some unfortunate chance that not one digital climatologist has felt curious enough to 'go and see' what a climate devoid of the greenhouse effect would be like, and conduct some thought experiment upon it?

Yves Lenoir. *Climat de Panique*, Favre, 2001.

Originally, the greenhouse effect was known as the 'hot-house theory', since the atmosphere was thought to behave in the same way as the glass in a greenhouse. The glass allows through the 'luminous heat' (solar radiation), but the 'dark heat' (infrared) cannot pass out through the glass, and the heat builds up, raising the temperature within the greenhouse. However, the atmosphere and the greenhouse are not directly comparable, as a glazed-in volume, or glasshouse, retains heat because of the absence of convection and advection, rather than through effects of absorption through glass and re-emission, as Wood pointed out in 1909 (in Jones and Henderson-Sellers, 1990). But the term 'greenhouse effect', accepted through use, is now general, to the detriment of the more appropriate term 'atmospheric effect'. In any case, it is too late to correct the expression, which has entered into everyday vocabulary; the essential point is that we should agree on a definition. Trace gases, known as greenhouse or emissive gases, absorb infrared radiation emitted by the surface of the Earth, by the atmosphere itself, and by clouds. They re-emit it into the surrounding gases. Therefore, these gases trap heat within the surface–troposphere system and re-transmit it: the natural greenhouse effect (cf. IPCC, 2001, glossary).

If the quantity of trapped heat increases, the temperature may rise, and this is what we call the enhanced greenhouse effect, also known as the anthropic greenhouse effect since it is thought to be a result of human activities. It is precisely this enhancement of the natural greenhouse effect which is nowadays considered to be responsible for 'climate change'.

The scenario of the so-called 'greenhouse effect' or 'global warming' is based upon the following postulates:

- the greenhouse effect warms the atmosphere;
- human activities produce an excess greenhouse effect;
- a resulting increase in temperature has been observed ... and attributed (with no formal proof) to this excess greenhouse effect; and
- consequently, and to put it briefly, humanity is responsible for the increase in temperature.

These points, upon which the whole scenario is founded, will be evaluated and discussed at an appropriate juncture. Obviously, we are not claiming to deal exhaustively here with this constantly aired subject. Furthermore, it is useful to bear in mind Kondratyev's words (2002, 2003): '*the greenhouse hypothesis of global warming appealed to a necessity of studying the "atmosphere–ocean–ice cover–biosphere" climatic system, taking account of the entire complexity of feedbacks between its interactive components*'; so artificially isolating this phenomenon means that it acquires a disproportionate importance. In this chapter, which deals with the greenhouse effect, we first examine the 'classic' version, now claimed to be 'set in stone': the version which has been progressively imposed and is now the one upheld by the Intergovernmental Panel on Climate Change (IPCC). Then, we will underline the major uncertainties specifically associated with the role of water vapour and clouds, and finally, the radical points which call into question the very notion of the greenhouse effect.

The source of the energy which drives the climatic system is the radiation from the Sun. This energy is constantly being absorbed, transformed, redistributed, dissipated, and renewed. All meteorological phenomena (transfers and transformations of energy, atmospheric circulation, weather and disturbances, climates) are consequences of the processes by which this initial solar energy is utilised.

5.1 PROCESSES OF RADIATION

Solar radiation propagates in straight lines and lies essentially in the visible, emitted at short wavelengths, and in the near infrared (thermal IR) (i.e., long waves). The radiation balance is the difference between, on the one hand, that part of incident solar radiation absorbed by the atmosphere and the Earth's land or ocean surface (incoming flux), and on the other hand, the direct (reflected) and infrared radiation sent by the surface and the atmosphere into space (outcoming flux). In a balanced Earth–atmosphere *ensemble*, received and emitted fluxes would have the same value. The mean value over one year, per one square metre and per twenty-four hours of incident solar radiation, is 342 W/m^2, and that of the planetary counter-radiation is 235 W/m^2, to which must be added the 107 W/m^2 directly reflected into space (Figure 6).

- The short-wave solar flux is reflected (as a function of the albedo of reflecting bodies), refracted (by diffraction or scattering), or absorbed and transformed

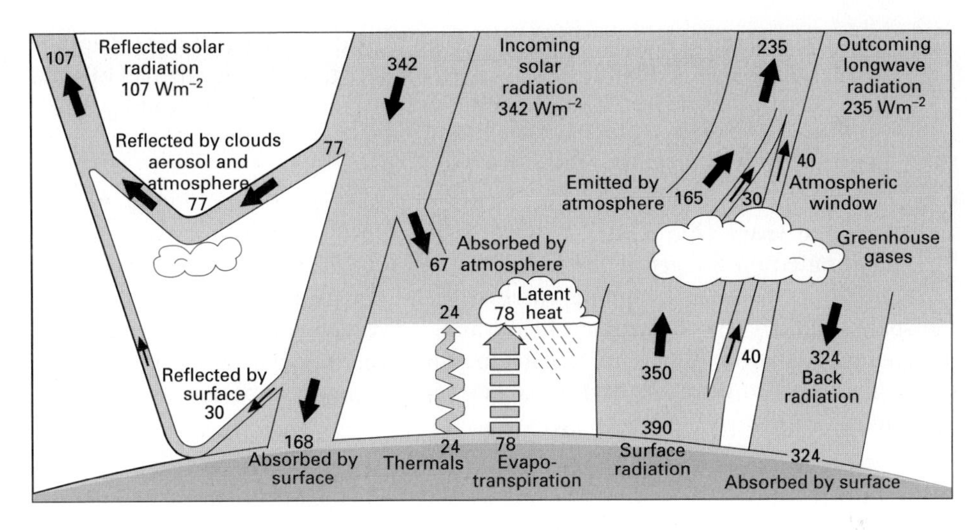

Figure 6. The Earth's annual and global mean energy balance.
From Kiehl and Trenberth (1997) in IPCC (2001), *The Climate System, Scientific Basis.*

into heat. About 31% (107 W/m^2) of the incident flux is directly reflected by cloud tops and aerosols (77 W/m^2) and by the ground (30 W/m^2) into space. The rest (69% or 235 W/m^2) is absorbed by the atmosphere (67 W/m^2 or 19% of the received flux) and by the Earth's land or ocean surface, which receives about 49% of the incident flux (168 W/m^2). Since the atmosphere is almost transparent to visible short-wave radiation, with the exception of stratospheric ozone for ultraviolet radiation, it is the Earth's surface which receives most of the radiation in almost equal proportions of scattered radiation (28%) and direct radiation (26%).

- The Earth's surface, which receives directly two and a half times as much energy as does the atmosphere (49% as against 19%), re-emits energy upwards (terrestrial counter-radiation or TCR) as long waves in the infrared and by transfer of perceptible and latent heat.
  - Because of the 'opacity' of certain components of the atmosphere to IR radiation, most of the radiation (350 W/m^2) from the Earth's surface is absorbed. As a function of the wavelength of the re-emitted radiation, the flux is either directly lost to space in the band between 8 and 12 microns (40 W/m^2 through the so-called 'atmospheric window'), or absorbed by trace gases such as water vapour, carbon dioxide (CO_2), methane (CH_4), nitrogen protoxide (N_2O), halocarbons (CFC), and ozone (O_3). The major components of the air, nitrogen (N_2, 78.08%) and oxygen (O_2, 20.95%) play almost no part in these thermal exchanges.
  - Energy may also be transferred by non-radiative processes (convection) in the form of *perceptible heat* (24 W/m^2), which accompanies thermal convection

directly, and *latent heat* (78 W/m^2) through the intermediary of water vapour. *Vaporisation heat*, held in reserve, is released over time during changes of state (condensation of water). The contribution of latent heat to the heating of the atmosphere is of the same order of importance as the direct absorption of short-wave solar radiation (78 and 67 W/m^2).

- The atmosphere re-radiates in its turn; this 'celestial' re-radiation is emitted either into space (235 W/m^2) or towards the Earth's surface (324 W/m^2). The global radiation budget at the top of the atmosphere is therefore in a state of equilibrium between incoming energy (342 W/m^2) and energy sent back into space, directly by reflection (107 W/m^2), to which we can add planetary counter-radiation (235 W/m^2). Any modification in either solar or infrared radiation will disrupt the energy balance, and any such imbalance is known as 'radiative forcing'.

5.2 THE NATURAL AND ENHANCED GREENHOUSE EFFECT

The Earth's surface receives more short-wave energy directly from the Sun than the atmosphere itself, and therefore represents – paradoxically – the source of energy. This means that the Earth's surface is of capital importance in climatology, especially in questions involving all phenomena of thermal origin, particularly in the lower layers: depressions, anticyclones, convection or subsidence of air, and meridional exchanges (Chapter 8). It re-emits long-wave energy (IR) and thereby warms the atmosphere. The trace gases, which absorb and re-emit the infrared radiation, are also found in the lower (densest) layers. So heat is conserved in the lowest 5,000 m of the atmosphere. The temperature decreases with altitude, until a mean temperature of −58°C is reached at the level of the tropopause. The Earth's surface, at a mean temperature of 15°C, is therefore warmer than it would be if the atmosphere did not exist. This enormous gain in temperature constitutes, in the 'classic' mould, the natural greenhouse effect.

It is characteristic of trace gases that they are transparent to short-wave visible light, but opaque to most of the long-wave infrared radiation re-emitted by the Earth. The capacity for absorption of each of these gases is a function of its particular characteristics as they warm up and re-radiate in all directions. Radiative forcing in the direction of the ground adds to heat received directly from the Sun, leading to an increase in temperature. Any augmentation in the concentration of emitting gases will therefore tend to increase the 'opacity' of the atmosphere. This extra heating is what is understood by the term 'the enhanced greenhouse effect', nowadays (wrongly) shortened to 'the greenhouse effect'.

- Water vapour is the principal greenhouse gas, although it constitutes only 0.3% of the atmosphere. Its distribution is extremely uneven, both geographically and as a function of altitude, and its lifetime in the atmosphere is brief. Through the mechanisms of evaporation and precipitation, the renewal of the global water

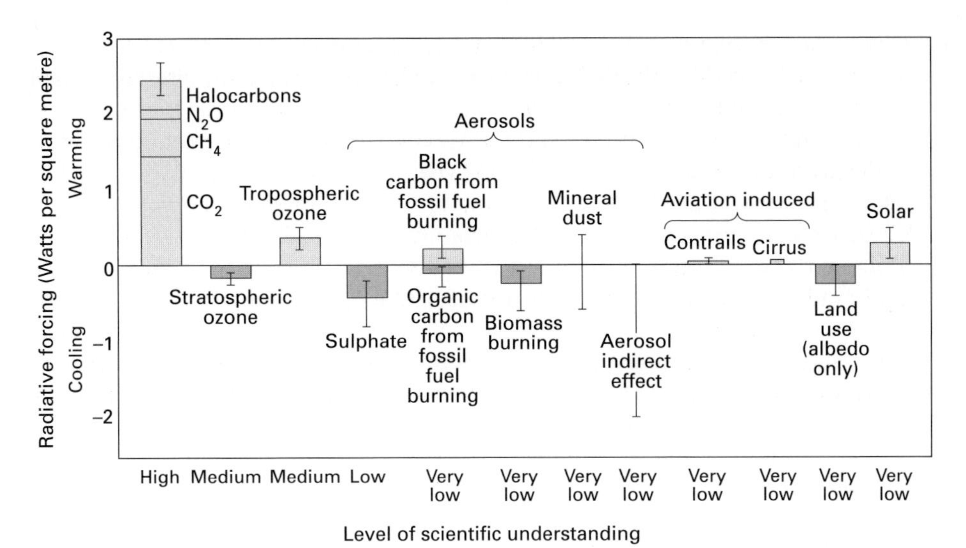

Figure 7. The global mean radiative forcing of the climate system for the year 2000, relative to 1750. These radiative forcings result from changes in atmospheric composition, modification of surface albedo, and variation in the output of the Sun. Rectangular bars represent estimates of the contributions of these forcings (warming or cooling, in W/m^2). Effects of water vapour, clouds, and volcanic events are not mentioned. The vertical line indicates a range of estimates.
From IPCC (2001), *The Climate System, Scientific Basis*.

potential takes less than two weeks. The action of this fundamental emissive gas is (oddly) not represented in Figure 7.

- Carbon dioxide, present at 370 ppmv (0.037%) in the atmosphere, is responsible for 39% of the natural greenhouse effect. Since the end of the 19th century, the amount of CO_2 has risen by 32%. **Emissions due to human activity represent only one-twentieth of the natural additions of carbon to the atmosphere**: three-quarters of this contribution comes from the combustion of fossil fuels, and the other quarter from deforestation and agricultural practices. The radiative forcing due to CO_2 since 1750 represents 1.4 W/m^2 (cf. Figure 7).
- Methane, at 1.8 ppmv in the atmosphere, has a lifetime of the order of 10 years and absorbs 43 times more infrared radiation than CO_2. Its origins lie in the decomposition of organic matter, the combustion of fossil fuels and biomass, and fermentation (animals, rice-paddies, household waste tips). The amount of methane has risen by 151% and its radiative forcing is estimated at 0.5 W/m^2, rising through induced effects to a value of the order of 0.7 W/m^2 (i.e., to half that of CO_2).
- Nitrogen protoxide at 0.3 ppmv, is very stable, with a lifetime of the order of a century. It has increased by 16%, and absorbs 253 times more radiation than CO_2. Its origin is both natural and anthropic, arising from nitrogen-based fertilisers, stock-breeding, and industry.

- Ozone, at 0.03 ppmv, is very stable, and has increased by 35%. In the troposphere it is formed by photochemical recomposition, mainly of pollutants found in urban atmospheres. It is a screen against both incident ultraviolet radiation and infrared radiation from the surface. Its contribution to the greenhouse effect is therefore of the order of 0.35 W/m^2. The diminution in stratospheric ozone between 1979 and 2000 is held to be responsible for a deficit of –0.15 W/m^2.
- CFCs (chlorofluorocarbons, or halocarbons) are of solely human origin. Their ability to absorb infrared radiation is 10,000 times greater than that of CO_2. Their radiative forcing reaches a value of 0.35 W/m^2 (Figure 7).

Together, these greenhouse gases (not including water vapour) are considered responsible for a radiative forcing of some 3 W/m^2 since 1750 (i.e., of the order of 1% of the energy received from the Sun).

Aerosols, especially those of human origin, entered the equation in the IPCC Second Report (1995), and were held to be responsible for a negative forcing of 0.5 W/m^2, lessening the said increase in temperature. So sulphur-based aerosols might counteract the warming associated with CO_2, and warming would therefore be due to the other greenhouse gases.

It is estimated that direct radiative forcing is: for sulphates (SO_2), –0.4 W/m^2; for aerosols originating in biomass combustion, –0.2 W/m^2; for organic carbon-based aerosols from fossil combustibles, –0.1 W/m^2; and for black carbon-based aerosols (soot) also from fossil combustibles, +0.2 W/m^2 (IPCC, 2001). Aerosols vary constantly and regionally, and react rapidly, their lifetimes being relatively short. The rate of forcing by aerosols is not at all well known, and modellers '*all agree that the aerosol forcing is more uncertain than any other feature of the climate models*' (Singer, 1999).

Together, Figures 6 and 7 summarise the most commonly encountered notions regarding the greenhouse effect. This should not, however, be taken to mean that the concept upon which 'global warming' is based is definitely established on some undeniable bases: all the figures quoted are to some extent contested, and '*the level of scientific understanding*' (Figure 7) is generally thought to be '*very low*'. Indeed, few values are derived from direct measurements (particularly from satellites) and come mostly from calculations, often approximate, which differ according to the studies. The procedures and respective responsibilities are open to discussion, in the main because the preceding schemas represent phenomena which evolve in vertical spaces which are theoretical and stable: no discontinuities or shear effects occur between the surface and the upper layers, and there are no horizontal exchanges advecting energy. This tends to suggest that these schemas do not really equate with reality.

We should also be looking at the actual contributions from greenhouse gases, at the major role of water vapour and clouds, and also at the very principle of the greenhouse effect, all of them ambiguous and contentious: more on these later.

5.3 WATER VAPOUR

The term 'atmosphere' comes from the Greek *atmos*, signifying vapour. This etymology recalls the unique role of water vapour in the climatic system, linking surface and atmosphere in the water cycle. Moreover, '*its radiative effects are the major factor in the atmospheric greenhouse effect*' (Elliot and Gaffen, 1995). Theory suggests that the global climate is quite sensitive to the least change in humidity at every level of the atmosphere, though observations verifying these hypotheses are few and far between. However, '*radiosonde observations over the past few decades suggest increases in tropospheric water vapour, globally and regionally*' (Elliot and Gaffen, 1995).

Amounts of water vapour are essentially incapable of regulation, in the geographical sense, the major source being above the oceans. Its distribution is also unequal at varying altitudes. '*Nearly half the total water in the air is between sea level and about 1.5 km above sea level. Less than 5–6% of the water is above 5 km, and less than 1% is in the stratosphere, nominally above 12 km*' (AGU, 1995). Its concentration varies by several orders of magnitude in just a few kilometres. Theoretically, a rise in temperature increases the ability of the air to hold water vapour. This means that if there is warming associated with CO_2, it will be approximately doubled by the retroactive effect of water vapour. But water vapour condenses, its concentration depending on its saturation vapour pressure, so any increase will not be as great as might be expected.

In the lowest 1 or 2 km, water vapour forms very readily as temperature rises, but its behaviour is by no means regular, especially as far as aerological stratification is concerned. Also, since the cycle of water vapour within clouds, and more especially within convective systems, is also poorly understood, water vapour is not correctly dealt with in global climatic models. The result is that there is great uncertainty about the real importance of water vapour. Although its contribution to the greenhouse effect is very important, it has not been precisely measured: estimates, of the order of 65% on average, actually range from 55% to 95%, and if clouds are brought in, even to 97%!

What does the IPCC have to say on this subject in its last Report (2001)? In the *Summary for Policymakers*, water vapour is not included in the list of greenhouse gases, and neither does it appear on page 6 (figure 2: *changes in atmospheric composition*), page 7 (*concentrations of atmospheric greenhouse gases*), page 8 (which shows Figure 7 as above, with no water vapour shown), or page 12, which proclaims that '*human influences will continue to change atmospheric composition*'. But on page 13, we read that '*global average water vapour concentration and precipitation are projected to increase ...*'. So here, the IPCC is 'obliged' to bring in water vapour, not because of some real interest which it may hold, but because an increase in water vapour is thought to explain increased levels of rainfall (cf. Chapter 11). Even the possible role of water vapour in the predicted rise in temperature is not mentioned!

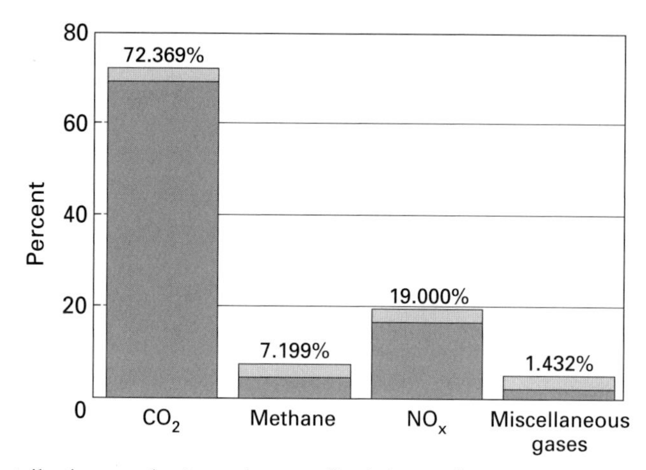

Figure 8. Contribution to the 'greenhouse effect' (natural and man-made sources – water vapour not included).
From Hieb (2004).
Note: methane, NO_x, and miscellaneous gases are adjusted for heat retention characteristics relative to CO_2.

Neither is this the case in the IPCC *Scientific Basis* (chapter 1, *The Climate System*), where we read: '*these so-called greenhouse gases, with a total volume mixing ratio in dry air of less than 0.1% by volume, play an essential role in the Earth's energy budget. Moreover, the atmosphere contains water vapour (H_2O), which is also a natural greenhouse gas . . .*'. This 'also', here, is ridiculous! Again, it seems that water vapour appears as an afterthought, having almost been omitted! Did someone simply forget? Or was it deliberately overlooked?

We should now re-estimate the respective importance of the different greenhouse gases. At a preliminary estimate, leaving aside water vapour (as IPCC regularly does), carbon dioxide is far and away the principal greenhouse gas (72.37%), the other emissive gases contributing about 28% (Figure 8).

The enormous 'dominance' of CO_2 is obviously much diminished when the contribution of water vapour is taken into account. Now, the proportions are (Figure 9):

Water vapour (WV)	**95.00%**
Carbon dioxide (CO_2)	**03.62%** vs. 72.37%
Methane (CH_4)	**00.36%** vs. 07.10%
Nitrous oxide (N_2O)	**00.95%** vs. 19.00%
CFCs and other gases	**00.07%** vs. 01.43%

So water vapour represents **95% of the greenhouse effect**, and the possible influence of the other gases is down to 5%! If we consider the real atmosphere, carbon dioxide, from natural and anthropic sources, the gas upon which the warming scenarios are based, represents no more than 3.62% of the greenhouse effect (i.e., 26 times less than water vapour!)

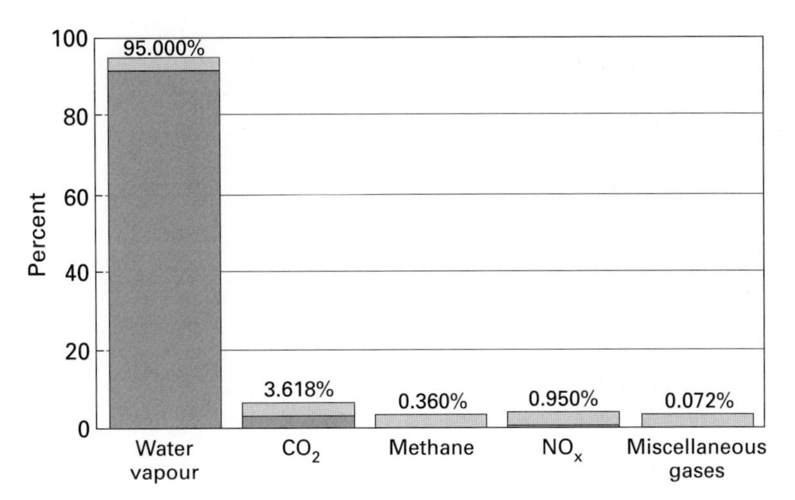

Figure 9. Contribution to the 'greenhouse effect' (natural and man-made sources – water vapour included).
From Hieb (2004).

The IPCC's outrageous overestimation led Hameranta (moderator of *Climate Sceptics*, 2004) to point out that '*water dominates, the influence of CO_2 is minuscule, unimportant!*' Now, nearly all water vapour (99.999%) comes from natural sources. So the real role of humanity in the greenhouse effect must be viewed with a sense of proportion. By bringing in the product of the adjusted contribution of CO_2 and the percentage of anthropic CO_2, Hieb (2004, figure 10) *obtained 0.117%* of the greenhouse effect. This insignificant value represents the contribution to the greenhouse effect from CO_2 originating in human activity!

Adding the contribution of the other gases to that of CO_2 (Figure 10), we now see that **human activities are supposed to be responsible for only 0.28% of the greenhouse effect**!

The 'insignificant' or even ridiculous estimate of Man's supposed responsibility will immediately draw down criticism. There do, of course, exist other distributions, for example, that calculated by Kiehl and Trenberth (1997), who suggested a figure of 60% for the supposed contribution of water vapour (corresponding to the current estimate of 100 W/m^2 out of 160 W/m^2 (i.e., 62.5%)), 26% for carbon dioxide, 8% for ozone, and the remaining 6% for nitrous oxide and methane. This is a lower figure (by one-third, cf. Figure 9) for water vapour, but it still does not permit us to 'forget' water vapour as part of the greenhouse process!

Water vapour is indeed much more abundant and better able to retain heat than carbon dioxide, and it may therefore be, as Broecker wrote, '*the only atmospheric component capable of warming and cooling the Earth*' (in Carlowitz, 1996). On a scientific level, then, we should first of all not falsify the presentation of the greenhouse effect by deliberately omitting its principal player. We should calculate and provide, in this field, indisputable proportions in order to be able honestly to estimate any human responsibility (if indeed this exists as regards the climate).

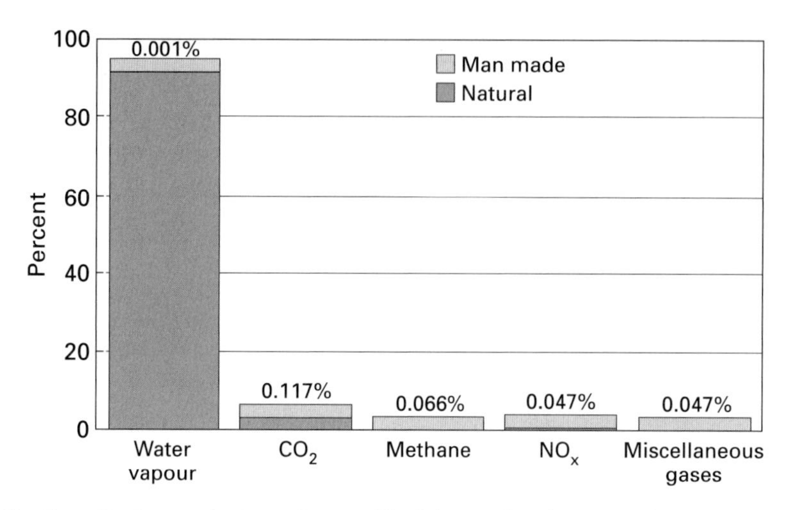

Figure 10. Contribution to the 'greenhouse effect' (natural and man-made causes – including water vapour).
From Hieb (2004).

It should also be pointed out that emissive gases '*water vapor, CO_2 and O_3 also absorb short-wave radiation*' (IPCC, 2001, *Scientific Basis*, chapter 1). This means that an increase in these absorbing gases will not necessarily enhance the greenhouse effect, which 'by definition' involves infrared radiation; it could however constitute a potential source of retroaction if water vapour increases. The share of the respective contributions, from direct solar radiation (short wave) and from terrestrial infrared radiation, is unknown.

These reflections on the major role played by water vapour, the '*greatest source of uncertainties*' (Keller, 1999), lead us to reconsider many aspects of the water cycle, especially in its vapour form. It will be necessary to analyse in depth its true contribution to any actual warming, past or future, and the evolution of its concentration and distribution. How important are its transfers (particularly in the lower layers where exchanges are concentrated)? What about its vertical distribution and the causes and processes of its evolution? The role of water vapour, which is already considerable, is further enhanced through the action of clouds.

5.4 CLOUDS

Clouds affect the radiation budget in the same way as greenhouse gases, but their effect is complicated by the fact that they reflect incident solar radiation, while their formation releases latent heat. Indeed, clouds may either reflect short-wave solar radiation from their summits, or absorb it; they also absorb and re-emit long-wave radiation from the Earth's surface. In the former case, they cool the surface by intercepting the solar flux, and in the latter they warm the lower layers. The relative contribution of these effects depends on the height, type, density, and optical and radiative properties of the clouds. Low-altitude clouds' effects are mostly reflective in nature, while those at high altitude have a much greater ability to absorb. Satellite observations suggest that clouds generally tend to cool the climate (Ramanathan *et al.*, 1989; Arking, 1991), but their effect depends on latitude: cooling is considerable over oceans at middle and high latitudes, with values up to 100 W/m^2, but over tropical regions maximum values reach 50 to −100 W/m^2. Anomalies remain, however, and so '*the role of clouds in modifying the Earth's radiation balance is well recognized as a key uncertainty in predicting any potential future climate change*' (Wielicki *et al.*, 1995).

These radiative properties of clouds are dependent on the evolution of water vapour, droplets, ice particles, and atmospheric aerosols (cf. IPCC, 2001). Processes within clouds are of prime importance in determining, through models, the radiative effects and thermal outcomes. However, although progress has been made in representing these processes, '*clouds remain a dominant source of uncertainty, because of the large variety of interactive processes which contribute to cloud formation or cloud–radiation interaction: dynamical forcing – large-scale or sub-grid scale, microphysical processes controlling the growth and phase of the various hydrometeors, complex geometry with possible overlapping of cloud layers*' (IPCC, 2001). The extent of these uncertainties is such that their representation in climatic models, where physical processes have to be parameterised, is an approximation. The IPCC itself admits that the spread of 1.5°C–4.5°C in the estimates of temperature rise if CO_2 levels double is precisely due to '*the interaction of model water vapour feedbacks with the variations in cloud behaviour among existing models*' (IPCC, 2001). This 3°C margin of error is evidence of the important part approximation plays in the prediction of temperature.

5.4.1 Negative or positive action of clouds?

Another example of this is the old debate, with diametrically opposite conclusions, which was already in train when the IPCC was founded. Its subject: do water vapour and clouds exercise a negative retroaction (a thesis supported by Lindzen, 1990), or a positive one, as is more widely believed nowadays (Cess *et al.*, 1990; Cess, 1991)? This fundamental question, altering the sign pertaining to the action of clouds, was not resolved by the Second Report of the IPCC, nor by the Third: '*because of the fact that climatic models involving clouds and precipitation are particularly complex, the precise extent of this retroaction – a crucial phenomenon – remains unknown*' (IPCC, 2001). Soon and Balunias (2003) confirmed that the situation has not fundamentally improved, given that '*the parameterisations of cloud microphysics and cloud formation processes, as well as their interactions with other variables of the ocean and atmosphere, remain major challenges*'. Therefore the figures associated with the action of clouds in Figure 6 (reflection and radiation) are only estimated, and the resultant energy budget carries an unwelcome and not inconsiderable margin of error. This is why Randall *et al.* (2003) could predict that '*at the current rate of progress, cloud parameterization deficiencies will continue to plague us for many more decades into the future*'.

Is this because the atmosphere is usually considered as homogeneous by the models, with, for example, surface temperature and the temperature in the free troposphere considered as coupled? Would this also suggest that the movement of water vapour between the surface and the upper levels might equally undergo (according to the theory) some continuous adiabatic decrease? This would seem to be the case with ascending movements (which, moreover, liberate latent heat), the convective cells of localised storms, the 'heat towers' of tropical cyclones, and the vertical structure of the meteorological equator (VME), for thermal or dynamical reasons, as high as the tropopause or even beyond it. But it is not always the case. The supposed coupling is not always observed, the best example being over those vast areas lying beneath a horizontal aerological discontinuity. The lack of continuity in exchanges between the lower layers and higher altitudes imposed by stratification causes two types of decrease, below and above the discontinuity, following first of all at the moist adiabatic rate, and then the dry adiabatic rate (or vice versa).

What is more: is atmospheric circulation, so often mentioned by the IPCC ('*the atmospheric water vapour content responds to changes in temperature, microphysical processes and the atmospheric circulation*' – 2001, *Scientific Basis*), really taken into account? For example, is the Trade Inversion (TI) considered, extending as it does across immense areas of the tropical oceans, where precisely evaporation is at its most intense? This inversion in the lower layers, marked by thin stratiform cloud formations, blocks or at least strictly controls the upward dispersion of the water contained within the lowest stratum of the trade wind. This water is channelled by the trades towards the updrafts of the ME, or diverted towards higher latitudes by the low-pressure corridors on the leading edges of Mobile Polar Highs (MPHs) (cf. Chapter 8). Fruitless attempts directly to determine the effect of water vapour by correlating surface variations and variations in long-wave outward flux (Raval and Ramanathan, 1989) show that these aspects are not necessarily taken into account.

Now, are the imperfect postulates currently in use all confirmed? We might for example estimate the value of latent heat released into the atmosphere by studying precipitation: a mean global value of 984 mm suggests a mean figure of 78 W/m^2 (Kiehl and Trenberth, 1997) as in Figure 6. Is such a relationship applicable to all scales of phenomena? For example, how reliable is the formula which states that '*where there is an increase in the amount of water vapour, there will be more clouds, and more rain*'? It has never been proved, there being no automatic link between water vapour concentration and rainfall levels (cf. Chapter 11). Can it really be claimed that '*an increase in temperature means that the ice sheet will melt, meaning that less solar radiation will be reflected with a resulting increase in humidity: therefore more rain will fall, leading to wider snow cover in cold areas*'? Does such a (utopian?) stairway of 'reasoning' reflect the reality of phenomena? It all stems from a very schematic way of looking at phenomena *in situ* (cf. Figure 6), even though precipitable potential is rarely used up *in situ*, but rather advected across great distances: water falling at high latitudes originates far away. Simplistic links like these are leading us, in the context of an enhanced greenhouse effect, to assume an increase in the intensity of perturbations and even in tropical cyclones (cf. Chapter 11)!

5.5 IS THERE REALLY A 'GREENHOUSE EFFECT'?

Until now we have considered the greenhouse effect scenario to be likely, though with the imperfections and uncertainties we have highlighted.

But it seems that there have been severe criticisms of the very principle of the greenhouse effect, even to the extent that, in the words of Thieme (2002): '*the greenhouse gas hypothesis violates fundamentals of physics*'. Such a statement merits our attention.

As support, the greenhouse effect scenario uses the examples of other planets, which serve as test beds to validate models. For example, Venus is considered to exhibit 'a quintessential greenhouse effect', with its surface temperature of 458°C, originally attributed to the carbon dioxide which represents 95% of its atmosphere (Courtin *et al.*, 1992). However, it seems that this temperature is the result of Venus' colossal atmospheric pressure, 92 times that of the mean pressure at the Earth's surface. Now the infrared absorption of a gas increases with pressure, and so the resulting temperature depends mainly on pressure.

Thieme (2003) took this as his initial hypothesis when he expounded the following arguments:

1. In the Earth's troposphere temperature decreases with altitude as far as the tropopause (i.e., in line with decreasing pressure); the rate of temperature decrease is moreover a function of air humidity, being greater in dry air than in damp air.
2. Atmospheric counter-radiation (or 'back radiation', the return of the flux emitted by the Earth's surface) can be caused only by reflection, but '*the CO_2 share in our atmosphere cannot cause reflection in any way*'.
3. If the radiation emitted by the Earth's surface were absorbed by the atmosphere, the absorbing air would warm up, and the initial structure of the air would be modified, most noticeably in its vertical temperature, density, and pressure profiles. Warm air tends to rise, following the basic principle of thermal convection: air warms up when in contact with the surface, and rises, transporting heat to upper levels. But air expands with altitude, and cools ... The greenhouse effect (i.e., the return of heat downwards) does not occur; instead, updrafts are transformed into horizontal advections.
4. One essential precondition for any type of heat transfer is that '*the emitter is warmer than the absorber*'. Now, if temperature decreases with altitude, a return transfer to the Earth's surface by way of the CO_2 is impossible. The IPCC notes, on this subject, that '*infrared radiation emitted to space originates from an altitude with a temperature of, on average, −19°C, in balance with the net incoming solar radiation, whereas the Earth's surface is kept at a much higher temperature of, on average, +14°C*' ...

These arguments, among others, lead to Thieme's statement that '*atmospheric back radiation is an arbitrary construct evoked to explain the temperature observed at the Earth's surface*', a conclusion supported notably by Von Storch *et al.* (1999), for whom this arbitrary postulate is not confirmed by observations, and is but a simple construct of the imagination. These critics do not stand alone. So, in Figure 6, the downward transfers, supposed to represent atmospheric back radiation, have no existence in reality. What is more, the schema does not represent the horizontal compensatory movements which occur in the general circulation, a fundamental omission to which we shall return later (Chapter 8).

5.6 THE GREENHOUSE EFFECT 'COOLS' THE ATMOSPHERE

It was Lenoir who asked whether, if the greenhouse effect involves 'warming', the absence of this effect would (logically) entail 'cooling'? He considered the theoretical case of a planet Earth with no greenhouse effect, which would have the following characteristics:

- at the surface, wide thermal contrasts would exist in time and space, and the surface temperature at the equator would be 90°C at midday;
- the ground would be cooler than the air just above it, at all times and in all places; and
- atmospheric circulation would be nil at altitude, and reduced only to daily slope effects near the surface (Lenoir, 2001, p. 46).

This was so unexpected that Lenoir expressed surprise that '*not one digital climatologist has felt curious enough to "go and see" what a climate devoid of the greenhouse effect would be like*'. Nature itself provides other examples: the Moon (cf. Lenoir, 2001), or Mercury with its almost total lack of atmosphere; the surface temperature on Mercury's daytime side is between 200° and 430°C, while at night it is between −150°C to −200°C.

The 'zest' which the greenhouse effect brings to the atmosphere entails considerable and important modifications to the terrestrial climate:

- thermal contrasts (and therefore gradients) are set up within the atmosphere;
- differences occur in thermodynamic potential leading to aerological circulation;
- there is an intensification of exchanges between the surface and atmosphere, with a consequent reduction in thermal contrasts at the surface (nights less cold, polar regions less cold, tropical regions less hot);
- the greenhouse effect cools the atmosphere, whose radiative budget is always negative: a deficit enhanced by energy lost by the ground in the form of perceptible and latent heat; and
- the ground becomes warmer than the air, on a global scale (Lenoir, 2001, p. 50).

So the introduction of greenhouse gases provokes a veritable climatic 'revolution': the resulting 'cooling' is not the least of its consequences! Lenoir examines the figures usually presented for determination of the Earth–atmosphere energy budget (Ramanathan *et al.*, 1989). Let us consider those supplied in Figure 6 (p. 79), which are not very different:

- The radiative energy flux received by the atmosphere comprises that absorbed at short waves (67 W/m^2) and the infrared re-emission from the surface (390 W/m^2), totalling 457 W/m^2.
- The radiative energy flux emitted by the atmosphere comprises the infrared flux emitted into space (235 W/m^2) and towards the surface (324 W/m^2), totalling 559 W/m^2.
- The energy lost by the atmosphere is greater, by 102 W/m^2, than the incoming flux. The greenhouse effect therefore tends to cool the atmosphere.
- The radiative deficiency of the atmosphere is compensated for by convection (perceptible heat: 24 W/m^2) and the liberation of latent heat (78 W/m^2), totalling 102 W/m^2 (Lenoir, 2001, pp. 50–51).

5.6.1 Vertical and horizontal exchanges

As a consequence, the atmosphere constantly loses energy, while the Earth's surface warms up (Ramanathan *et al.*, 1989), and the ground is warmer than the air (Lenoir, 2001; Thieme, 2002). This is the classic mechanism inducing convective instability and turbulence. Evaporation and condensation of surface water further complicate heat exchanges. Non-radiative transfers (i.e., the provision of direct heat (via turbulent transfer) or differentiated heat (latent heat liberated during condensation)), compensate for the thermal deficit of the atmosphere. In other words, Figure 6, presented by the IPCC as the unquestionable argument for the reality of the greenhouse effect, precisely proves that **the 'greenhouse effect' by itself does not warm the atmosphere**! This is a point which must be highlighted: **the balance of the radiation budget depends principally upon direct convective heat transfers from the surface towards higher levels**, above all **through the intermediary of the water cycle** (now assuming greater importance), transferring the latent energy stored during evaporation, and liberated in updrafts.

There is another important consequence: the greenhouse effect is also 'the source of the wind', since, without the atmosphere, there can be no thermodynamical activity (Lenoir, 2001, p. 51). The existence of thermal gradients means that exchanges are set up and less insolated areas enjoy a greater input of energy at the expense of warmer regions. **The most decisive consequence is therefore the creation of atmospheric circulation**. At high latitudes, the temperature of low-level air falls markedly, receiving less solar energy, and little from the ground, but the air continues to dissipate its energy. Cooling, it becomes denser and migrates towards tropical latitudes (in compact masses, cf. Chapter 8), and the departure of this low-level air is compensated for by the arrival of warm air from lower latitudes.

These meridional exchanges, which transfer enormous amounts of energy and create aerological stratification and perturbations (and thereby, updrafts and vertical exchanges of heat), seem not to have been properly considered in the establishment of the radiation budget (Figure 6) and the processes determining surface temperature.

5.7 THE IPCC 'WATER VAPOUR FREE' GREENHOUSE EFFECT

As has already been pointed out, the IPCC ignores the influence of water vapour in its *Summary for Policymakers.* The predicted scenarios for the 21st century anticipate emissions and concentrations of CO_2 or SO_2, but not of water vapour. Climatic models project the reactions of climatic variables, including temperature, as a function of the evolution of emissive gases, but water vapour is not counted among these. Notwithstanding, according to '*global simulations obtained through models, concentrations of water vapour should increase ... during the 21st* century' (IPCC, 2001). Apparently, though, this can be of little consequence for the radiation budget, since the subject is not mentioned!

Let us now look at the *Technical Summary* of Working Group 1, by definition scientific since it is not submitted for political approval: it was '*accepted, but not approved in detail*'. It is however supposed to offer '*a complete, objective and balanced exposition of the question*'.

This summary indicates that data from the most reliable observations show a general increase in water vapour at the Earth's surface, and in the lower troposphere, during recent decades. Figure *TS* 7b (IPCC, 2001, p. 33), on the subject of hydrological indicators, specifies:

- for the lower stratosphere, an increase in water vapour of 20% since 1980;
- for the upper troposphere, no noticeable trend worldwide since 1980;
- but for the upper troposphere tropics (between 10°N and 10°S), a 15% increase;
- for the troposphere, increases in many regions since about 1960; and
- large increases generally in water vapour at the surface in the northern hemisphere.

Figure *TS* 7b (IPCC, 2001, p. 33) also mentions a 2% increase in total nebulosity above the oceans (since 1952) and over land masses (during the 20th century).

Chapter 7 of the *Scientific Basis* (IPCC, 2001), based on the work of Cess *et al.* (1990), Hall and Manabe (1999), Schneider *et al.* (1999), and Held and Soden (2000), again mentions the amplification effect: '*the water vapour feedback acting alone approximately doubles the warming*'.

We understand the primary importance of water vapour. It acts in three ways, by absorbing short-wave radiation, by absorbing long-wave radiation and being a major emissive gas, and as a vector of latent heat through meridional transfers and updrafts. Its importance has been reaffirmed above. We know that '*if water vapour increases and is distributed to higher, colder altitudes, less heat is radiated to space and thus climate warms*' (Del Genio, 2002). We know that water vapour has definitely shown a (sometimes marked) increase in its concentration. It is also thought that '*if you change the water vapour concentration by 30%, you can change planetary temperatures by 5–6°C*' (Broecker in Carlowitz, 1996). We also know that, without water vapour, '*carbon dioxide alone is not a very effective greenhouse gas*' (Courtin *et al.*, 1992), and that intense transfers of perceptible and latent heat have taken place over recent decades (Chapters 12 and 13).

It might well be expected that the IPCC would take a good look at the consequences of such increases in water vapour for the energy budget, consequences not only possible but inevitable; these increases are already observed, or envisaged. But nothing is mentioned ... save a presumed increase in rainfall, as a result of a hypothetical (unverified) link between precipitable potential and supposedly precipitated water (cf. Chapter 11).

It is thus manifestly clear that *in the IPCC's atmosphere, there is no water*!

The *Technical Summary* (IPCC, 2001, p. 45) notes, however, that '*the increase in the water vapour content of the atmosphere is one of the principal retroactions at the origin of the strong warming predicted by climatic models ...*'. At last! The importance of water vapour has finally been acknowledged. But there is more: '... *reacting to an increase in CO_2 concentration*'! In other words, water vapour evolves because the CO_2 evolves! Is it really so important to hang on to the CO_2 at all costs? Fortunately, the IPCC calls this report 'objective'. The word 'objective' is doubtless meant to describe the 'goal' towards which the IPCC is striving, but it certainly does not refer to scientific objectivity. We ought here to review the hierarchy of factors, by briefly recalling the effect of water in the determination of climatic characteristics, particularly temperature.

5.8 THE EFFECT OF WATER

The so-called 'greenhouse' effect is fundamentally a 'water' effect, with other emissive gases possibly playing a part. The presence or absence of water determines the characteristics of the climate, and especially its thermal behaviour. Figure 11 deals with this fundamental mechanism.

- If water is absent (which it never is, totally), intense warming by day and a considerable decrease in temperature by night mean a wide daily temperature range (Figure 11(a)). This thermal behaviour is experienced over dry land masses, over ice-covered ground or in the 'pure' air far from cities or in mountainous areas.
- If water is present, (Figure 11(b)), the incoming flux is absorbed while evaporation consumes vast quantities of energy. The temperature cannot therefore rise steeply, and any thermal convection is considerably reduced, especially over oceans. Water vapour stores energy, while long-wave emissions from the ground are intercepted. Night-time temperatures fall moderately, and the diurnal thermal range is narrow. This thermal behaviour is experienced over

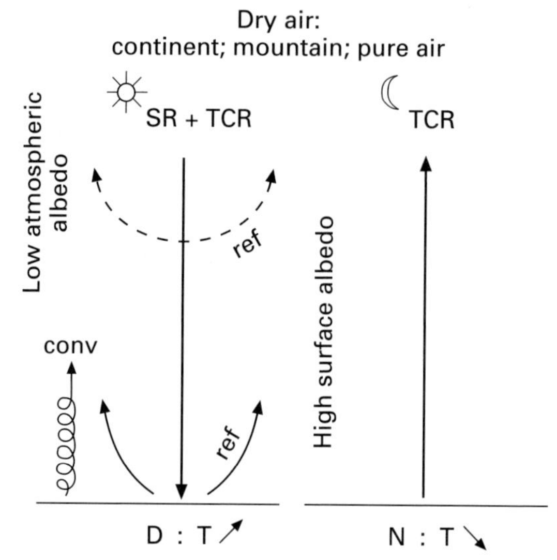

Large thermal amplitude
Very limited greenhouse effect

Moist air:
ocean; dense forest; polluted dust-laden air

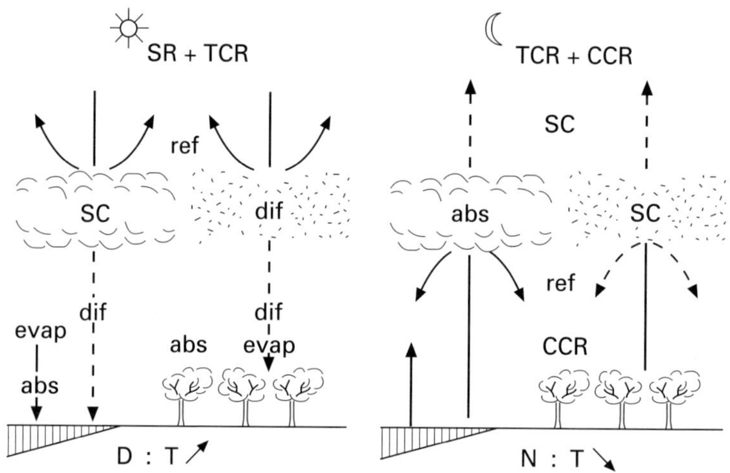

Small thermal amplitude
Intense greenhouse effect

SR : solar radiation (SW)
CCR : celestial counter-radiation (LW)
SC : scattering
evap : evaporation
TCR : terrestrial counter-radiation
abs : absorption
ref : reflection
conv : convection, turbulence

Figure 11. Water effect on thermal behaviour: (*top*) continental; (*bottom*) oceanic.
From Leroux (1996, 1998).

water and in coastal areas, but also in forested regions where it is hyper-oceanic to the extent of nullifying any daily contrasts, as happens in the Congo or the Amazon. These characteristics are also found to some degree in dust-laden atmospheres, or the polluted, 'dirty' air around large conurbations.

With altitude, the thermal gradient is determined by water (and by pressure). If the atmosphere is dry the decrease attains 10°C/km (dry adiabatic rate), and if humid, this falls as low as 5°C/km (moist adiabatic rate). Rising air causes cooling and condensation, and the liberation of latent heat which slows the cooling and encourages updraft, without which the vertical development of cloud formations could not proceed.

5.9 SUMMARY: A FRAGILE SCENARIO

The concept of the 'enhanced greenhouse effect' seems at first sight to be scientifically supported, and, a priori, shows that emissive gases originating from human activity are capable of changing the climate. However, this scenario remains highly speculative, with uncertainties so numerous that the very notion of a greenhouse effect is seriously called into question (cf. Dietzer, 2001; Thieme, 2003; Hoyt, 2004; De Laat and Maurellis, 2004). It is readily understandable that scientists have for more than a century hesitated to go down this road, and the question still remains: why did they speak out with such assurance on this subject during the 1980s? Did they really feel that some point had been reached where major uncertainties had been eliminated? Were they urged on by the promising futures offered by climatic models? Or were they indeed swayed, as were the politicians, by the rapid uptake of such things as acid rain or the famous 'ozone hole'? **The very foundations of the house of cards known as the 'greenhouse effect scenario' are still far from firm, to say the least**.

The very tendentious presentation of the greenhouse effect scenario by the IPCC in its reports, constantly highlighting minor emissive gases such as carbon dioxide, shows a deliberate insistence upon human responsibility. But this too still appears uncertain, and is plainly insignificant in comparison with **the fundamental role of water vapour**. Analysis of this role has been, however, carefully avoided. Indeed, the role of water vapour is not even brought in, except in a minor capacity, as a factor in the amplification of the action of CO_2.

So we must shake off this unfounded obsession with the anthropic greenhouse effect, and reconsider the problem of climate change in a different way, re-establishing the proper hierarchy of phenomena and giving the 'water effect' the major climatic importance which it deserves. What is it that can increase the concentration of water vapour? Can water vapour evolve by itself (i.e., without being merely a consequence of an increase in CO_2)? Why has water vapour already increased? Has this increase already had some consequence for the evolution of temperature? What can explain its dispersion towards higher levels? What are the implications on the dynamics of circulation?

Again, because evoking the greenhouse effect deals with only a small part of the problem, what proof have we that the enhanced greenhouse effect is really causing temperatures to rise? The mean global temperature curve is constantly put forward by the IPCC (Chapter 10), but has it any real climatic value? And is it really correlated with the greenhouse effect (i.e., physically rather than just statistically)? Is the greenhouse effect the only possible correlation? Answers to these questions are slow in coming.

6

Causes of climate change

Felix qui potuit rerum cognoscere causas.

Virgil (P. Virgilius Maro). *Georgics* (II, 489), 39–29 B.C.

We have no Theory of Climate for the moment or in the foreseeable future.

Hämeranta. Moderator of *Climate Sceptics*, 2004.

According to the models and the Intergovernmental Panel of Climate Change (IPCC), it would seem that the greenhouse effect, especially in its anthropic aspect, is responsible for all climatic changes. This is obviously a simplification, a caricature, because other factors are involved, and very much more effectively. These different factors act upon what is generally called the 'climatic system', a practical label but one which does not always imply the totality of the phenomena which need to be taken into account in our understanding of the dynamics of climatic changes.

6.1 THE CLIMATIC SYSTEM

The climatic system has five major components:

- *The atmosphere*. The most unstable and changeable component of the system: the modification of its constituents is considered to be the essential phenomenon in the greenhouse effect, thanks to the properties of its emissive gases, principally water vapour, to which must be added solid and liquid particles in suspension (aerosols), and clouds.
- *The hydrosphere*. All liquid water, including water underground; freshwater in rivers, lakes, and aquifers; salt water in oceans and seas (which are both sinks

for, and sources of, carbon dioxide); and water/vapour and/or liquid in suspension in the air.

- *The lithosphere.* Land masses and their distribution and relief (altitude and disposition), soils, volcanic and terrigenic dust in the form of aerosols.
- *The cryosphere.* Sea ice (icefields), ice on land (the *inlandsis* of Greenland and Antarctica, glaciers on mountains, permafrost), snowfields, and ice crystals in high clouds.
- *The biosphere.* On land and at sea, represented by vegetation, and particularly by extensive entities such as large areas of forest, not forgetting plankton fields.
- *The noosphere* (*noos*, intelligence) – which may be added to these major components. Representing the actions of the human race (though this does not always correspond to a definition of those actions).

These tightly interwoven components are influenced (or forced) by processes both internal and external. Internal processes are interactions affecting climate and depending on it, and external ones are factors which affect climate but are independent of it. To these more or less direct actions, we can add retroactive (feedback) processes. Positive feedback intensifies the original effect of a forcing event, and a negative feedback reduces it. As an example of positive feedback we may recall that an increase in water vapour is induced by a rise in temperature, and further encourages warming, water vapour being the principal emissive gas. However, an increase in water vapour content may also contribute to increased cloud cover, causing negative feedback (cf. Chapter 5). Figure 12 shows the great complexity of the climatic system, whose constituents are linked by physical, chemical, and biological interactions, on very variable scales of time and space.

This diagram (Figure 12), which is supposed to represent the *global climate system*, should not be taken to mean that there exists a *global climate*. There is no such thing. It would be a mere construct of the imagination. The definition of climate or, to be more precise, of climates involves a precisely defined geographical base. Also, because of its static character, this schema cannot convey the integral nature of the climatic system. In fact, it represents what one might call the 'Earth system', but does not express the climatic system as a whole, because this system is essentially one of movement. Indeed, together with stable or only slowly modified elements, easily represented on such a diagram, there are also changing elements displaying more or less rapid evolution. The ocean, for example, driven at the surface by the wind (drift currents), or by contrasts in density resulting from thermal gradients (density currents) or salinity (thermohaline circulation). Perhaps the best example, much more reactive and fast-moving, is the circulation of the atmosphere, with its meridional transfers of air and energy (Chapter 8). These extremely mobile elements, especially the air, cannot be represented in the schema of Figure 12. They are, in fact, often forgotten, particularly within the models, in spite of their essential role in determining the climate. We will expand upon their importance later (cf. Chapters 8, 9, 11, 12, and 13).

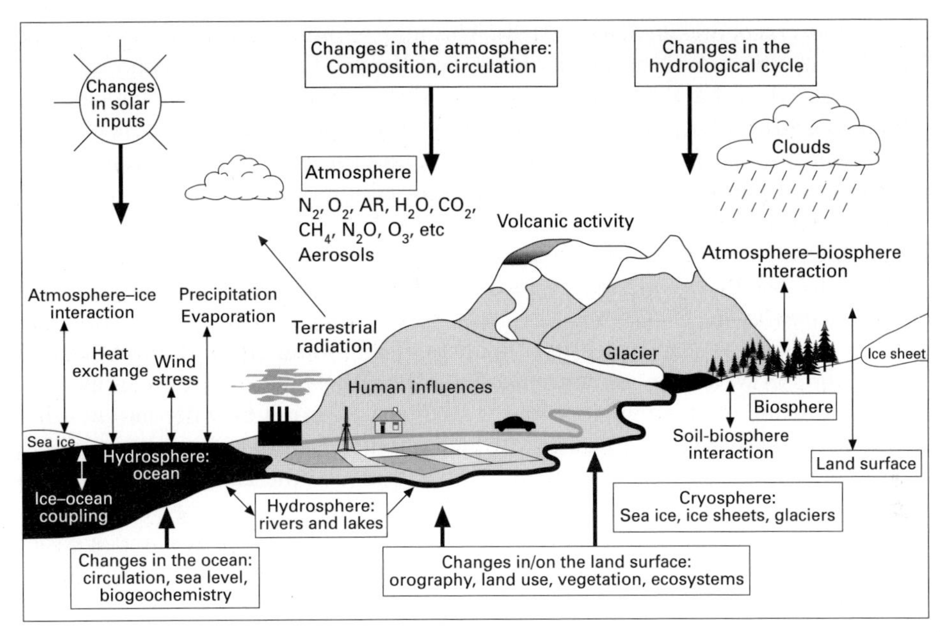

Figure 12. The components of the global climate system, schematic view. Processes and interactions are represented by thin arrows; aspects that may change are represented by bold arrows.
From IPCC (2001).

6.2 THE CAUSES OF CLIMATE CHANGES ACCORDING TO THE IPCC

6.2.1 The principle basis

In the very first paragraph of the *Introduction to the Climate System* (2001), the IPCC states: '*If one wishes to understand, detect and eventually predict the human influence on climate, one needs to understand the system that determines the climate of the Earth and of the processes that lead to climate change*'. A praiseworthy aim, on an obviously scientifically correct path (though one not always followed in the IPCC reports!). The IPCC goes on to say that: '*The climate of the Earth as a whole depends on factors that influence the radiative balance, such as, for example, the atmospheric composition, solar radiations and volcanic eruptions*'. It is also stressed that '*climate is determined by atmospheric circulation and by its interactions with the large-scale ocean currents, and the land with its features such as albedo, vegetation and soil moisture*' (IPCC, 2001).

What the IPCC goes on to say shows that, by omitting movements of the air and the oceans, it sees climatic changes as essentially driven by human activities. Almost its only concern seems to be the '*identification of a human influence on climate change*'

(IPCC, 2001, *Tech. Summ. Scient. Basis*). This human influence includes the burning of fossil fuels, the combustion of biomass, and the production of greenhouse gases and aerosols which have an impact upon radiative forcing. Also, changing land use methods (agriculture, irrigation, deforestation, and reforestation) affect the physical and biological properties of the Earth's surface, and the growth of cities leads to the formation of urban heat islands with very localised impacts (cf. Chapter 10).

Without dwelling again on the greenhouse effect (cf. Chapter 5), and the way it is proffered, exaggeratedly and in a very fragmentary fashion, let us not forget that, according to the IPCC itself, '*water vapour is the strongest greenhouse gas ... water vapour is central to the climate and its variability and change*' (IPCC, 2001). However, this does not seem to lead to a proper analysis of the 'water effect'. Remember also that variation in water vapour concentration is not solely dependent on an increase in carbon dioxide (and is therefore only a feedback effect): it may depend on other factors, particularly dynamic transfers (which are not mentioned).

Climate models are used to simulate and quantify the climate's response to humanity's present and (supposed) future activities, according to different scenarios (IPCC, 2001). Incidentally, it is worth mentioning that it is much easier to vary some parameter or other (e.g., emissive gas rates, albedo value) within the equations of models than it is to take account of other modifying factors, and their transmission via general circulation (cf. Chapter 8), which models cannot do.

6.2.2 Natural and ... 'supernatural' causes

Again it should be pointed out that the IPCC (2001) seems to recognise two different types of climatic variability. Variation can be the result of '*natural fluctuations of the forcing of the climatic system*', for example variations in the intensity of incident solar radiation, or modifications in the concentration of aerosols after a volcanic eruption. Also, '*natural climate variations can likewise occur in the absence of modification of external forcing, through the effect of complex interactions between the diverse components of the climatic system, particularly ocean–atmosphere interaction. The El Niño–Southern Oscillation (ENSO) phenomenon is an example of this natural "internal"* variability on interannual weather scales'. So it seems therefore (if the 'interaction' mentioned is only an 'internal' process) that there are variations in the climate ... because of the climate ... (i.e., with no apparent cause!) What might 'natural internal variability' be, if it is not responding to some (external?) forcing? Could there be 'natural' modifications, and also 'supernatural' ones not derived from fundamental, long recognised causes (or even causes as yet undervalued)?

This would seem to be the case, because alongside causes that were identified long ago, such as the Sun and volcanoes, we also find 'oscillations' promoted to real players at the forefront of the meteorological stage. Consider this: as a result of a '*quasi-periodically varying ENSO phenomenon, caused by atmosphere–ocean interaction in the tropical Pacific ... the resulting events have a worldwide impact on weather and climate*' (IPCC, 2001). So the ENSO phenomenon (cf. Chapter 13), not linked with any external forcing, and which the IPCC deems completely 'independent' of the climatic system, becomes a cause in its own right of world climate change (although it is obviously only a consequence, cf. Chapter 13)! Another example is that of the 'North Atlantic Oscillation (NAO)' which '*fluctuates on multi-annual and multi-decadal timescales, perhaps influenced by varying temperature patterns in the Atlantic Ocean*' and which '*has a strong influence on the climate of Europe and part of Asia*' (IPCC, 2001). Similarly, there is the 'Antarctic Oscillation' or the 'Arctic Oscillation' ... Here we have phenomena (cf. Chapters 12 and 13) whose origin is not well defined, and which are considered to be causes of climate change! Is this justifiable, or are these things promoted to the rank of 'cause' because of a too-frequent tendency to make a *deus ex machina* out of a phenomenon whose initial cause is unclear, or merely inexplicable, or whose exact place in the chain of processes (i.e., in general circulation) cannot be determined?

Confusion is therefore rife: there is polarisation on the subject of the greenhouse effect, inconsistent application of other climate change factors, and the classification as causal of phenomena whose mechanisms are not clear. So the IPCC's report cannot be said to allot the greenhouse effect its proper place compared with other factors. It is therefore necessary to look again at possible causes of climatic variations, and thereafter in any case to indicate mechanisms through which possible modifications brought about by these causes are distributed around the world by general circulation (cf. Chapter 8). Leaving aside the hypothetical greenhouse effect, the principal causes of climate change covered here are orbital variations in radiation, variations in solar activity, and volcanism.

6.3 ORBITAL VARIATIONS IN RADIATION

Seasonal and latitudinal variation in radiation is a function of the positions of the Earth with respect to the Sun. The polar axis, presently aligned with the star Alpha Ursae Minoris (the Pole Star), forms an angle of almost 66° 33′ to the plane of the ecliptic. The Earth–Sun distance changes by about 5 million kilometres between Earth's aphelion (when it is furthest from the Sun, near the summer solstice in early July) and its perihelion (when it is nearest the Sun, near the winter solstice in early January). But the orbital parameters of radiation are constantly changing, with variations taking place in the Earth–Sun distance and in the inclination to the ecliptic and orientation in space of the polar axis, which determines the precession of the equinoxes. These three parameters follow overlapping cycles of different lengths, and slowly modify conditions affecting the arrival on Earth of the Sun's radiation.

The astronomical theory of palaeoclimates, involving variations in the three aforementioned astronomical parameters, and the outcome of lengthy reflection about the genesis of glaciations discovered by Agassiz (1837), saw decisive progress with the work of Adhémar (1842) and Croll (1875) at the end of the 19th century, and thereafter especially through the work of Milankovitch (1924). Glacial periods had first of all been associated with a decrease in energy received at high latitudes in winter, with corresponding accumulation of snow. Milankovitch, however, thought that conditions promoting the formation of ice corresponded to a minimum of summertime solar radiation in the northern hemisphere, insufficient to melt the snow from the previous winter. None of these 'explanations', which consider phenomena *in situ*, addresses the dynamic of ice accumulation (cf. Chapter 9).

These early concepts, and also more modern concepts on this subject, offer only a static vision of phenomena; they take into account neither the necessity of enormous transfers of precipitable potential towards the poles, nor the dynamical factors needed to bring about such long-distance transfers. So Milankovitch deserves recognition not for an explanation of the mechanism of glaciations (which cannot be explained without recourse to the idea of general circulation, cf. Chapters 8 and 9), but for having revealed the causes of the long-term variations in insolation experienced by the Earth. The astronomical theory (sometimes known simply as the 'Milankovitch parameter') has been alternately recognised and discredited, as geological debates have continued. Then, in the 1970s, long chronological series based on the study of marine sediments and ice cores gave more credence to the theory: more recently, the Vostok cores from Antarctica (Petit *et al.*, 1999) have revealed four cycles, lasting more than 400,000 years, of eccentricity in the Earth's orbit, and lesser variations within those principal cycles. Periodicities of about 100,000, 41,000, and 22,000 years have therefore emerged. So, the astronomical theory is established as the fundamental explanation of palaeoclimatic modifications.

It should be remembered that, although these cosmic parameters explain past variations in the climate, comparisons with the present situation are possible only if we understand the geographical context of the periods in question. At the same time we should take into account distribution of land masses and their major mountain chains, which control units of circulation in the lower layers of the atmosphere, in a known context of general circulation (cf. Chapter 8).

6.3.1 Variation in the eccentricity of the Earth's orbit

The Earth's orbit is not circular, but elliptical, its revolution around the Sun being perturbed by the gravitational attraction of the other orbiting planets. The Earth is therefore not at a constant distance from the Sun, its orbital path changing from almost a circle to a more or less flattened ellipse. The term *eccentricity* describes the degree of deviation from a circular path, and is expressed as a percentage relating the minor axis to the major axis of the ellipse: the smaller the percentage, the more circular the shape becomes. Variation in the eccentricity, which has been relatively small since it has not yet exceeded 7% (A. Berger, 1992), shows a near-periodicity of about 100,000 years (or 100 kyr (kiloyears)), with variation between 90 000 and 120,000 years (90 kyr and 120 kyr). The last maximum of eccentricity, which was small (of the order of 2%), happened about 16 kyr BP, and the previous one dates back to about 125 kyr BP, with an eccentricity of about 4% (Figure 13).

The relatively slight eccentricity of the Earth's orbit has little effect on the total energy received throughout the year, the difference amounting to only 0.2%. However, seasonal contrasts are important, and if we add those based on other parameters, they are involved in a major cycle of climatic variations of about 100 kyr, which constitutes the '*principal cause of the quasi-periodic succession of*

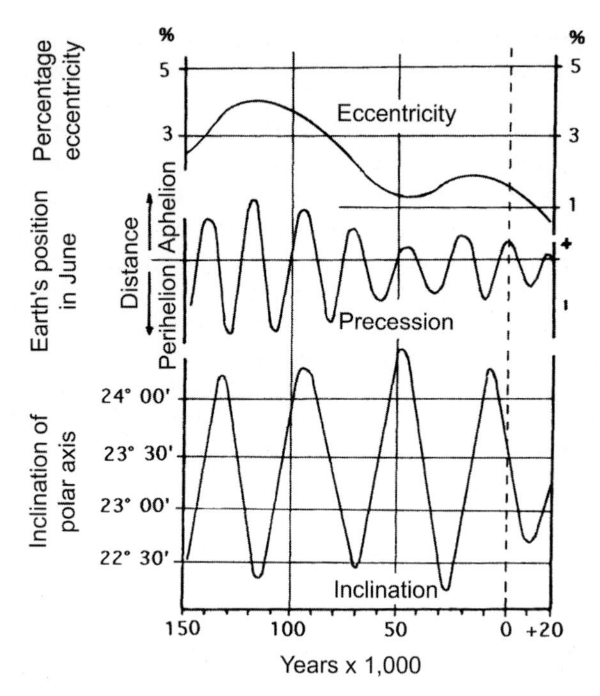

Figure 13. Variation in orbital parameters from 150 kyr BP to +20 kyr.
After Berger (1992).

glaciations on Earth' (Duplessy and Morel, 1990), in good agreement with recent values based on the Vostok cores.

6.3.2 Variation of the angle of inclination of the Earth's polar axis

The angle between the polar axis and the perpendicular to the plane of the ecliptic is moving very slowly towards 23° 26′ (in 1996 the value was 23° 26′ 22″), and decreasing by 0.5″ annually (Fong Chao, 1996). This angle varies through 3°, between 22° and 25°, over a quasi-period of 41 kyr. The last minimum of obliquity (involving an angle of 22° 10′), occurred about 28 kyr BP, and the last maximum (24° 30′) occurred about 9 kyr BP (Figure 13). The polar axis has been slowly moving back ever since. A decrease in the angle, with the axis moving towards its most upright position, reduces seasonal contrasts, with milder winters and cooler summers. Conversely, when the angle is larger and the axis more tilted, contrasts are more marked. The winter hemisphere is relatively 'farther' from the Sun and is therefore colder, the summer hemisphere being warmer as it is 'nearer' the Sun.

Variations are identical in the two hemispheres, but consequences differ according to latitude, especially in the tropics and in polar regions:

- In the tropical zone, variations due to axial inclination are comparatively weaker, but the latitude of the two tropics moves from 22°N and S to 25°N and S, bringing about a reduction or extension of the astronomically defined tropical zone. Briefly, the latitude of the tropic, cosmically defined, is determined as follows: 90° minus the angle formed by the polar axis and the plane of the ecliptic. The latitudinal belt lying between the tropics, and within which the Sun will reach the zenith, thus varies by 6° in latitude, with a width of from 44° to 50°. This difference means an extension in latitude of nearly 700 extra kilometres above which the Sun may pass at the zenith, which may seem a small distance, but is by no means inconsiderable if we consider: the total surface in the tropics capable of receiving optimal (zenithal) solar radiation; the dynamic of the meteorological equator (ME); and the reduction in the resultant amplification of monsoon circulation (cf. Chapter 8). About 9 kyr BP (Figure 13), the tropical zone (in the strictly astronomical sense) stretched across 49° of latitude (24° 30′ × 2), but in our epoch this is reduced to 46° 54′, and will continue to shrink (as the amplitude of zenithal movement decreases).

- The effects of obliquity are amplified in polar regions, where the latitude of the Polar Circles (today at 66° 33′, and moving towards 66° 34′) oscillates between 65° and 68°, reducing or extending the area where 24-hour darkness in winter may occur. During summer, a maximal tilt will increase polar warming, and the difference between the two extreme positions of obliquity means a difference of 14% in energy received in high latitudes in summer. Conversely, when the axis is at its most upright, the energy received by the poles in summer is at its least, a situation pertaining at about 28 kyr BP, when the inclination was close to 22° 10′ (Figure 13).

The cumulative effect of the total quantity of energy received simultaneously by the tropical and polar zones is particularly striking:

- When the axis is more upright (i.e., not much inclined) the tropical zone is smaller and the zone of the 'polar night' correspondingly larger. The total energy deficit is considerable, especially near the poles where the winter deficit is greater and summer insolation correspondingly reduced. This was the situation at about 28 kyr BP.
- When the axis is more tilted, the tropical zone is larger and the zone of the 'polar night' correspondingly smaller. The total energy surplus is considerable, and the polar thermal deficit is less in the winter, with summer insolation correspondingly increased. This was the situation in the period centred on 9 kyr BP.

Remember that the axis is currently moving to a more 'upright' position (i.e., there is a very gradual reduction in the tropical zone and a no less gradual extension of the 'polar night' zone. But the timescale is very long (estimated at about 10,000 years according to Figure 13).

6.3.3 Variation of the orientation of the polar axis (precession of the equinoxes)

The Earth is not a sphere, but an oblate spheroid, with a slight bulge at the equator due to rotation. The gravitational attraction of the other bodies in the solar system causes a slow gyration of the Earth's axis, an oscillation like that of a spinning top.

The polar axis does not therefore always point towards the same place among the stars, and the Pole Star (on a line projected from the polar axis northwards) is at present Alpha Ursae Minoris; 4,000 years ago the Pole Star was Alpha Draconis, and in 12,000 years' time it will be Vega, in the constellation of Lyra.

This oscillation gradually displaces the positions of the solstices and equinoxes on the ellipse travelled by the Earth, and changes that moment in the year when it reaches its furthest point (perihelion) and its nearest point (aphelion) with respect to the Sun. The combination of orbital parameters gives a mean value for the periodicity of the precession of the equinoxes of 22 kyr.

The outcome of this periodicity is a swing, in opposite directions, of the polar axis: thus, 11 kyr ago (Figure 13), at the summer solstice, the Earth was at perihelion (unlike today when it is at aphelion), whilst the winter solstice coincided with aphelion (perihelion today). In the northern hemisphere therefore, summers were warmer but winters colder; in the southern hemisphere summers were cooler, as a result of the increased solar distance, but winters were less cold, the distance being shorter. Seasonal contrasts were therefore less pronounced in the southern hemisphere, but very marked in the northern hemisphere.

6.3.4 Orbital parameters and the evolution of insolation

The role of the astronomical factor in long-term insolation, and in climatic variations, is undoubted. The 100-kyr cycle is obvious, and is apparent in all long-term curves. Since it is seen to be the same all over the planet, this cycle explains the synchronous character of major changes in the two hemispheres. But factors within the three periodicities combine differently and inertial factors must also be taken into account, for example, the accumulation and melting of continental ice (cf. Chapter 9).

Let us take the example of the last major cycle of glaciation, as illustrated in Figure 13:

- The period centred around 125 kyr BP, when eccentricity was 4%, obliquity nearly 24°, and summer insolation in high latitudes more than 13% greater than at present, constituted the interglacial Eemian period, which was somewhat warmer than the present.
- The thermal deficit which followed, at high latitudes in the northern hemisphere, from 115 kyr BP (with eccentricity still pronounced, but with low obliquity (22° 24′), and summer at aphelion), marked the beginning of the last Würm glaciation. Through successive stages, with the maximum deficit centred around 28 kyr BP (see above), this led to the Last Glacial Maximum (LGM) which ended at about 15 kyr BP.
- Around 10 kyr BP (Figure 13) a remarkable situation pertained, with northern summer occurring at perihelion (11 kyr BP) and the polar axis strongly tilted (9 kyr BP) (i.e., with extension of the tropical zone and diminution of the 'polar night' zone (cf. above)). Because of glacial inertia (the time needed for the

melting of enormous ice sheets), the Holocene Climatic Optimum (HCO), the most recent equivalent to the Eemian Period, was at its height around 6 kyr BP.

- Since that time, conditions have gradually been becoming less clement, and as a function of orbital parameters, 'the cooling which began 6,000 years ago will continue for another 5,000 years' (Berger, 1992), then evolving into glacial conditions.

Although these orbital parameters are fundamental in the understanding of palaeoclimates, they occur too slowly to be considered in studies of present evolution. However, they enable us to see recent millennia, and our own epoch, as part of a general trend of lessening insolation.

Different combinations of parameters introduce shorter variations within the major cycles. Consequences vary according to latitude, being slight near the equator and increasing in magnitude towards the poles. Obliquity and precession have a great influence on insolation in polar latitudes (NB where the greatest variations are encountered); it is in these very regions that the Mobile Polar Highs (MPHs) originate, influencing all general circulation, which is accelerated (rapid mode) or decelerated (slow mode) as a function of the intensity of the thermal deficit in high latitudes (cf. Chapters 8 and 9).

6.4 VARIATIONS IN SOLAR ACTIVITY

Variations in the Sun's activity have been a subject of debate for a very long time, and there is an abundant scientific literature (cf., e.g., Waple, 1999). These variations crop up from time to time in the media to 'explain' meteorological anomalies. The Sun 'goes crazy' and its 'wrath' is responsible for heatwaves, droughts, and even floods. There is an endless fascination with the subject, even though '*in spite of the massive literature, there is little or no convincing evidence of statistically significant correlations*' (Pittock, 1983) between solar activity and weather or climate (with just a few exceptions). Interest has recently been rekindled following new observational and analytical capabilities, new methods of investigating past solar activity, and new theories moving beyond the Sun itself to take account of variations in the flux of galactic cosmic rays (GCR).

6.4.1 The sunspot cycle and solar activity

Irregularities in the Sun's activity manifest themselves through the appearance of darker patches on the photosphere (the visible luminous surface of the Sun), linked to the Sun's magnetic field. The spots appear dark because they are at a lower temperature (about 4,500°C) than the rest of the photosphere (at about 6,000°C). Observations of sunspot activity since the beginning of the 17th century have revealed that the number of spots varies between sunspot minimum and sunspot maximum, following an average cycle of 11 years, with variations around this mean value of between 9 and 13 years (Figure 14). Also, cycles occur in pairs, making the

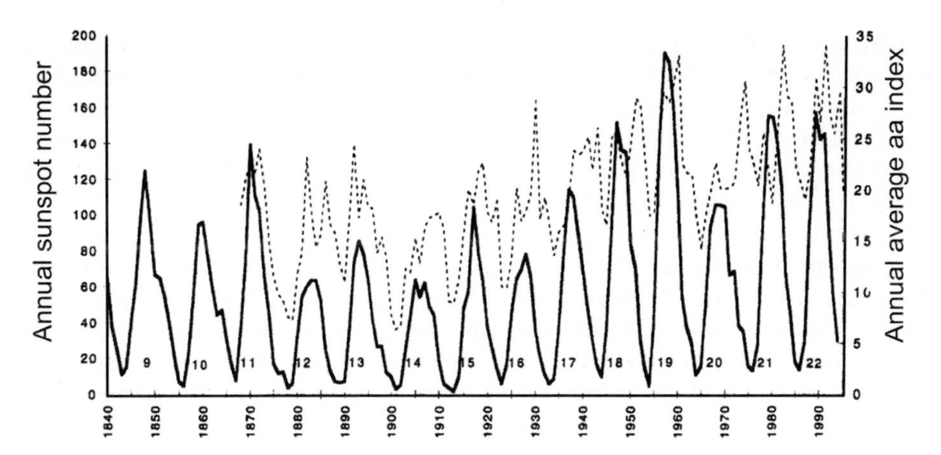

Figure 14. Annual sunspot numbers from 1840 to 1995 (thick line) and annual average geomagnetic index from 1868 to 1995 (dotted line).
After Joselyn *et al.* (1997).

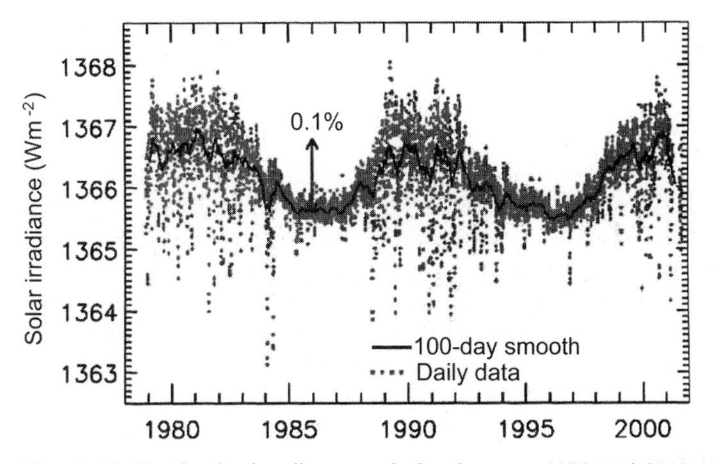

Figure 15. Total solar irradiance variation between 1978 and 2002.
After P. Foukal (2003).

periodicity 22 years, involving the reversal of the Sun's magnetic field. The maximum before last, in cycle 22, occurred in 1991, and the last minimum in 1997, when cycle 23 began, with a maximum in 2000–2001 (Figure 15). The next minimum is expected in 2007–2008.

These 11-year cycles are of unequal length. The values of the minima remain very similar to each other, but the maxima exhibit considerable variations. There is a relationship between the number of spots and solar activity: maximum solar radiation (the active Sun) occurs at times of spot maxima (Figures 14 and 15).

In the course of one cycle, variation in the intensity of radiation is never very great. Since 1978, variations in the intensity of solar radiation have been measured directly by satellites, and results have shown that the amplitude of variation in an 11-year cycle is only 0.1% (Foukal, 1994, 2003; Figure 15 above). The amplitude of variations within a cycle attains 0.3% over short periods during phases of increased activity (around maximum), and is very small at times of least activity (around minimum). Cycles involve all solar activity, with maxima corresponding to periods of intense solar wind and greatly increased ultraviolet radiation.

It is possible to infer past solar magnetic activity (and associated illumination) by analysing isotopes formed by the interaction of radiation and atmospheric molecules. Production of carbon-14 (^{14}C) from nitrogen-14 (^{14}N) in the upper atmosphere varies with the intensity of solar radiation. Small variations in amounts of ^{14}C fixed by trees and corals are revealed by isotopic analysis of their carbon content. Analysis of cosmonucleides in polar ice can also be used to estimate abundances of beryllium-10 and chlorine-36. Such studies have led to an appreciation of variations in solar activity over several thousands of years, and cycles of 11 years, and longer cycles, have been confirmed, as has the existence of periods of reduced solar activity. Examples are the Wolf Minimum of the 13th century, the (more pronounced) Spörer Minimum (15th century), the Maunder Minimum (17th century), further evidence of which is the almost total absence of sunspots between 1645 and 1715, and the (less pronounced) Dalton Minimum at the beginning of the 19th century.

6.4.2 Solar activity and climate change

The impact of variations in solar activity on the way climate evolves is still constantly discussed (cf. Landscheidt, 1998, 2000–2003; Benestad, 2004). Simple models have been put forward: a 1% change in the solar constant would entail a variation in the Earth's surface air temperature of 0.6°C (a value which corresponds to the presumed secular rise in global temperature). If we consider only what has happened in the course of the last century, then cycle 19, peaking in 1958, showed the highest values (Figure 14), and the amplitude of cycles 14 to 18 has progressively widened (radiation having apparently increased by 0.6% between 1910 and 1960). Cycle 20 saw a decrease in amplitude, which picked up again during cycles 21 and 22. It is impossible to ignore parallels with global thermal evolution in recent times. The marked rise in temperature since the beginning of the 20th century (mainly between 1910 and 1940) seems indeed to be due to solar 'forcing' (Tett *et al.*, 1999), whilst the *contemporary climatic optimum* of 1940–1960 corresponds to cycles 18 and 19, followed by a change in the 1970s (cycle 20) and, since, by a new rise in temperature coinciding with a new increase in the length of the cycles. Co-variation of the length of cycles and 'global temperature' is therefore particularly narrow, and even more so than the co-variation (or supposed link) between the evolution of 'global temperature' and values for the concentration of CO_2 (Chapter 10).

Just when it seemed that everything had been said on the subject of the supposed influence of the Sun, an 'amazing', nay 'dazzling' association was recently revealed between the length of the solar cycle (between 9.7 and 11.8 years) and temperature anomalies in the northern hemisphere between 1860 and 1990 (Friis-Christensen and Lassen, 1991). The amplitudes of cycles and of solar activity correspond: when the

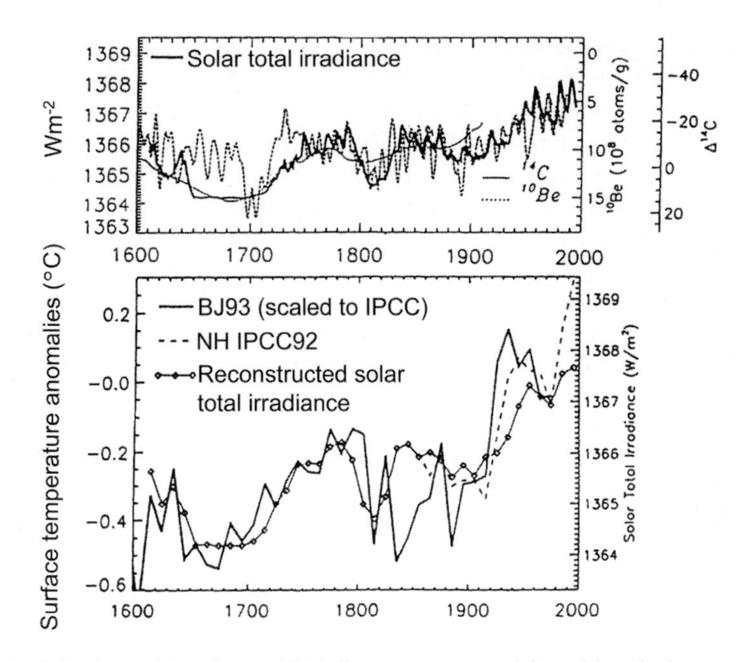

Figure 16. (*top*) Sun's total irradiance (thick line: reconstructed from historical sunspot group numbers; thin line: deviation of carbon-14 abundance from its long-term trend in tree rings; dotted line: trend for beryllium-10 in polar ice cores). (*bottom*) Representative summer surface temperature for the northern hemisphere, extrapolated to 1600 (thick line: from Bradley and Jones; dashed line: IPCC northern hemisphere temperature in the common period; diamond line: decadal solar irradiance, from Lean *et al.*, 1995).
After Pang and Yau (2002).

cycle is short, solar activity is high, and when the cycle is long, activity is lower. Cycles were long (more than 11 years) at the end of the 19th century and at the beginning of the 20th century (with associated low temperatures), and then they became shorter (of the order of 10 years) until 1940–1960 (the period of the *modern climatic optimum*); then came a slight lengthening of the cycles until 1970 (with slight cooling), and finally a renewed reduction in length to below 10 years, along with the higher temperatures of recent decades. So here is another surprising co-variation, even more remarkable than that involving the 'greenhouse gases', so often invoked by the IPCC.

Similarly, and rather later (Figure 16), Pang and Yau (2002) considered that, despite the fact that '*the long-term variations account for less than 1% of the total irradiance, there is clear evidence that they affect the Earth's climate*'. The period known as the 'Little Ice Age' seems to have been associated in its early phases with the Maunder Minimum: an estimated 0.24% reduction in total radiation would have meant a fall in global temperature of 0.46°C (Lean and Rind, 1994). Another minimum, the Dalton, occurred between 1795 and 1825, and was shorter and less severe. Since then, irradiance has gradually increased, and we are now experiencing the *Modern Maximum*, which Lean *et al.* (1995) claim is responsible for half of the rise in temperature since 1860.

We can only be surprised by the 'spectacular' character of links established between, on the one hand, sunspot cycles and their amplitude and length, irradiation values, the prolonged maxima and minima of solar activity (and even the magnetic index, the resultant of the Earth's field and the field induced by the solar wind), and, on the other hand, the secular evolution of mean global temperature (the IPCC curve or estimates). Few co-variations are so neat or closer than that involving CO_2 and temperature. Beryllium studies in polar ice indeed show that the period of enhanced solar activity of the last 60 years is unique in the last 1,150 years. We may be right to think, like Sala and Chiva (1996), and there are other recent examples, that in the Mediterranean part of Spain '*the appreciable "natural rise" in temperature, after correction for the effects of urbanisation, may be attributed to solar activity*'. In the same way, the contribution of the Sun was judged to be comparable or even greater to that of the greenhouse effect by, for example, Baliunas and Soon (1995), Dietzer (2000), Shaviv and Veizer (2003), Landscheit (2003), Jaworowski (2004), Hoyt (2004), and many others. Criticism is multiplying of the exaggerated importance ascribed to the greenhouse effect, and we see an increasing acceptance of the role of the Sun as an essential factor in climatic evolution.

Demonstrative as these relationships are, should they be seen, in the same way as the well-known IPCC 'relationship', as co-variations, or as real, proven cause-and-effect relationships? It is difficult to make up one's mind in the absence of an identified physical mechanism. Such remarkable relationships pose the essential question, before any statistical analysis takes place, and because of the multiplicity of steps involved between causes and effects: can a relationship between solar activity and an isolated climatic element (here, temperature, and on a mean and global scale) appear in such an immediate fashion?

The results above raise first of all problems of estimation:

- Estimation of the variation in solar activity itself, which is of narrow amplitude; though is this still true when we consider that the climatic system functions on less than half a billionth of the energy emitted by the Sun?
- Estimation of the real climatic value of the 'reconstituted' IPCC thermal curve.

In any case, the co-variation of solar activity and 'temperature' is much narrower than that of CO_2 levels and 'temperature' (the CO_2 curve climbs in a surprisingly regular fashion, but the thermal curve shows fluctuations: cf. Chapter 10)!

6.4.3 New approaches...

Now, how does this solar variability manifest itself? As well as emitting visible radiation, the Sun also produces the *solar wind*, a powerful flux of energetic charged particles (protons, neutrons, and electrons). It is diverted, unlike the Sun's light, which travels in straight lines, by the Earth's magnetic field into space, essentially around the magnetopause, which is at a distance of 65,000 kilometres. Nevertheless, because of the lines of force of the magnetic field, which in the present era 'enter' the planet at the North Magnetic Pole and 'leave' at the South Magnetic Pole, the *polar cones* are not well protected from these emanations, whence the occurrence of polar aurorae. The ejections are at their most intense, as solar wind storms, when solar activity is at its greatest. It appears that the effects of solar activity work upon ozone, as the most important modifications involve ultraviolet radiation, which may represent 32% of the changes (Haigh, 1996) affecting ozone chemistry, essentially above the poles. The maximum concentration of ozone is indeed reached during times of maximum solar activity (Angell, 1989). During these maxima the Sun in fact emits intense ultraviolet radiation which affects ozone levels, actually created above the tropics and transferred to the poles by way of the stratosphere. Could these particles be contributing to the seasonal variation in ozone concentrations above the poles, and could they, taking into account aerological links and the decisive part played by the polar regions in determining meridional transfers, explain a possible relationship with climate? This is a relationship that remains to be proved.

Svensmark and Friis-Christensen (1997) and other astrophysicists (see Lenoir, 2001) seem to have come up with the 'missing link' between the Sun and climate, opening a new avenue of research. This showed that cloud cover could be influenced by the intensity of galactic cosmic rays (GCR). An increase in cosmic radiation, modulated by solar magnetism, which varies by about 15% over a cycle, leads to an increase in cloud cover (low cloud) of 3–4%. This increase in cloud brings about a fall in temperature, so the more intense the GCR, the less warm it gets. The amount of GCR also depends upon solar activity: the more magnetically active the Sun, the more intense the solar wind; and the result on Earth is less GCR received, reducing the density of low cloud so that the temperature rises. Forcing due to the Sun is equal to $1.5\,W/m^2$, the same value attributed to CO_2 ... Physical mechanisms have been proposed in order to explain how cosmic rays can affect cloud density, but in-depth investigation is still needed (Carslaw *et al.*, 2002; Benestad, 2004). New approaches are thus being made, and some as Kirkby *et al.* (2004) even supposed that cosmic rays may command glacial cycles. It would be therefore ill-advised to take no notice of them, and to be as insistent, as the IPCC is, on the near-exclusive role of the greenhouse effect ...

To complete the picture, mention should be made of the increase in the radioactivity of the air due to human activities. This can also be instrumental, '*in a way analogous to the effect of cosmic rays, in the formation of clouds*' (Lenoir, 2001).

A re-examination of statistical relationships as a function of physical mechanisms has yet to be undertaken before we can make objective statements about the reality of links between the solar cycle and climatic parameters (especially changes in temperature). But the upsurge of interest in this subject, once thought to be exhausted, is very encouraging, especially if we determine variations in insolation as a function of latitude. In particular, the trend arising from a linear interpolation by Kukla *et al.* (1992) has shown a decrease in annual insolation north of latitude 65°N during the period 1945–1986. Borisenkov *et al.* (1983) also predicted that northern hemisphere insolation from 1800–2100 would decrease in high latitudes, but increase in middle and low latitudes. Given the dynamical importance of high latitudes, driving and feeding general circulation (Chapter 8), the kind of evolution proposed is particularly worth noting.

6.5 VOLCANISM AND CLIMATE

The idea that there may be a link between volcanism and the climate is a very ancient one (cf. Leroux, 1998, 2001). Many authors have discussed the link between major volcanic eruptions and marked drops in temperature, using data from recent times, and also looking back through geologic time. Ancient eruptions have even been seen as factors in Quaternary glaciations. Undoubtedly the most earnest student of the subject, having had the most influence upon later studies, has been Lamb (1970, 1972, 1977) who suggested the dust veil index (DVI) to characterise how dust veils from volcanic eruptions influence climate. Very many studies, now supported by satellite imagery, have shown that volcanic aerosols play a major part in causing temperature change, especially in the short term.

6.5.1 Volcanic emissions and ejecta: silicates and sulphates

Volcanoes, either active or extinct, are a continuous source of gases, which may or may not accompany lava outflows; half of this outgassing probably occurs at the mid-oceanic ridges. Volcanoes also throw into the atmosphere:

- water vapour;
- sulphur compounds (mostly sulphur dioxide, SO_2);
- carbon dioxide (35–65% of the CO_2 needed to balance out the deficit of the ocean–atmosphere system, according to Gerlach (1991)); and
- chlorine (36 million tonnes annually in years without major eruptions, according to Maduro and Schauerhammer (1992)).

A large eruption, sending considerable amounts of chlorine into the stratosphere, is accompanied by a decrease in ozone levels. For example, Mount Erebus, in Antarctica, in continuous eruption since 1972, emits more than 1,000 tonnes of chlorine a day (370,000 tonnes in a year), and is a major contributor to ozone reduction above the South Pole. After the eruption of Pinatubo in 1991, measurements showed a decrease in the ozone layer of about 5–6% above northern hemisphere tropical latitudes (where the cloud first spread). There were decreases of 3–4% over mid-latitudes, and of 6–9% in high latitudes (Mahfouf and Borel, 1995) as a result of the concentration of aerosols in the polar vortex.

The meteorological effects of volcanoes depend upon the density, extent, and duration of atmospheric veils of aerosols. Aerosols of non-volcanic origin (dust, and sand particles from deserts), which generally rise to only modest altitudes, remain in the troposphere for relatively short periods. They are washed out of the atmosphere by rain, or driven towards the poles, where there is evidence on the ice sheets of the

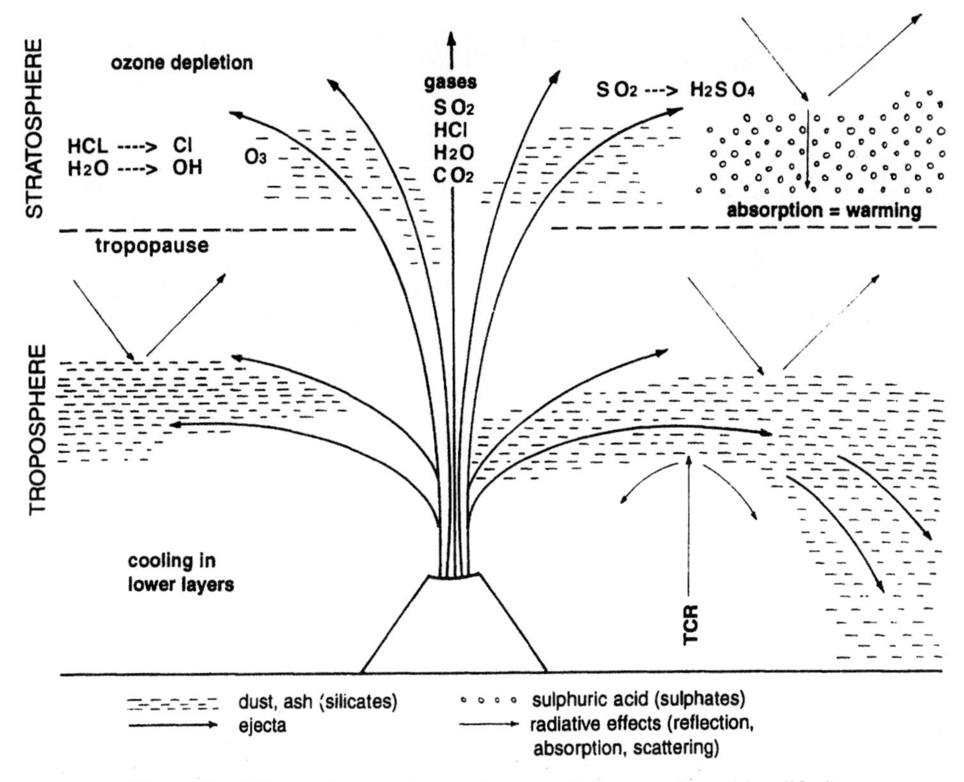

Figure 17. Effects of volcanic eruptions on the atmosphere (simplified).

large amounts transported. Because volcanic aerosols are projected up into the stratosphere, they remain in the air for much longer. These aerosols are of two types (Figure 17):

- Silicate aerosols, which consist of the ash and dust from pulverised rock and are produced in abundance during the more explosive types of eruption, often form veils of fine dust. When material is projected into the stratosphere, long-lasting veils are formed, and the finest particles may stay aloft for more than 10 years.
- Sulphate aerosols originate from the transformation of sulphur compounds (sulphur dioxide, SO_2) into sulphuric acid (H_2SO_4) through contact with atmospheric water. The size of these sulphuric acid droplets (less than 1 μm) allows them to remain in suspension in the stratosphere for several years, and there is a permanent aerosol layer at the 20–25-km altitude. We now know that these sulphate aerosols, by reason of their high reflectivity as droplets, have a greater thermal impact than do ash and dust. Satellite observations of the eruption of El Chichón in 1982 supported this view.

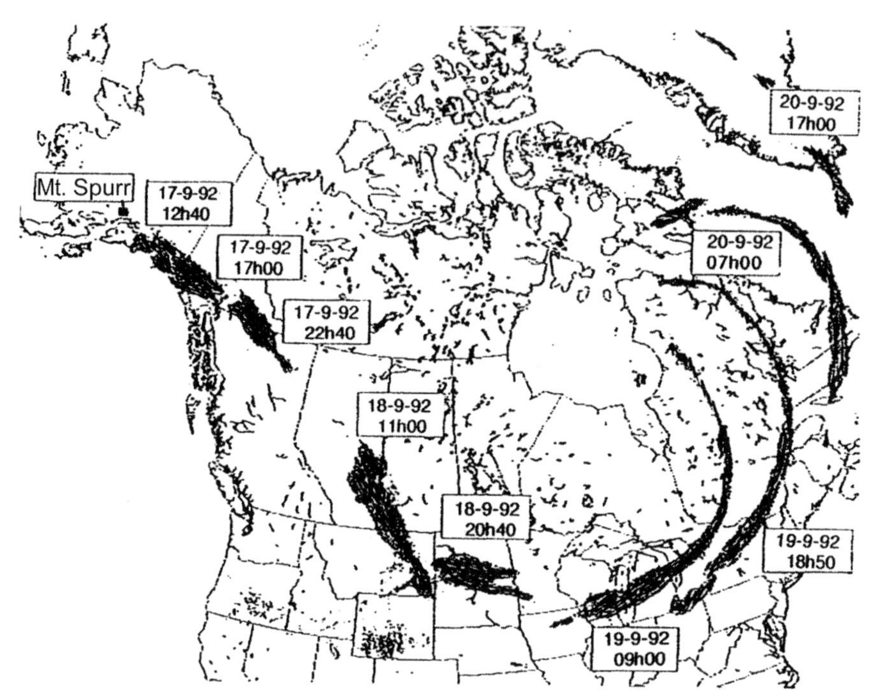

Figure 18. Transport of a dust cloud from Mount Spurr (Alaska) on the leading edge of an MPH over North America, 18–20 September 1992.
After Schneider (1994).

The way in which aerosols are dispersed depends upon the location at which they are ejected:

- In the tropical zone, where the tropopause is higher, and where ejecta can reach a height of 20 km, dust is distributed both northwards and southwards and the spread can be planet-wide. Dispersed by high-level winds, material takes from 2–6 weeks to spread around the planet in lower and mid-latitudes, and thence towards the poles. Material from the eruption on 14–15 June 1991 of Mount Pinatubo in the Philippines, at 15°N, reached a height of up to 35 km. It took 21 days to travel round the globe, according to satellite observations. Two months after the event, there were aerosols above 42% of the Earth's surface between 30°N and 20°S (Bluth *et al.*, 1992). These aerosols later gradually drifted northwards. Two years afterwards, whilst tropical latitudes were back to normal, northern higher latitudes were still optically affected, mainly because of reinforcement by aerosols from the eruption of Mount Spurr in Alaska (Figure 18).
- At higher latitudes, where the tropopause is lower, particles reaching an altitude of more than 10 km form a circumpolar dust ring. Cyclonic circulation on the leading edge of MPHs soon transports the material towards the poles. Figure 18 shows the spreading across Greenland and the Arctic of a dust cloud which was ejected on 17 September 1992 by Mount Spurr.

In both cases, whether or not distribution is general, volcanic aerosols more or less rapidly arrive in polar latitudes. They join forces with terrigenic aerosols lifted from the surface (Tegen *et al.*, 1996); the increasing amount of dust in suspension (Andreae, 1996) forms a polluted layer above the Arctic, especially in winter. This *Arctic haze* is as big as the continent of Africa (Shaw, 1995). However one looks at it, such a concentration of aerosols above the poles can only increase the thermal deficit.

6.5.2 Radiative and thermal effects of aerosols

The main effects due to aerosols from volcanoes or from the Earth's surface involve the scattering of solar radiation into space and towards the Earth. Major dust eruptions are more likely to cause such effects, but more modest eruptions may also contribute if the ejected magma is rich in sulphur.

Effects on solar radiation have been measured ever since the eruption of Krakatoa (near Sumatra, Indonesia) in 1883. The aerosols from this eruption reduced direct solar radiation by 20–30% for a few months, though there was a certain compensatory effect from scattered radiation. The explosion of Gunung Agung (Bali, Indonesia) in 1963, then called '*the eruption of the century*' because of the quantity of ash sent up into the stratosphere, led to a 24% reduction in direct radiation, but compensatory scattering effects brought this down to only 6% of total radiation; it took 13 years for the volcanic dust to disperse. After the eruption of El Chichón (Yucatán, Mexico) on 28 March 1982, the planetary albedo showed an increase of the order of 10% (Halpert *et al.*, 1993). There was also a decrease of 25–30% in direct solar radiation lasting several months after the eruption of Mount Pinatubo in June 1991 (Dutton and Christy, 1992). The effects on solar radiation are greater in high latitudes, since its path through the aerosol layer is longer because of the lower angle of incidence.

Thermal effects involve the stratosphere, where the absorption of part of the radiation causes warming. The undoubted effects on the Earth's surface temperatures may be of lesser scope, but they are more important in climatic terms. In the northern hemisphere, generally, an eruption will cause a drop in temperature of the order of 1°C, the fall occurring 2 or 3 months after the eruption, whence the relevance of the season during which the volcanic event happens (Self and Rampino, 1988). The eruption of Tambora (on the island of Sumbawa, Indonesia) on 11 April 1815, one of the greatest volcanic cataclysms, caused mean temperatures in the winter of 1815–1816 to be 1.5°C lower than normal (Chenoweth, 1996). 1816 was called '*the year without a summer*', particularly in North America, where, from June onwards, four major cold snaps brought severe frost and snowfalls, New England being the hardest hit. Krakatoa, in 1883, erupted one-tenth as much ash as Tambora, and caused global cooling estimated at 0.5°C. Gunung Agung, 1963, brought about a fall in mean annual temperatures of 0.4°C in the northern hemisphere, but 1.3°C for latitudes between 60°N and 90°N from 1963 to 1965, the area covered by Arctic ice sheets increasing by 1.5 million km^2. El Chichón, 1982, brought about surface cooling estimated at 0.5°C for 5 months after its eruption. Pinatubo's

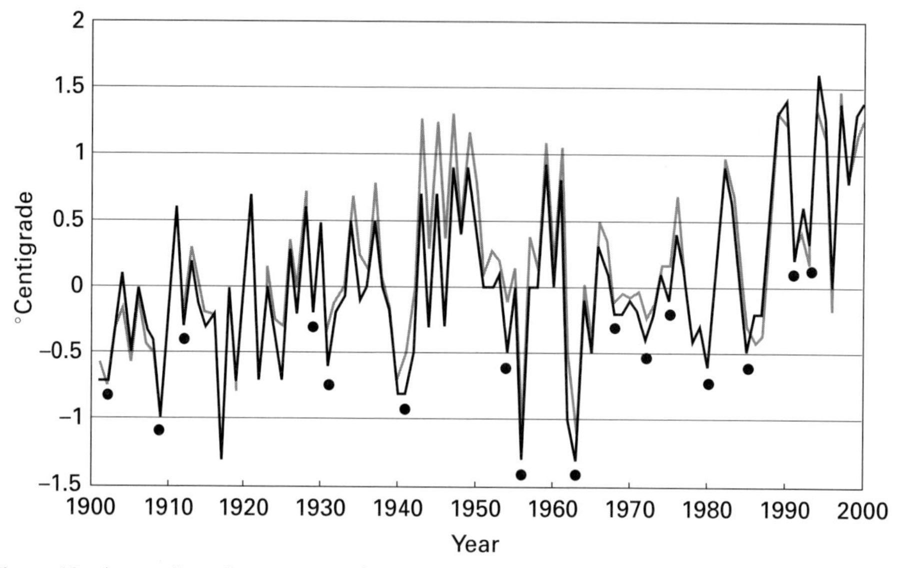

Figure 19. Anomalies of mean annual temperature in France from 1901–2000, related to the period 1961–1990, from Jones *et al.* (2001): thin grey line; from Météo-France, 2002: thick black line from Moisselin *et al.* (2002). Black dots: outstanding volcanic events.

eruptions caused, at the end of 1991, a reduction in temperature in the troposphere of between 0.5° and 0.7°C. In 1992 anomalous below-average temperatures still pertained over North America and northern Asia, down by −3° to −4°C from January to March near Greenland (Halpert *et al.*, 1993).

The involvement of volcanism in the lowering of temperatures therefore seems undeniable, as Figure 19 shows for the 20th century for France. Rapid drops in mean temperatures, some more pronounced than others, coincide fairly closely (in the same or the following year) with these outstanding volcanic eruptions:

- 1902: Santa Maria (Guatemala), Mount Pelée (Martinique), Soufrière (Guadeloupe).
- 1907: Ksudach (Kamchatka).
- 1912: Katmai (Alaska).
- 1928: Paluweh (Indonesia).
- 1929: Reventador (Ecuador).
- 1932: Cerro Azul (Argentina).
- 1942–1947: Hekla (Iceland).
- 1954: Mount Spurr (Alaska).
- 1956: Bezymianny (Kamchatka).
- 1963: Gunung Agung (Indonesia).
- 1966: Awu (Indonesia).
- 1971: Hekla (Iceland).
- 1974: Fuego (Guatemala).

- 1980: Mount St Helens (USA).
- 1982: El Chichón (Mexico).
- 1985: Nevado del Ruiz (Colombia).
- 1991: Pinatubo (Philippines).
- 1992: Mt Spurr (Alaska)...

This co-variation is far too striking to be a mere coincidence. On the scale of mean annual values, according to Veyre (2000), the fall in temperature is of the order of 0.5–1°C for minimum temperatures, and of the order of 1–2°C for maximum temperatures. As far as seasonal variation is concerned, the consequences are less marked in the summer, and much more marked in the winter, outstanding examples being 1956, and 1963, when the eruption of Gunung Agung caused a negative anomaly of about 4°C.

One might reasonably claim that volcanism should occupy a primary place among the factors marshalling climatic evolution in the far and near past, but analyses are still needed (Robock, 2002). An objective classification of volcanic events is fairly difficult to establish.

Indices have been proposed, but they do not all use the same parameters. Lamb's (1970) DVI is based on estimates of quantities of ash emitted, measured by its effects on the optical properties of the atmosphere. The volcanic explosivity index (VEI) (Newhall and Self, 1982) essentially considered the geological characteristics of the magnitude of the explosion. The ice volcano index (IVI) (Robock and Free, 1995), involves the measurement of acidity in the ice of the *inlandsis*, mainly in Greenland. This acidity reflects the amount of volcanic aerosols in the atmosphere. Each of these indices offers a classification of volcanic events, but none has any absolute climatic significance. Many eruptions have gone unreported, and even as recently as the 1970s, sulphur concentrations in the stratosphere went unexplained. In another example, the IVI departs from the other indices by ranking third in importance the minor eruption of Hekla (Iceland) in 1971, after the Katmai (Alaska) event of 1912 and the eruption of El Chichón (Mexico) in 1982. The Hekla event throws new light upon the climatic shift of the 1970s, marked principally by the onset of the Great Sahel drought (Chapter 11), and the El Niño episode of 1972–1973 (Chapter 13).

To establish the relationship between volcanoes and climate, we still need to determine what dynamical links there are between ejecta and thermal modifications. We can use the case of '*the year without a summer*', 1816, to study the climatic consequences of the eruption of Tambora in April 1815, a volcanic event which occurred moreover at the time of the 'Dalton Minimum' of solar activity (Figure 16). The effect of the much-increased polar thermal deficit was a strengthening of MPHs (Chapter 8), bringing much colder weather. More powerful MPHs led to an acceleration in meridional exchanges, and to much more intense perturbations. More vigorous meridional exchanges in the northern hemisphere (Catchpole and Hanuta, 1989; Rampino, 1989; Chenoweth, 1996; Soon and Yaskell, 2003) resulted in continual descents of cold air masses and snowfalls over North America (and also in China and Japan). There was an unprecedented area of ice cover in summer in Hudson Bay and eastern Canada, and abnormally high pressure at the location of the 'Azores' Anticyclonic Agglutination, evidence of reinforcement from MPHs. At the same time, lower pressures were recorded in the Icelandic sector, the deeper depressions associated with stronger MPHs (cf. Chapters 9 and 12). Cold air arrived more frequently in summer around the Mediterranean (Chapter 11) ... Phenomena like these, characteristic of a *rapid mode* of general circulation, accompany all major volcanic events to a greater or lesser extent.

Volcanism is therefore another cause of climatic variation, aperiodic and of short duration, and characterised by rapid falls in temperature. Its influence may be felt over longer periods if volcanic events succeed each other closely, or if, during relatively 'quiet' periods, lowered air turbidity, with fewer aerosols, encourages temperatures to rise. However, climatic consequences are neither immediate nor uniform. Response time is conditioned by the length of time it takes aerosols to be transported towards the polar regions, adding to the thermal deficit; MPHs, reinforced, then take up the baton, spreading the climatic consequences. The real importance of volcanism, which, like solar activity, requires further investigation, would therefore seem very largely underestimated.

6.6 CONCLUSION: THE GREENHOUSE EFFECT IS NOT THE CAUSE OF CLIMATE CHANGE

The possible causes, then, of climate change are:

- well-established orbital parameters on the palaeoclimatic scale, with climatic consequences slowed by the inertial effect of glacial accumulations (Chapter 9);
- solar activity, thought by some to be responsible for half of the 0.6°C rise in temperature, and by others to be responsible for all of it, which situation certainly calls for further analysis;
- volcanism and its associated aerosols (and especially sulphates), whose (short-term) effects are indubitable; and
- far at the rear, the greenhouse effect, and in particular that caused by water vapour, the extent of its influence being unknown (Chapter 5).

These factors are working together all the time, and it seems difficult to unravel the relative importance of their respective influences upon climatic evolution. Equally, it is tendentious to highlight the anthropic factor, which is, clearly, the least credible among all those previously mentioned.

The IPCC (Figure 20) does this, though. Its simulations and descriptions of recent climate change tell us that '*the indication that the trend in net solar plus volcanic forcing has been negative in recent decades ... makes it unlikely that natural forcing can explain the increased rate of global warming since the middle of the 20th* century'.

- Figure 20(a) is supposed to demonstrate, using simulations from models, that net natural forcing (i.e., solar plus volcanic) has been negative over the last four

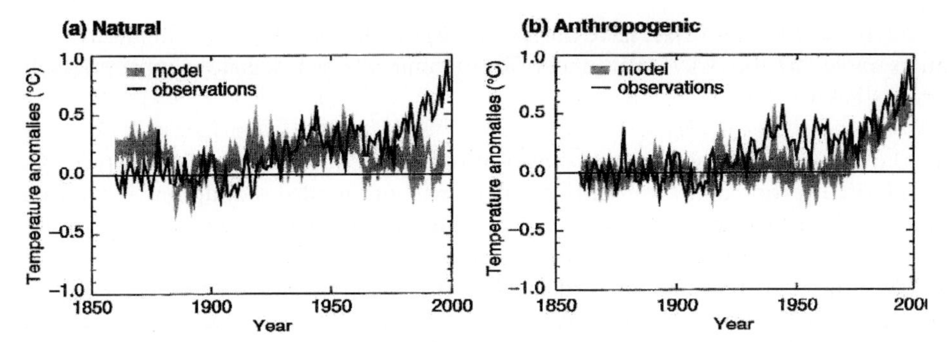

Figure 20. Simulations of annual mean surface temperature anomalies relative to the 1880–1920 mean, from the instrumental record (thermal reconstructed temperature), compared with simulations with a coupled ocean–atmosphere climate model. (a) With natural forcings only: solar radiation and volcanic activity; (b) with anthropogenic forcings: greenhouse gases, change in stratospheric and tropospheric ozone, and an estimate of sulphate aerosols.
From IPCC (2001).

decades approximately, which has tended to lower global temperatures. However, are these forcings on the same scale? Can they be associated, given that volcanic forcing is discontinuous and of limited duration? Moreover, the references cited by the IPCC (2001) in this 'attribution', basing '*the volcanic forcing on Sato et al. (1993) and the solar forcing on Lean et al. (1995)*', do not take into account the recent developments discussed above, thereby ignoring the influence of the 'Svensmark factor', thought to amplify solar forcing by a factor of four (Dietzer, 2000).

- As for Figure 20(b), it is supposed to demonstrate the preponderant, even absolute, influence of the greenhouse effect as opposed to natural causes (Figure 20(a)), after and despite a century of their parallel evolution, in determining recent thermal changes. It would seem that these changes are the result of anthropic forcing (and nothing else) through the greenhouse effect, ozone and sulphate aerosols. The association of these parameters seems unwarranted when discussing human intervention, since the Sun's activity largely determines ozone levels, and sulphate aerosols are the result of volcanic activity! Why not deal with just greenhouse gases here? Doubtless because '*at the current level of understanding, global environment change resulting from increasing atmospheric CO_2 is not quantifiable*' (Soon *et al.*, 2001).

So the problem remains in its entirety: furthermore, the curve on Figure 20 for global temperature (marked 'observations'), upon which the model curve is superimposed as 'visual' proof in Figure 20(b), has no recognised climatic significance (cf. Chapter 10).

Figure 20 does, however, have the advantage of being symptomatic in nature. It is an example, in more ways than one, of shortcomings which it is now appropriate to discuss below:

- First of all it reveals the partisan spirit of the IPCC, seeking always to promote 'formal' proof of the superiority of its still unfounded standpoint by relying upon the final part of the curve, all the while failing to underline that this presumed 'relationship' is (oddly) a feature of recent decades only;
- It shows that the method used (that of the model) is of questionable scientific value, and is thought to be a *sine qua non*: the simulations are deemed to provide 'indisputable' answers to questions of climate (cf. Chapter 7); and
- Causes are discussed, but the dynamic of the transmission of climatic changes is never made clear. There is indeed no point in discussing causes, if one ignores the modalities of the sequences of events marshalling climatic outcomes within the general circulation of the atmosphere (cf. Chapter 8).

We shall return later to Figure 20, with special attention to the last part of the curves, after an analysis of the dynamic of meridional exchanges, the vectors of climate change (Chapters 12 and 13).

7

Models and climate

Dynamical meteorology had developed concepts more global than that of the particle in order to understand atmospheric evolution. It was in this spirit that computerised prediction was born (...)

The power of computers increased ... and the basic equations could be dealt with directly. It was just a question of numerical analysis and of the potential for calculation (...) So we have returned, through digital processing, to the reductionist concept of the particle as used by Richardson, even with climatic modelling ...

Guy Dady. Analysis of *The End of Certainty*, by Ilya Prigogine,
in *La Météorologie*, No. 33, 2003.

'Models' are seen as absolute references: they pronounce; they predict; they confirm or deny; one might even say that they decide! Did the modellers not announce the Apocalypse to come, and orchestrate the catastrophism of the IPCC?

'The' model inhabits the realm of the supernatural. Shrouded in the mystery of its mathematics and physics and its association with the fascinating subject of computers (and those that it uses defy the imagination with their power), it realises to some extent humanity's ancient dream: a Godlike weather machine! The very name 'model' suggests perfection, the ideal, the incontrovertible example.

There cannot then be any doubting the model, as it is obvious that they cannot be wrong. Unbelievers are guilty of lese-majesty; detractors commit sacrilege. O heresy! The model has spoken: full stop! 'They' forewarn us of global warming and its *cortège* of catastrophes, so it must be true, no question of doubt. Therefore, models represent endorsements at the highest level. The IPCC and the media bobbing in its wake (together with many a scientist) 'take refuge' within their predictions: Delphic oracles, guarantees of the highest order, the supreme diagnostic sacrament!

The faith we put in models is obviously due to their image: a very attractive one but also very ill defined and immaterial, bordering on the mystical. Such faith is moreover directly proportional to our lack of knowledge about their true nature and their ability to represent reality. The everyday person, the media, and even some scientists who do not specialise in the field, are, in their ignorance, quite '*amazed by such science and technology*'. The result is blind faith, and belief in the vast and limitless, nay, divine possibilities of the model. The modellers, who are not all meteorologists or climatologists (some far from it), are the new alchemists with their philosopher's stone, today's grand masters of modern magic. Many people think this, including some of the modellers themselves who, in their certainty, forget that the age of the Humanists is over, and that it is nowadays not possible to embrace all science. Each of us is necessarily inexpert in many fields. The community of modellers is relatively restricted and for obvious reasons will not itself seek to dismantle such a fine myth. Essex and McKitrick (2002) decry this way of thinking, equating the model to '*the Enchanted Computing Machine (ECM) that can magically cope with all the details needed to compute all of the theory while securing all the necessary initial data to implement it ...*': meaning that, to the modellers, '*accurate prediction is just a matter of having a big enough machine and enough data*'. Echoes of the perpetual quest for the funding of state-of-the-art machines.

But we are not discussing only the quest for more 'gigaflops'. Even though models are undoubtedly a considerable intellectual advance, they are not magical; neither are they infallible modern seers, as people so often (erroneously) believe. It will be necessary to provide here a summary presentation of numerical climate models in a general way, and not in the mathematical and physical detail which is the province of the specialist. One would need to be a modeller to give an exhaustive technical presentation, and what is really necessary, climatologically speaking, is to know about what goes into the computer and what comes out of it. As we all know, you do not have to be a top racing driver or a first-class mechanic to appreciate a Formula One car: it is judged concretely by its performances. Since the 'global warming' scenario is a product of models, the main thing is to be able to judge the validity of their results, comparing the computer predictions with actually observed climatic facts.

7.1 NUMERICAL MODELS

Modelling, or the calculation of atmospheric evolution through the use of equations of fluid mechanics and thermodynamics, is already an 'old' story: since the beginning of the 20th century. It was V. Bjerknes who claimed in 1904 that '*if, as any person who reasons scientifically must think, atmospheric phenomena develop from those preceding them according to precise laws... the laws which govern how one state of the atmosphere develops from another must be known with sufficient accuracy*'. This postulate of the continuity of phenomena is the very foundation of modelling.

It was Richardson who took on Bjerknes' ideas and attempted to resolve the equations involved in predicting weather via numerical process. In 1922 he was thwarted principally (though not uniquely) by the fact that the means for such calculation were lacking. He imagined a weather forecasting 'factory' with an army of human calculators: at least 64,000 of them to '*speedily grasp the evolution of the weather*'. This 'factory' foreshadowed the technology which later emerged. The first electronic computer, *ENIAC* (1946), developed primarily by von Neumann, was used for the first time in 1950 by Charney, Fjörtoft, and von Neumann for weather forecasting. As Coiffier (1997) so rightly commented, later came '*the time of the "experts"*' and progress was continuous, but in a numerical rather than a meteorological way as computer technology made its rapid strides. Moreover, weather forecasting merged into climatic prediction, because, in the words of Rochas and Javelle (1993), '*climate is simulated using the same models as those used for forecasting the weather*', though of course the scales of space and time were different.

7.2 THE ARCHITECTURE OF MODELS

A numerical model of the climate is a representation of the physical and mechanical processes governing the evolution of the climatic system. The clear aim is to solve the '*equations describing the behaviour of the atmosphere, i.e., to determine the future values of its characteristic features from initial values known from meteorological observations*' (Coiffier, 2000).

7.2.1 The elementary cell

To achieve this aim, the architecture of the model is schematically as follows:

- Space is divided into boxes or cells, within whose confines calculations are carried out, with transfers of mass or energy taking place between cells. Climatic variables are considered to be uniform and are defined at a point within the cell. This technique of reducing space to cells or to elementary points at the heart of the cell (the 'scale of the particle') is called the '**reductionist method**'.
- The model is based on a grid defining the dimensions of each elementary box both horizontally and vertically:
  - The number of boxes on a plane is determined by a horizontal grille on its surface whose intervals are from 200–500 km (usually 250). This spatial resolution may be reduced for greater working precision, especially for meteorological prediction, on a smaller field within the general framework (the 'encased' model whose intervals may be as little as 10 km).
  - The horizontal mesh is complemented by a vertical one which divides the atmosphere into (approximately) 1-km slices, on levels varying in number from 10 to 20. To increase accuracy, cells in the denser lower layers are often more shallow than those in the upper layers.
- Time itself is divided into segments of half an hour or one hour, sometimes more. The characteristics of a given cell at a given time are deduced from the characteristics of that cell as observed at earlier times (combining observation and calculation).

- The model's 'limit conditions' (i.e., at the limit of the atmosphere) are also specified. At the top of the atmosphere the incident solar flux will be involved; at the base, the characteristics of the Earth's surface (e.g., albedo or exchanges at the ocean–atmosphere interface). The model may also be extended below level zero into the ocean (the ocean–atmosphere coupling model) on the same pattern and using the same methods.

So, **the fundamental element of the model is the elementary cell**. In each cell, **at a particular point and at a given moment**, the essential climatic variables are defined. These variables, depending on and representative of the fluid state considered, are: wind, defined via its three components, two horizontal (latitudinal and longitudinal) and one vertical; temperature; specific humidity; and pressure. All are expressed as functions of time or space. Meteorological observations supply horizontal wind speed (the vertical movement being inferred from the horizontal), atmospheric pressure, temperature and humidity, at the surface and at altitude. Secondary data, such as vertical movement (see above) or rainfall totals, may be deduced from primary variables. Entering the observational data into each cell determines its initial state (i.e., the starting point of the exercise) which is represented by the initial values of all the variables at all the points of a network extending throughout the atmosphere. From this initial state, all successive states will be calculated. This operation, aiming to define the state of the atmosphere at a given moment, is known as an *objective analysis*. The sum total of the observational data (requiring vast computer memory) is considered to be the '*climatological*' part of the model, though the definition of this 'climatology' is that long used by meteorologists: the collecting and storage of observational data.

7.2.2 The fundamental equations

The laws governing atmospheric (and oceanic) phenomena are represented by well-known mathematical equations, some of them two centuries old and bearing the names of illustrious scientists such as Newton, Hadley, Simon, Coriolis, Lagrange, Navier, and Stokes. The equations express the relationships between the different variables and simple notions such as the conservation of mass, energy, and momentum, laws of fluid mechanics, radiative transfer, and thermodynamics ... Or, more simply put, they describe how variations in pressure cause the wind to blow, how energy is absorbed, how temperature varies, how water is evaporated at the surface ... These equations, the very same as those used by Richardson, are known as 'primitive equations'.

- Equations 1 and 2 describe the horizontal movements of the air, with latitudinal and longitudinal components linked to the forces exerted upon the air by the Earth's rotation, by horizontal pressure gradients, and by retardant forces such as friction and the dissipation of energy through turbulence.
- Equation 3 describes the vertical movement of the air under the influence of

forces exerted by gravity, by vertical pressure gradients, by the Earth's rotation, and by the constraints of friction and turbulence.

- Equation 4 correlates modifications in the density and speed of the air in order that mass be conserved.
- Equation 5 (thermodynamic) correlates the supply of heat to a volume of air by radiative and convective processes with the resulting variations in temperature and pressure.
- Equation 6 (state) correlates the pressure, density, and temperature of the air.
- Equation 7 represents evaporation, condensation, and precipitation of moisture, the total mass of water being conserved; the term 'warming' from equation 5 is here modified to include the supply of latent heat.

These general equations are the object of a certain number of simplifications as a function of the spatial and temporal scales involved. These 'primitive' equations comprise the *dynamical* part of the model.

So models are based on the laws of physics, laws which are considered 'proven', though '*theoretical laws, which assume a universal character, become empirical relationships as soon as they are adapted to describe the Earth*' (Fellous, 2003). For effects of complex physical processes not explicitly dealt with by the primitive equations and occurring on a smaller scale than that envisaged by the model, other terms must be formulated in a simplified way, in order realistically to reproduce the evolution of the atmosphere. These physical processes, whose resolution can only be approximate due to their complexity or because of problems of scale, are the object of parameterisations, which determine '*their mean effect (in a statistical sense) on the dynamic variables of the model*' (Coiffier, 2000). Also parameterised are the effects of solar radiation, of exchanges with the ground, of condensation and convection. These other calculations comprise the *physical* part of the model, complementing the *dynamical* part derived from the equations.

The modification of one variable, or forcing, brings about variation in other parameters, firstly at a given point (i.e., within the box being considered), then step by step from one cell into others: 8 cells on each horizontal plane around the cell in question, and 2 cells (or, 'at the limit', one) above and below that cell in the vertical column. Then, another 8 cells are involved, and so on ... So, in order to describe changes, equations must be integrated in steps starting from an initial point at a given moment, then recalculated again and again for all directions and for every hour, if that is the chosen time interval. Obtaining the new values gives access to the next stage. Reducing the size of the cells (high resolution) will certainly increase accuracy in defining them, but this considerably lengthens the time needed for the calculations.

So a numerical model is therefore '*a large collection of computer programs solving by repetition the equations introduced into it*' (Fellous, 2003). This is the very 'factory' imagined by Richardson. We must already remember that in that factory, the 'research centre' or 'brain' in which the science is done (i.e., the elementary cell) occupies a very much smaller space than the vast halls full of enormous machines (all the cells) processing and producing data (the *climatological* area), constantly and tirelessly repeating the same operations (the *dynamical* area), and then calculating new data from those produced in an absolutely mechanical fashion. It is obvious that all this repetitive and purely technical part will not introduce errors (except through the hardware), since processing at this level is extremely efficient. As a result, the quality of what comes out depends only on what goes in (i.e., **on the quality of the definition of the phenomena in the elementary cell**).

7.2.3 The types of models

There are several types of model, and their complexity has increased in step with our growing computing power (cf., e.g., Mason, 1979; Dickinson, 1986; IPCC, 2001; Weart, 2004).

A simple numerical model confines itself to the calculation of a small number of carefully chosen quantities. In order to do this, it simplifies certain parts of the problem, or ignores them, for the sake of exploring specific aspects with greater accuracy. These models represent certain mean climatic features quite well. For example, a 1-D model which treats the atmosphere as a simple vertical column may be used to give an idea of the mean effects of radiative transfers, from the surface to the upper layers in the column in question. With a 2-D numerical model, it is possible, for example, to represent a differential thermal field on a horizontal plane, or atmospheric movements resulting from average pressures across bands of latitude.

Complex models incorporate as fully as possible, though usually with some simplification, all available knowledge on atmospheric thermodynamics and physics, and also take into account atmosphere–ocean exchanges. The first ('coupled') model of this type dates from 1975 (Manabe and Bryan). A 3-D numerical model, known as a General Circulation Model (GCM), treats the atmosphere as a vast, turbulent, and rotating fluid, heated by the Sun and exchanging heat, humidity, and momentum with the land masses and oceans beneath. These oceans are of comparable complexity to the atmosphere. The model embraces monthly and seasonal variations in solar radiation, ocean surface temperatures, ice cover, the nature of the ground surface, etc ... GCMs require computers of enormous capacity, very costly not only in monetary terms but also in time and effort. It takes several years to set up a GCM.

All models will obviously have many things in common, but they differ considerably in the mathematical techniques involved, in the efficacy of their calculation, in their definitions of limiting conditions, and in the parameterisation of physical processes, particularly in the case of phenomena too small to be resolved within the framework of a model.

7.3 LIMITATIONS OF MODELS

Models, then, are magnificent, complex, and expensive intellectual constructs, and as they have continued to improve, '*confidence in the ability of models to predict the future climate*' has grown, as the IPCC (2001) points out. But there are still disagreements between the predictions of the models and observations; even if they *are* extraordinary tools, they remain tools. The modellers themselves say that '*uncertainties remain high*' (Beniston *et al.*, 1997). The authors also remark that the '*uncertainties in high-resolution simulations need to be quantitatively communicated to the users of these data*', a recommendation often overlooked by both the authors and the users!

Models have yet to achieve the desired state of perfection, and so, as Essex and McKitrick (2002) stated: '... *measured against reality, GCMs are primitive and childlike, but they are the best we have and they are precious things*'. Leaving aside purely technical problems of hardware and software, the imperfections of models have many causes: these involve the observational data, the physics of phenomena, the capabilities of the machines, and the appropriate structuring of models (i.e., the very logic involved in modelling). There is no point in listing now all the imperfections or shortcomings of models: others have already done this in much greater detail than is appropriate here. A few essential observations on these points are all that is needed.

Uncertainties arise first of all with the data required to define the initial state. Problems exist with the heterogeneous spatial distribution of observational data: there are vast areas, mostly oceanic but also on land, which offer no data or only data of questionable quality. Also, certain types of data are absent or have to be calculated from other data, an example being the vertical component of air in motion, which is too slight to be measured. What is more, although the ocean seems to be more complex than the atmosphere, it is much less well known in terms of observations.

Difficulties present themselves with physical schemas of parameterisations, which offer inadequate simulations of the mechanisms of retroaction between different elements of, for example, the genesis of precipitation, orographical effects, or biosphere–atmosphere links. One example of a serious difficulty concerns the process of the formation of clouds, and their radiative properties. These are not convincingly linked to other meteorological processes, since the modellers follow simple but not necessarily satisfactory physical arguments. Moreover, some processes remain unresolved, and/or are deemed to be insignificant (in the absence of a more acceptable alternative).

Shortcomings are inherent in models, and particularly so in their internal workings. The horizontal dimension of their mesh (resolution) does not allow for the treatment of reduced scale meteorological phenomena (which will 'slip through' the mesh), examples being thunderclouds or localised storms, and regional representation of climatic outcomes. The difficulties encountered in the inclusion of clouds (because of their size) into models are well known: a cumulus formation may be less than one kilometre across, while organised cloud systems may have widths of less than a hundred kilometres. Such a dimension may, however, be encountered in the case of a powerful phenomenon, such as the squall lines of tropical Africa. The depth of a layer of stratus cloud may be less than the interval between two levels.

So the very size of the basic cell, the resolution of the equations within each cell, and the fact that all values are supposed to be constant for each cell are tolerable hypotheses from the point of view of climate, but only for certain elements: temperature, pressure, wind ... if normal conditions apply; but not if there is some marked change in the nature of these climatic elements. These initial hypotheses become fanciful if phenomena of limited extent are involved. These include clouds, rainfall, relief, vegetation types, etc., and they are not correctly represented within the models (Fellous, 2003).

7.3.1 '*Sui generis*' shortcomings...

We recall Bjerknes' initial hypothesis: '... *an atmospheric state develops from the preceding one*'. Although this physically based principle of continuity has been followed to the letter ever since, is it really always applicable to meteorological phenomena?

- *Yes*, if the connection is able to assert itself without difficulty from one cell to the next, horizontally and vertically. An example might be that of the modification of solar radiation, passing down through space from cell to cell in the vertical column until it reaches the surface and increases or decreases its temperature. In the cell immediately above the ground, and as a function of seasonal variation in insolation, the temperature will be higher (in summer) and lower (in winter) in concert with the apparent motion of the Sun. So the process of radiation is easy to understand, winters are colder than summers, it gets hotter in the tropics than at the poles, and higher up in the atmosphere it is colder ... simple stuff! In any given year, the thermal pattern should unfold much like this, progressively and without any great surprises, with each day adding or subtracting its tenth (or fraction of a tenth) of a degree ... But, as everyone knows, temperatures do not behave like this day after day in any one place (except, perhaps, in the heart of the Congo forest, closer to the equator, under the tree cover). Cold and warmth succeed each other irregularly and brusquely, with factors other than solar radiation coming into play ...
- *No*, because transmission from cell to cell cannot take place freely in the real atmosphere. The initial hypothesis assumes a homogeneous atmosphere and an uninterrupted continuum, allowing constant exchanges between cells: the atmosphere is not allowed any aerological or orographical discontinuity, nor any stratification or shear between fluxes and air masses, either horizontally (so there can be no opposing advections) or vertically (no inversions of temperature, wind, humidity, or vertical movements ...). Can such a hyper-simplified atmosphere really be that of Earth?

Figure 21 shows exactly, with a French example, that the evolution of temperature is not a steady process. June 2004 heralded, as expected, the coming summer. Around the middle of June, France was enjoying, under a high-pressure area, what weather forecasters and the media here like to call in their colloquial expression '*the return of*

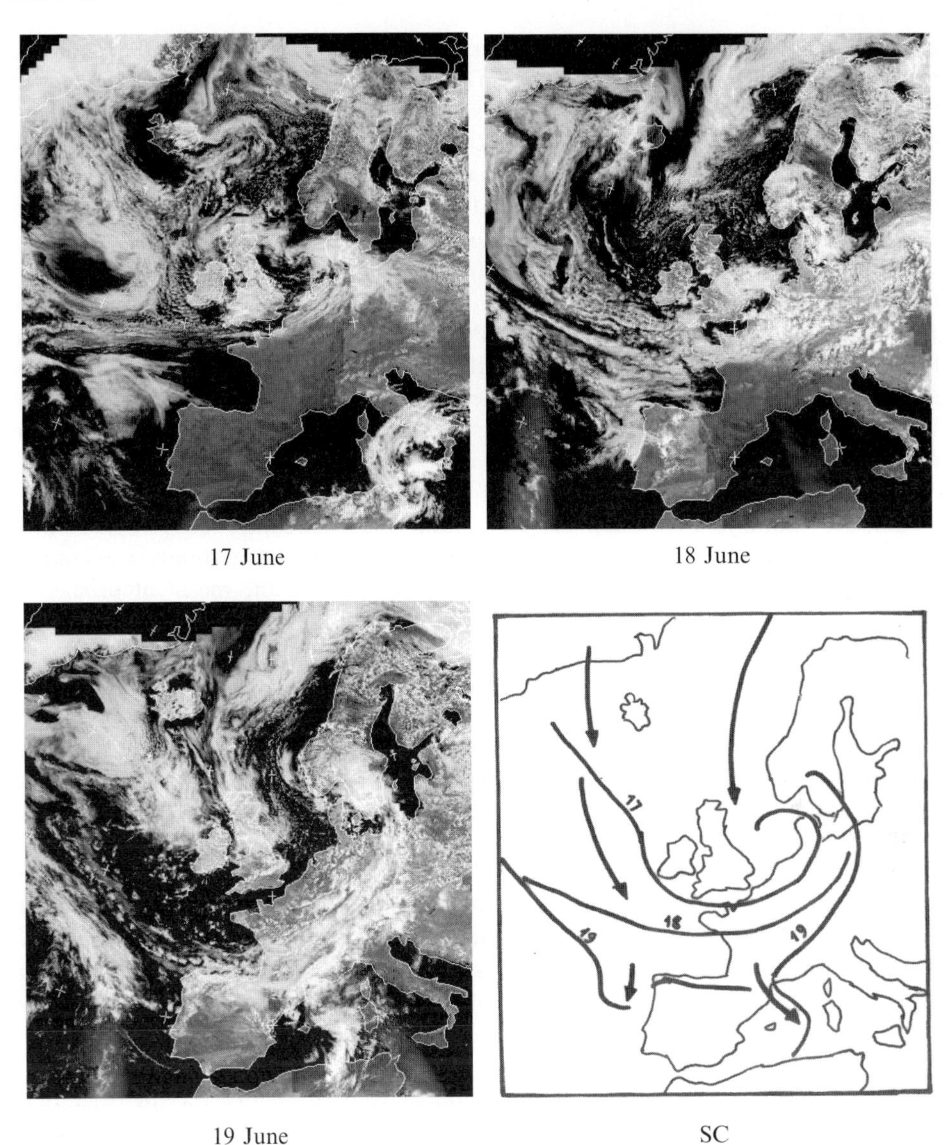

Figure 21. Cold air of a Mobile Polar High (MPH) following a Scandinavian path invades western Europe and reaches the Mediterranean. 17–19 June 2004 (visible, NOAA, 14.00). Summary chart (SC): arrows – path of cold air coming from the Arctic Ocean and Greenland; thick line – leading edge of MPH (with date indicated).

the Azores High', synonymous with fine weather. Daytime temperatures were already exceeding 30°C. Locally, in towns, they were even reaching 33°C. In the south of France, night-time temperatures held up at over 20°C. The TV screens showed beaches crowded with happy bathers, and fires were already scorching the parched earth of the Midi. Well-known politicians were visiting hospitals and retirement homes under the invited gaze of television cameras, doubtless atoning for negligence and insufficient action during the heat crisis of August 2003. All the talk was of air-conditioning, and Météo-France took to the media with its message of '*be on the alert for the heatwave*!'. Summer was upon us, and the solstice would soon set the seal on it. Could things be otherwise, given the logic of radiation? Then, on 17 June (Figure 21), a mass of cold air (a Mobile Polar High (MPH)), which was already moving down across the British Isles from the Arctic, having passed between Greenland and the Scandinavian mountains, began to invade northern France. On 18 and 19 June, the Arctic air of the MPH, reinforced by newly arrived air direct from Greenland, crossed France. It brought cold and rain. Blocked by the Pyrenees and the Alps, it fell upon the Mediterranean in the form of *tramontane* and *mistral* winds. Temperatures fell by about 10°C, sometimes even by 15°C. It was no longer summer (briefly)!

The sudden change in the weather of 17–19 June, and the irruption of cold air across the country: were they present within the elementary cells representing mainland France in the models for the preceding days? Apparently not, since, normally, such rapid incursions of cold air 'pass through the mesh' of predictive models. Three-quarters of MPHs come in mainly from the west, from America, and (statistically) the likelihood of a direct southward (Scandinavian) trajectory into France is actually relatively small in summer, because MPHs are normally less intense at that time of the year.

So here we have an extremely paradoxical situation. It is claimed that models can embrace extremes, linking '*the particle to the planet*' and giving a global view of the climate from an initial, elementary base. The popular image of '*a butterfly flapping its wings somewhere ...*' gives credence to the interdependence of all climatic phenomena across the planet ... but such a (theoretical) process is possible only in a static atmosphere in which nothing interrupts the continuity of step-by-step 'interrelations', and there are absolutely no 'compartments' in any direction! This is an imaginary atmosphere dreamed up in a laboratory, belonging to what Lenoir (1992) called a '*video-planet Earth*': immobile, arrested, with changes from cell to cell taking the allotted time – hardly the Earth's atmosphere (cf. Chapter 8)!

Let us now return to solar radiation and its effects. Radiation having organised the thermal field and thereby the pressure field, circulation is created as gradients appear and the structure of the troposphere is established, with all its discontinuities. So the effects of radiation are differentiated, transported, transformed, and combined with other (principally geographical) factors. Consequently, it is pointless to try to describe temperature changes with reference to radiation in a fixed theoretical vertical column: such a thing almost never exists.

Models are therefore fundamentally, *sui generis*, incapable, because of the very principle upon which they rely, of giving a world climatic picture: the atmosphere is not a homogeneous whole, and a 'global' climate has no existence in fact, by reason of the great diversity of climatic behaviours.

7.4 MODELS AND TEMPERATURE

It is said the 'global warming' scenario is a product of models. Consider the following: Arrhenius (1903) suggested that, if CO_2 levels doubled, the average temperatures would rise by 5–6°C. Möller (1963) gave the first estimate based on a 1-D atmospheric model: a rise of 1.5°C if levels went from 300 to 600 ppmv. Manabe and Wetherald (1967) gave an estimate of possibly 2.4°C. In 1975, with a 3-D model (GCM) now available, they suggested a rise of 3.5°C in the troposphere, and a fall in temperature in the stratosphere. In 1980, the first Villach Conference gave a figure of 1.5–3.5°C. In 1981, Hansen *et al.* predicted a rise of 4°C, claiming that, by the end of the 20th century, anthropogenic warming would exceed the natural variability caused partly by volcanic aerosols and the atmospheric dust veil. Ramanathan *et al.* (1983) showed the importance of clouds (both their types and their radiative properties) in the modelling of atmospheric processes. By 1984 Washington and Meehl had already announced an increase of 6°C. In 1985, the second Villach Conference foresaw a rise of 1.5–4.5°C. In 1990, the First IPCC Report gave 1–3°C; the Second, in 1995, 1.5–3.5°C; and the Third, in 2001, opened it out: from 1.4–5.8°C.

So, a century after Arrhenius, we arrive at exactly the same prediction. In 1903, however, there were no models. Conclusion: in spite of what is constantly claimed, **the presumed 'global warming' scenario owes nothing to models**; on the contrary, it is they who owe much to that scenario!

7.4.1 Models and 'global warming'

Estimates of the hypothetical warming supposedly originating from models boils down in the end, in spite of its sophisticated calculations, to a simple rule of three, involving:

1 the present level of CO_2;
2 a supposed future level, usually times two; and
3 the corresponding temperature.

Thanks to a single schematic equation, '$1CO_2/T/2CO_2$', a simple relationship involving a modification in radiative flux and a temperature anomaly, Dickinson (1986) proposed that, if CO_2 levels doubled from 300 to 600 ppmv, temperatures would rise by 4°C, and if other emissive gases were considered, this would increase to 5.5°C. We can perform a quick and simple calculation: between 1900 and 2000, CO_2 levels increased by about 70 ppmv, the presumed increase in temperature (given the mean curve of 'reconstituted' temperatures) being from 0.6–0.8°C. This gives a figure of roughly 0.1°C per 10 ppmv. The IPCC (*Summary for Policymakers*, 2001, figure 5b) supposes possible CO_2 levels for 2100 of 550–900 ppmv, a rise in one century of 190–640 ppmv. This would roughly (as a 'book-keeping hypothesis') correspond to a warming of 1.9–6.4°C. The 'rough and ready' calculation is not very different from the IPCC's forecasts, especially if we remember the margin of error: from 1–3.5°C. This shows the pointlessness of using such a profusion of computing power! Moreover, the comparison between, on the one hand, the supposed increase in temperature of 0.8°C (margin of error: 0.2°C) for the one and a half centuries between 1860–2000, a period which includes the Industrial Revolution, and on the other hand, the forecasts for the current century (reckoned to be 10 times greater) is further evidence for the trite nature of the 'equation' used by models.

It is, then, elementary and very simplistic: the models perform no better than Arrhenius, or a 'back-of-envelope' estimate, or a simple 1-D model. According to Le Treut (1997), '*an ever increasing number of ever more sophisticated models all agree that temperatures are rising*'. This forecast is held to be a '*remarkable fact*' and proof of the ability of models to predict the future. However, it is a mere statement of the obvious – could models based on radiation (the simple 'rule-of-three' kind, as already seen) predict anything other than *warming*? If any one parameter is increased, then a rise will be the result: 'elementary, my dear Watson'! It is not therefore necessary to proceed via models, unless perhaps one is trying, under their 'incontrovertible' scientific umbrella, to impress the public and the non-climatologists who form the great majority of the IPCC's members.

One way in which models might demonstrate their usefulness would be to piece together an explanation of the way temperatures evolved during the 20th century, with a rapid rise from 1920 to 1940, and cooling from 1940 and 1970. This has yet to be done! Other problems remain: for example, the evolution of temperature on the regional scale, especially in high latitudes.

7.4.2 Models and temperature at high latitudes

One very hypothetical aspect is that of the (presumed) considerable temperature increase (up to 10–12°C) in high latitudes, paradoxically in the winter at either pole. Globally, these high values would have a marked influence on the predicted mean thermal trend, since tropical regions are supposed to experience little change.

Arrhenius made such a prediction: '*the effect of these changes will be at its maximum in the vicinity of the poles*'. We had to wait until 1975 for Manabe and Wetherald to state that their model showed greater warming in the Arctic than in the tropics. According to them, '*simple reasoning*' would predict that '*not only does a more active circulation transport more heat towards the poles, but less snow and less ice mean more solar radiation absorbed by the land and the sea*' (in Weart, 2004). However, things are not so simple: why should circulation be more active, given that warming actually slows circulation (cf. Chapter 8)? The resulting circulation would, on the contrary, be slower and less heat would be transported! Why should there be less rainfall if circulation accelerates, which entails an increased transfer of potential precipitable water? What is more, the energy supplied by the Sun at a low angle is never very great. So the aforementioned '*simple reasoning*' is not as reliable as might be claimed, since its propositions do not reflect the facts, or are contradictory. All in all, the 'reasoning' is flawed. This does not mean however that the models do not subscribe to it!

In fact, for what physical reason should higher latitudes warm up so much? Do we observe a more intense terrestrial counter-radiation in these regions, more especially in winter, when there is no insolation? Should we go against logic and suppose that the greenhouse effect is greater near the poles, though the water vapour content is less and the cold waters around the edges of the ice also act as effective sinks for CO_2? Could the presumed increase in temperature, which cannot be the result of phenomena *in situ*, be caused by intensified meridional transfers? We know for a fact that in cold periods, exchanges are more vigorous, whilst in warm periods they are considerably slowed (see Leroux, 1993: the slow mode of general circulation). Is it because cooling during glacial periods was more marked in polar latitudes that, conversely and contrary to all reason, there should also be greater warming there?

The last IPCC Report (2001), which also foretold greater warming in northern areas of North America and Asia, put forward the following 'scientific' argument: '*... snow and ice reflect the Sun's light, so less snow means that more of the Sun's heat is absorbed, bringing about warming ...*', and therefore '*... warming of more than 10°C is forecast for parts of northern Canada and Siberia in winter*' (sic!). A new 'peak' of scientific insight: though the Sun here seems to be shining during the polar night of winter ... and this must have been approved by the IPCC's experts!

Is the physical *non sequitur* of these phenomena aggravated by an illogical technique whereby, as the meridians converge towards the pole, the elementary 'boxes' become smaller and smaller? The fact remains that the accelerated 'warming' at high latitudes is neither observed nor explained (cf. Chapter 10). There has already been criticism of the inappropriateness of models: '*current climatic models do not correctly integrate the physical processes affecting the polar regions*' (Kahl *et al.*, 1993). But nothing seems to have changed, in spite of the crucial importance of high latitudes in driving general circulation (cf. Chapter 8).

Models are no better at predicting the evolution of temperatures on the regional scale. According to the IPCC, '*regional temperature values could be appreciably different from the global mean, but it is not yet possible to determine these fluctuations accurately*'. This would seem to imply that the mean value was known *before* the local and/or regional ones used to determine it! Here we have a novel method of calculating the mean! This simply suggests that the temperature prediction was carried out using a simple 1-D model, employing an elementary vertical column to represent global temperature, but not taking account of the whole atmosphere, which would require a 3-D general circulation model. Moreover, is it true to say, as the IPCC does, that '*it is not possible to determine*' regional changes, while all that is necessary is simply to observe them? Is it merely because the models are unable to reproduce these behavioural differences? Can they really do it? Do they have at their disposal a coherent schema of general circulation, which would allow them to take account of regional disparities (Chapters 8, 12 and 13)?

7.5 MODELS AND PRECIPITATION/PERTURBATIONS

Climate evolves as the weather evolves, self-evidently, and precipitation is particularly a sign of perturbed weather, resulting as it does from a breakdown in the equilibrium of water concentration in the air.

7.5.1 Models and precipitation

The predicted warming will bring in its wake a modification of the hydrological cycle: '*mean rates of both evaporation and precipitation are expected to increase by 2–3% for each degree of global warming*' (WCP, 1988), and consequently, '*the total volume of precipitation should increase*' (IPCC, 2001). On what argument is this prediction based? As with temperature, the forecast for precipitation is based upon schematic reasoning, its foundation being '*the relationship between evaporation and surface temperature ... a well established one, confirmed by all models*' (EOS, 1995). This reasoning is extremely simplistic and linear: temperature rise (T) = more evaporation (ev) = increase in water vapour content (precipitable water potential) = more rain (R)! This is the 'schematic' equation 'T/ev/R', or, even more concisely, 'T/R'! The last part of this relationship may also be expressed thus: '*(deep) convection will be stimulated (...) if water vapour is supplied through strong evaporation or a marked convergence of moisture*' (Coiffier, 1997). It is simple, but, we are assured, 'physically based': '*the underlying physics on this is well established*' (EOS, 1995) and '*all the models confirm this relationship*'. The fact that all the models use it does not however mean that it is scientifically sound!

It is well known that the existence of precipitable water potential is only one of the conditions necessary for pluviogenesis (cf. Chapter 11). Nowhere, however, is any direct relationship observed between precipitable water potential and actual precipitation! For rain to fall, it is not necessary merely to have water vapour present, or even evaporation to add to it: if such were the case, it would rain all the time in the wet tropics and on the shores of the Red Sea, where moisture is not in short supply. Similarly, there would be incessant rain in summer over the Mediterranean basin, where precipitable potential is at a maximum. It comes, then, as no surprise that certain models (using this elementary equation) make it rain heavily (and 'logically') in the Sahara, since it is hot there! So a restriction is imposed on the respective values in this relationship (i.e., rainfall and evaporation), with an emphasis on the following secondary schematic estimate: 'evaporation > rain = drought'. So it is hoped that both floods and drought might be predicted ... but only by abandoning the first relationship ...

Although precipitable water potential is a necessary factor, it is not the principal one in pluviogenesis. With some localised exceptions, there is always enough water vapour in the air (at hand or transferable) to support rainfall. Even over the Sahara, the air, so they say (as a joke), holds as much water vapour as London's during one of its pre-1960s fogs. And this is true, but the deficit in saturation is very great indeed there, and aerological structural conditions (stratification) are drastically different. The process of pluviogenesis requires much more complex conditions than in the schematic, elementary (and erroneous) relationship of 'T/R', or 'evaporation/rain' used by the models.

So the mediocrity of the models' results is apparent: '*Elevated temperatures will bring about a reinforcement in the hydrological cycle, increasing the risk of droughts and/or floods in certain places and a possible diminution in the severity of these phenomena in other places ...*' (IPCC, 1995). Can this be called prediction, when everything or nothing, aggravation and diminution, droughts and floods are imagined, with no actual locations specified? An identical 'prediction', worthy of the Pythian oracle, reappeared in 2001: '*The total volume of precipitation should increase, but, locally, trends are much less certain ... there is as yet no sign of either an upward or a downward evolution in soil moisture world-wide*' (IPCC, 2001). Kukla (1990) had already pointed out the poor performance of models in reproducing the pluviometric field. This is still the case, and will long remain so all the while the associated 'logic' is based on such a reductionist and erroneous relationship, taking no account of links with perturbations in the weather.

7.5.2 Models and perturbations

It is surprising that **the prediction of precipitation has been dissociated from that of weather**, since precipitation is closely linked with perturbations, being their most symptomatic manifestation. It would be wise, therefore, to analyse the evolution of the weather, but can models do this? There is a simple answer: no.

Meteorological models are incapable of predicting the weather for more than 2 or 3 days ahead. Beyond that, the margin of confidence is only three-fifths, or two-fifths: say, a one in two chance of being right (i.e., 'it will rain/it will not rain'). At this level of 'accuracy', we are no longer forecasting!

Climatic models first of all predicted more clement weather in the future, in the First IPCC Report of 1990: '*mid-latitude storms will also weaken or change their tracks, and there is some indication of a general reduction in day-to-day variability in the mid-latitude storm tracks in winter in model simulations, though the pattern of changes varies from model to model*' (IPCC, 1990). Météo-France confirmed this hypothesis in 1992: '*Mid-latitude storms ... are the result of temperature differences between the pole and the equator ... as this difference narrows because of warming ... mid-latitude storms will weaken*'. Again, in 2000, Planton and Bessemoulin of Météo-France affirmed that: '*climate change as simulated by numerical models will manifest itself as a reduction in the north–south temperature gradient at lower atmospheric levels ... the effect of this reduction will be an attenuation of the atmospheric variability associated with depressions, since instabilities, especially over the North Atlantic, are strongly conditioned by the intensity of the temperature gradient*'. So, a rise in temperature should be accompanied by less vigorous MPHs, a decrease in meridional exchanges of air and energy (slow mode of circulation, Chapters 8 and 9), and, in temperate and polar latitudes, by a reduced thermal contrast between fluxes. There is no need of models to deduce this, as the seasonal contrast between the mildness of summer and the rigours of winter show us (Leroux, 1993). A 'warm' scenario brings with it therefore a reduction in the intensity of meridional exchanges and more clement weather.

Nevertheless, the IPCC (2001) announced '*changes in atmospheric moisture, thunderstorm activity and large-scale storm activity*'. It also predicted greater tropical cyclonic activity, on the basis of the '*warm water/cyclone*' link, with warmer oceans stimulating more activity ... Again, this simplistic link is deemed to be relevant (Chapter 11)! Is this happening because the weather itself gives the lie to the predictions made by models in 1990, predictions which are logical consequences of the greenhouse effect scenario? Is it simply because it is now necessary (even by backtracking) to 'exploit' dramatic events and in spite of everything to stick to reality? Bizarrely, we now see (merely opportunist?) predictions of exactly the *opposite* of what was forecast in 1990 (and confirmed several times since) – with catastrophist pronouncements much echoed by the media, who do not express any surprise at this *volte-face*, catastrophe being obviously much more satisfying than tranquillity!

Anyway, is it really the models which are now predicting this evolution in the weather? This is what the IPCC has to say: '*The frequency and intensity of extreme meteorological conditions such as storms and hurricanes could be changing. However,* **models cannot yet predict how.** *The models used to study climate change cannot in themselves simulate extreme meteorological conditions*' (IPCC, 2001). This is very clear: **models cannot predict the evolution of the weather**, and are therefore incapable of foreseeing either calmer or more violent perturbations. But this does not prevent the IPCC (2001, see above) from announcing 'changes in storm activity'!

Climate models are therefore unable to demonstrate a presumed link between the greenhouse effect and the evolution of weather:

- For reasons connected with the scale of phenomena, an example being: '*very small-scale phenomena, such as thunderstorms, tornadoes, hail and lightning ... not simulated in climate models*' (IPCC, 2001).
- For larger scale phenomena also, because their causes are still not fully understood and are still debated (cf. Chapters 8, 11, 12, and 13).

Why, then, is the 'authority' of models still constantly evoked? And how is it that we can be told that the conditions 'could change' (a truism, since weather changes all the time!) if the models are incapable of predicting this? As Kondratyev (2003) underlined, it is all to do with the necessity of 'sticking to reality': '*the results of numerical climate modeling that substantiate the "greenhouse global warming" hypothesis, are nothing but an adjustment to observational data*'. The 'predictions' (wrongly attributed to models) of supposed catastrophic weather to come have no real basis, and moreover no basis in models, since '*simulated analyses of extreme phenomena are still at a very early stage, especially as far as the trajectories and frequency of storms are concerned*' (IPCC, 2001). The so-called 'scientific guarantee' of the model announcing an increase in meteorological mayhem does not therefore exist. The constant reference to models by the IPCC, scientists, and the media when discussing this subject is a mere swindle.

The question, then, is particularly confused, and we come back to the fundamental but still vexed question: that of the dynamic of perturbations, and, importantly:

- the absence of a link between general circulation and those perturbations (Chapter 8); and
- between perturbations and precipitation (Chapter 11).

Models are used sometimes as arguments, sometimes as justification, sometimes even as an excuse! But lacking adequate input, these models cannot show the relationship between the greenhouse effect and the evolution of the weather. If we are incapable of predicting the weather, how can we claim to be able to predict future rainfall? **The catastrophist weather predictions of the IPCC are therefore completely without foundation** (i.e., without scientific support). This situation reveals limited knowledge of the dynamic of weather, mainly but not exclusively in the mid-latitudes.

7.6 THE IMPERIALISM OF MODELS

Models reign supreme in the discipline of climatology and at the IPCC. Dady, who brought meteorological modelling to France, thought that '*the drift towards reductionism (...) is dangerous, not only because it interprets reality imperfectly, but also because it is totalitarian*' (in *Le Monde*, 24 February 1995): the use of models has come to dominate, with its 'scientific totalitarianism', climatological, and meteorological practices.

7.6.1 Models and climatology

Climatological analysts tend nowadays to consider that the (obligatory) way of the model will lead to analysis of the climate. Consulting synoptic charts (or, even better, preparing them oneself), radiosonde exploration, perusing meteorological data, studying satellite images, trying as a matter of priority to understand the way weather works, establishing the dynamics of phenomena, proposing hypotheses and having them verified by calculation ... all these things now seem out of fashion. Nowadays, the so-called *diagnostic* studies invariably establish 'teleconnections', but do not try to show any possible causal links between the parameters analysed, and therefore, as Fontaine put it in 1990, method '*is no longer interested in physical reality*'.

It is not difficult to 'establish remote links', which is what 'teleconnections' really implies. For example, sea surface temperatures in the North Atlantic and the rainfall in the Sahel are 'connected', even though the precipitable water potential of this part of the Atlantic has almost no chance of being advected across Africa, because of the trajectory of the maritime trade wind. Statistical relationships, acting at a great distance, have also been proposed for the El Niño–Southern Oscillation (ENSO) and precipitation on regional and global scales (Ropelewski and Halpert, 1987) ... in spite of these facts:

- the phenomena in question are influenced by totally different factors;
- they are separated by thousands of kilometres and belong to specific units of circulation;
- great mountain barriers intervene, blocking any lower layer interaction between those units; and
- there is, as a consequence, really no physical justification for a 'connection' between them.

These diagnostic analyses are of little use, and serve rather to divert research away from truly interacting mechanisms, or 'proxiconnections'. What benefit can be derived from the remark that years when the Indian monsoon is below normal are associated '*with relatively high potential over the Caspian Sea ... and with troughs over eastern Canada or the Yellow Sea*'? Canada, influencing monsoons in India? Or, to give another example, the abnormally dry character of the West African monsoon has been associated with '*anticyclonic blockage over north-west Europe, and ... a deepening of geopotential lows centred over the Aleutians and northern Siberia*' (Fontaine, 1990). Are areas of low pressure over the Aleutians really influencing the African monsoon, which comes from the south? These long-distance links may well fire the imagination, but they are of no interest in the study of rainfall in India or the Sahel, given that, on the one hand, these parameters have no link in a coherent schema of general circulation (cf. Chapter 8) and on the other hand, no precise analysis of the actual conditions pertaining to pluviogenesis in India and Africa has been carried out. Consequently, **these 'teleconnections' do not further by one iota our understanding of the dynamic of perturbations and pluviogenic processes**, whose real mechanisms are not integrated into the models.

7.6.2 Models and meteorology

In the field of meteorological forecasting, on a daily basis, models allot secondary importance (sometimes even less) to synoptic analysis, in spite of all the progress made in observational techniques. According to a Météo-France forecaster in Toulouse, models are relatively reliable on only two days out of three, but '*the problem is that these models work a little like "black boxes": the observations are fed in at one end and they spit out the predictions at the other. As a result, we no longer try too much to give a descriptive explanation of atmospheric mechanisms, since the model is supposed to give the answer*'. So forecasting is becoming automatic, performed in the dark, with the forecaster allowing the machine to do the thinking ... and to come up with ready-made answers. Now, this might be just about acceptable if the answers were right! And this applies even more for long-term, medium-term, or seasonal forecasting which is far from being an exact science. This results in '*the meteorological services working exclusively (or nearly so) on modelling, while nobody tries to explain how the climatic system functions globally*'.

So, when an unexpected event occurs, for example the heatwave of August 2003 in western Europe, the meteorologists cast around within their stock of ready-made (but inappropriate) 'recipes' to try and 'explain' the heat, and they quite simply forget to analyse what has actually been observed (cf. Section 11.3.1: on the heatwave, or the 'dog days').

At the beginning of the last century, meteorologists had attempted to set out all-embracing concepts, for example the mass theory of the Polar Front, or the dynamic theory of high-altitude jet streams, seeking a better understanding of the reality of phenomena (cf. Chapter 11.1.1: schools of thought). However, since the middle of that century, these endeavours have been derailed with the rise of modelling. Modelling does not in fact feel the need to use such concepts, and claims to have been freed from them. This is absolutely untrue. The concepts are everywhere in the equations and the paramaterisations which they exclusively use. Meteorological analysis has thus moved progressively away from reality, meaning for example that the appearance of the really useful tool of satellite observation to investigate and understand phenomena has gone almost unnoticed and has not posed its fundamental and necessary questions. For 40 years now, satellites have presented evidence that circulation does not work in a continuous and linear fashion, or step by step, but obeys the impulsion of great mobile air masses (MPHs: cf. Chapter 8) which bring about a number of structural modifications. But the models see nothing of this; and the eye of the observer (with its 'subjective' analysis) is held to be less reliable than the so-called 'objective' analysis of the model!

As a result, the mechanisms of perturbations, and especially those in middle latitudes, are still the subject of much debate (Chapter 11), and meteorology does not possess an agreed picture of general circulation, a situation which might appear absolutely incredible to non-climatologists. Having an understanding of general circulation would mean that the reality of meridional exchanges would be recognised, and perturbations would be integrated into that circulation, with meteorological phenomena unambiguously occupying their true places (Chapter 8). This is certainly not the case at present: as already mentioned, any phenomenon may be fitted anywhere into the chain of processes, as cause or as effect, at the beginning or at the end; it seems not to matter, as the model does not carry the organisational schema or the direction of the links. So there is no identified starting point, and no recognised first cause in the circulation. No necessary moving parts, no specific structures: and, in the final analysis, no resemblance to reality!

7.6.3 Models and ways of thinking

With swelling numbers of model-minded 'technicians' (i.e., numericists), usually non-meteorologists, specific ways of thinking have evolved, mainly since the time of return to the 'primitive equations' in 1956 (Coiffier, 1997).

The idea of the continuity of phenomena from one cell to the next leads to the notion of the general interdependence of phenomena. So two parameters (or more, if available data permit) may be associated. For example, in the case of the Sahel drought, 'links' may involve sea surface temperatures in the Atlantic or the Indian

Ocean, and rainfall (Chapter 11). However, squall lines, the inclined structure of the meteorological equator (IME) or pulses within the monsoon (i.e., real rainmaking conditions, cf. Leroux, 1983, 2001), will be absent, as these are not 'model-friendly'. Similarly, there is the 'association' between sea surface temperatures and episodes of heat and drought in North America, which does not take account of real meteorological factors. This mindset allows us to roam freely (and illogically) about the planet (since everything is propagated from cell to cell), and so El Niño becomes a phenomenon with global implications (cf. Chapter 13), without the slightest physical link being evoked, or sought, or demonstrated a fortiori.

The impossibility of taking the reality of phenomena into account, with the consequent need to parameterise, leads to over-simplification, or just out-and-out simplism. Complicated structures and mechanisms, or particular conditions inducing rain, are out of the question: a simple link has to be found, some short-cut or 'wrinkle', an 'equation' which can be used at the level of the individual cell. For example, the rule-of-three equation 'CO_2/temperature' ('$1CO_2/T/2CO_2$'), the 'evaporation/rainfall' relationship ('T/ev/R'), the 'sea surface temperature/cyclonic activity' relationship ('SST/cycl'), or 'SST/R' for rainfall in the Sahel ... the real question is whether the real nature of the phenomena is actually there within these types of 'relationship'.

Models give neither the direction nor the meaning of the links which they establish. The term 'link' or 'relationship', so constantly employed, must be clearly defined. A simple example will suffice. Consider the case of two neighbours, Smith and Jones. They both leave their respective homes at 7.30 a.m. Someone observing these departures in the course of a year will infer a definite connection, with a high degree of confidence. Can there be a cause–effect relationship between the two departures? Certainly not, since it is sometimes Smith and sometimes Jones who leaves first; and on one or more days Smith came out, but Jones (being unwell) did not (or vice versa). One departure is not contingent upon the other. The two variables are completely independent. Then, Smith gets into his car, but Jones boards the bus, and they go off in different directions to their different jobs. So though there is a strong correlation between the two morning departures, it is *only a statistical* one (i.e., accountable); we have here a *co-variation*, the cause of which lies outside the terms of our analysis: the external 'physical' reason for the departures and their near-coincidence is that the two neighbours, like countless others, have to go to work, a cause which is not taken into account in the parameters analysed, and so not studied.

This confusion involving correlations, or co-relations, or co-variations, is always present. As a result, meteorological phenomena around the world, far apart and very different in nature, with absolutely no connection between them, are blithely 'correlated'! It does not seem to matter that the relationship has no physical meaning. Is not the main point of the exercise to present 'statistically significant' results? The *real* question is: does the claimed link have any meteorological significance? Far from it, if the analysis is undertaken 'blind', with no *previous* evidence of some physical link which the calculations seek to verify. What is certain is that **a 'link' which is not properly verified cannot be used to show that the correlation 'explains' a phenomenon**, even if there is a high degree of confidence; it explains nothing as long as the physical link between cause and effect remains unidentified and undemonstrated. The story comes to mind of the scientist who tried without success to find a link between sunspot numbers and rainfall; later, however, he discovered, in a light-hearted study, an amazing relationship between sunspots and the annual highest scores at his golf club! It is always possible to unearth 'statistical links'; but the same is not true of physical ones.

The weather has always had something of a mysterious aura. Was it not marshalled by the gods themselves? Today's array of mathematical, statistical, and computerised practices has progressively eroded interest in meteorological analysis, and forecasting the weather has assumed a new kind of 'magical' character. Dady (1995) denounced a veritable '*intellectual and universal blocking in meteorology*'. The causes:

- a falling off in conceptual meteorological research, trying to understand phenomena in their entirety;
- the (totally illusory) claim of modellers to be able to work outside these concepts; and
- the disdain for the analysis of real facts.

This attitude has engendered a near-refusal to consider what is concrete. Phenomena are not carefully distinguished from one another, and essential and identifiable entities such as MPHs, the trade inversion (TI), and the double structure of the ME (inclined and vertical) are not evoked or well described. It all seems to indicate that the quest for a faithful description of the climatic system is not the major aim of the modellers.

Consequently, meteorology generally has been in a veritable **conceptual impasse for fifty years**, mostly because of the principles, and almost exclusive use, of models.

7.7 CONCLUSION: MODELS ARE GREATLY OVERESTIMATED...

Numerical models have been around for a long time, since the beginning of the 20th century. Despite all the progress made in computerised techniques, V. Bjerknes' initial concept (1904) remains unquestioned. Which immediately raises a fundamental question: does that concept still apply, and can we, from the basis of the elementary cell, encompass all meteorological phenomena and the totality of the climatic system?

It is said that models originated the global warming scenario. In fact, they neither originated it (we have to go further back to Arrhenius at the beginning of the 20th century), neither do they perpetuate it, having done no better than Arrhenius in the field of prediction. And it has taken them 100 years to arrive at the same estimates, with no extra proof of the reality of the greenhouse effect having a determining influence on climate.

It is also said that models forecast an increase in rainfall and extreme weather. They are in fact incapable of so doing, for, as the IPCC admits, while basing its announcements of worsening weather on them, they are not up to the simulation of extreme phenomena. How could they visualise an increase in rainfall if they cannot predict the evolution of the perturbations which are the very cause of that rainfall?

If models cannot reconstruct the evolution of the climate during the century which has just come to an end, how can they claim to be able to predict the climate a hundred years hence? Are they serious? To establish their credibility, they may look back and 'reconstruct' the climate of 1900, since the result is immediately verifiable! But it would seem easier for them to make predictions for the climate during the coming century, since none of us can check their validity! When they look at the climate of 2100, we should bear in mind the reservations expressed by the modellers themselves: '*Uncertainties are still great ... the changes associated with different parameterisations are of the same order of scale as the errors in the model*' (Beniston, 1997), and so '*the accumulation of uncertainty factors makes the detailed prediction of future climatic evolution illusory at present*' (Le Treut, 1997). What could be clearer? So why do these same modellers forget their own reservations? Why do they so doggedly preach a scenario which they cannot, by their own admission, predict?

The forecasts/predictions of the models are considered to be the fine fruits of an accomplished science of meteorology. However, they can favourably impress only the ill-advised, since they imply that the modelling of meteorological phenomena has come into its own and that the workings of general circulation are exactly understood. This is far from true, as we will see below. In fact, the predictions reflect the approximations, rash simplifications, inconsistencies, and contradictions of a meteorological discipline stripped of its concepts and shackled by its old dogmas, anxious to affirm what it cannot demonstrate. Climatology has, as yet, no real 'models', in the proper sense of indisputable references.

The 'reductionist' view, criticised by Dady in the introductory quotation of this chapter, has considerably shrunk our vision of phenomena. It would seem that the modelling community suspects this: '*a major effort will be to tackle the question of how the uncertainty in climate projections is related to the limitation in the model's representation of physical and dynamical processes*' (*Executive Summary*, Community Climate System Model (CCSM), 2004, www.ccsm.ucar.edu). But future improvements in the field of the atmosphere will involve new treatments of clouds, solar radiation, water vapour, terrestrial radiation, aerosols, atmospheric chemistry ... (Merilees, 2003) (i.e., they still remain at the level of the basic cell). Hameranta (2004) stated: '*It's interesting to see what they recognise they don't know yet, but more interesting is to see what they don't recognise they don't know yet*'. Nothing is in fact planned which will widen the framework and embrace phenomena in their entirety, allotting them their rightful place within general circulation.

The concentration by models on the greenhouse effect, and principally its anthropic aspect, eclipses or minimises other possible factors in climate change, such as water vapour, cloudiness (Chapter 5), orbital parameters, solar activity, volcanism, atmospheric turbidity (Chapter 6), and urbanisation (Chapter 10), and especially the dynamic of meridional exchanges (Chapter 8), parameters which are not expressly accounted for by models.

8

The general circulation of the atmosphere

This theory had radically changed our vision of the Earth, and, more than this, our way of practising geology, to such an extent that its emergence gave rise to bitter and acrimonious scientific debates, often too heated ...

The more novel an idea is, the more its power to shock, and the more it upsets those whose reputations have been established elsewhere, and those whose intellectual comfort has been troubled by its emergence. Originality is a prized virtue, provided that it is not too disturbing. Beyond a certain threshold, any bold innovation will be met with marginalisation, or even sacrificial reaction.

Claude Allègre. *Histoires de Terre*, Fayard, Paris, 2001.

Fedorov stated the fundamental principle in 1979 during the First World Climate Conference in Geneva: '*Variations in the climate are the effect of changes in the general circulation of the atmosphere, and also, no doubt, in the general circulation of the oceans*'. Later, the Intergovernmental Panel on Climate Change (IPCC) declared: '*Climate is determined by atmospheric circulation and by its interactions with the large-scale ocean currents and the land with its features*' (in IPCC, 2001, *Glossary*). Everyone, a priori, agrees on this point: the general circulation of the atmosphere is considered to be the vehicle for climatic variations, and all known (and predicted) variations are supposed to be analysed particularly by models within the framework of general circulation.

8.1 CLIMATE IS DETERMINED BY ATMOSPHERIC CIRCULATION

This means (or should mean) that climatology is (or should be) already in possession of a coherent schema of the general circulation of the atmosphere, applicable on all spatial and temporal scales. That schema explains (or should explain) how the 'atmospheric system' works, why and how it changes, and how the climatic consequences of previously established causes are transmitted through the links of general circulation. Also, such a schema must (or ought to) enable us to put each element into its correct place in the logical chain of phenomena, since each element, individually identified, is necessarily involved in the *ensemble* of mechanisms through links of causality. Logically, this is the way things *should* be, but is it the way they are?

Our thoughts turn immediately to models (cf. Chapter 7), and more especially to 3-D numerical models, more specifically known as General Circulation Models (GCMs). Are they really an indispensable tool? **The answer, unfortunately, is no.** In the words of Lindzen, '*models do not begin with a scheme of the general circulation; the general circulation is supposed to emerge as part of the solution*' (pers. commun., 12 January 2004). So the circulation is 'supposed to emerge'. It is not even a certainty! Surprising, but there it is.

8.1.1 Absence of a general schema

So, to use a concrete image, the climatic model is like a huge building site from which some future (and hypothetical) structure will arise: all manner of building materials, great and small, have been sorted and laid out in countless little piles on different parts of the site (the basic cells), but there is no general plan about this, no attempt to relate any element to any others. Each pile on the grid is seen only in the context of its neighbours: it is like an enormous 'Lego' kit with no user instructions. The architect can see no further than the basic cell, and cannot know what final form the building (the model) will take. The general layout and logic of the architectural whole will emerge by themselves (it is supposed), from the equations linking the basic cells. The only framework is that suggested by the points on the grid, but how each point, or set of points, relates to the whole is unspecified.

This shows the pointlessness of interrogating such a model as to how meteorological phenomena are governed by some initial cause (unless it can be defined previously), since the main thrusts and chains of activity are not identified.

What is more, without a general, coherent schema, it is impossible to appreciate the real importance of any one element in the context of the whole, and allot it its exact place in the unravelling processes. To take another image, let us compare this climatological reasoning with the assumption that, in the case of a car, the component (out of a multitude of possibilities) that makes the vehicle move is the radiator fan (!). This fan, which immediately suggests itself as the equivalent of an aircraft propeller or a ship's screw, is situated at the front of the car (rather like the propeller). When it is not rotating, the engine is not running; if the car accelerates, it moves faster. It seems to have all the properties of a 'factor', and might easily be mistaken for the 'engine' itself ...

But this is approaching the problem from the wrong direction. We can describe the fan (its colour, diameter, the width of its blades, what it is made of, etc.) as accurately as possible. We can observe and analyse, to a high degree of accuracy, what it does (direction of rotation, speed, ratio of its turns to those of the wheels, or to the speed of the vehicle), considering the multiple co-variations and statistical correlations involved. But this will still not reveal the cause of its motion, even though the existence of the fan belt might suggest a possible cause of the observed effect. Even then, only our knowledge of the way in which everything works will confirm that the fan has no part to play in the vehicle's actual motion, but belongs at the end (not at the start) of a chain of events and is but a minor consequence of it. The engine will continue to run even if the fan is disconnected. Our approach may be crude and erroneous, but it is made possible, as in many other cases, by the absence of a general schema in which the true place and function of the fan are made clear.

Unfortunately, climatology uses similarly futile reasoning when it suggests, for example, that:

- droughts, heatwaves, floods, and extreme meteorological events are the 'results' of the greenhouse effect, though no causal links have been established;
- the Sahel drought, or the 'Dust Bowl', are caused by temperature changes in the oceans, though the root cause of the changes is neither known nor investigated; and
- El Niño is a kind of *deus ex machina* causing numerous, totally different events worldwide, though its real nature and exact place in the chain of events are not known: it is only a consequence (like the motion of the car's fan) further along the chain. We shall come back to this later in Chapter 13.

How has such a situation come about? Dady (2001) underlined that the advent of modelling had spelt the end of working with concepts 'beyond the particle', in an effort to 'understand atmospheric evolution'. Ever since, for 50 years, climatology has been in a real conceptual *impasse* as far as general circulation is concerned. What is the first cause in the workings of general circulation? What are the causes of its variations, its structure and components, its modalities and organisation? What is the precise place of perturbations and their integration into the general dynamic? Where do the processes transmitting climatic modifications belong, whatever their timescale? What is the state of the debate on these questions, and many others? We need first of all to look back at past efforts to construct a complete and coherent image of atmosphere general circulation.

8.2 A BRIEF HISTORY OF THE CONCEPTS OF GENERAL CIRCULATION

Since the beginnings of meteorology, a description of general circulation has been sought, but this fundamental aim remains to be achieved. Let us examine the principal efforts.

The great voyages of the 15th century led to a progressive description of the 'Brave West Winds' of temperate latitudes, which brought Columbus back from America, and the tropical trades and monsoons. The first chart of winds by Halley in 1686, showing the trades between 30°N and 30°S, was used to elaborate the first theory of the dynamics of wind: the so-called 'equatorial chimney'. The first theory of general circulation saw Hadley (1735) assert the primacy of tropical heating: warm air rose at the equator, streamed at altitude towards the poles and returned to the tropics via the lower layers, drawn in by the relative vacuum of the 'chimney': general circulation comprised two convection cells, one in each hemisphere. Hadley claimed that the direction in which the trades blew was the result of the difference in the speeds of rotation of the equator and the poles: air coming from the pole is slowed by the Earth's surface, which moves fastest at the equator, and therefore seems to come from the east. By the same reasoning, air moving towards the pole travels eastward faster than the Earth's surface, whence the existence of westerly winds in mid-latitudes.

8.2.1 Birth of the tri-cellular model of circulation

A century later, in 1835, the mathematician Coriolis showed that the trajectory of any object moving across a rotating body will describe a curve at all points on that body's surface, independently of the starting point and of the direction of the object, once set in motion. The oceanographer Maury carefully collated from ships' logs a considerable amount of documentation concerning sea and air currents, and revealed the existence, previously unsuspected, of zones where mean atmospheric pressure remained relatively constant: low-pressure zones near the equator and the poles, and high-pressure zones at about 30° north and south. In 1855 he put forward a plan of general circulation incorporating these pressure zones, with two cells in each hemisphere, and air rising at the equator and subsiding at 30° in each hemisphere (already the 'Hadley cells'). But he also showed a current from the tropics to the poles in the lower layers, rising at the poles, which were calmer zones, like the equator.

Ferrel, a teacher, drew upon the principle set out by Coriolis and the observations of Maury, and in 1856 proposed a schema of general circulation marshalled by the geostrophic force, based on three circuits. This tri-cellular model would influence meteorology for a long time to come. In the subtropical (now 'Hadley') cell, equatorial air rises, then descends at about 30°N and 30°S to return to the equator. However, all the air does not return towards the equator: some travels towards the poles at low level, but at about latitude 60° it encounters cold air leaving the poles, and rises, moving back in upper levels towards 30° and falling again. This forms the second circulatory cell (later known as the Ferrel cell). The third cell is the polar one: cold air moves away from the pole, warms up, and at about latitude 60° it rises and returns towards the pole. Notice that, at the extremity of the temperate cell, we have a meeting of 'warm' air from the south and 'cold' air from the north: cold air which, in the polar cell, is nevertheless capable of rising on its journey back to the pole!

Ferrel's tri-cellular schema gave (provisional) answers to questions of the day, such as that of the existence of the equatorial calms feared by sailors, at the junction of two tropical cells. Subsidence of the air at latitudes 30°N and 30°S, with high-pressure belts, explained the great continental deserts and the oceanic tropical calms known as the 'horse latitudes'. At polar latitudes, the Coriolis effect caused prevailing easterlies, and for the same reason the equatorial (trade) winds were westerlies.

The convergence, at about 60°, of polar air and warm air from the temperate zones formed the great depressions and anticyclones of middle latitudes.

Ferrel's model was hailed in his day as a great breakthrough, and the call of the tri-cellular model of circulation still echoes loudly in today's meteorology. However, it has its shortcomings, and many people have tried to improve upon it. For example, Thomson's schema of 1857 caused Ferrel to modify his own in 1889. The cells became far less individualised, and the polar easterlies disappeared, to be replaced by the 'polar calm', but the 'tropical calms and dry belt' and the 'doldrums and equatorial rain' were retained.

8.2.2 Improvements of the tri-cellular model of circulation

The end of the 19th century saw the advent of new representations from Guldeberg-Mohn (1875) and von Helmholtz (1888), a forerunner of the Bergen school, who laid down the principle of the conservation of energy and the vortex theorems. In the early 20th century there were further contributions from Margules (1903), Hasselberg-Sverdrup (1914), Ekner (1917), I. Bjerknes (1923), Bergeron (1928), Dedebant and Wehrle (1933), and Rossby (1941) (in Olcina and Cantos, 1997). Margules described the structure of a discontinuity within a fluid in rotation (i.e., a front). J. Bjerknes (1923) kept the trade wind, the counter-trade, and subtropical subsidence, but between latitude 30° and the pole he introduced the Polar Front, with families of perturbations and sporadic winter irruptions of polar air into lower latitudes (foreshadowing Mobile Polar Highs, MPHs). Dedebant and Wehrle (1933) expounded the theory of differentiated rotation: briefly put, the atmosphere does not move *en bloc* like the Earth, but is separated into rings of different velocities as a function of latitude and altitude, with easterly winds in the equatorial ring and westerlies in the polar rings, where their speed increases up to the tropopause.

Rossby, another member of the Bergen school, did not follow Bjerknes' schema, and, like Bergeron (1928), he proposed a tri-cellular circulation model (1941) with general circulation explained through combinations and adjustments of forces, the dynamical factor taking precedence over the thermal factor. The success of this model owed much to the status of Rossby, who had been in the USA since 1926 and had founded the Meteorology Department at the Massachusetts Institute of Technology (MIT). In its main features, this model is actually quite close to that of Ferrel, with three cells in each hemisphere: the Hadley cell, the Ferrel (with exchanges between these two cells) and the polar; a new feature was the Polar Front, a continuous separation between polar air and air flowing from subtropical high-pressure areas of essentially dynamical origin. The Ferrel cell, also essentially dynamical, is supplied with energy mobilised by enormous Norwegian 'cyclones'. This model does not include jets, but in 1947 Rossby overturned concepts when he stated that the origin of temperate-latitude perturbations (the Polar Front) lay in the high-altitude jet stream.

In 1921, Defant had explained that exchanges took place in steps rather than in a continuous current from the pole towards the equator, and that these exchanges were mechanical and carried out within large-scale perturbations (anticyclones and moving lows), which create widespread turbulence. The progress of families of cyclones brought about step-by-step meridional exchanges. Palmen (1951) took up this idea, and attempted to reconcile matters: he considered the polar zone to be a zone of mixing, and he introduced jets and connected the temperate and tropical cells. During the 1950s, Ferrel's model was judged to be unrealistic by the meteorological community, as it contradicted observations. Since then, Palmen's model (which nevertheless owes much to Ferrel's), with later slight modifications (Palmen and Newton, 1969), is considered to be the best elaborated of the schemas of general circulation.

Modelling has not brought further decisive progress in the interpretation of general circulation. Phillips' attempt in 1956 at simulation of the movement of the atmosphere in one hemisphere, in response to the controversies of 1940–1950, did not produce the desired results. Lewis (1998) commented: '*The experimental design was bold*', but '*the simplicity of the model dynamics exhibited an almost irreverent disregard for the complexity of the real atmosphere*'.

It must be pointed out that, without special instructions, models based on radiation can only deal with the more or less direct consequences of temperature differences (on pressure gradients or circulation) from cell to cell. So they can only describe a unicellular circulation in each hemisphere, from the pole towards the equator in the lower layers, and in the opposite direction at altitude, in fact reproducing Hadley's initial 'equatorial chimney' schema of 1735.

Representations of general circulation therefore remained at a certain point, and for about 50 years Ferrel's tri-cellular model (1856), as revised by Rossby (1941) and Palmen (1951, 1969) has been the preferred doctrine. Although this model is now seen as '*a vast oversimplification*', it still provides a useful conceptual tool (Henderson-Sellers and Robinson, 1989). In France, it continues to be reproduced and taught, especially in universities (cf. Beltrando and Chémery, 1995). Météo-France, the French Meteorological Society (SMF) and the Laboratoire de Météorologie Dynamique (LMD) still consider that the general circulation of the atmosphere comprises '*three meridional cells in each hemisphere, from the equator to each pole. These are the Hadley cell, by far the largest and most active, the Ferrel cell and the Polar cell*' (De Félice, 1999, cf. site `www.smf`).

8.3 INSUFFICIENCIES IN THE REPRESENTATION OF CIRCULATION

Although Ferrel's legacy has been amended, it does not even approach an explanation of the true nature of meridional exchanges. The principal shortcomings are as follows:

- In the Polar cell (direct sense) 'warm' air rises and cold air 'falls', conforming to thermodynamics. But it is difficult to understand how air flowing from the pole might have been 'warmed' on reaching latitude 60°, and be warm enough to rise and return to the pole! This Polar cell is improbable. Also, *polar easterly winds* are not observed. What is more, if this happened, it would never be cold outside the polar regions! We know that this cell has been re-named, but to call it now a 'mixing zone' does not tell us much more about the nature of meridional exchanges.

- In the temperate or Ferrel cell (indirect sense) the air which is moving back towards higher latitudes is progressively cooled, but it still rises at around 60° latitude! It is supposed that the movement is brought about by the other cells: a rather simplistic viewpoint. This cell, which works contrary to physical principles, is also therefore quite improbable!
- In the Hadley cell (direct sense) equatorial updrafts are certainly observed, but the subsident aspect (which is supposed to create deserts) does not involve the surface, because at the latitude where subsident movements happen, we find not only the Sahara, but also the West Indies and the Yucatán!
- The tri-cellular schema segments the exchanges and thereby does not envisage long-distance transfers, nor can it explain waves of heat and cold; also, it does not allow for thermal equilibrium in the Earth–atmosphere system. The 'cellular' concept might well have been credible in the days when distances seemed so great and zones so well separated, when tropical weather was 'independent' and red rains across Europe were known as 'rains of blood' ... though nowadays we all know that the red dust comes from the Sahara, just as those living around the Gulf of Mexico know that the *nortes*, like the winter rains of the Cape Verde Islands, are of far-away 'polar' origin!

Ferrel's schema and those that followed it were based on observations: initially those of Maury, and mainly of the mean pressure field:

- with low pressure corresponding to convergence and upward movement; and
- with high pressure corresponding to divergence and downward movement.

With these principles in mind, it was easy to imagine the three cells; but these schemas established 'truths', often still considered to be incontestable postulates. For example:

- The meteorological equator (ME), axis of the so-called 'equatorial' lows, is a location of updrafts, a notion implied within the generic term 'Intertropical Convergence Zone' (ITCZ). This is not always true, especially in the case of the inclined meteorological equator (IME), which is not necessarily characterised by vertically ascending air.
- Anticyclonic areas are created by air descending through the whole depth of the troposphere. Chen *et al.* (2001) mention the ambiguities which surround this idea: '*The low-level tropical anticyclones in particular continue to be an open question*'. So '*tropical high-pressure areas*' (e.g., the 'Azores' or 'Hawaiian' anticyclones), defined using mean pressures, are generally considered to be 'permanent centres of action'. This way of thinking has caused an incredible **confusion of scales. A 'centre of action', defined using mean pressures (i.e., statistically), has no real existence on the scale of real (synoptic) weather**. This

confusion is still very much alive, and falsifies the perception of phenomena, masking the real origin of these anticyclonic cells.

The representativeness of mean wind values does not have the same meaning in the tropical zone and in the extra-tropical zones:

- in the tropics, mean values are representative of the real wind field (i.e., with the direction of the synoptic wind deviating very little from the vector of the mean resultant wind); and
- outside the tropics, the continual passage of MPHs causes the wind constantly to change its direction completely, so the 'mean direction' of the wind has no real synoptic meaning.

Therefore it is impossible to represent tropical and extra-tropical circulations in the same way in a schema of general circulation (which necessarily deals with means): any streamlines shown will be representative of reality only in the case of tropical winds.

Let us also mention that perturbations, and especially those of middle latitudes, are not integrated into the dynamic of circulation. For example, '*the most significant feature of the general circulation of the atmosphere in temperate latitudes is the existence of prevailing westerly winds. Superimposed upon these are perturbations which can mask the dominant character of these winds*' (Rochas and Javelle, 1993): meaning that perturbations in temperate areas, appearing *ex nihilo*, are not integrated into the westerly flow, and 'graft' themselves onto the circulation without really becoming an integral part of it even if they are moving in the same direction!

As a consequence of these preceding points, meteorology, and therefore models, do not really possess a coherent picture of general circulation. To a non-climatologist, this might seem unbelievable, but it is a fact that cannot be ignored, if we are discussing the prediction of climatic changes; because **no parameter, and no climatic region, can evolve independently,** since everything is more or less closely connected!

8.4 THE CONCEPT OF THE MOBILE POLAR HIGH

The first time that the necessity for a new conception of general circulation came to me was during research into tropical meteorology. For example: observations of variations in temperature, pressure, and the speed of the trade wind showed that subsident air was quite incapable of instigating the rapid accelerations and sudden cooling in that flux. The preparation of 250 mean meteorological charts at different levels for tropical Africa and the description of the vertical structure of the troposphere (Leroux, 1983) showed equally that the troposphere over the tropics is not homogeneous, but highly stratified, with distinct horizontal discontinuities such as the Trade Inversion (TI) and the IME. It seemed moreover that the tropical dynamic was closely associated with the extra-tropical, represented by MPHs arriving at

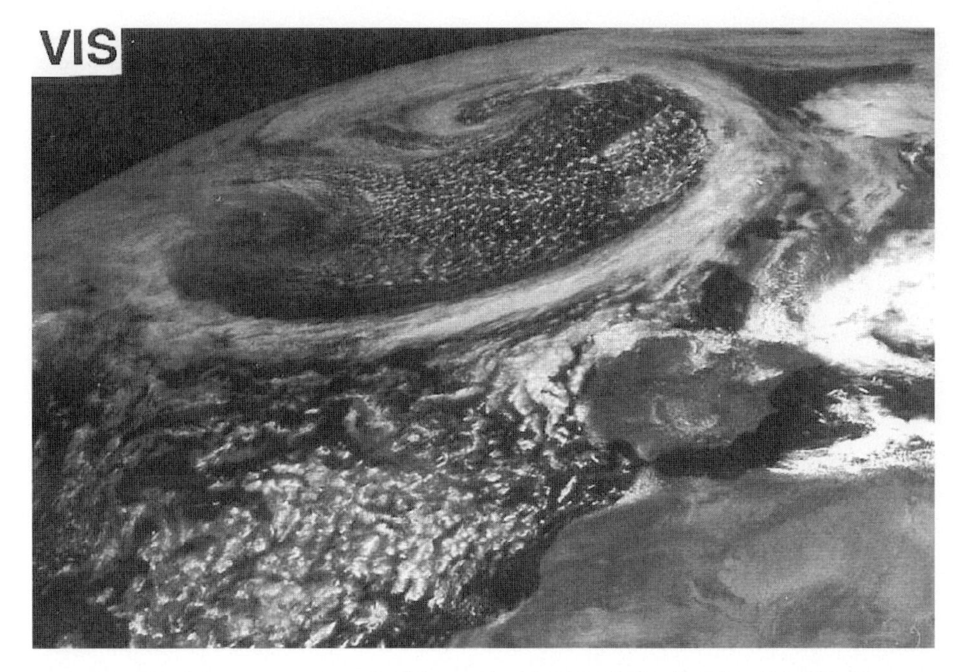

Figure 22. Cloud pattern connected with a typical MPH (28 April 1986, 12 h UT, Meteosat, visible). Cold Arctic air flows across the North Atlantic Ocean in the form of an MPH in the lower levels off the southern part of Greenland. On the leading edge of the MPH, a warm cyclonic air flow carries the subtropical water potential toward high latitudes and especially toward Greenland. A previous MPH now supplies the Atlantic trade flow, blowing southward off the coasts of the Iberian Peninsula and West Africa.

tropical margins and feeding trade wind circulation, and thereafter monsoon circulation, in the lower layers (Figures 22 and 23).

It was then necessary, by tracing the MPHs back to their source, to make a day-by-day analysis, using synoptic charts and satellite images, of the appearance, motion, transformation, and (supposed) 'disappearance' at the edge of the tropics of the 'motors' of circulation, the constantly renewed MPHs. A manual count for the whole of the northern hemisphere based on the *European Meteorological Bulletin* served chiefly to establish the trajectories and frequencies of MPHs for the period 1989–1993 (Guimard, Mollica, Moreau, de La Chapelle, and Reynaud of LCRE). Additional regular computer-based analysis has been done by Favre (for the North Pacific) and Pommier (North Atlantic), as part of their theses at the Climatology Laboratory (LCRE), Lyon.

8.4.1 Characters and structure of MPHs

MPHs (Leroux, 1983, 1986, 1996a, 2000) are in the main directly and visibly responsible for variations in pressure, wind speed and direction, temperature,

Figure 23. 8 July 2004, Goes 10, 18 h UT, visible. Five MPHs, separated by bands of clouds, are present on this picture of the south-eastern Pacific Ocean. On the right-hand side, an MPH has reached the South American relief (visible on the extreme right) and supplies the Easter Anticyclonic Agglutination and the maritime trade. South of this first MPH, another is rushing down from Antarctica. In the centre of the picture, one whole MPH is well outlined by interior anticyclonic rotation and by the cyclonic circulation around it; however, its north-western part is already opened to feed a maritime trade flow blowing north-westward. On the left-hand side of the picture, one (partly visible) MPH is followed on its southern side by another one rushing from Antarctica.

humidity, and cloud and rain amounts in extra-tropical areas – they are indirectly, and to a lesser extent, responsible for them in tropical areas. They are therefore responsible for the perpetual variations in the weather, and for the variability of the climate, on all timescales.

The thermal deficit which is ever-present at high latitudes, and at its greatest in winter, is responsible for the cooling and subsidence of the air above the Arctic/Greenland and Antarctica. As the descending air falls into step with the Earth's rotation, it will reach a critical mass and detach itself, rather like a drop of water, moving away from the pole in the lower layers as a *mobile lenticular body of dense air*, approximately 1,500 m deep and of the order of 2,000 to 3,000 km in diameter.

Figure 24, based on the *Meteosat* image from 26 April 1986 and the corresponding synoptic chart, shows in simplified form the association of surface wind and pressure fields in an MPH. In form, an MPH is a mobile combination of an anti-

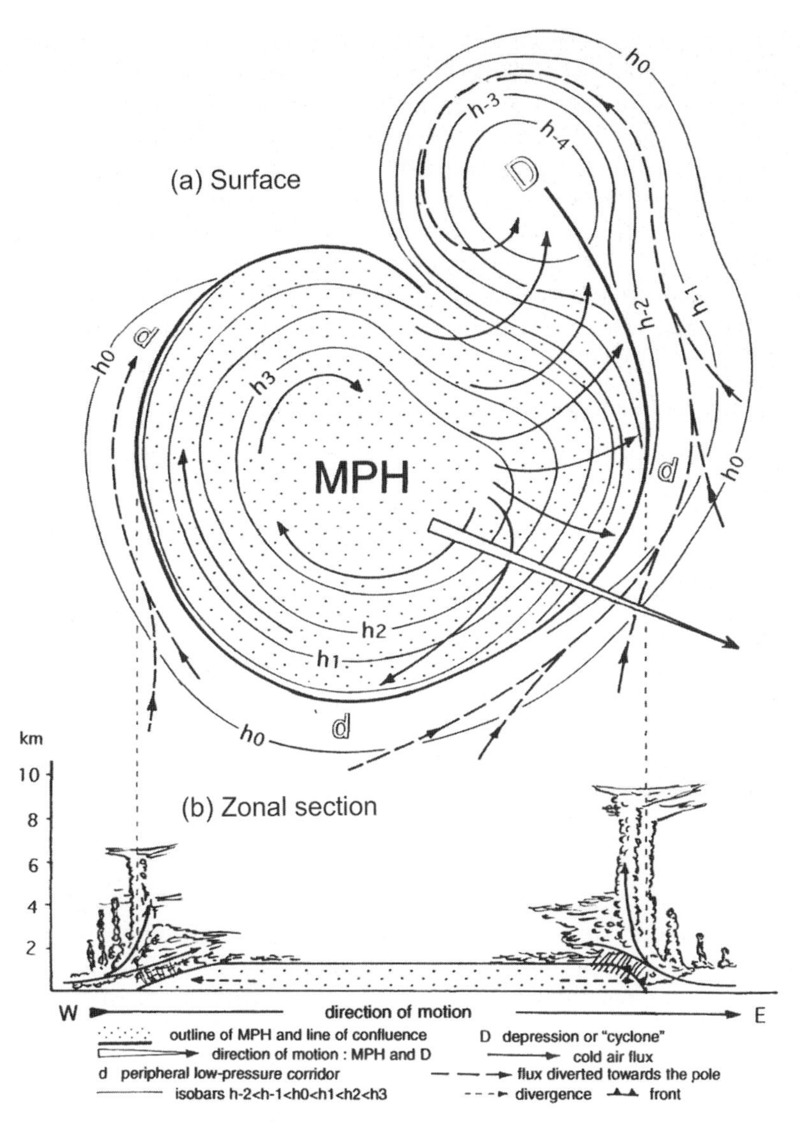

Figure 24. An MPH: (a) northern hemisphere surface pressure and wind fields; (b) vertical structure of an MPH and associated clouds.
From Leroux (1986, 2000).

cyclone (cold air: the MPH itself, the 'motor' of the whole unit), a peripheral low-pressure corridor, and a closed depression (of warm cyclonic air), both closely associated with the anticyclone and owing their existence and characteristics to it. The terms 'cold' and warm' as applied to air can have either an absolute or a relative value. The displacement *en masse* of an MPH cannot be related to the actual direction in which the wind blows: within the MPH it is anticyclonic, and in the peripheral low-pressure corridor and/or the closed depression, cyclonic.

The weather associated with MPHs in polar and temperate latitudes is a function of the respective densities of the air within the MPH, and the surrounding air. All the while the MPH's air remains cold, and therefore denser, the surrounding (relatively or absolutely) warm air will be lifted, and the MPH will be surrounded by more or less dense cloud formations (Figures 22 and 23). In winter, MPHs are more powerful and their trajectories more meridional, intensifying meridional exchanges and transfers of energy (especially perceptible heat and subtropical, or even tropical, latent heat). As a result, the peripheral low-pressure corridors and the closed depressions (cyclones) are deepened, updrafts (dynamical convection) are more vigorous, and the weather 'worsens', with increasingly frequent storms (cf. Chapter 11).

8.4.2 Trajectory and formation of Anticyclonic Agglutinations

The trajectory of MPHs from the poles towards tropical margins is determined by:

- The dynamic nature of the MPHs themselves, with generally NW to SE trajectories in the northern hemisphere and a more or less pronounced meridional component, moving closer to the tropics in winter as the MPHs' vigour and speed increases.
- Relief of more than 1,000 m (the approximate mean depth of MPHs), especially continuous mountain ranges. Relief channels some or all of the mass of the MPH, imposing trajectories and determining the units of circulation in the lower layers.

The gradual slowing of MPHs, the intersecting of their trajectories, and the effects of relief cause them to merge, and low-pressure corridors and cyclonic circulation between MPHs will diminish and disappear, with the formation of Anticyclonic Agglutinations or AAs (Figure 25). These AAs may become 'permanent' (mostly over oceans to the west of mountain ranges), they may be seasonal, or appear only in winter (and not only over land); or they may be occasional and of variable duration (cf. Chapter 11).

Trade wind circulation is born of these agglutinations of MPHs on tropical margins, which are veritable 'buffer zones' of general circulation within which anticyclonic rotation gradually asserts itself. The trades will possibly evolve as monsoons on crossing the geographical equator. Pulsations in the fluxes of trades and monsoons deflect the MPHs which are constantly arriving at the AA (Figure 25).

8.5 UNITS OF CIRCULATION IN THE LOWER LAYERS

The lower layers of the troposphere are of primary importance in climatology. They are the densest layers, and contain nearly all the water vapour and other greenhouse gases. Also, circulation in the lower layers is much more complex than that of the upper layers, because of:

- the Earth's surface warming the atmosphere;

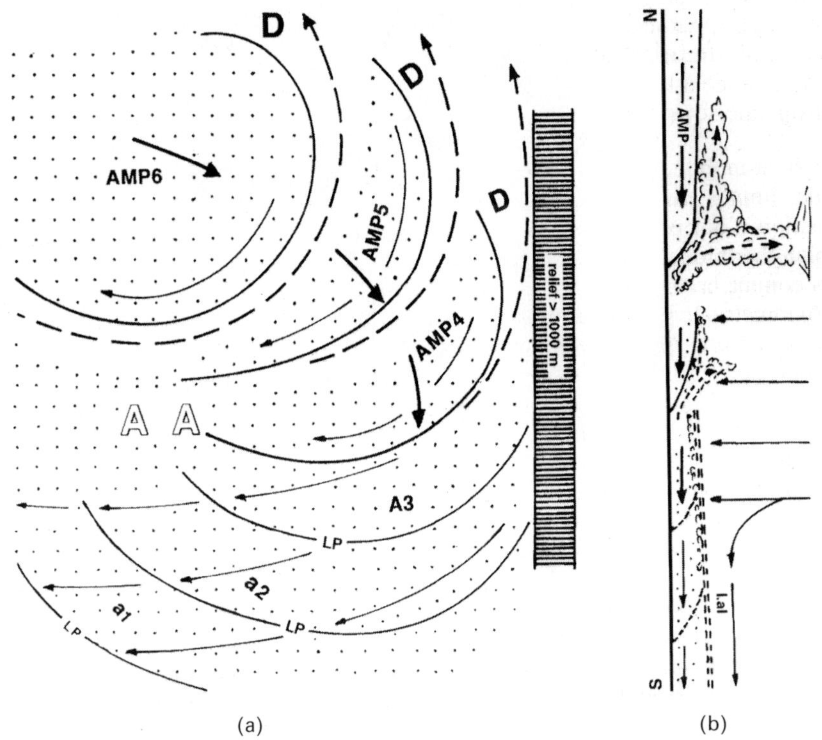

Figure 25. Formation of an AA, by merging of MPHs and birth of trades. (a) pressure and wind fields at the surface (a1, a2, a3: serial number of successive MPHs); (b) meridional vertical cross section corresponding to surface pattern (a) (AMP = MPH, LP = line of pulsation in the trade, I.al = trade inversion or TI).

- differences in the substratum;
- the thermal behaviour of oceans and land masses;
- vertical upward and downward movements caused by the surface (compression, turbulence, convection);
- thermal gradients arising from differential warming;
- effects of attraction caused by deep tropical thermal lows; and
- differences in altitude (especially the presence of high mountain masses).

Circulation in the lower layers is marshalled by MPHs, 'motors' of the general circulation, and by the geographical factor (namely mountain ranges). The permanent thermal deficit at the poles, and the resulting subsidence, cause a lenticular mass of cold air, initially about 1,500 m deep, to move off into the circulation at the rate of about one a day from either pole. As they travel and spread, these vast discs become shallower. Therefore, mountain chains of the order of 1,000 m high will present an almost insurmountable barrier to this cold air (though this is not the case with the warm cyclonic air associated with MPHs, which can cross such obstacles). Great mountain walls such as the Rockies and the Andes, or the line of

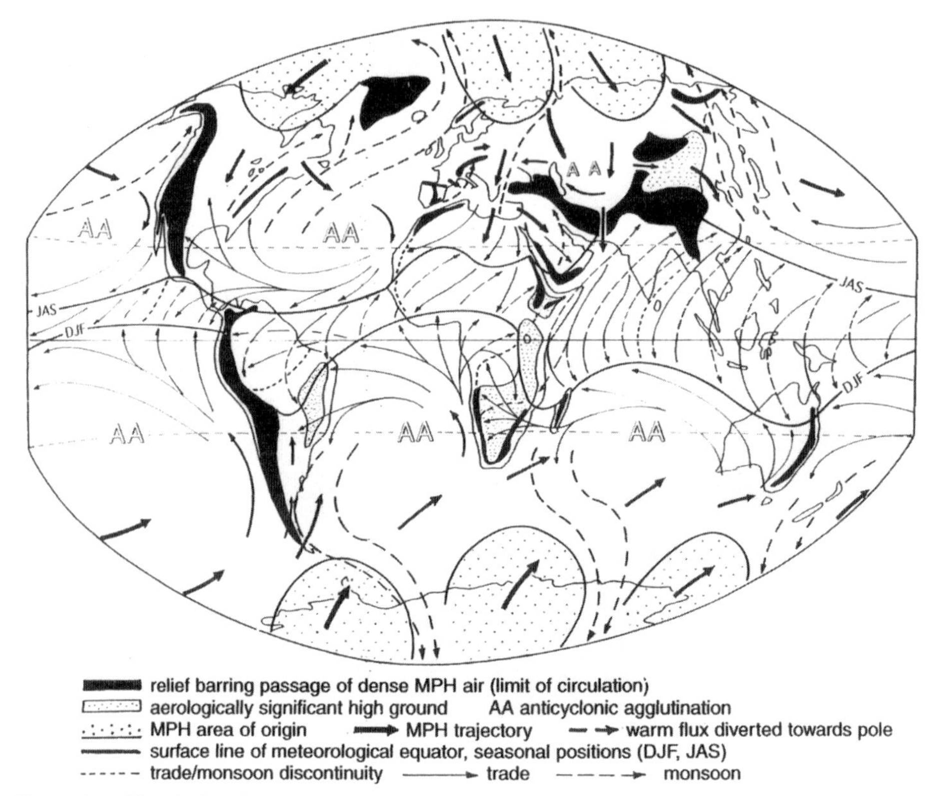

Figure 26. Circulation in the lower layers (diagrammatic), showing the six main aerological units determined by MPH dynamics and relief.
From Leroux (1996a).

highlands from the Pontus, the Caucasus, and the Zagros chain, right down into the Himalayas and Tibet, are absolute barriers to the dense air carried in MPHs, which are diverted in their entirety. Continuous mountain ranges, even if they are not quite as imposing as those already mentioned, cause the agglutination of MPHs, and by impeding the free passage of air, they determine the courses of vast units of circulation.

The vast spaces within which circulation occurs are fairly distinct, if extendable, and there may be 'traffic' to varying degrees between these spaces. Six aerological units are recognised, three in each hemisphere (Figure 26):

- North America/North Atlantic/western Europe;
- Northern and central Europe/Mediterranean/Middle East/northern Africa;
- East Asia/North Pacific/western North America;
- South America/South Atlantic/western and central Africa;
- Southern and eastern Africa/Indian Ocean/Australia; and
- Eastern Australia/South Pacific/western South America.

The dynamics of two of these spaces will be discussed in Chapters 12 and 13.

Each aerological unit of the lower layers has its origins at the pole, with relief intervening sooner or later to channel MPHs and lock them into their proper units of circulation. Within each unit, cold air advected by MPHs describes immense, elongated 'figures of eight', with air arriving from the pole and then travelling from west to east with a meridional component roughly down the middle of the unit. Next, an AA forms at the eastern edge, the trade circulation picking up tropical heat and water, and air returning to the pole by way of the cyclonic circulation on the leading edge of new MPHs. The latitudinal extent of each unit is limited by the surface line of the ME, their maximum extension occurring in winter when MPHs and their associated circulation are at their most vigorous.

In the oceans, circulation at the surface is controlled by the circulation of air (cf. Chapter 14). The driving force of marine circulation is yet again the MPH, which by pressure and its very size brings surface water eastwards in mid-latitudes, to more southerly latitudes in winter and more northerly latitudes in summer, in the northern hemisphere. The encounter with the land mass to the east of the unit, beneath the AA, divides marine circulation, with anticyclonic rotation drawing subtropical currents towards the equator. These currents store energy which will be sent back towards the temperate zone on the western edge of the great oceanic gyres thus created. This (slower) circulation effects in this way between 20–25% of the meridional heat exchanges. As a consequence, its intensity varies with that of the circulation of the air, and by extension, with the vigour of MPHs, more rapid in winter (or long-term cold periods) and slower in summer (or long-term warm periods).

8.5.1 Dynamical unicity and climatic diversity

In each unit of circulation spatial diversity is wide, and on the dynamic scale we can distinguish some very distinct regions:

- those from which cold air preferentially departs (i.e., near the pole (with little warm air advection from the south observed at the surface));
- those directly upon the trajectory of MPHs: regions which experience alternating circulation – cyclonic (ahead of MPHs) or anticyclonic (beneath MPHs), with western edges of units preferentially (on average) transferring cold air, while eastern edges experience intense updrafts of warm air;
- those away from the usual path of MPHs (i.e., in areas of associated depressions and warm cyclonic fluxes), though the passage of MPHs is not completely out of the question;
- those located beneath a remarkably stable AA, though this stability is only relative, since variations in strength and north–south migration are nevertheless observed; and
- those beneath a trade wind circulation extending the AA, possibly transformed as a monsoon. In this case, in the tropical zone, the eastern edge will exhibit both cool air and cool water, while the western edge is warm.

While some regions experience more or less constant conditions (especially if the influence of high mountain chains is felt), others show continual variations on a daily, seasonal or even interannual basis. For example, in the zone between the two extreme seasonal positions of the ME (Figure 26), the trades and the monsoon blow alternately. Other regions lie at the junction of two meteorological spaces. One of these is western Europe, which is influenced alternately by the Atlantic and the Eurasian units of circulation. The latter is responsible for very cold periods, especially in the winter.

In spite of the diversity of these geographically scattered units, **within each aerological space the initial dynamic is the same**, dictated to various degrees by the same MPHs, and so all the parameters of the unit are interdependent. Within the space, even if the logic of reactions seems to differ and climatic outcomes vary, there exists a **general co-variation** of climatic parameters. Another aerological space will have its own dynamic, organised by its own MPHs and the particular geographical conditions affecting its circulation. Interactions involving separate units in the same hemisphere may arise from communication between more or less compartmentalised units, or from the **conditions prevailing at the starting point of the circulation (i.e., at the pole concerned).**

8.5.2 The fundamental questions

The way in which circulation in the lower layers is organised calls into question many points which had been thought (and which are still thought by some) to be solidly established.

- The origin of the circulation does not lie in updrafts resulting from heat in the tropics, but is rather caused by the thermal deficit, and the variations in that deficit, over the polar regions. Because of this deficit, new 'discs' of cold air are constantly being injected into the circulation, with varying degrees of energy, thereby encouraging the return of warm air towards the poles.
- Circulation is not a continuous process (i.e., flowing step by step as in models), but is constantly renewed as cold, lenticular air masses (MPHs), transporting their inherent thermal characteristics (and others picked up along their paths), perturb the circulation of higher and middle latitudes. Then, calmed, they supply tropical fluxes. The arrival of these enormous volumes of 'new' air (or, more correctly, **recycled via the poles**), orchestrates the intensity of meridional exchanges, both warm and cold.
- The polar/temperate zone is not, strictly speaking, a 'mixing zone', but a zone of rapid meridional transfers, where the notion of 'mean winds' has no climatic meaning.
- Cyclonic circulation to the fore of MPHs effects an intense transfer of water vapour, and, consequently, of energy, most but not all of which originates in the tropics. The majority of the water vapour is transported in the lower layers (i.e., the lowest 1,500 m), as Peixoto and Oort (1983) underlined, remarking especially that '*the transport of water vapour is clearly influenced by topography*'.

- There is *no break in the circulation within each aerological space, and no interruption between temperate and tropical circulation*, from the pole to the ME, and therefore there are **no closed cells in the lower layers**.

Figure 27 is an expressive illustration of this continuity. It shows an encounter between an MPH and the Great Escarpment at the edge of southern Africa, on 1–7 August 1999. The MPH is slowed and divided by the relief, with most of the air it carries remaining in the South Atlantic, where an AA forms (statistically defined via mean pressures as the 'St. Helena anticyclone'). Now a trade wind, it is accelerated along the coast of Namibia along the base of the western escarpment, which, at a mean height of 1,500 m, remains above the cold air and prevents it from flowing onto the southern African plateau. The maritime TI generates low, thin, stratiform clouds. The air of the MPH spreads progressively northwards and westwards, crossing the Angolan highlands and entering the Congo basin. On 5–6 August, it crosses the geographical equator and assumes a monsoon trajectory (Atlantic monsoon) moving into western Africa on 7 August. Over the ocean, that part which is now the maritime trade nears the coast of Brazil on 6–7 August. At the base of the higher eastern escarpment (the Drakensberg range), the southern fraction of the MPH, now fairly small, reaches Natal on 3 August, and southern Madagascar on 5–6 August. Therefore, the cold air of the MPH moves onto the continent by way of the anticyclonic rotation, entering through the valleys of the Limpopo and the Zambesi, and supplying the continental trade, which flows westward across the plateau; and then, after crossing the Namibian escarpment over the Atlantic, above the TI topping the lower layer of the maritime trade. On 7 August, a new MPH has arrived off southern Africa. A long band of cloud, oriented NW–SE on its leading edge, is evidence of the transportation of warm air back towards the pole. This new MPH feeds the maritime trade in turn, and then the subsequent monsoon. So the intensity of tropical circulation is determined by the frequency and power of the MPHs (i.e., by the polar thermal deficit). This is a fundamental aspect of the tropical dynamic.

- There are definite hiatuses between different units of circulation, and in the lower layers within which most air moves. So there is actually no 'single' general circulation, but specific circulations integrated into general circulation. The notion of globality and the interdependence of all phenomena, as presented by the models, is a long way from being a fact.
- AAs, or 'subtropical highs' (Figure 25), are present in the lower layers, with no intervention from upper levels. Now, subsident air is still almost universally recognised as '*causing the presence of the great continental deserts of subtropical regions: in Africa, the Sahara to the north and the Kalahari to the south; in America, the Mexican desert and the Atacama; the Gobi in Asia* (sic!)*, and the Australian desert*'! (Fellous, 2003). This is wrong for a number of reasons, not least of which is that the air subsidence in question occurs only above AAs, and cannot therefore reach the ground!

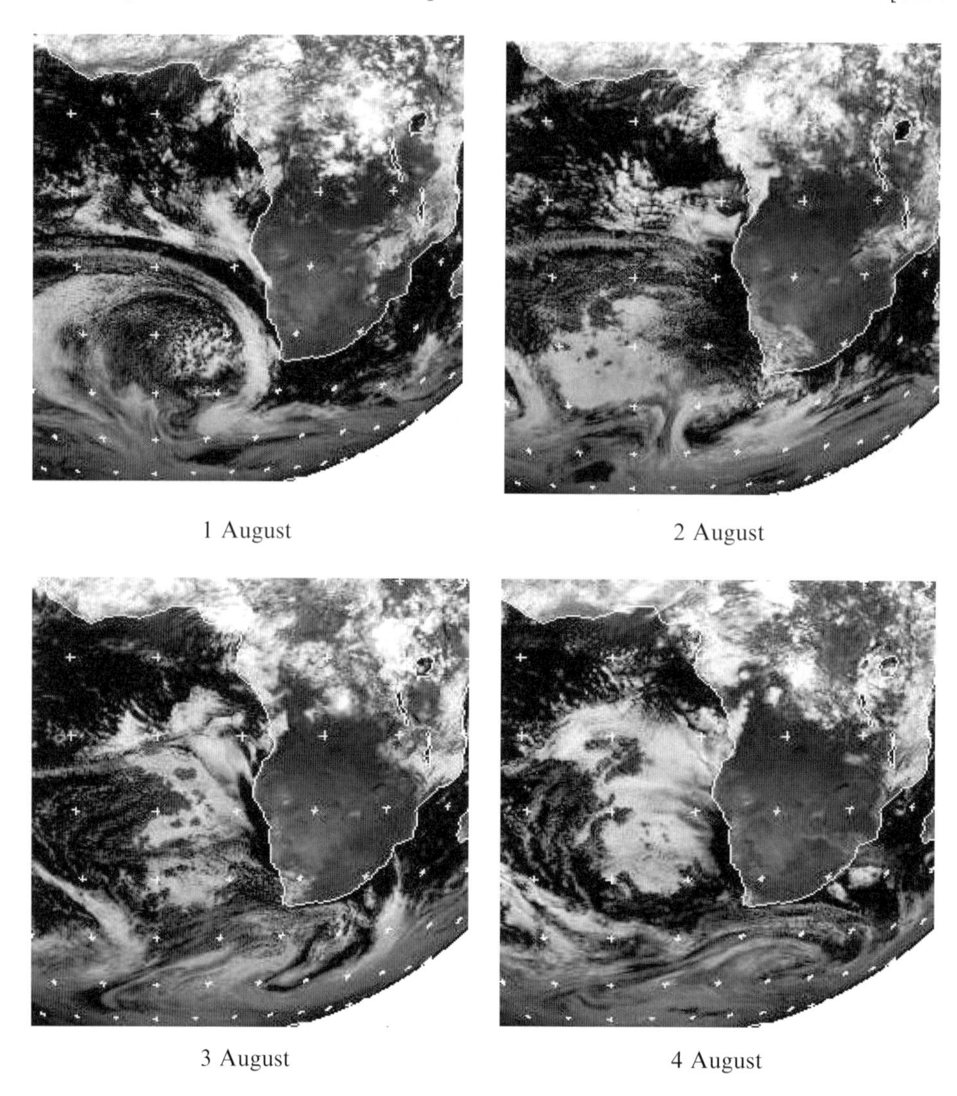

Figure 27. Supply of tropical circulation, trade, and monsoon by MPHs (1–7 August 1999, *Meteosat*, visible, 12 h, from Satmos–Eumetsat). Summary chart (SC) (path of the MPH's air from 31 July to 7 August): unbroken line – leading edge of MPH (or part of MPH); broken line (PL) – pulse line; GE – Great Escarpment around the highlands of Southern Africa.

- The geographical factor is normally not often taken into account, especially by models. Its importance is considerable as far as circulation is concerned, however, with relief being a main factor. For example, the Rockies, stretching from Alaska to southern Mexico, divide North America into two practically distinct units as far as cold air in the lower layers is concerned. Similarly, to the east of this barrier, water vapour is advected essentially from the Atlantic.

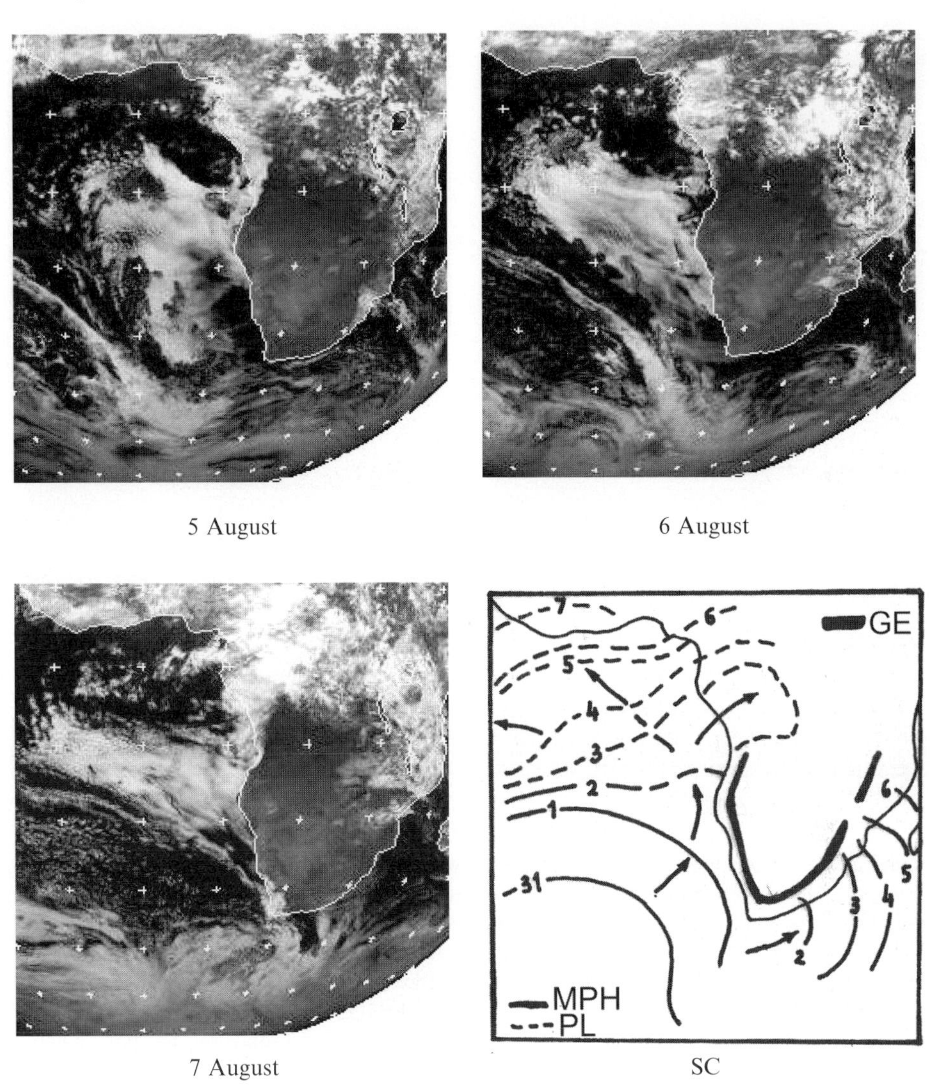

5 August

6 August

7 August

SC

The Andes do the same with the cold air and water vapour of South America. Again, the Himalayan–Tibetan mountains form a fundamental climatic boundary, denying MPHs access to the Indian subcontinent. The role played by relief is immediately obvious in the case of the high mountain chains, but more modest relief can also have a comparable effect. For example, the Great Escarpment around southern Africa, rising to about 1,500 m in the west along the Namibian coast, is able to split the flow of MPHs into both the Atlantic and Indian Oceans (Figure 27). Also, the mountains of the Iberian peninsula (Meseta, Sierras) force MPH air around them onto the western

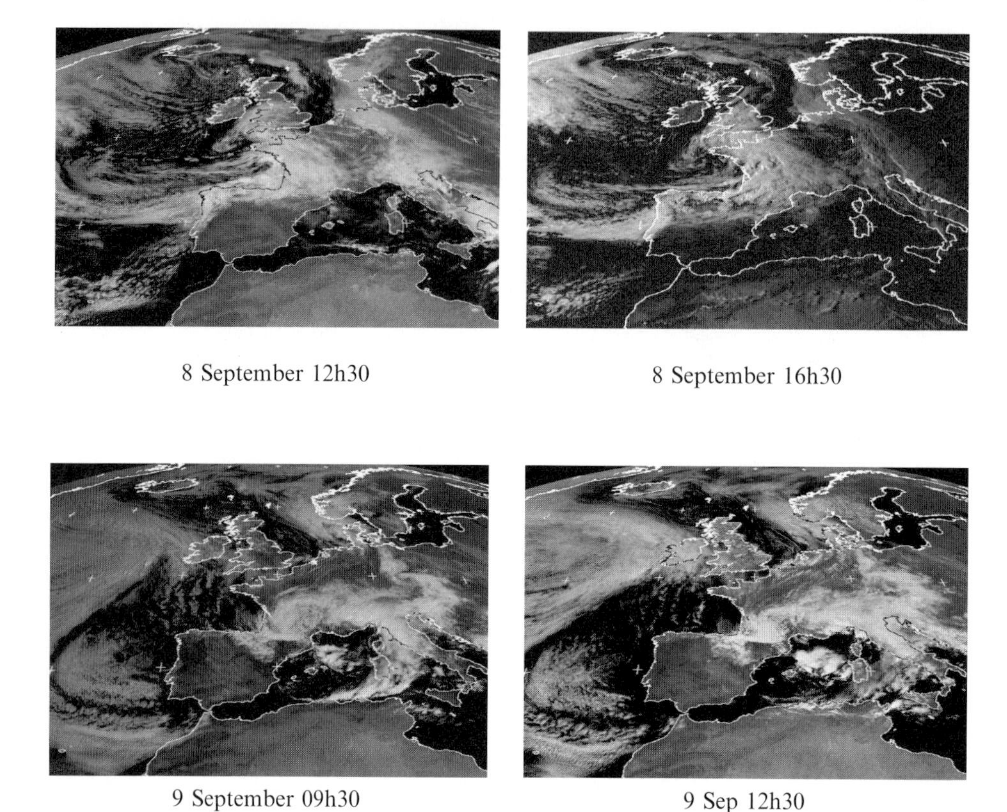

Figure 28. The influence of relief on lower level circulation (8–9 September 2003, *Meteosat*, visible). A large MPH moves across Western Europe. The east–west chain of the Cantabrian Mountains and the Pyrenees halts the cold air over the Bay of Biscay and the Aquitaine Basin. The leading edge of the MPH runs southward along the western side of the Iberian Peninsula, and eastward to reach the Mediterranean through the orographic funnel between the Pyrenees and the Alps. During the night of 8–9 September and the morning of 9 September, the MPH has passed along the northern side of the Alps toward Central Europe, and has passed round the Meseta high plateau and the Iberian Sierras. It invades the western Mediterranean area, following three paths: one through the French funnel (as *tramontane* and *mistral*), another (less intense) through Catalonia, between the Pyrenees and Celtiberian Sierra (*cierzo*), and the last (but not the least) between the Sierra Nevada and the Atlas. The Atlas range is itself impassable to the cold air of the MPH as far as Tunisia and the Gulf of Gabes. The MPH now supplies, on the western side of the Atlas, the Atlantic maritime trade, and on the eastern side of the Atlas, the Saharan continental trade (which will become the *harmattan* over western Africa).

Mediterranean, and the Atlas Mountains, extending E–W, prevent MPHs from directly penetrating North Africa west of Tunisia (Figure 28). The climatic consequences of these particular configurations are in the final analysis very significant (Chapter 11).

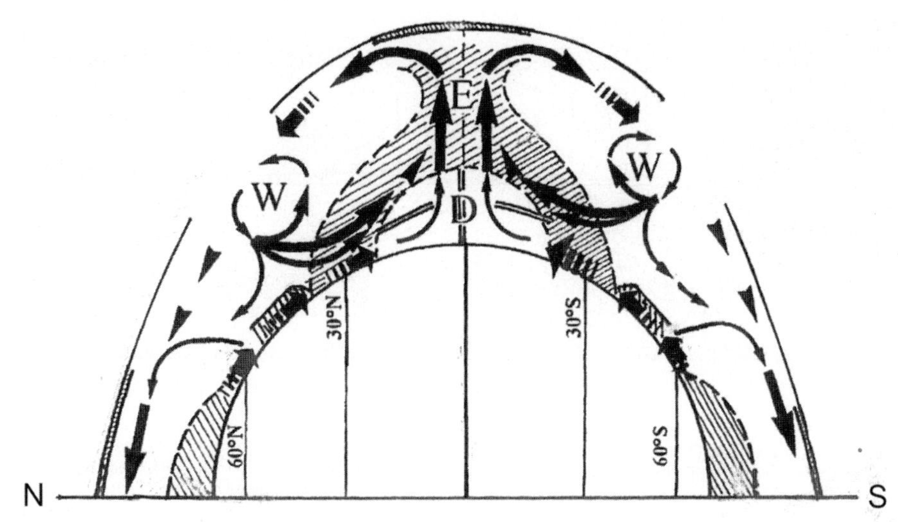

Figure 29. Troposphere: mean general circulation. Météo-France has adopted this scheme of general circulation, which has been taught at the French National School of Meteorology since 1992 (cf. Bonnissent, 1992), and published in *Météorologie Générale et Maritime,* Cours et Manuels No. 14, figure 9.7, p. 83 (Ecole Nationale de la Météorologie, Météo-France, Toulouse, 2001).
From Leroux (1983, caption: see figure 30).

8.6 GENERAL CIRCULATION IN THE TROPOSPHERE

The complexity of circulation in the lower layers is the result of the interference between the geographical factor (and particularly its orographical aspect) and the thermal factor, which causes the formation of the disc-shaped MPHs and of tropical thermal lows over continents. The influence of these factors is necessarily attenuated and finally negated at altitude. Then circulation becomes simpler, while the air becomes less dense. Eventually, only the major zonal currents remain, extra-tropical westerlies and tropical easterlies, concentrated into jets at about the altitude of the tropopause (Figure 29).

The starting point for circulation is at the poles, where MPHs form. The polar thermal deficit thereby provides the driving force for general circulation.

As MPHs move through the middle latitudes, they create, in the manner of a snow plough, two important displacements, horizontal and vertical:

- The diversion of warm air back towards the poles (Figure 24(a)) in the lower layers on the leading edge of the MPHs, and above them. This returning air will supply future MPHs.
- The lifting of warm air on the leading edge of MPHs (front and closed depression), and the liberation of latent energy (Figure 24(b)).

In those latitudes where intense vertical transfers of air take place, we also observe accelerations in the westerly (jet) circulation. These accelerations are strongest in

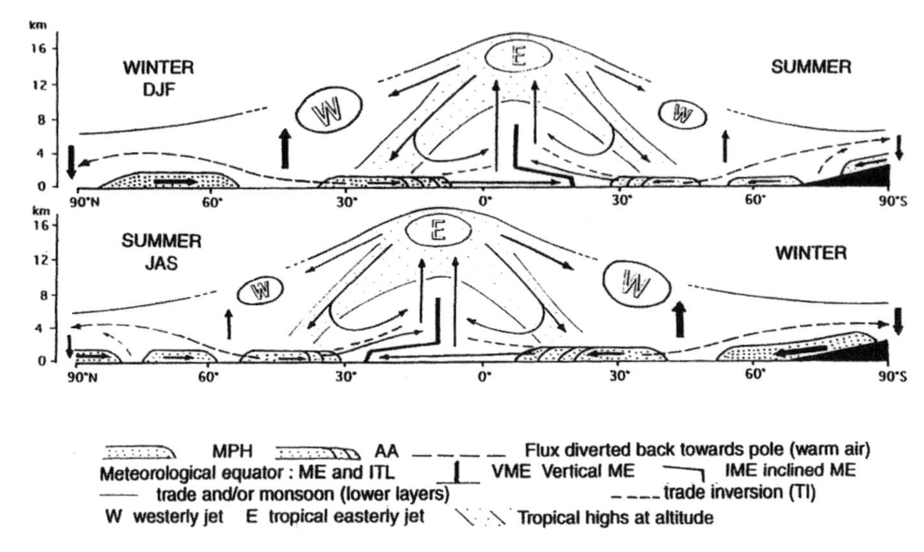

Figure 30. General circulation in the troposphere (vertical sections) according to seasons. From Leroux (1983, 1996a).

winter and weakest in summer as the power of MPHs and the intensity of updrafts vary. They are shifted nearer to the tropics in winter than in summer, because of the difference in the latitude reached by MPHs (Figure 30). The jets are consequences of lower layer phenomena, and especially of considerable vertical updrafts of air and energy caused by MPHs (and not vice versa).

AAs, formed by MPHs, are found only in the lower layers. From them springs the trade circulation, or more properly, the lower stratum of the trade, originally (within the AA) of the order of 1,000 m deep. The lower stratum of the trade becomes progressively warmer, and slowly spreads and deepens, becoming laden with water vapour. It meets the trade or the resulting monsoon (Figure 30) coming from the opposing hemisphere, along the ME.

Along the ME, updrafts, with deep convection, become general (Figure 29), for thermal but especially dynamic reasons, connected with the confluence/convergence of the circulatory hemispheres, impelled initially from the poles (cf. Figure 27).

The upward movement of air at the heart of the tropical zone has two major consequences:

- It supplies the Tropical Easterly Jet (TEJ), the vigour of which varies with the intensity of the updraft.
- It raises pressure at higher levels, forming Tropical High Pressure areas (THPs) which enclose the tropical zone in an inverted 'V' configuration (Leroux, 1983). These highs drive circulation in the direction of the poles, but the geostrophic force will not permit meridional exchanges across such a distance. Obeying mechanical laws, air at higher levels is rapidly drawn down towards the

surface, the descents tending to close off the Hadley cells at around latitudes 30° north and south.

The downward movements do not however reach the surface, as the lower layers are already occupied by the AAs and/or the lower stratum of the trade wind issuing from these same AAs (Figure 25(b)). A fundamental discontinuity is thereby created, between, on the one hand, subsident warm, dry air above and, on the other hand, the AA, or the lower stratum of the trade wind below, as it warms, spreads and possibly gains moisture. This discontinuity is the **Trade Inversion (TI)**, the climatic consequences of which are essential within the tropical zone as it is a horizontal, unproductive discontinuity (i.e., discouraging the vertical development of cloud formations). This is also the case in extra-tropical zones, as this discontinuity firmly dictates how water vapour is utilised, hindering its upward dispersion and concentrating it in the lower stratum of the trade.

Then, the remaining subsident (Hadley cell) fluxes may take two possible directions, one back towards the ME above the lower stratum of the trade in the middle layers, and the other towards the temperate and then polar zones around and above the MPHs, thus closing a circuit initiated at the pole (Figure 29).

8.6.1 Seasonal variation in general circulation

Variation in the Sun's radiation associated with its zenithal movement causes seasonal modifications in the intensity of meridional exchanges (Figure 29). All latitudes are affected to a greater or lesser extent, but it is the variations associated with the polar thermal deficit which have the most important effect on circulation, as they determine the dynamic of MPHs.

The winter thermal deficit increases the vigour of MPHs, intensifying confrontations in mid-latitudes, strengthening AAs, and accelerating the trades. Circulation is accelerated throughout the atmosphere in winter. The more meridional trajectory of MPHs displaces the most intense vertical transfers, and the corresponding westerly jet is in its turn shifted towards the tropics, now moving at maximum speed (twice that of the jet in summer).

The reinforced trade edges the ME in the direction of the summer hemisphere, and develops into the monsoon as it crosses the geographical equator. This crossing of the equator, requiring the support of the initial trade, is enhanced over tropical land masses in summer by deepening thermal lows which draw in monsoon air. The winter meteorological hemisphere therefore spills over to some extent into the summer hemisphere with its transequatorial monsoon fluxes (Figure 26).

Impelled by the trades, the monsoons are drawn into thermal lows; these are phenomena of the lower layers. Because of this, only the lower part of the ME is influenced by this 'overspill'. As a consequence, the ME possesses two vertical structures:

- In the middle layers, which are little (or not at all) influenced by lower layer phenomena, the vertical meteorological equator (VME) and, above it in the upper layers, the TEJ, take part in the overall shift of general circulation, similar to that of the AAs. The VME is the axis towards which easterly fluxes move in the middle layers (the upper strata of the trades).

- In the lower layers, the IME trespasses far into the opposite hemisphere, causing the trans-equatorial monsoon flux to move beneath the trade blowing in that hemisphere (Figure 30). The superposition of a trade upon a monsoon makes the IME a fundamental discontinuity, unproductive since the two fluxes have different origins, characteristics and directions.

Displacement continues as the seasons succeed each other: southward during northern winter, and northward during southern winter. In southern winter, the fact that the Earth is at aphelion reinforces the southern thermal deficit, which is already enhanced by the altitude of Antarctica (Figure 30). This situation encourages the spreading of the southern meteorological hemisphere, the incursion of well-supplied monsoons and the northward migration of the surface line of the ME, amplified by deep thermal lows over northern land masses (Figure 26).

8.6.2 Partitioning and stratification in circulation

As we saw in Chapter 7, modellers see the troposphere as uniform, homogeneous, and smooth, with neither partitioning nor discontinuity. Lindzen (pers. commun., 12 January 2004) wrote: '*there is a difference between sharp gradients and discontinuities, and it is the latter that do not exist*'. This 'ideal', but hypothetical, atmosphere has little in common with the real thing, which is, on the contrary, rigorously organised: it is separated into near-autonomous units of circulation and possesses vertical and horizontal discontinuities with distinct identities.

Circulation is first of all separated out in the lower layers (Figure 26). One of the primary discontinuities is relief, forming barriers at various altitudes and blocking the cold, dense air of MPHs. Consider, for example, both North and South America, where the west coasts have climatic characteristics distinctly different from those of eastern areas: MPHs of different origins come from different directions, as does precipitable water. The Great Escarpment around the southern African plateau prevents the maritime trade, blocked at the Namibian coast, from penetrating inland until it passes beyond the highlands of Angola, to flow into the lower lying Congo basin (Figure 27 above; Leroux, 1983).

The partitioned circulation thus created will possibly flow further into the tropical zone as a Trade Discontinuity (TD), a division between two trade wind circulations. Thus, in southern Mexico, the evolved Atlantic trade, rounding the Sierra Madre mountains, meets the nascent Pacific trade, which is denser, and rises above it (Figure 26). Similarly, south of the discontinuous barrier of the Cantabrian mountains, the Pyrenees, the Meseta, the Iberian Sierras, and the Atlas range in Morocco, a TD separates the Atlantic maritime trade from the continental Saharan trade which is energised by MPHs moving into northern Africa via the eastern Mediterranean basin. The warmer and lighter continental trade, with its load of dust, then passes above the cooler and denser maritime trade (Figure 26 above; Leroux, 1983).

Acknowledging the role of relief would preclude inanities like this one from Météo-France (Cours et Manuels No. 14, 2001, p. 152), on the subject of the Asian 'winter monsoon': '*the masses of cold air expelled from this thermal anticyclone experience considerable compression (a foehn effect) beneath the wind of the Himalayan and Tibetan relief*'. This seems to suggest that, contrary to all that is known about density, the very cold, dense, and thin layer air of the MPHs moving across China (Figure 26) manages to rise several thousand metres in order to cross the Himalayas and descend into the Indian subcontinent! This kind of 'explanation' is absurd, for reasons which ought to be borne in mind by those proposing so many similar dynamical 'links': the Pacific and the Great Plains of America?

In extra-tropical zones, the passage of MPHs sets up a synoptic and discontinuous stratification around and above the MPHs. The mobile inversion of wind, temperature, and humidity marking the top of the MPH separates fluxes of different origins, directions, vertical movement, and characteristics. Situated at an altitude of about 1,500 m in the vicinity of the pole, it progressively sinks as the MPH moves towards the tropics. The low-pressure corridors between MPHs possess no such inversion, and here, updrafts dominate in the cyclonic circulation.

The stratification becomes permanent and continuous within the AAs. An inversion at about 1,000 m separates air advected by MPHs from the subsident air above. This inversion, in concert with the anticyclonic character of the AA, is particularly unproductive (i.e., it inhibits the vertical development of cloud formations). It extends from the AA further into the trade circulation, becoming the equally unproductive TI. Beneath this horizontal discontinuity, the turbulent lower stratum warms up and may become moist (over the ocean) or drier (over land). Above, the subsident upper stratum is warm and dry. Between the two, the TIs mark the ceiling for thin stratiform clouds when the trade is a maritime one, or the upper limit of the concentrations of dust lifted by turbulence within the lower stratum of the continental trade.

The IME is another stratified, unproductive structure. The superposed fluxes (easterly trade above westerly monsoon) are of different origins, directions, and characters. Beneath such a structure, which may extend for several hundred (or even a thousand) kilometres (cf. Figure 26), precipitable water potential advected by the lower layer monsoon is scarcely exploited, with rain being normally absent (Leroux, 1983).

8.7 CONCLUSION: GENERAL CIRCULATION IS PERFECTLY ORGANISED

Whatever the modellers, with all their skills, might say, it is not the model which is right, but reality, the only possible reference. But **the models do not represent this reality**. **General circulation is complex, partitioned, stratified ... but perfectly organised**. When meteorological phenomena are labelled 'chaotic' (a comforting assertion for peace of mind), it is normally a sign of a (deliberate?) lack of understanding of this rigorous organisation. The fact is that 'chance' plays a small part, and introducing it is often the resort of unavowed ignorance.

So, when the supposedly 'unruly' climate is discussed, it is usually because those discussing it do not appreciate the '*rules*' (i.e., the rigorous mechanisms determining, localising, and characterising climates). Questions such as, '*Can the climate be turning on us*?' (Bard, 2004) are absurd, echoing the catastrophism of 'media weather', and serve only to point up an ignorance of the way in which climate works. 'Turning' suggests that the causes, mechanisms, or even the direction of general circulation really could, for some reason or other, abruptly change, or go 'turning' into reverse!

The general circulation of the atmosphere is rigorously organised, is always subject to the same physical principles, and always functions according to the same mechanisms (in well-defined geographical conditions). Its variations are therefore not variations in its nature, but are the result of variations in its intensity.

Before we claim that there is some relationship between two parameters, or, worse still, that one is the cause of the other, we must understand and recognise the respective places of each of these parameters, and the sequences linking them within the context of general circulation. We cannot say that something is tied to something else without having first established the reality of the physical link between them. No matter how sophisticated the statistical analysis, it is completely worthless climatologically if it is performed 'blind', with no basis in any proper meteorological analysis; observation and, above all, proof of the reality of physical links are indispensable.

Even if the schemas presented above, involving MPHs, are as yet incomplete, they represent the most realistic version (i.e., the version closest to meteorological reality), because they are based faithfully on direct observation. Let those who contest them, or query the role of MPHs at the origin of general circulation, supply proof that this concept does not conform to observed reality. Let them indeed suggest an alternative, indisputable concept, equally well supported, and based on the facts as they are observed!

The concept of the MPH as applied to general circulation has the advantage of representing, in the field of current research on this subject, the only schema embracing the initial cause of circulation, and the cause of its daily, seasonal, and indeed palaeoclimatic variations. It offers a complete and coherent overview of the dynamic of meteorological phenomena, encompassing all events, normal or extreme.

It is applicable to all scales of intensity, time, and space. This concept explains everything, leaving nothing out: **a concept that does not explain everything explains nothing**. This is why we shall constantly come back to it in the course of this book.

Part Two

The lessons of the observation of real facts

9

The observational facts: Past climates

In spite of everything, no climatic model has ever been able to simulate the snowfall, ten times greater than today's, in Canada. The growth of the great glaciers therefore represents a truly drastic change in the conditions then prevailing in North America, and the onset of glaciation still wears its cloak of mystery.

J.C. Duplessy and P. Morel. *Gros temps sur la Planète*, O. Jacob (ed.), 1990.

In its chapter entitled *The Evidence from Past Climates* (UNEP/WMO, 2002), the Intergovernmental Panel on Climate Change (IPCC) reminds us of the role alloted to palaeoclimatology:

- '*Past natural climate changes offer vital insights into human-induced climate change*'. This assertion allows one to underline the relationship between greenhouse gases (GG) and temperature, and of course echoes the insistence that humans are responsible for the enhanced greenhouse effect, all totally in keeping with IPCC policy, and so there is nothing new about it. However, when the relationship between greenhouse gases and temperature (GG/Temp) is evoked, we should also be careful to evoke an opposite relationship (Temp/GG) in order to test the reality of some supposed physical link. Is the increase in greenhouse gases a cause, or on the contrary, a consequence? What is the significance on the palaeoclimatic scale (just as on the seasonal one) of the more or less close co-variation of CO_2 and temperature?
- '*Studies of past climates (palaeoclimatology) give a sense of the scale of future changes*'. This extract seeks to put climate change in the long-term context, giving a scale for possible variations, with the past throwing light on possible future scenarios. Moreover, such studies '*provide a crucial check on scientists*' understanding of key climate processes and their ability to model them' (IPCC, 2002). So the reconstruction of past climates will serve to reveal and explain fundamental climatic mechanisms, and as a consequence the ability of models to

represent climatic reality will be tested, as will the validity of their projections for the climate of the future. This is an essential, indispensable procedure. Now, the question arises: how well do we know the mechanisms of the past, especially those of glaciation and deglaciation, and will they really help us to understand present-day phenomena?

The analysis of observational facts on the palaeoclimatic scale therefore assumes particular importance, especially as a basis for the verification of any long-term (supposed) relationship between the greenhouse effect and temperature.

9.1 PALEOCLIMATOLOGY AND THE GREENHOUSE EFFECT

The Vostok ice core research from the Antarctic offers evidence of the way climate has evolved over the past 420,000 years approximately (Figure 31). This remarkable analysis, which represented our deepest view into the past until the drilling of the Dome Concordia ice core of 2003, demonstrated quite unambiguously that orbital parameters of radiation influence the climate (Muller and MacDonald, 2000). The Vostok research revealed a succession of four cycles of the order of 100,000 years,

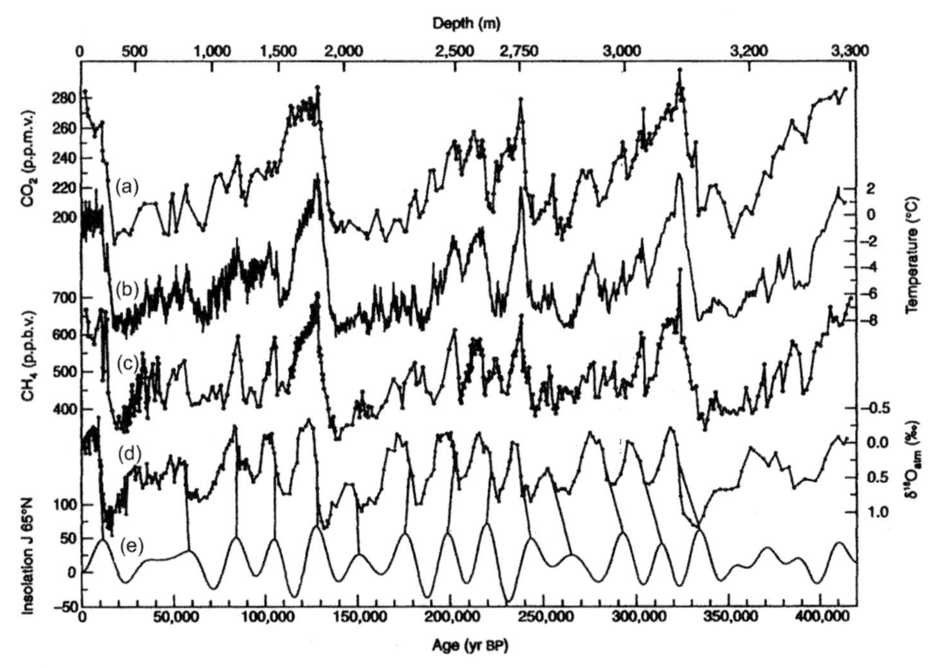

Figure 31. Vostok time series of: (a) concentration in CO_2; (b) isotopic temperature of the atmosphere; (c) concentration in CH_4; (d) $\partial^{18}O_{atm}$ (i.e., global ice volume changes); and (e) mid-June insolation at 65°N (in W/m^2).
From Petit *et al.* (1999).

associated with the variation in the eccentricity of the Earth's orbit; while within each great glacial/interglacial cycle, there are variations of shorter duration, associated with both the change in the Earth's axial tilt and the precession of the equinoxes. This work represented the full vindication of the concepts proposed by Milankovitch in 1924, verifying that **the primary cause of this evolution is external to all the parameters analysed**.

Co-variation is consistent in the case of changes in: CO_2 and CH_4 concentrations; deuterium excess (related to the temperature of that part of the ocean where the initial evaporation occurred); volume of ice; calcium content; dust ... and the corresponding temperatures as deduced from the analysis of water isotopes. This co-variation is the result of 'forcing' from outside the Earth itself (cf. Chapter 6). All the parameters vary in concert, and are therefore statistically correlated: $r^2 = 0.71$ between CO_2 and temperature, and $r^2 = 0.73$ for CH_4. But the evolution of temperature on this timescale is the consequence of variations in insolation at high latitudes. What then, is the significance of the conclusion: '*we confirm the strong correlation between atmospheric greenhouse gas concentrations and Antarctic temperature*' (Petit *et al.*, 1999)?

What indeed does the term 'correlation' mean here? If there is a physical link, it implies that the variation in the content of CO_2 and CH_4 has determined the evolution of the temperature. If this were the case, it would still be necessary to demonstrate, before all else, the causes of the variations in greenhouse gases, and to explain their periodicity, and this has not been done. Quite obviously, this is a spurious link, since **the cause of the variation in temperature is external**! To make the claim that '*changes in greenhouse gas concentrations may have helped to amplify ice age cycles*' (IPCC, 2001) does nothing to alter the nature of the initial problem; it simply transforms a cause thought to be primary (greenhouse gases controlling the temperature) into a secondary one (greenhouse gases contributing to the enhancement of the greenhouse effect). One cannot in fact help noticing '*the parallelism between variations in air temperature and the atmospheric content of greenhouse gases*' (Masson-Delmotte and Chappellaz, 2002). There may be parallelism to a greater or lesser degree, or concomitance, or coincidence, or simultaneity, but there is certainly not a physical correlation, and neither has it been demonstrated. To be valid, any supposed relationship would have to be analysed in both directions!

Now, in its turn, palaeoclimatology reveals the unacceptable confusion, involving both the past and the present, between statistical correlations, physical links, and mere co-variation, in contempt of the physical reality of climatic phenomena.

9.2 PAST AND PRESENT LEVELS OF GREENHOUSE GASES: ARE THEY COMPARABLE?

The evolution of the climate in the past evokes a (fallacious) argument often put forward by partisans of the greenhouse effect. This systematic *parti pris*, which goes against the very spirit of research, has developed within the IPCC and even finds official expression, for example in France by the Centre National de la Recherche

Scientifique (CNRS). In 2002 the CNRS awarded its gold medal to Lorius and Jouzel for having '*provided evidence of the link between levels of greenhouse gases and the evolution of the climate*', stating that '*their work has contributed to our appreciation of the potential influence of human activity on the future evolution of the climate of the planet*'. The medal in question was justly deserved, given the pioneering work done by Lorius, and the remarkable results obtained in the field of palaeoclimatology (or, to be more precise, in isotopic analysis); but the motive behind the award, based as it was on hasty conclusions, was simplist and misguided, though it did reflect the spirit of the times! The real problem, however, lies elsewhere: the fundamental question is whether past and present levels of greenhouse gases are directly comparable, and whether the supposed link is a real one.

9.2.1 Evolution during the last millennium

The IPCC has published, in its *Technical Summary* (figure 8), WG1, *Sc. Basis* (IPCC, 2001), a curve representing the evolution of the CO_2 content of the atmosphere over the last thousand years, including both estimated and directly measured levels from, respectively, ice and the atmosphere. The IPCC states that '*over the millennium before the industrial era the atmospheric concentrations of greenhouse gases remained relatively constant*'. From the year 1000 until 1800 (Figure 32), carbon dioxide levels as estimated from the ice remained in fact quite close to 280 ppmv, which would seem to indicate that there was little change in temperatures for eight

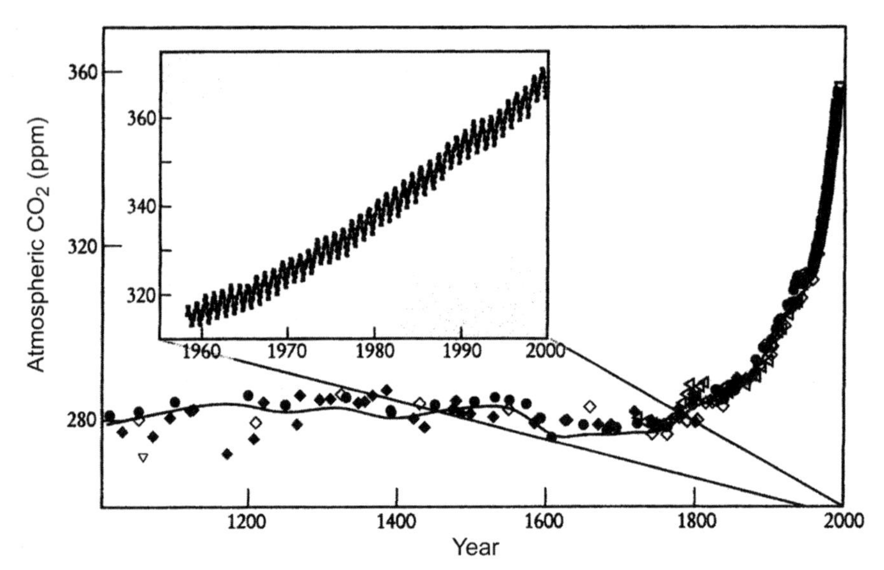

Figure 32. Carbon dioxide levels during the last millennium. Data before 1958 from measurements of air bubbles trapped in ice cores (from several sites in Antarctica); data since 1958 from Mauna Loa, Hawaii.
After Sarmiento and Gruber (2002).

centuries. Both the Medieval Warm Period (MWP) and the Little Ice Age (LIA) would seem to have occurred while the maximum variation in CO_2 levels was only about 4 or 5 ppmv! It must be remembered that, if we give credence to these values for carbon dioxide, we accept *ipso facto* that greenhouse gases have had absolutely no effect on climatic evolution over the last thousand years, since it must have been particularly uniform! However, ice-based measurements suggest a rapid increase since about 1850, from 280 to 310 ppmv, in accordance with the marked rise shown by atmospheric measurements both at the South Pole and on Mauna Loa (from 310 to 370 ppmv). The recent rise in temperature is therefore 'triggered' by the increase in carbon dioxide levels: this might seem to 'work' for the last century, but it does not seem to have applied during the preceding eight hundred years! On the climatic scale this comparison raises another problem, this time of magnitude: is the supposed evolution of the climate since the end of the LIA of the same order of magnitude as the evolution of the climate since the year 1000 (cf. the 'hockey stick' diagram, in Chapter 10)?

Let us look again at Figure 32. Doesn't the shape of the curve representing carbon dioxide concentrations simply reflect the fact that atmospheric measurements and ice-based estimates are fairly similar only at depths relatively close to the surface, given that the air is 'locked' into the ice only below depths of 80–100 m? Compression and the passage of time diminish CO_2 content and apparent variations are 'smoothed out'. I wrote to two laboratories where research in palaeoclimatology is carried out, the LSCE *(Laboratoire des Sciences du Climat et de l'Environnement*: Climate Science and Environment Laboratory); and the LGGE (*Laboratoire de Glaciologie et Géophysique de l'Environnement*: Glaciology and Environmental Geophysics Laboratory), and asked them this fundamental question: '*Can measurements taken from ice and from the atmosphere really be compared*?'. I received no definitive reply, though the LSCE claimed that such a comparison could probably be valid for very recent periods, cross-checking with current measurements (i.e., only at levels near the surface).

9.2.2 Evolution during the Holocene period

Also, what are we to make of variations in carbon dioxide levels during the Holocene period? '*CO_2 levels varied little during this period (~20 ppmv). From a maximum at the beginning of the Holocene, levels decreased until about 8,000 years ago, and then came a slow rise until pre-industrial values were attained*' (Flückiger *et al.*, 2002; www.lgge, 2004). So a margin of only 20 ppmv is held to be an 'explanation' of the Holocene Climatic Optimum (HCO) and the abrupt episodes of cooling which punctuated the Holocene, its climate being, however, '*less stable than previously thought*' (deMenocal and Bond, 1997); it is also claimed that '*the global carbon cycle has not been in steady state during the past 11,000 years*' (Indermühle *et al.*, 1999).

Also, what are we to make of the 100 ppmv (i.e., from 200 to 300 ppmv) said to 'explain' the formidable glacial/interglacial turnabouts (cf. Petit *et al.*, 1999)? A change of about 100 ppmv is supposed to have caused a fall in temperature in the

Antarctic of more than 10°C (Figure 31(b)), or, globally, a fall of 6°C. A maximum value of 300 ppmv is, in the same way, supposed to 'explain' interglacial periods with temperatures higher than today's. This slight range, relative to the amount of CO_2, is in the final analysis of the same order of magnitude as recent variation (from 290 to 370 ppmv). Can we compare, climatically speaking, this variation of 100 ppmv over 420,000 years to the rise of more than 80 ppmv which has taken place over the last one-and-a-half centuries, and which is supposed to 'explain' the 0.6–0.8°C of the thermal curve used as a reference by the IPCC? To recapitulate, briefly and schematically:

- A variation of 10 ppmv is supposed to correspond to a change (in the past) of about 1°C in the Antarctic, or 0.6°C in mean global terms.
- The same 10 ppmv variation is supposed to correspond to a change (in the present) of 0.07–0.09°C.
- If we applied the 'logic of past times' (0.6–1°C per 10 ppmv) to the present, the recent rise in temperature should have been of the order of 6–8°C!
- If we applied the 'logic of present times' (0.07–0.10°C per 10 ppmv) to the past, thermal variations over the last 400,000 years should not have exceeded 0.7–1°C!

Obviously, all this doesn't really mean anything at all, except in that, if we retain only the CO_2 levels, the last 150 years have experienced thermal, not to say climatic, changes, of the same order of magnitude as the last 420,000 years! Isn't this astonishing? There is nothing in common within these 'correspondences', and the very idea of some relationship between greenhouse gases and temperature loses all meaning, especially if we consider the enormous disproportion in scale between the climatic events of the glacial periods, and those of today. It only remains to estimate whether the evolution of temperature in the past is itself comparable to the recent mean thermal reference curve, the real climatic significance of which has not been demonstrated (cf. Chapter 10).

It is therefore incontrovertible that the CO_2 levels of the past (i.e., estimated), and present-day CO_2 levels (directly measured), are not comparable from the climatic point of view.

Recent information, from the work of Jaworowski (2004), gives good confirmation of this statement. '*Glaciological studies are not able to provide a reliable reconstruction of CO_2 concentrations in the ancient atmosphere.*' This inability stems from the fact that '*the ice cores do not fulfil the essential closed system criteria ... One of them is a lack of liquid water in ice, which could dramatically change the chemical composition of the air bubbles trapped between the ice crystals*' (Jaworowski, 2004). So a score of physicochemical processes act to considerably alter the chemical composition of the air trapped in the ice, and, as a consequence, '*CO_2 concentrations in the gas inclusions from deep polar ice show values lower than in the contemporary atmosphere.*' Therefore, it cannot with any certainty be claimed that '*present-day levels of CO_2 and CH_4 ... are unprecedented during the past 420 kyr*' (Petit *et al.*, 1999), a formula reiterated by the IPCC: '*Today's CO_2 concentrations have not been exceeded during the past 420,000 years*' (IPCC, 2001). Such a statement is simply wrong, because it is without foundation!

Therefore, using evidence from palaeoclimatology to claim that the climate in the past and the climate in the future are directly comparable, and that they are controlled by the greenhouse effect, is an ideal booby trap for the non-climatologist and, *a fortiori*, for the journalist and the average citizen who are not familiar with Milankovitch: and what is more, 'bad' chemistry is no basis for 'good' climatology!

9.3 THE RELATIONSHIP BETWEEN TEMPERATURE AND GREENHOUSE GASES

Figure 31, along with other isotopic analyses, poses another crucial problem. In what sense does the link between greenhouse gases and temperature operate? In order for the greenhouse effect to control temperature, the amounts of emissive gases would have to increase first, for some as yet unstated reason, a reason which would oblige us to suppose that greenhouse gas amounts vary according to a 100,000-year cycle! Now, ice core samples have definitely shown that changes in temperature occur before changes in the levels of emissive gases: as Barnola *et al.* (1987) described it, '*the temperature signal decreasing before the* CO_2 *when proceeding from the interglacial to the glacial period*'. So '*the* CO_2 *decreases generally following the temperature decrease in the Antarctic, with a delay of as long as 4,000 years*' (cf. www.lgge, 2004). In the same way, during deglaciation (e.g., that of 240,000 years ago), warming precedes the increase in CO_2: '*a delay of about 800 years seems to be a reasonable time period to transform an initial Antarctic temperature increase into a* CO_2 *atmospheric increase*' (Caillon *et al.*, 2003). Are we now saying instead that temperature directly determines the levels of greenhouse gases? The answer is still no.

CO_2 concentrations fall during periods of glaciation, slowly decreasing towards levels of 180–200 ppmv (Figure 31). A fundamental question arises from this: '*Where does the carbon dioxide go during glaciation?*' (Bopp *et al.*, 2002). One answer has been sought in the role of vegetation, both on land and in the sea, mirroring seasonal CO_2 evolution during the year (cf. Figure 32, insert), the seasonal amplitude being of the order of 6–7 ppmv. In the northern hemisphere the maximum level is reached in May, and the minimum occurs in September–October as a function of the cycle of vegetation, with CO_2 being absorbed in the spring and summer and restored to the atmosphere in autumn and winter. But variations in the continental biosphere cannot really explain palaeoclimatic evolution, because, during cold periods, the great reduction in biomass considerably reduces the stock of carbon which can be held by vegetation.

So we look to the oceans, which contain 50 times more carbon than the reservoir of the atmosphere. It is usually thought that, during periods of glaciation, it is actually the oceans, and more particularly those of the southern hemisphere, which regulate the carbon cycle, because:

- cold water absorbs far more CO_2; and
- '*accumulation of carbonate and carbonaceous matter increased during the glacial age due to intensified sink water formation*' (Olausson, 1985).

Many hypotheses have been put forward in an effort to explain this evolution, both physical (mixing of surface, mid-layer, and deep waters) and biological (involving the fixing of carbon by plankton). There is, however, still no agreed explanation. In the words of Stephens and Keeling (2000), '*also lacking is an explanation for the strong link between atmospheric CO_2 and Antarctic air temperature*'; and in the opinion of Bopp *et al.* (2002), '*at the moment none of them manages to explain the totality of the 80–100 ppm difference between glacial and interglacial periods*'.

CO_2 concentrations increase during interglacial phases, attaining (presumed) levels of around 280–300 ppmv, at a quite remarkably rapid rate when compared to the slow rate of decrease during colder periods (Figure 31). There is still a delay, though, all the while the temperature is changing, since a latency period is necessary for the vegetation to be able to spread back into its former territory (with methane levels closely reflecting changes in the global biomass), and for the oceans to become a source of CO_2, notably in equatorial warm water areas.

So here we have a complex question, and the possibility presents itself that there is actually no link, not even a differentiated one. '*The marked similarity in temporal variations*' is not observed, for example when '*a large portion, about 7°C, of the isotopic temperature decrease leading to the glacial conditions takes place between 132 and 117 kyr, a time during which no significant CO_2 changes are observed*' (Barnola *et al.*, 1987).

This question is therefore of very great importance, but it remains unresolved as yet, for past as well as present times. In the case of the present, a similar question arises, as far as atmospheric CO_2 is concerned: '*the terrestrial and the marine environments are currently absorbing about half of the carbon dioxide that is emitted by fossil fuel combustion*' (Schimel *et al.*, 2001). But, '*at present, no global oceanic model exactly represents the variations in CO_2*' (Metzl, 2002). Consequently, if greenhouse gases are not controlling the temperature, or, as Caillon *et al.* (2003) put it, 'CO_2 is not the forcing that initially drives the climatic system', then it would also appear that the temperature does not (directly) control the evolution of greenhouse gases.

On the palaeoclimatic scale, greenhouse gas concentrations are therefore not a cause, but indubitably a consequence (like other parameters) of a dynamic that still remains largely undefined. **The so-called greenhouse effect scenario has not been established as a long-term phenomenon**. This being the case, an analysis of past climates which supposes them to be a 'laboratory' for the investigation of present-day phenomena will not be able to supply the information desired. What, then, is our comprehension of the great climatic mechanisms involved, and more especially our comprehension of the dynamic of an atmosphere which could account for the considerable upheavals in the climate of the past?

9.4 MODES OF GENERAL CIRCULATION

In an effort to explain the phenomena of the past, such as the evolution of the carbon cycle (which we have already mentioned), hypotheses have been proposed; among others, there is one which aims to throw light upon the absorption and restitution of carbon by the oceans. One of the premisses of this hypothesis is reckoned to be that, '*in a glacial period, precipitation diminishes* ...' (Paillard and Parrenin, 2004). This is an application of the equation 'T/R', which means in this case 'low temperatures = low rainfall' (cf. Chapter 7). If this were so, how could those enormous conglomerations of ice have been built up (see the introductory quotation)? Here, then, is a hypothesis that gets off to a shaky start. The argument should, if nothing else, be based from the outset upon coherent schemas. There are those who try to see an explanation in oceanic circulation: '*only deep circulation changes could be the cause of the CO_2 variations*' (Barnola *et al.*, 1987). Since oceanic circulation is largely marshalled by atmospheric circulation, let us put this reflection into the widest context, that of general circulation.

Working on the basis of seasonal variations in circulation (Figure 30), which are orchestrated by the intensity of the polar thermal deficit, and also, of course, reasoning on the basis of actually observed facts and wide-ranging reconstructions such as CLIMAP, COHMAP, IPCC–*Scient. Basis*, etc. (cf. Leroux, 1983, 1993a, 1994c, 1998, 2001), we can propose the existence of two modes of general circulation, a rapid mode and a slow mode.

9.4.1 The rapid mode of general circulation (cold scenario)

Here, the simple initial postulate is that, in one of the meteorological hemispheres in winter, phenomena are intensified as a result of the polar thermal deficit (Figure 30). A rapid mode of general circulation therefore combines, schematically, two meteorological hemispheres in winter (i.e., a global situation corresponding to a strong thermal deficit), all the year round and simultaneously (with more or less marked deviations), at high latitudes in both the north and the south (Figure 33). The general tone of the climate is a cold one, even in summer, and seasonal thermal contrasts are much reduced as cold tends to dominate.

The major traits of this rapid mode of general circulation are:

- Mobile Polar Highs (MPHs) are powerful, deep, and of vast dimensions, and they maintain considerable coherence, retaining their original low temperatures across longer distances. They transport a larger amount of cold polar air, they move more rapidly, and their trajectories are more meridional, so that they drive more deeply into tropical margins.
- The greater dynamism of MPHs causes an intensification of the cyclonic circulation at their leading edges, and as a result there is a much-enhanced transfer of energy from the tropics towards the poles. This transfer involves perceptible heat and latent heat through the intermediary of water vapour, vigorously diverted towards higher latitudes. Remember that the more meridional trajectory of the MPHs means that they are able to both reach and divert warmer tropical or subtropical air, with its supply, over the oceans, of a richer precipitable water potential.
- This transfer towards the poles may also involve continental air, and therefore much drier conditions are favoured as Anticyclonic Agglutinations (AAs) are

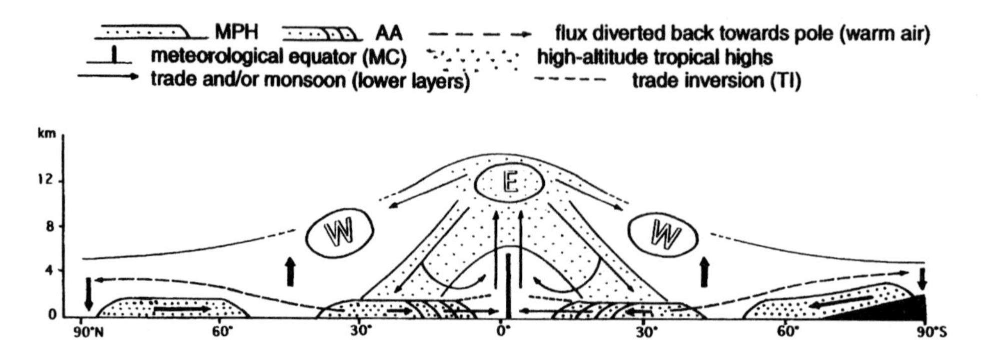

Figure 33. Rapid general circulation mode connected with a strong polar thermal deficit.

extended and the unproductive character of the Trade Inversion (TI) is reinforced. This continental air can carry much greater quantities of dust. The greater densities of dust in ice core samples for colder periods from Greenland and the Antarctic testify to the more vigorous nature of this transportation, typical of greatly accelerated fluxes in the lower layers and at altitude.

- Disturbances in middle latitudes are more violent, because of accentuated thermal contrasts and the greater vigour of the MPHs, which leads to more powerful updrafts (with deeper and wider depressions). Westerly jets, abundantly supplied by these enhanced updrafts, are in turn accelerated and displaced in the direction of the tropics.
- AAs, fed by strengthened MPHs, become in turn more vigorous and spread further northwards and southwards over both oceans and land masses. Over the land, continental AAs are much stronger in winter, bringing about stable anticyclonic conditions which last for a long time. These continental anticyclonic situations are just as frequent in the summer, and are accompanied by dry conditions (cf. Chapter 11). Generally speaking, in winter as in summer, AAs form at more tropical latitudes, their stability is reinforced and their unproductive character (or inversion) covers a wider area, especially in the tropical zone where the desert extends towards the equator.
- Tropical circulation, which is powerfully supplied by MPHs, is much accelerated, but the trades cover a relatively smaller area; their energy may be handed on to monsoon fluxes extending from them, and these too will be more limited in their range.
- At the heart of the tropical zone a paradoxical situation occurs: fluxes are more vigorous and faster, but the area across which they sweep is reduced because of the northward and southward *rapprochement* of MPHs and the resulting contraction of the tropical zone. Another factor is that of the strong opposition set up by contrary fluxes from the other hemisphere, which hinder the migration of the meteorological equator (ME) (Figure 34(a)), less strongly drawn, at the surface, by less deepened thermal depressions.
- The structure of the ME, with its limited range of migration, becomes closer to

(a) Cool global situation: rapid circulation mode

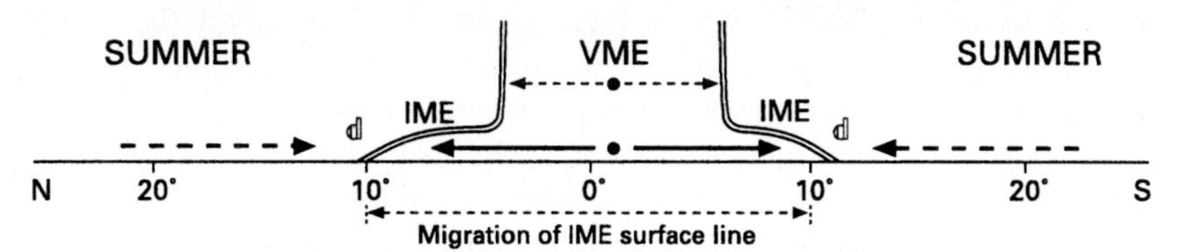

(b) Warm global situation: slow circulation mode

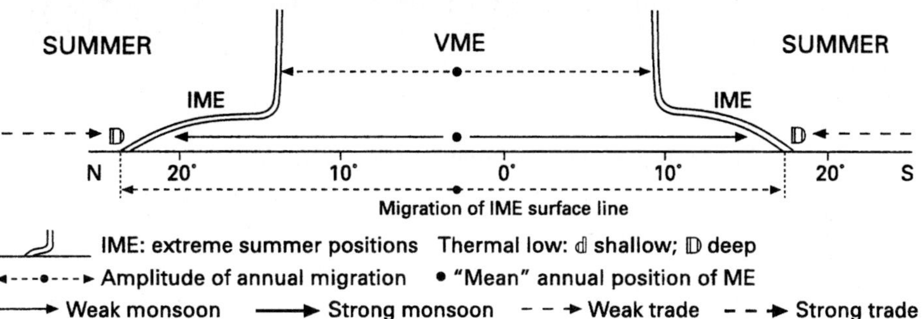

Figure 34. Differential seasonal migration of the ME's vertical structure: (a) associated with a rapid general circulation mode; (b) associated with a slow general circulation mode.

the Vertical Meteorological Equator (VME). The likelihood of any extension of its inclined structure (IME) is reduced by the dynamic nature of the opposing flux and the weaker continental thermal depressions. The monsoon circulation which is characteristic of the IME's structure is also considerably reduced in its amplitude, not straying far from the equator, especially in Africa and Asia (Figure 34(a)).

- The migration of tropical pluviogenic structures is limited to a narrow zonal belt, notably in the case of the VME, which remains relatively close to the equator and is associated with regular and abundant rainfall. Within the structure of the IME, now moving across a much more limited area, the same narrow range applies to squall lines, beneath which rain is light and uncertain.
- The water vapour of the tropics is normally exported in two ways: either to the fore of MPHs, or following the updrafts of the ME (Chapter 8). The greater power of MPHs progressively encourages more intense transfers towards the poles, at the cost of diminished precipitable water potential over the tropical zone, where rainfall is now generally in short supply.
- The polar thermal deficit is further aggravated in the northern hemisphere by the extent of the land masses, sea ice, and the *inlandsis*. The relative strengthening of the northern meteorological hemisphere means that it trespasses upon its southern counterpart, and the ME is therefore shifted southwards all along its line (Figure 34(a)).

Oceanic surface circulation (cf. Chapter 14) is in turn everywhere accelerated, impelled by colder MPHs, which exert greater pressure upon the water, and by trade winds which are now, like all the other fluxes, more rapid. The great gyres are now closer to the geographical equator, and ocean currents are accelerated, in particular those cold currents which flow towards the heart of the tropical zone along the eastern edges of oceans, where upwellings are also more vigorous. Currents of density at high latitudes are reinforced by lower temperatures and by the growth in area of icefields, and more CO_2 is absorbed and stored, meaning that there is a lesser concentration of CO_2 in the atmosphere. With this generally increased circulation in the oceans, there is a much greater intensity of thermal transfers in the water, as well as in the atmosphere.

9.4.2 The slow mode of general circulation (warm scenario)

In either of the meteorological hemispheres in summer, phenomena are less intense as a result of the diminution of the polar thermal deficit (Figure 30). A slow mode of general circulation therefore combines, schematically, two meteorological hemispheres in summer conditions (i.e., there is a global situation corresponding to a moderated thermal deficit), all the year round and simultaneously (with more or less marked deviations), at high latitudes in both the north and the south (Figure 35). The general tone of the climate is a warm one, with mild winters, and seasonal thermal contrasts are wider as meteorological phenomena have greater freedom of movement.

The major traits of this slow mode of general circulation are:

- MPHs are less powerful, more shallow, and of smaller dimensions, and they are less coherent as they move along their trajectories, warming more rapidly. They have a smaller amount of cold polar air to transport, and their trajectories are less meridional, so they do not go beyond tropical margins.
- The reduced dynamism of MPHs causes an attenuation of the cyclonic circulation at their leading edges, and as a result there is a reduced transfer of energy from the tropics towards the poles. The supply of perceptible heat and latent heat is favoured, but the intensity of the transfer is slowed. The less meridional

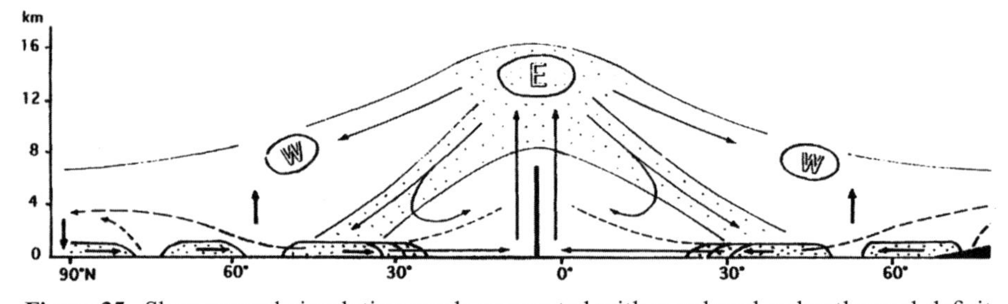

Figure 35. Slow general circulation mode connected with a reduced polar thermal deficit.

trajectory of the MPHs means that they are unable either to reach or divert the maximum possible precipitable water potential.

- When transfer towards the poles involves continental air, less dust is transported because of the deceleration in fluxes, both in the lower layers and at altitude, and because of the reduced extent of regions dried out by winds now reduced in strength.
- Disturbances in middle latitudes are more clement, because of attenuated thermal contrasts, and the reduced vigour of the MPHs fuels less powerful updrafts (with shallower depressions). However, a stormy character may be more prevalent as a result of thermal convection and the greater availability of energy. Westerly jets, less well supplied by reduced updrafts, are less rapid, and displaced away from the tropics.
- AAs, fed by the weakened MPHs, exhibit reduced pressures and have smaller areas over both oceans and land masses. Over the land, continental AAs may form only in winter, with stable anticyclonic conditions being less frequent. Generally speaking, these weaker AAs form at latitudes more removed from the tropics. They are less stable, and their unproductive character (beneath the inversion) covers a smaller area.
- Tropical circulation, less generously supplied by MPHs, is decelerated, but the trades cover a relatively larger area, as do the monsoon fluxes extending from them.
- At the heart of the tropical zone, a paradoxical situation still occurs: fluxes are slower, but the area across which they sweep is much wider because of the northward and southward movement of AAs away from each other, and the resulting dilation of the tropical zone. Another factor is that of the relative lack (or even the complete absence) of opposition from contrary fluxes originating in the other hemisphere (Figure 34(b)).
- The ME, its range of migration having broadened, now possesses a very obvious double structure. The VME migrates beyond 12–15°N and S, and the zonal band of abundant rainfall is thereby widened. Meanwhile, the lower layers of the IME may impose their structure beyond the tropics, drawn mainly by much deeper continental thermal summer depressions, and also because of the less dynamic nature of the opposing flux. Trans-equatorial monsoon circulation is considerably enlarged, being found far from the equator, and bringing with it its load of moisture from the oceans, deep into tropical continents, especially Africa and southern Asia (Figure 34(b)).
- The seasonal migration of tropical pluviogenic structures covers a wider area, in the case of both the VME and the structure of the IME, which benefit from a richer precipitable water potential, delivered more evenly over a larger area (Figure 34(b)).
- The existence of less vigorous MPHs reduces the intensity of transfers of tropical energy towards the poles, and the tropical zone is able to conserve most of its precipitable water potential, which is transported into the interiors of tropical continents by amplified monsoon airstreams.

- Now it is the turn of the southern meteorological hemisphere to trespass upon its northern counterpart, its encroachment being more extensive over northern tropical continents, where it is drawn in by deep continental thermal summer lows in the lower layers. The ME is shifted to the north of the geographical equator all along its line (Figure 34(b)).

Oceanic surface circulation, impelled by less dense MPHs and slower trade winds, is, in turn, slowed. The great oceanic gyres are now further from the geographical equator, and ocean currents driving them are decelerated, especially along the eastern edges of oceans, where upwellings are also less vigorous. Currents of density at high latitudes are less marked, absorbing and storing less CO_2 and thereby increasing the concentration of this gas in the atmosphere. The whole of the circulation in the oceans is slowed, and thermal transfers are diminished.

Remember that the repercussions of the modifications dealt with in Figures 33, 34, and 35 take place principally within the lower layers of the troposphere. It will therefore be useful to complement them with (horizontal) schemas of circulation for the six aerological units defined in Figure 26 (Chapter 8). This will not be done now, but later when we come to examine recent climatic evolution in certain sectors such as the North Atlantic and the North Pacific (Chapters 12 and 13). As an aid to comprehension, it will now be possible to refer if need be to these complementary schemas of slow and rapid circulation in the lower layers.

9.4.3 The aerological dynamic of past climates

Variations in insolation at high latitudes are responsible for the changing modes of general circulation. Figures 33 and 35 offer a variety of responses to questions raised by the observation of past climates, and more importantly they offer insights into:

- Variations in the intensity of meridional exchanges of both air and energy, and more particularly the modification of wind speeds and the amounts of water vapour transferred. Tropical latitudes experience a very limited amplitude in variations of insolation, and still retain an abundant precipitable potential, exportable even when it is reduced during cold periods.
- The propagation of climatic modifications, especially from middle and high latitudes towards tropical latitudes by uninterrupted circulation, with relatively moderate tropical thermal variations advected by meridional circulation.
- Changes in the intensity of perturbations, especially in middle latitudes, with rapid circulation characterised by violent weather during cold periods, while during warm periods slow circulation is characterised by clement weather. The ability of MPHs to capture tropical energy, to the advantage of the higher latitudes, is to a greater or lesser degree enhanced (a fundamental point).
- The changes in the extent of the tropical zone, as it expands and shrinks between AAs as they move towards and away from each other is a function of their size. The freedom of movement of tropical fluxes and discontinuities is closely controlled by this ever-changing configuration. The ME shifts more or less

markedly, and readily or reluctantly deploys its vertical, pluviogenic structures to greater or lesser effect. Here is the explanation of the paradox involving strong tropical fluxes accompanying an attenuated amplitude in the migration of the ME, and, conversely, weaker tropical fluxes accompanying an increased amplitude in the migration of the ME, associated with widespread trans-equatorial monsoon circulation (Figure 34).

- Changes in the speed of ocean circulation, especially near the surface, varying with their impulsion by MPHs, which direct ocean surface currents and also the intensity of upwellings, while cold polar air affects the intensity of density currents.

These, then, are the principal points involving MPHs and general circulation as applied to palaeoclimatology (Leroux, 1983, 1993a, 1994c, 1998, 2001).

9.5 MODELS AND PALEOCLIMATOLOGY

9.5.1 The modelling contributions to the reconstruction of past climates

- Possible changes in general circulation are always being discussed, for example, by the IPCC: '*current evidence indicates that very rapid and large temperature changes, generally associated with changes in oceanic and atmospheric circulation, occurred during the last glacial period and during the last deglaciation*' (IPCC, 2001, *Sc. Basis*, chapter 2). But these changes in atmospheric circulation are not specified. Indeed, how could they be, since the models do not integrate the general circulation and are therefore not able to take it into account? Consequently, phenomena are generally considered *in situ*, as is evident from a previously quoted precondition: '*in a glacial period, precipitation diminishes ...*' (Paillard and Parrenin, 2004). This 'relationship' is nothing more than the application, at the level of the basic cell of the model, of the simplist principle 'T/R': 'cold = less evaporation = less rain' (cf. Chapter 7); and it does nothing to advance the arguments beyond the hypotheses already formulated by Adhémar or Milankovitch.
- The important part played by the polar regions in driving the aerological dynamic also goes unrecognised. For example, Figure 31 shows the curve for insolation in June at latitude 65°N (Petit *et al.*, 1999). Why choose 65°N if we are discussing the Antarctic? Is it simply because '*according to Milankovitch, it is the high latitudes of the northern hemisphere which ... are the most susceptible to changes in insolation*' (Paillard and Parrenin, 2004), and because '*the amount of sunshine received at 65°N is traditionally invoked as the preponderant factor in the forcing leading to glacial conditions*' (`www.lgge`, 2004)? Are we, then, dispensing with individualised meteorological hemispheres, and is there no ME (a fundamental discontinuity)? Is circulation in the Antarctic perhaps controlled from the North Pole(!), and not by the aerological dynamic of the three southern units of circulation which meet at the South Pole (cf. Figure 26)?

- And what of the classic undulatory concept of 'westerly waves', giving priority to the upper layers, and supposed to control phenomena in the lower layers (cf. Chapter 11)? This concept led the COHMAP group (1988) to suppose that the westerly jet might be divided into two by polar relief, with one branch passing to the north of the Arctic Circle, and the other to the south of the *inlandsis*. What physical law might explain the presence of a jet near the pole, when in winter or during cold episodes the westerly jet is strongly shifted towards the tropics (cf. Figure 33)? How, in physical terms, can the *inlandsis* divert the main branch of a jet situated very high above it, and what is the meteorological significance for the climate of North America of the existence of a jet in the upper layers to the south of the *inlandsis*?
- Do we still suppose that a '*stable anticyclone ... with easterly surface winds*' (COHMAP, 1988) could maintain its existence above the frozen polar relief? This kind of stability is obviously inconceivable: no such anticyclone could possibly maintain itself there, given the extremely mobile conditions observed above Greenland and the Antarctic, with violent catabatic winds accompanying the departure of every MPH (Figures 22 and 23).
- What has become of the classic concept of subtropical anticyclones, seen as a 'barrier', or an 'anticyclonic rampart', allowing or denying the passage of meridional exchanges? This is in fact tantamount to saying that '*frequent meridional exchanges imply a weakening of anticyclones*', and conversely that '*less frequent meridional exchanges imply a reinforcement of tropical anticyclonic cells*'. This leads to the conclusion that, 18 and 12 kyr ago, the Hawaiian AA was '*a weaker subtropical high*' but, by contrast, 9 and 6 kyr ago it was '*a stronger subtropical high*' (COHMAP, 1988). Unfortunately, this is quite the opposite of what is actually observed in the real dynamic (cf. Figures 33 and 35). Knowing as we do that 'subtropical anticyclones' are still very much with us in some minds, and are thought to be associated with subsidence (cf. Chapter 7), it is easy to understand why the link between temperate and tropical phenomena is still (very) difficult to establish.
- And what might we say about the association: '*stronger Aleutian low*'/'*weaker subtropical high*', established for the Pacific of 18 kyr ago (COHMAP, 1988)? This 'relationship', which contradicts reality, highlights a lack of knowledge about the true nature of the dynamic link between these two entities. This is still the case, not only for the North Pacific, but also for the North Atlantic (cf. Chapters 12 and 13).

9.5.2 The present state of the question

Where do we stand, 10 years after the COHMAP conclusions? The article by Bartlein *et al.* (1998), on the subject of the simulation of the palaeoclimate of North America over the last 21 kyr, is one of many examples which offer rapid insights. Several of the points covered above are mentioned:

- '*displacement of the jet stream by the Laurentide Ice Sheet to the south*', even though we still do not know how the *inlandsis* manages to divert the jet, and what impact this diversion might have upon glaciation;

- '*generation of a glacial anticyclone over the ice sheet*', the 'glacial anticyclone' which, it will be remembered, cannot maintain its existence, and which (if it did exist) would prevent advection onto the *inlandsis* of precipitable potential moving up from the south (given the direction of rotation of the associated winds);
- '*strengthening of the eastern Pacific and Bermuda high-pressure systems in summer as the ice sheet decreased in size*': high-pressure systems growing stronger, even though they should quite obviously be doing the opposite: the vigour of anticyclonic cells actually diminishes as the *inlandsis* shrinks during periods of warming (Chapter 13); and
- '*development of a heat low at the surface*' in the south-western USA, a supposed thermal low which does not exist nowadays on this continent, as this type of depression is a normal feature of tropical climes!

No mention is however made of the actual circulation of the lower layers, and neither is anything included on the importance of relief (indeed considerable in this context), on the origin and transportation of the water which will form the *inlandsis*, or on the mechanisms of the weather and their evolution. Moreover, in spite of the fact that modelling prides itself on using no concepts, at least two accepted concepts appear here: the westerly jet and subtropical anticyclonic cells!

So it cannot really be claimed that things have moved on: this is far from being the case. Models still display the same faults inherent in their conception:

- the consideration of phenomena *in situ*;
- simplistic relationships within their basic cells (e.g., temperature/rain, temperature/pressure);
- statistical correlations being equated with physical links;
- averaged and static vision;
- reference to outmoded concepts;
- a lack of individualisation of the principal agents of the dynamic, of circulation, and of transfers (especially transfers of water in the atmosphere);
- a lack of appreciation of the real processes of weather; and
- a faulty perception of the *ensemble* of phenomena acting within a well defined aerological framework.

Even if we disregard manifest errors, it would therefore seem that no decisive progress has been made by numerical models in the explanation of past climate changes. Bard (2004) echoed this sentiment when writing about climatic transitions: '*In spite of the use of ever more sophisticated models in recent years to study the amplitude, duration, and initial conditions of these sudden transitions, these questions remain open, and debate among the modellers is lively*'. But is it possible to picture reality, if no account is taken of that reality itself as one proceeds, and still avoid fundamental contradiction (inevitable anyway)? Can one keep falling back on ever more remote 'teleconnections', or evade the issue by seeking to show that infinitesimal density differences in seawater are able to trigger climatic catastrophes (Chapter 14)?

It is manifestly impossible to make progress without restoring phenomena to their rightful places within the context of the general circulation of the atmosphere, on condition that we impose a coherent schema equally applicable to phenomena both past and present.

9.6 GLACIATION AND DEGLACIATION

In 1990 Duplessy and Morel reminded us that '*the onset of glaciation still wears its cloak of mystery*'. Nothing has changed: the Laboratory of Glaciology in Grenoble added its judgment: '*How does the Earth move into its periods of glaciation? This question has interested the scientific community for many years, but it remains an enigmatic one*' (`www.lgge`, 2004). What is more, the question will continue to inhabit the realms of magic rather than science if we persist in imagining, a century after Milankovitch, that '*when insolation is diminished in summer, snow which has fallen in the winter lies unmelted, and accumulates year after year ... this leads to the building of great ice caps*' (Paillard and Parrenin, 2004)! How long would it take to build such an ice cap, even if the suggested mechanism were capable of doing it? How does such an ice cap come to be, especially if, as is supposed, 'precipitation diminishes' during cold periods?

9.6.1 Onset of glaciation?

The question – and it is a crucial one – of the origin of the water which forms the *inlandsis* is often debated, but without consensus. An example is that of the Barents Sea ice sheet, an extension of the Fennoscandian *inlandsis* stretching to the Svalbard archipelago. It was 3,000 m thick, and formed relatively recently, originating about 25 kyr ago and reaching its maximum size 20 kyr ago (Siegert, 1997). With others, Hebbeln *et al.* (1994) expressed their wonder: '*Such a rapid growth of a large ice sheet requires significant amounts of moisture, but the origin of this moisture has been unclear*'. This origin has been sought in the vicinity of the *inlandsis*, but since the Norwegian Sea is almost always covered in ice, the authors suggest that '*seasonally ice-free waters were an important regional moisture source for the Barents Sea ice sheet*'. Is this proposed local summertime source really sufficient, given the poor evaporation rate in cold air, especially deep within a period of glaciation? An accumulation of this depth, in such a short time, cannot be countenanced unless there is a powerful advection of large amounts of precipitable water from far to the south, channelled in this case north-eastwards between the ice sheets of Greenland and Fennoscandia (cf. Figure 41, p. 198).

Khodri *et al.* (2001) also sought the cause in the ocean (thermohaline circulation), and, always like Milankovitch, in the '*persistence of snow in summer*'. Their model also suggests that '*this enhanced the equator-to-pole thermal gradient in summer, particularly over the North Atlantic where subtropical latitudes are warmer and high latitudes colder*'. Here we have an enhancement of the '*equator-to-pole thermal gradient*' by the authors, though it should be a 'pole-to-equator thermal gradient'. A pressure gradient is set up that decreases as one goes further south, thereby preventing the establishment of any northward flux from the south! And such a transfer would still have to be established directly over such a great distance ... elementary! It is particularly surprising to read, a little further on: '*the summer increase of the equator-to-pole surface temperature gradient acts to enhance the annual northward transport of moisture by the atmosphere*' (sic!). Now we see a transfer – northwards(!) – which is supposed to take place in competition with the pressure field! This transfer, reckoned to furnish '*optimal conditions for delivering snow*' is unfortunately impossible, since the reasoning behind it is completely wrong, turning on its head the very simple principle of the force of the gradient! And, as if this were not enough, the authors also point out that (according to observations) '*precipitation is therefore enhanced over land*', but they associate the increased precipitation with '*a decreased Icelandic low ... (and with) weaker cyclonic activity*', which obviously runs counter to meteorological reality! This article, which oddly appeared in *Nature*, is symptomatic of efforts to simulate glaciation via models, which efforts serve to reveal the faults (*sui generis*) of the models themselves, to which we can add (as in the present case) a surprising lack of knowledge of the rudiments of meteorology.

9.6.2 Dynamical processes of glaciation

The building up of mountains of ice more than 3,000 m high requires the intervention of powerful meteorological phenomena, continually acting over thousands of years. The importation of extraordinary volumes of water, as water vapour, and then clouds, and finally as abundant falls of snow, which piles up and is stored as ice, results from **a very great acceleration of meridional exchanges**. The only climatic agents capable of organising this intense, repeated transfer, by reason of their dimensions, energy, and regenerative power, are the MPHs, and more specifically the cyclonic circulation engendered within the low-pressure corridors formed at their leading edges, bringing precipitable water potential gathered up in tropical latitudes and transporting it towards the polar regions. Climate models do not take MPHs into account, and therefore cannot simulate the dynamic conditions responsible for the formation of the *inlandsis*, as Duplessy and Morel (1990) point out in our introductory quotation.

Conditions pertaining to this transportation of air across the North Atlantic may be investigated with reference to the meteorological situations causing snowstorms along the eastern coasts of the USA and Canada (Kocin and Uccellini, 1990), and blizzards such as those of 1888 (Kocin, 1988) and 1993 (Forbes *et al.*, 1993), both labelled 'blizzard of the century'. Similarly, conditions have been very severe during many recent winters (January 2004 being a prime example). For instance, in Figure 36, we see the position on 5 February 1996 of the immense MPH that brought a record-breaking cold spell to North America. Temperatures fell to −51°C in Chicago, and to −15°C in Louisiana and Florida. On its leading edge, this MPH

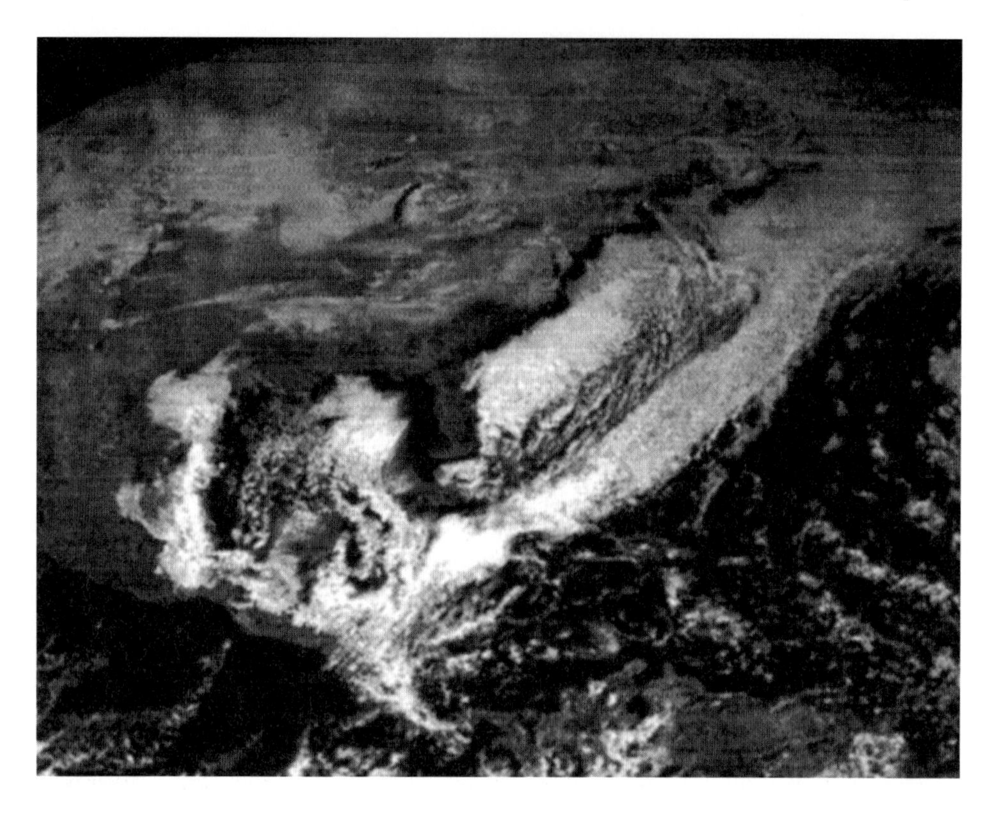

Figure 36. 5 February 1996 (18h48 UT, *GOES E*, visible). A vast MPH, reinforced by a second one which has already reached the Great Lakes, covers almost the entire American continent (east of the Rockies), the Gulf of Mexico, and the near Atlantic Ocean. It causes a strong and direct transfer, on its leading edge, of tropical energy towards the pole, from the Caribbean Sea to Greenland, the Norwegian Sea, and beyond.

caused a particularly intense, direct, and rapid transfer, restricted in the main to the lower layers, of perceptible and latent heat from the tropics in the direction of the north-eastern Atlantic.

The lower level of insolation in polar latitudes, a summer phenomenon (obviously not a winter one, strictly speaking, if the Sun does not rise) leads to the constant renewal of this type of situation during all seasons. The thermal deficit is accentuated, meaning more powerful MPHs: so the onset of glaciation takes place in the context of a (more and more) rapid mode of circulation (Figure 33). MPHs venture deep into the tropical zone to pick up their precipitable water potential, which is well conserved beneath trade inversions with their reinforced 'unproductivity'. The tropical zone is generally left short of precipitable water, and rainfall levels progressively decrease. However, the extra-tropical zones benefit from this water, though not in their totality, certain areas being favoured because of the trajectories found within the units of circulation; these areas preferentially harbour low-pressure areas at the north-eastern edges of the units (Figure 26).

The lower level of insolation in polar latitudes also means that the altitude at which precipitation advected by MPHs freezes becomes lower and lower – this is quite low down in these latitudes anyway, even in summer. As the ice massif gains height, dynamic updraft will be reinforced and the rain–snow boundary is reached sooner, increasing the proportion of snowy precipitation. However, above the 'optimum snowfall' level, the amount of precipitation is diminished. The accumulated ice, with its higher albedo and its radiative capability, encourages the cooling caused by the lower insolation, and increases the energy of MPHs, which are vigorously ejected. This in turn promotes the return of moist and warmer air associated with low pressure, and the intensity of the transfers increases. All the while, the ice stored within the *inlandsis* is depleting the meridional exchanges; as ever greater quantities of their immediately available water are extracted, sea levels progressively fall.

9.7 ANTARCTIC GLACIATION

The question of when the ice of the Antarctic began to form is not an easy one to answer, since '*the main mass of the Antarctic ice cap has remained unmelted ever since it was formed, 60 million years ago*' (Postel-Vinay, 2002) It may have happened even earlier, in aerological conditions very different from today's. Also, the extent of the ice sheet has seen little variation, not venturing far from the land mass beneath it during cold periods, which were characterised essentially by an increase in volume. The dynamic of the exchanges to and from the most southerly latitudes is fairly easy to schematise, since the Antarctic continent is centred upon the pole, and is isolated far from other continents, except in the case of the southern tip of South America. This geographical configuration, and the easy passage of MPHs with very great initial energy in both their low-pressure corridors and their associated closed Southern Ocean lows, form part of a regular evolution in tandem with cosmic parameters, as shown on Figure 31. However, the zonal band of heavy southern rains (on average greater than 1,000 mm annually) lies currently between about 40°S and 60°S, along the corridor into which the lows associated with MPHs move. Consequently, most of the rain falls into the sea, while the Antarctic continent, situated mostly within the 70°S circle, receives less than 300 mm annually, except in the north of the Antarctic Peninsula, where precipitation may attain values of 500 mm. Values decrease towards the interior of the continent, less so in western Antarctica, and more rapidly with altitude to the east of the Antarctic Cordillera; over most of the Antarctic Dome, values stay below 100 mm (rain equivalent).

The dynamic of the weather over the Antarctic ice cap does not exhibit uniform behaviour.

- The eastern Antarctic (to the east of the Antarctic Cordillera), which contains 85% of the volume of the ice cap, reaches a height of more than 4,800 m. '*There has not been any important modification in the thickness of the ice since the end of the last glaciation*' (Lorius, 1983). This part of the *inlandsis* has seen little in the way of variations on the palaeoclimatic scale (Bindschadler *et al.*, 1998). A kind of 'glacial immunity' is conferred upon it by its geographical position (and especially its great distance from South Africa and Australia), its latitude, and its altitude. Its altitude, with very low temperatures in the central part of the Antarctic Dome, assures a snow coefficient of 100%, but, as at Vostok, only moderate snowfalls of the order of 50 mm per year are possible, which are unrepresentative of the general pluviometric pattern.

- The western Antarctic, sometimes referred to as the West Antarctic Ice Sheet (WAIS), having attained its maximum volume about 20 kyr ago, is thought to have lost two-thirds of its ice mass during the last deglaciation (Bindschadler *et al.*, 1998). Most of the mass of this ice sheet, with its much more modest relief, sits on a rocky base, which is below sea level. The ice projecting beyond the land forms shelves, such as those found in the Weddell Sea and the Ross Sea. This part of the ice cap, which is the most vulnerable (because the ice is floating on water), is sensitive to variations in insolation, and grows during cold periods, with their more abundant precipitation, and shrinks during warm periods, when melting takes place and precipitation is less abundant. It would seem, however, that this area of ice has held its own during the last three interglacial periods: 125 kyr ago, when conditions were warmer than during the HCO; 220 kyr ago; and 320 kyr ago, with a possible fourth period 420 kyr ago (Postel-Vinay, 2002). This last period is thought to have been the warmest interglacial of the last 500,000 years.

9.7.1 The behaviour of the West Antarctic Ice Sheet

The unusual behaviour of the WAIS is worth noting. The dynamic here, typical of a rapid mode of circulation (cf. Figure 33), is first of all due to the lower altitude of the ice sheet, which makes it more vulnerable to variations in insolation and variations in the rain/snow boundary. But the principal factor is its geographical position, and specifically the position of the Antarctic Peninsula, which is really a prolongation of the Andean chain. From 55°S northwards, the Andes present a formidable barrier to the passage of MPHs. A large fraction of southern MPHs are therefore regularly channelled into the Pacific, and this stream of cold air moving northwards is responsible not only for the vigorous nature of the so-called 'Easter Island' AA, but also for the permanent spilling of the ME northwards across the geographical equator, and over the eastern Pacific (Figure 26).

Before the MPHs encounter the Andes, the progressive closing of the low-pressure corridor greatly accelerates the stormy north–north-westerly flux (known as the *Norte*, although it is a warm flux). This flux, originating in the Pacific, is driven energetically southwards along the leading edge of the MPH and also by the mountain barrier (Figure 37). The Antarctic Peninsula therefore lies right on the path preferentially taken by warm air advected towards the South Pole. The MPHs and/or fractions of MPHs which pass further to the south of the Andes, and

Figure 37. 19 January 2004 (*GOES E*, visible). The leading edge of an MPH encounters the southern part of the Andean Cordillera. The cyclonic warm airflow is strongly channelled southwards, between the MPH and the Cordillera. Then, after the encounter of the MPH with the relief, the cold air of MPH is channelled northwards.

continue their westward paths, also briefly cause warm Atlantic air to be diverted towards the Antarctic, especially into the Weddell Gulf, the far end of which is surrounded by high relief (Figure 38).

The western part of the Antarctic, and in particular the Peninsula and the Weddell Sea, therefore have a specific thermal pattern determined by the variations of polar insolation. An enhanced thermal deficit over the Antarctic ice cap increases the power of the MPHs which stream away from it, an effect heightened in the case of those MPHs which slide off its (more extensive and higher) eastern side. As a consequence, the volume of the returning southwards flux is increased, 'warming' the area of the Antarctic Peninsula. This increased delivery of warmer air, of precipitable water and of actual precipitation by the low-pressure area associated with each MPH (rapid scenario) helps to build the ice sheet (glaciation). Conversely, in the case of the slow scenario, the deceleration of the circulation causes less water to be delivered and encourages the melting of the ice cap (deglaciation), as a function of the variation of the level of freezing.

Figure 39 illustrates the variations in summer polar insolation during the final part of the last glacial cycle. The figure shows that insolation in the vicinity of the North and South Poles does not follow exactly the same pattern. Although evolution generally proceeds along the same lines, there are discrepancies in the case of high latitudes (White and Steig, 1998; Steig, 2001). In the vicinity of 85°S, minimum insolation occurred about 30 kyr ago, and the volume of ice was at its greatest

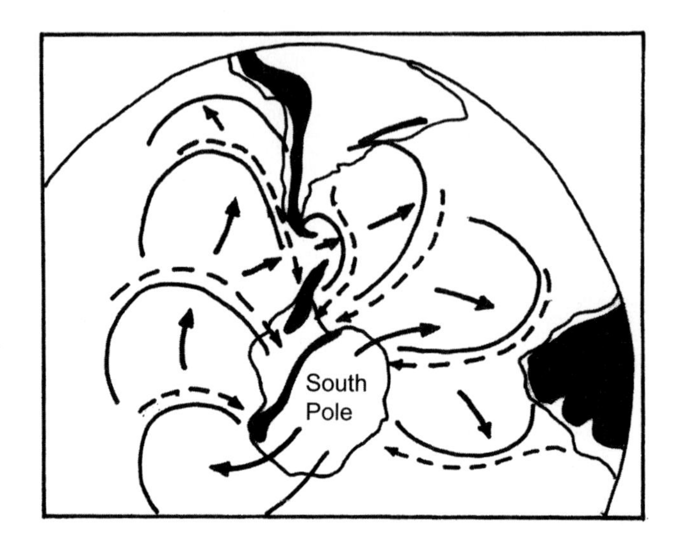

Figure 38. Dispersion of MPHs by the Antarctic Dome, and the dynamical influence of the Andean Cordillera on warm air advection towards western Antarctica, mainly the Antarctic Peninsula.

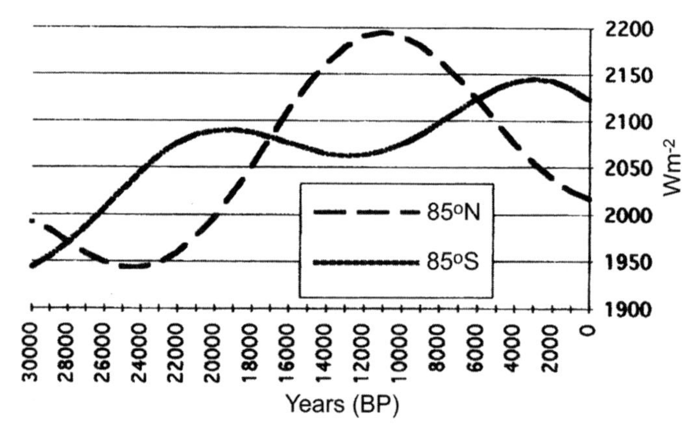

Figure 39. Summer polar insolation at 85°N and 85°S during the last 30 ky (sum of monthly averages of daily insolation in W/m^2).
From Davis (1988).

about 20 kyr ago, before a maximum insolation between 20 kyr and 3 kyr brought about rapid deglaciation (Figure 31(d)), mainly of the WAIS. Since about 2,000 years ago, there has been a slow decline in insolation in southern polar latitudes.

The palaeoclimatic dynamic of the Antarctic ice cap should serve as an example of the way in which past climates can be used to facilitate understanding of present-day phenomena (and vice versa). Unfortunately, this does not always seem to be the case, especially when a section of a glacier or an ice shelf breaks off and drifts away.

For example, before stating somewhat rashly that an iceberg originating from the Larsen ice shelf (March 2002) was '*the most visible sign of the warming of our planet*' (in *Le Monde*), the so-called 'scientific' journalists, some 'scientists', and the IPCC might have benefited from a close study of the dynamic of the past. We shall return to these matters in Chapter 14.

9.8 GLACIATION IN THE NORTH

Although the glaciations in the south and in the north were roughly synchronous, they were certainly not identical in character. Neither was their geographical spread the same, since one was limited to the Antarctic continent, while the other extended widely into continents near the Arctic, but the *inlandsis* did not cover the pole itself (cf. Figure 41, p. 198). Their respective dynamical characters were also different: the southern dynamic was particularly simple, while encounters between MPHs and mountain masses (of both rock and ice) created a more complex situation within the northern dynamic.

9.8.1 The Greenland inlandsis

There have certainly been variations in the volume of the *inlandsis* of Greenland, but in spite of its latitude, it has been preserved during the Quaternary period. So Greenland has always featured in the *décor* of air circulation in the North Atlantic aerological space, and is still pivotal. Since this island, lying between 82°N and 61°N, is quite 'off centre', relatively speaking, *vis-à-vis* the North Pole, its ice should by now have melted, not least in its southern areas which are at the same latitude as Scandinavia. The reason for the longevity of the ice is not a directly climatic one, but rather an orographical one. The ice is spread across a basin surrounded by mountains. These mountains:

- on the one hand prevent direct contact between the ice and the sea, thereby preventing large-scale 'calving'; and
- on the other hand, because of the altitude of the ice surface (on average, above 2,000 m), ensure temperatures low enough to preserve most of the ice, with a snow coefficient of almost 100%.

Greenland's great extension in latitude means that, in the south, precipitation exceeds 1,000 mm per year, while in the north, the mean annual figure is only about 100 mm, a roughly south–north decline, matching in direction that of the height of the snow–rain boundary. Consequently, climatic conditions are not homogeneous, and ice core samples from different parts of Greenland are not (in terms of latitude) immediately comparable in significance. Neither are they comparable *a fortiori* to the Vostok samples, since there is a general synchronicity but no precise simultaneity (White and Steig, 1998).

9.8.2 Dynamics of the northern glaciation

In the northern hemisphere, the presence of land masses and the disposition of high relief create individualised entities of circulation in the lower layers (Figure 26). The trajectories of the MPHs within these entities determine the directions of diverted cyclonic fluxes, and thereby the locations where precipitation and possible accumulation of ice will be heaviest. In the North Pacific, the Rockies channel the precipitable water potential towards Alaska (Figure 40). In the North Atlantic, air moving up from the south is directed into Canada, Greenland, the Norwegian Sea, and Scandinavia (Figure 36). In northern Eurasia, the limit of the ice of western Siberia (Thiede and Mangerud, 1999) is determined by the impoverishment of the precipitable water

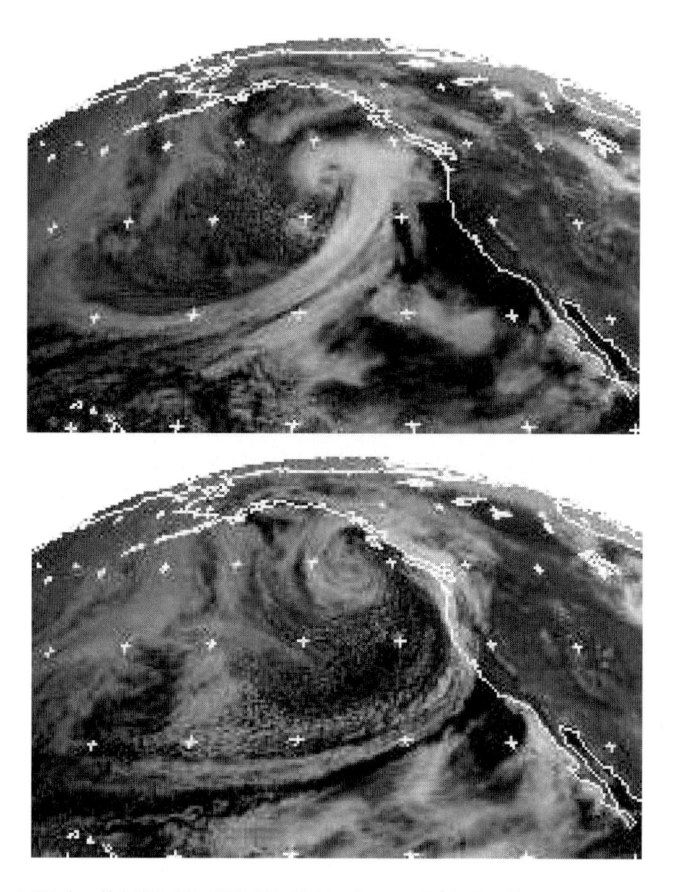

Figure 40. (*top*) 26 April 2000 (*GOES W*, 18 h). (*bottom*) 27 April 2000 (*GOES W*, 18 h). An MPH, moving across the Pacific Ocean from the Bering Strait area, creates on its leading edge a strong cyclonic, warm, and wet low-level airstream, towards the western side of the Rockies. The leading edge of the MPH runs southwards, while the connected low runs towards the Gulf of Alaska.

potential reaching it from the Atlantic by way of the Norwegian Sea and western Europe, and from the Mediterranean across central Europe. The impoverishment in the latter case is due to the presence of the Caucasus–Zagros–Himalaya mountain barrier and the 'continentalisation' of the air that occurs after the relief has been crossed. Variations in polar insolation determine the vigour of MPHs and the resulting intensity of this circulation (cf. Figures 33 and 35).

The Eemian interglacial period, which lasted from 130 kyr ago until 120 kyr ago (calendar years), was, compared with the present era, a time of greater summer insolation (13% more) in northern high latitudes, and higher temperatures (at least 2°C higher). The Eemian was also warmer than the HCO (around 6 kyr ago). The ensuing deterioration in the climate unleashed the most recent (Würm) glacial period, which lasted for about 100,000 years and was marked by '*a succession of advances and retreats of the icefields*' (Duplessy and Ruddiman, 1984). This period was at its height about 20 kyr ago. About 115 kyr ago, insolation in northern high latitudes (with northern summer at Earth's aphelion) was 9% less than it is today, and the ensuing lower temperature was the first act (Cortijo *et al.*, 1994). MPHs were thereby strengthened, resulting in an intensification of the transportation of tropical energy towards the more favoured areas of each unit of circulation. A consequence in Scandinavia, for example, was a rise in air temperature (Mangerud, 1991), and sea surface temperatures were 1 or 2°C higher than today's (Ruddiman and McIntyre, 1979).

Then the weather became colder, and more and more violent, and there was abundant precipitation at high latitudes, increasing in volume as the lowered snowline encouraged more and more snowy precipitation. The ice cover already present in Greenland and on Ellesmere Island spread across Baffin Land and the Labrador plateaux. Because of the trajectory (similar to today's) preferentially followed by MPHs across North America between the Rockies and Greenland, the earliest invasion of the ice took place across eastern Canada and Greenland (where advections of tropical air came to a halt). At the same time, intense streams of warm, moist Pacific air along the Rockies towards Alaska led to the amassing of ice on those mountains (Figure 40). Northern Eurasia was invaded later, the ice spreading from the Scandinavian highlands. Cooling in the polar regions was progressively amplified by the increased albedo of the ice, enhancing the radiative effects of reduced insolation. Slowly, the ice of the *inlandsis* became thicker, and gradually, the channel between the Canadian Arctic and the Atlantic was closed. MPHs slid both northwards, into the Arctic, and southwards, into the Atlantic, from these new, impressive ice mountains, which reached heights of 3,800 m in Canada, and 2,500 m in Scandinavia, northern Russia, and western Siberia.

9.8.3 Glacial relief and MPHs

The slow build-up of these imposing mountains of ice gradually altered patterns of circulation, and especially the directions in which MPHs moved out. The narrowing of the corridor between the Laurentide ice cap and the Rocky Mountains caused a progressive acceleration of the cold air of MPHs moving from the Arctic towards the

Figure 41. Glacial topography and dynamics of MPHs during the Last Glacial Maximum (LGM).

Gulf of Mexico, until the corridor was sealed. Other passages also became blocked, the first being that between Greenland and Baffin Land (via the arc of Ellesmere), and later, the gap between Scandinavia and Scotland: an almost continuous barrier was erected between the Arctic and the Atlantic. The ice attained its greatest mass around 20 kyr ago, or approximately 4,000–5,000 years after the minimum of insolation which occurred at latitude 85°N (about 24 kyr ago, Figure 39). At the time of maximum glaciation, the Arctic Basin was divided from the Atlantic by an impassable 'chain' of ice mountains, with only one narrow corridor left, opening into the Norwegian Sea (Figure 41): powerful MPHs, truly 'polar' in nature and therefore very cold, moved out across eastern Siberia, China, and the Pacific. In the case of the Pacific MPHs, southerly advections on their leading edges resulted in a much greater accumulation of ice along the Rockies and on the western *inlandsis* (Figure 40). In America itself, the ice spread beyond the Great Lakes, as far south as latitude 40°N (the latitude of Valencia, in Spain, or Naples, in Italy). Just as happens nowadays in Greenland, 'American' MPHs, in this case not really 'polar', moved off the icy highlands, which formed a ridge at about 50°N (Figure 41).

Cooling increased the pressure of MPHs upon the surface of the sea, accelerated the winds, enlarged the icefields and increased the density of cold waters: this could only lead to an intensification in ocean circulation. As evidence of this, we may cite the acceleration of the cold, driving Canaries and California Currents, and the increased vigour of their associated upwellings. In turn, the Alaskan Current, the Gulf Stream, and the North Atlantic Drift/Norwegian Current (all warm currents) were necessarily accelerated. As happened with warm air during the onset of glaciation, warm waters were also vigorously fed into the Gulf of Alaska and towards the Norwegian Sea, at least early on, before a general cooling took hold with a considerable extension of the ice pack. This intensification of circulation, the cold in high latitudes and the proximity of the *inlandsis* and the icefields, all enhanced the absorption of carbon dioxide by dense currents. The lower CO_2 concentrations (simple consequence) were accompanied (in time) by lower temperatures.

9.9 DEGLACIATION IN THE NORTH

Insolation in high northern latitudes gradually improved from 22 kyr ago, increasing by 13% to reach its culmination at about 11 kyr ago at 85°N (Figure 39). Around 10 kyr ago, a remarkable situation existed (cf. Chapter 6): simultaneously, northern hemisphere summer occurred at the Earth's perihelion (11 kyr) and the Earth's axis was strongly tilted (9 kyr). There was abundant precipitation during the first phase of deglaciation, associated with still vigorous MPHs. Warming elevated the rain–snow boundary so that precipitation was increasingly in liquid form, and, as every mountain dweller and winter sports enthusiast knows, snow and ice lose the protection of their albedo when rained upon, and melt very rapidly as the comparatively warmer water flows across them. As the 'American' MPHs departing from the *inlandsis* gradually lost their strength, the intensity of meridional exchanges was correspondingly slowed, as the water potential advection, bringing about a progressive decrease in rainfall and cloudiness, as insolation increased. The thickness of the *inlandsis* rapidly decreased – melting 10 times faster than it had formed. However, deglaciation was not a steady process. It was interrupted by intervals of (sometimes severe) cold, as insolation increased. We can mention the two main intervals, the first between 12.7 and 11.5 kyr, and the other between 8.4 and 8.0 kyr ago, and there were other more short-lived events. As a consequence of glacial inertia, the most favourable period of the HCO did not occur until 6 kyr ago (i.e., 5,000 years after the maximum of insolation) (Figure 39). Northern polar insolation has been decreasing since 11,000 years ago, and at 85°N, its value is lower today than it was at the beginning of deglaciation.

The retreat of the ice caps created new opportunities for the formation and movement of MPHs. These conditions throw light upon the main fluctuations of deglaciation associated with increased insolation. As the height of the ice mass decreased, catabatic winds were slowed and 'American' MPHs became less energetic (and relatively less cold). Conversely, however, corridors appeared one by one through the glacial relief, renewing communication between the Arctic and the Atlantic, and allowing colder Arctic MPHs to move across America (advected cooling). Access to the Norwegian Sea (Figure 41) increased between 17 and 15 kyr ago, as the ice retreated between Svalbard and Scandinavia, and a further corridor was open between the Scottish and the Scandinavian ice caps (Siegert, 1997). The so-called 'Scandinavian' trajectory followed by MPHs now gave direct access to Europe, and probably contributed to the first cold episode of the Older Dryas.

9.9.1 (Some) severe cold returns

From about 15 kyr ago, a passage began to form between the glaciers of the Rockies and the Laurentide ice field (see `ngdc.noaa.gov/paleo/pollen`, 2004). Through this corridor, which widened only very slowly (it was still a narrow defile about 14 kyr ago), Arctic MPHs were soon vigorously streaming southwards past the Rockies. These MPHs, colder than their 'American' ice field counterparts, brought on the cold episode of the Younger Dryas (12.7 kyr to 11.5 kyr ago), abundant evidence of which is to be found '*from regions in and around the North Atlantic*' (Rodbell, 2000). This rapid return to cold conditions even as insolation was reaching its maximum (Figure 39) can be explained only in dynamical terms. The coming of these powerful Arctic MPHs, channelled along the eastern side of the Rockies and towards the Gulf of Mexico, saw a short-lived return of the glacial dynamic (i.e., an upsurge in the importation of water from the tropics and a return to very rainy conditions). This was a warm, localised episode, followed by a further cold period when the Laurentide ice field briefly advanced. Rainfall during this warmer episode increased by 50%, and the weather was violent in character; the temperature in southern Greenland rose by 7°C (Dansgaard *et al.*, 1989) as warm air was advected from the south. A similar situation occurred in Canada's maritime provinces (Mott *et al.*, 1986). In the interior of Greenland, the accumulation of snow had doubled by the end of this episode (Alley *et al.*, 1993). This warming was succeeded by marked cooling across North America, mainly to the south of the Laurentide ice field, and in Europe, where glaciers began to advance again (Anderson, 1997). In northern Africa, partly influenced by the North Atlantic aerological unit, the Younger Dryas spelled the end of the wet 'wild Nile' episode, and dry conditions briefly reasserted themselves (Leroux, 1994c).

The opening up of the corridor allowing some Arctic air to flow across America also gave some relief from the cold to Asia, and explains the time lag, which is often mentioned, between Atlantic and Pacific phenomena. In central China, the onset of the Younger Dryas saw an 'abrupt reversal' of conditions, and put an end to the accelerated deposition of *loess* brought from the moraines along the leading edge of the Eurasian *inlandsis*, which was in rapid retreat, melting back to Scandinavia by 11 kyr ago. The *loess* was channelled preferentially towards the Gobi (as the 'yellow wind') through the sill of Dzungaria (Figures 26 and 41). This wind dynamic gave way to an episode when the deposition of dust was reduced and soils were formed in rainy conditions (Zhisheng *et al.*, 1993; Ding *et al.*, 1998). Less powerful MPHs meant that moist air made its way to the interior, a prelude to the re-establishment of the Chinese summer monsoon.

Conditions of severe cold returned again between 8.4 kyr and 8.0 ky ago. About 10 ky ago, although the corridor along the foot of the Rockies was now well open, there was still a large cap of ice centred on Hudson Bay, though much larger than the Bay itself. By 9 ky ago, this ice still covered the entire Bay, Baffin Land, and the Labrador plateau, but by 8.2 ky ago the cap was breaking up: Hudson Bay was ice-free and substantial deposits persisted only in the Labrador area and Baffin Land. Now, Arctic MPHs were able to move off eastwards, an important consequence being that it now became possible for cold air to flow into the Davis Strait and Baffin Bay. Greenland, the north-eastern Atlantic, and Europe then became rapidly colder (Alley *et al.*, 1997; Von Grafenstein *et al.*, 1998; Barber *et al.*, 1999). In Saharan Africa, the HCO (from 9 ky ago until 6 ky ago) was also interrupted by a short, cooler period, separating the warm, moist Tchadian and Nouakchottian episodes (Leroux, 1994c); conversely, the summer monsoon was re-established in China.

9.9.2 Is the key in the water, or in the air?

In any discussion of the weather, it would seem normal to include a mention of aerological effects, but this has not, however, always been the case in the analysis of the above developments. For example, on the subject of the Younger Dryas: '*There has been a certain consensus around one relatively simple explanation: a considerable inflow of freshwater from an unidentified source*' (Grousset, 2001). Certainly, water from the glacial Lakes Ojibway and Agassiz flowed at this time into Hudson Bay, and thence into the Labrador Sea, but can it therefore be claimed that '*this cooling event was forced by a massive outflow of freshwater from the Hudson Bay*' (Barber *et al.*, 1999)? For how long did this freshwater flow last? And during what proportion of the whole cold episode did this occur? According to Clarke *et al.* (2003), the '*inflow ... might have occurred in less than a year*', and the '*result was a northern hemisphere temperature drop of about 5°C for about 200 years*'! Can water draining from a lake be responsible for such effects (and on such disproportionate scales) for two centuries? Is it on the same scale of phenomena as the climatic consequences which reached even as far as northern Africa? Especially if we consider the 'vehicle' which had such a far-reaching influence: '*A sudden increase of freshwater flux from the waning Laurentide ice sheet reduced sea surface salinity and altered ocean circulation*' (Barber *et al.*, 1999). Can a reduction in salinity, localised and short-lived (though its duration is not specified), be responsible for such climatic consequences?

There is a recent trend in thinking which allots a fundamental role to the thermohaline circulation in the dynamics of the climate. First of all, this ignores the fact that ocean circulation is a *consequence*, driven by the circulation of the air and by density factors. Also, although the density of the water is certainly affected by its salinity, it is also very much dependent upon its temperature. What does this density difference really amount to? Is the slight difference susceptible to the effects of the kind of temperatures encountered at these latitudes? Does this mean that a lack of saltiness does not allow the water to cool, and that it remains 'warm' and stays near the surface? What has become of the cold currents flowing out of the

Arctic Ocean, and especially the East Greenland Current with its polar water? Where, too, is the warm water of the North Atlantic Drift, continually arriving from the south – where has it gone? Also, where did the 'severe' cold spell come from, since we are discussing '*the most abrupt and widespread cold event to have occurred in the past 10,000 years ...*' (Barber *et al.*, 1999)? How did its influence spread as far afield as the *Sahara*? Fortunately, the authors themselves recognise that their '*scenario oversimplifies ocean circulation and climate boundary conclusions*'! That is the least one can say! Let us add moreover that such an interpretation is not shared by everybody: Veum *et al.* (1992) remarked that '*warming apparently took place at times of both strong and weak thermohaline circulation*'. They pointed out particularly: '*we find that rapid changes in thermohaline circulation cannot account for the transient return to a cooler climate during the Younger Dryas episode*'. We shall go further into this later (Chapter 14).

The climatic oscillations which characterised the last glaciation, such as the so-called 'Heinrich Events', the post-glacial cold spells mentioned above, and the 'Dansgaard–Oeschger Events', have been interpreted in many different ways. Proposed causes include: the contribution of freshwater flowing from lakes and rivers, meltwater from the ice or from icebergs, ice surges and earthquakes, etc.; but alongside these 'fresh and salt water solutions' (or rather, instead of them), we find no 'aerial' explanation! This is really too much! **When discussion centres on the weather, it is obviously first and foremost to the aerological dynamic that we should turn**, to try to unravel meteorologic–climatic phenomena.

This omission is symptomatic of a climatology lacking in explicatory concepts: into the gap thus created leap, by default, other disciplines, first among them oceanography. What is needed is a truly climatological approach to the analysis of the dynamic of deglaciation, with reference to variations in insolation at high latitudes, the changing intensity of circulation (MPHs, returning fluxes), the evolution of glacial relief and/or the thermal and aerological consequences of volcanic activity. Such is the approach of Soto (2005), currently working thesis at the Climatology, Risks and Environment Laboratory (LCRE).

9.10 CONCLUSION: THE GREENHOUSE EFFECT DID NOT CONTROL PAST CLIMATE

Evidence left by past climates is of major importance in our attempts to understand the climate of today. Although we have devoted a long chapter to this aspect, it can still be only an incomplete *résumé*, with the amount of research available, and being done, in the field of palaeoclimatology. We have concentrated on high latitudes because of their importance for general circulation, often at the expense of tropical regions (cf. Leroux, 1983, 1994c, 2001). So only the essential ideas have been brought out here, but they have numerous lessons for us, the main ones being as follows.

- The levels of greenhouse gases in the past (which are only estimated), and their present-day levels (which have been directly measured), are **not comparable from the climatic point of view**. It is therefore particularly rash to claim that CO_2 levels, for example, have never been as high as they are today: there is no absolute certainty on this point.
- **The so-called 'greenhouse effect' scenario cannot be fitted into a long-term context.** *We see from past climates that temperature depends on neither CO_2 nor CH_4 levels, since* it *changes first, under the influence of external causes, and greenhouse gas levels change only later.* This increase or decrease in greenhouse gases is therefore a consequence. Greenhouse gas levels have their place in the radiation budget, enhancing or attenuating the greenhouse effect as part of normal, everyday processes, but this cannot be seen as having a fundamental effect: greenhouse gases certainly do not cause climatic evolution.

The climate has not evolved in the past because of greenhouse gases, and there is consequently no scientific reason for it to evolve because of greenhouse gases either now or, a fortiori, in the future.

- Probable scenarios for the future are based upon what we know from the past, which gives clues to their likely characteristics and amplitudes, and the ways in which they change. With this in mind, we can say that recent climatic evolution (and especially that observed in the North Atlantic and North Pacific spaces) shows with no shadow of a doubt that we are now living in **the initial phase of a glaciation**. This is manifesting itself in one region through warming, lower pressures and heavier rains, while at the same time in other regions there is cooling, higher pressures, and less rain ... We shall come back to this in Chapters 12 and 13.
- Applying traditional mechanisms (currently being proposed to explain meteorological phenomena) to palaeoclimatology **does not really allow us to take account of the considerable modifications which have occurred in the past**. This has already been underlined (Duplessy and Morel or LGGE) in explanations of glaciation or deglaciation. Broecker (2000) made much the same point when discussing thermohaline circulation: '*Nobody has yet been able to describe the workings of those mechanisms which, in both the ocean and the atmosphere, are capable of bringing about such changes*'. This means, therefore, that either the behaviour of the atmosphere was different in the past compared with the present (which is absurd), or the proposed mechanisms are in error (which seems much more likely). It is quite obvious that the atmosphere does not behave in a whimsical manner, constantly changing its character according to places, seasons, or epochs, or simply for some bizarre reason, which may vary from analysis to analysis. Consequently, studying the past shows us that the schemas, concepts, and explanations commonly used nowadays in meteorology are not applicable to all scales of space and time: they are therefore marred by errors, or downright wrong.

- Glaciologists and oceanographers search for explanations in water, both fresh and salt; strangely, though, nobody looks to the air for a climato-meteorological explanation. Atmospheric processes, which are the first things to examine in any attempt to explain the weather, are strangely absent from explanations. **Looking to the past confirms therefore that meteorology is now in a conceptual impasse**. Into the yawning gap thus opened swarm experts from other disciplines, who, in spite of their efforts, are not, and cannot be, at the centre of the debate, which is primarily concerned with the atmosphere.
- Far from demonstrating our 'skill at modelling phenomena', the study of past climates rather shows up the **weaknesses** (sui generis) **of the modelling process**, prime examples being the unacceptable confusion between correlation and co-variation, and/or the consideration of phenomena *in situ* (i.e., on the scale of the basic cell). Models do not reproduce the dynamical conditions of either glaciation or deglaciation phenomena observed and verifiable in every detail. They even introduce (sometimes gross) errors! So what significance should we ascribe to predictions made by these same models of 'the climate in 2100', predictions all the more gratuitous because nobody can (or will be able to) verify them?
- Even if some aspects of past climates are still unclear, and ambiguities remain, our study of them shows that **the dynamic of the climate is rigorously organised**. We read about the 'collapse of the climate', the 'breakdown in the climatic machine' or the climate 'turning', but such terms are a perversion of language. The dynamic of the weather and of the climate always follows the same principles, in both the past and the present, and any modifications are not of their nature, but of their intensity. So in order to discuss their 'breakdown', we first of all have to know how they really work: what are the real mechanisms? Before summoning up catastrophist scenarios (such as Bard's '*Can the climate be turning on us?*' (2004)), more suitable for the tabloid press than for scientific debate, we should be pondering the lessons of the past!
- Palaeoclimatology shows us that the concept of the MPH and its application to general circulation, which explains particularly the 'mysterious' mechanism of glaciation, can help us to understand the key mechanisms of the climate within a defined context of circulation. It can be applied on all spatial and temporal scales, and reveals the rigorous organisation of the dynamics of climate.

10

The observational facts: Present temperatures

The fallibility of methods is a valuable reminder of the importance of skepticism in science ... organized and searching skepticism as well as an openness to new ideas are essential to guard against the intrusion of dogma or collective bias into scientific results.

National Academy of Sciences. *On Being a Scientist: Responsible Conduct in Research*, 1995.

The 'global warming' scenario is based upon the postulate that greenhouse gases cause temperatures to rise. This temperature increase is vouched for by the Intergovernmental Panel on Climate Change (IPCC) via the curve of mean global ('reconstituted') temperature anomalies for the period 1860–2000, republished annually by the World Meteorological Organisation (WMO) in their *WMO Statement on the Status of the Global Climate* (`www.wmo.ch`). Since the appearance of its Third Report, the IPCC relies equally upon the 'curve' of (estimated) temperatures for the last millennium to support its postulate. This is the 'hockey stick' curve, so called because of its shape, which, it is claimed, shows that '*the late 20th-century warmth was unprecedented over at least the past millennium*' (IPCC, 2001).

The first of these curves, supposed to provide 'proof positive' of warming, has not demonstrated a climatic significance; and the second curve, which is even less convincing, has aroused considerable debate, to the extent that its scientific integrity has even been called into doubt.

It will therefore be of use to examine these curves, and to ask ourselves about their real climatic significance. In the interests of continuity, given the subject of the previous chapter, let us first consider the evolution of the climate over the last 1,000 years.

10.1 THE EVOLUTION OF TEMPERATURE DURING THE LAST MILLENNIUM

In its First Report (1990), the IPCC published a temperature curve (Figure 42) for the period beginning as long ago as 900 AD. This curve reflects the existence of two

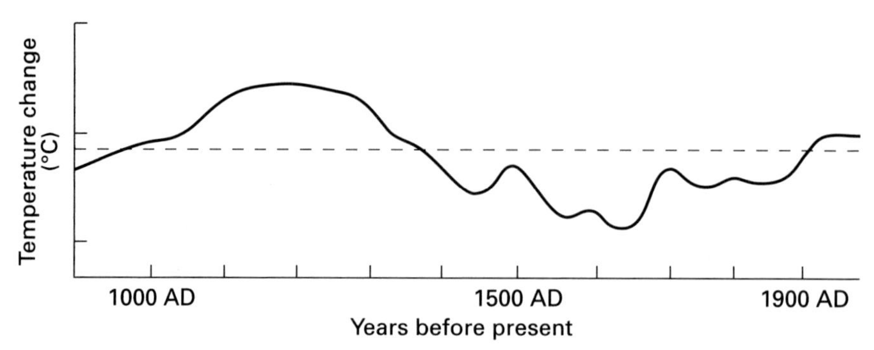

Figure 42. Schematic diagram of temperature variations in the last 1,000 years. Dashed line: conditions near the beginning of the 20th century.
From Folland *et al.* (1990).

important and contrasting periods, the 'Medieval Warm Period' (MWP) and the 'Little Ice Age' (LIA). The evidence of the curve is very clear: the MWP lasted roughly from 1000 to 1300, and was '*exceptionally warm*', especially in western Europe, Iceland, and Greenland. According to Folland *et al.* (1990), '*this period of widespread warmth is notable in that there is no evidence that it was accompanied by an increase of greenhouse gases*'. So the present-day period, which might be said to represent a sort of 'return to normality' after the rigours of the LIA, was preceded by an even warmer period which owed nothing to the presumed 'anthropic greenhouse effect'. This is obviously quite inconvenient for those who would, at any price, put both greenhouse gases and humankind in the dock! Is this why the first curve to call into question the existence of the MWP was so readily adopted by the IPCC?

10.1.1 The 'hockey stick'

In the words of Daly (2001), '*one scientific coup overturned the whole of climate history*'. The curve by Mann *et al.* (1998, 1999), 'miraculous' indeed for the IPCC, did away in one fell swoop with the MWP and the LIA, and replaced them with a more linear trend, very lacking in contrasts (Figure 43). This curve in fact resembles that of CO_2 concentrations in Figure 32, with levels so unvarying that they seem, as has already been pointed out, quite improbable (Chapter 9). However, this time the curve shows a slight downward tendency, which, it is claimed, represents the LIA. For dates before 1900, temperatures are estimated mostly from studies of tree rings (the trees in question being located high up in a region of western North America); corals; ice cores; and historical records. From 1900 onwards, the 'official' IPCC temperature curve is suddenly grafted on (cf. Figure 46, p. 212). Here we have the same procedure as was followed in Figure 32, juxtaposing estimated and measured values for CO_2 concentrations; and the same question has to be asked: 'are the two data series, whether they represent CO_2 or whether they represent temperatures, truly comparable, and can they be fitted together in this way?' This 'apples and oranges' procedure recalls closely that followed by Hansen in 1988, unleashing the

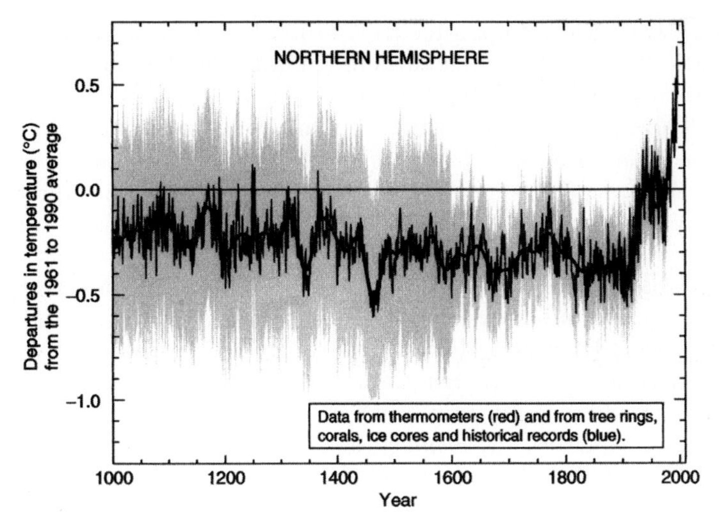

Figure 43. Variations in surface temperature over the last millennium (1000–2000): the so-called 'hockey stick' diagram (the end of the curve, from 1860–2000, titled 'red' on the original figure, comes from observed temperatures: see Figure 46).
After IPCC (2001), from Mann *et al.* (1998, 1999).

'greenhouse panic': the end of his curve (Chapter 2, Figure 4, p. 33) juxtaposed a thermal value established over five months with mean annual temperature values, a procedure devoid of any scientific rigour!

Thus constructed, the 'hockey stick' curve shows (and spectacularly!) that '*the rapidity and duration of warming during the 20th century were much greater than in any of the nine preceding centuries*'. It is a crude trick, which seems to have begotten a mania, but it would seem that it still 'works'! The views of Mann *et al.* (1998) were in fact immediately endorsed by the IPCC, which in so doing quickly forgot its previous reports, and the curve appeared in the Third Report as an additional 'proof' of the exceptional nature of recent climatic evolution. It became '*one of the great propaganda icons of the United Nations climate change machine*' (Corcoran, *Financial Post*, 13 July 2004). Of course, for the IPCC (2001), the MWP now became *ipso facto* less warm ('*... appears to have been less distinct, more moderate in amplitude*'), and similarly, the LIA '*can only be considered as a modest cooling of the northern hemisphere*' (IPCC, 2001). When one considers the habitual and justifiable reluctance of scientists to take on board new ideas, one can only marvel at the speed with which the IPCC 'changed its clothes'. Such haste might even seem suspect. We are certainly no longer moving in the realms of science here!

This 'warming' may well have been without precedent during the last 1,000 years, but this seems not to be long enough for the IPCC! The period in question had to be extended: from the original 600 years (1998), through 1,000 (1999), and finally to two millennia, according to Jones and Mann (2004), who pushed the starting point back by 1,800 years, and even beyond, though, obviously, the same conclusions were reached. And the conclusions will always be the same, as long as the thermal curve used relies upon the trick of mixing incompatible data: one can blithely go back in time as far as one cares to, knowing that the curve will terminate every time with the 'reconstituted' temperatures (Figure 46, p. 212)! And of course, the culprit will always have to be identified, in line with the dogma of the IPCC, in words like: '*modeling and statistical studies indicate that such anomalous warmth cannot be fully explained by natural factors, but instead, require a significant anthropogenic forcing of climate*' (Mann *et al.*, 2003), or again, with greater affirmation: '*only anthropogenic forcing of climate, however, can explain the recent anomalous warming in the late 20th century*' (Jones and Mann, 2004). This, however, is to go beyond simple research into the evolution of temperature, and on into the realms of propaganda, or what Essex and McKitrick (2002) called '*nescience*': they justifiably asked '*whether statistical methods can detect a human influence on climate*'.

Though past climates in no way show a link between greenhouse gases and temperature (cf. Chapter 9), the *montage* of the 'hockey stick', and its partisans, provide the IPCC with its 'ultimate weapon'!

10.1.2 The IPCC ... against ... the climatologists

The controversy generated by the '*infamous hockey stick*' (Daly, 2001) is still very much with us, and rightly so, as the IPCC rejects out of hand existing, patient research and the findings of much serious investigation, based on historical records and archaeological, botanical, and glaciological work from many different parts of the world. These studies are the fruit of long years of research by experienced scientists, and we need only cite here some of the best known contributors: Lamb (1965, 1977, 1984); Mayr (1964); Le Roy Ladurie (1967); Alexandre (1987); Grove (1988) ... the list is very long. The IPCC thus demonstrates the shortness of its memory, and/or at least the limits of its 'climatological culture'!

During the MWP, between 1000 and 1300, the Arctic icecap melted back considerably, and was not present in the vicinity of either Iceland or Greenland between 1020 and 1200. In southern Greenland, the mean annual temperature was between 2–4°C higher than it is today. A stormier period set in after about 1250, and sea travel became more difficult. Around 1340, the Vikings' sea routes lay further to the south, and after 1410, communication by sea ceased. In central and western Europe, the climatic optimum occurred between 1150 and 1300, and the limits for crop-growing and viniculture edged northwards by 4 or 5 degrees. The '*gentle twelfth century*', with its mild winters and dry summers, gave way to the '*golden age*' (according to a Scottish history) of the thirteenth, the most beneficent of all in terms of weather, good harvests and increasing trade. North America also experienced a warm and relatively dry period, though after 1300, conditions became cooler and wetter. A similar example around this same time was that of sub-Saharan Africa, where rain was abundant, and the great and prosperous Sahelian/Sudanian empires flourished between 1200 and 1500 (Mauny, 1961; Toupet, 1992).

The decline into the LIA saw an increase in storm activity, heavy rains, snowfalls, and waves of cold. Also observed, according to the IPCC (1990), were '*extensive glacial advances in almost all alpine regions of the world*'. Cold did not dominate continuously throughout this period, but '*the Little Ice Age was probably the coolest and most globally extensive cool period since the Younger Dryas*' (IPCC, 1990), and climatic conditions were particularly severe between 1550 and 1850.

Looking at past research (even if very briefly) is one way of demonstrating first of all the negative stance of the IPCC, which sees nothing wrong in 'invalidating' former studies in a cavalier fashion, and in contradicting itself from one report to the next. Also, looking at these past studies underlines the unrepresentative nature of the curve adopted by the IPCC. Daly (2001) was quick to point out the '*falsification of climatic history*' with the help of 14 eloquent examples from all over the world. Soon, Baliunas, Idso, and Legates (2003, 2004) analysed the results of more than 240 studies completed within the last four decades, and concluded that '*the 20th century is neither the warmest century, nor the century with the most extreme weather, of the past 1,000 years*'. They also stated that '*clear patterns did emerge showing that regions worldwide experienced the highs of the Medieval Warm Period and lows of the Little Ice Age, and that 20th century temperatures are generally cooler than during the medieval warmth*'.

McIntyre and McKitrick (2003) used the same data as Mann *et al.* (1998), and caught them in a trap of their own making. In particular, they showed that '*the data set of proxies of past climate used in Mann, Bradley, and Hughes (1998) for the estimation of temperatures from 1400 to 1980 contains collation errors, unjustifiable truncation or extrapolation of source data, obsolete data, geographical location errors, incorrect calculation of principal components, and other quality control defects*'. Their conclusion was quite categorical: '*The major finding is that the values in the early 15th century exceed any values in the 20th century. The particular "hockey stick" shape ... is primarily an artefact of poor data handling, obsolete data, and incorrect calculation of principal components*'. Temperature estimates, when corrected by McIntyre and McKitrick (Figure 44), provide proof that the current period is certainly not the hottest for a thousand years!

The recent study by J. Luterbacher *et al.* (2004) might have thrown some light onto the debate, but it involves only the period from 1500 onwards, and therefore does not include the MWP. So, its conclusion (immediately made much of by the proponents of 'global warming') was, unsurprisingly, that '*the late 20th and early 21st-century European climate is very likely warmer than that of any time during the past 500 years*'. There is nothing new in this, since the period under consideration is the LIA, and so the study does not confirm the IPCC stance. The same artifice is seen in the final part of the curve: the 'official' IPCC curve for 1901 onwards (Figure 46, p. 212) is tacked on to the estimated temperatures from 1500–1900. It is interesting to note that the coldest winter in the series is that of 1708–1709; the severity of this season drew comment from Réaumur in the *Mémoires de l'Académie des Sciences*, 1709 being called '*the year of the great winter*'. Summers in Europe between 1530 and 1570 were slightly warmer than those of 1901–1995, on average, and the summer of 1902 was the coldest along the curve, whilst that of 2003 was the warmest.

There are other studies which look back further. They are far too numerous to be exhaustively listed here, but let us take one as an example. Using isotopic analyses of ice from Greenland (GISP2), Grootes *et al.* (1993) confirmed (Figure 45) the

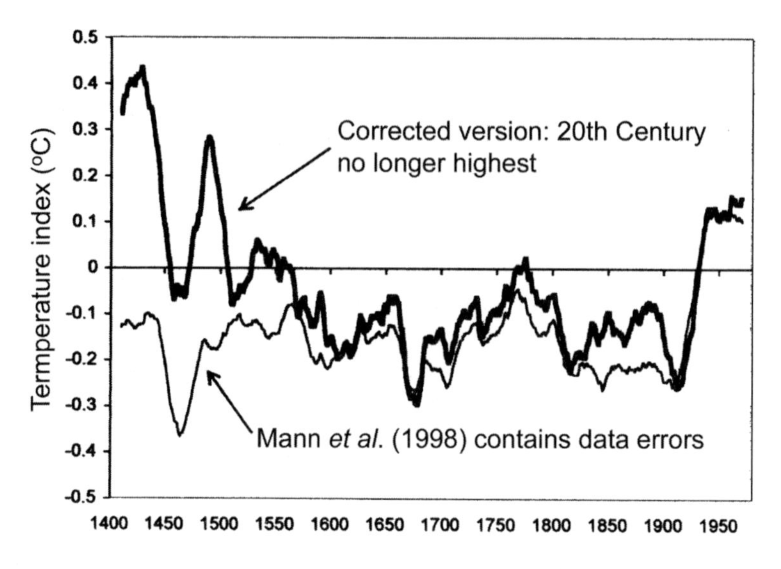

Figure 44. Temperature anomalies index (°C) for the northern hemisphere (1400–1980), average temperature reconstruction.
(*bottom*) From Mann *et al.* (1998). (*top*) From McIntyre and McKitrick (2003).

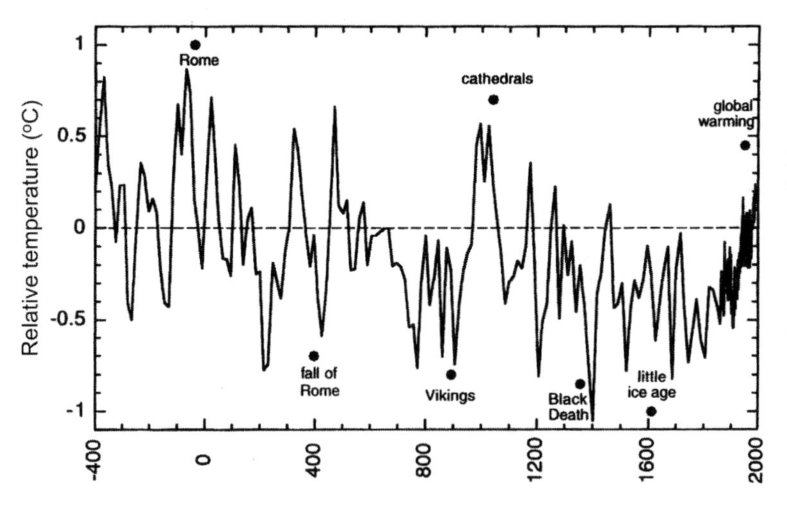

Figure 45. Variations of temperature over the last 2,400 years.
From Grootes *et al.* (1993).

existence not only of the LIA (and the two peaks within it, one in about 1700 and the other in about 1400) but also of the MWP, whose opulence favoured the building of cathedrals. These periods are also confirmed by the Antarctic 'Dome C' core (Benoist *et al.*, 1982). Before the periods in question, the Vikings' expansion into France has been associated with low temperatures, especially in northerly latitudes.

The triumphant years of the Roman Republic and Empire, however, corresponded to a warm period centred on the beginning of our era (Figure 45).

A recent analysis of winter temperatures in central China (Ge *et al.*, 2003) confirms, for a period of 2,000 years, the climatic evolution represented by Figure 45. '*From the beginning of the Christian era, climate became cooler at a rate of 0.17°C per century*': this was the cooling which spelt the end of the 'Roman Warm Period', and was the overture for the 'Dark Ages Cold Period', during which '*temperature reached about 1°C lower than that of the present*'. When subsequently the temperature rose, '*the peak warming was about 0.3–0.6°C higher than the present for 30-year periods, but over 0.9°C warmer on a 10-year basis*', and so the MWP '*was warmer than has yet to be experienced in modern times*'. From 1310 onwards, the temperature declined rapidly, and during the LIA, mean temperatures were '*0.6–0.9°C lower than the present, with the coldest value 1.1°C lower*' (Ge *et al.*, 2003).

10.1.3 Key problems raised by the 'hockey stick'

The so-called 'hockey stick' curve presents various problems as far as the estimation of actual climatic evolution is concerned.

- The first problem involves the greater part of the curve (i.e., the 'reconstituted' temperatures). As has already been stated, for example in the case of the levels of CO_2 concentration (Figure 32), the climatic significance of this reconstruction seems to be without foundation, and its adoption by the IPCC, in spite of the conclusions reached in many excellent studies, is hasty, to say the least, because it does not confine itself to strictly scientific preoccupations.
- The second problem is that of the juxtaposition of two data series which are not necessarily comparable. Also, why do Mann *et al.* choose to use the IPCC's thermal curve (Figure 46), even though its climatic validity has not been clearly demonstrated? Moreover, as Soon pointed out in 2004, why has the thermal curve climbed more and more steeply in recent years in articles by Mann *et al.*, with the maximum temperature anomaly going from 0.30°C in 2002, to 0.58°C in 2003? Quite an impressive leap, is it not? Singer (2003) worked with the post-1980 data (data which Mann *et al.* had considered not worth using), and proved thereby that the most recent temperature series are in agreement with those provided by satellites and weather balloons (see below), which indicate only a slight warming, and do not suggest that recent decades are the hottest for 1,000 years!
- The third point to be raised is that, according to the curve, the last hundred years appear to have been quite 'exceptional' in nature! In comparison, the last nine centuries (or an even longer period) seem quite bland (cf. Figure 43), and set against them, the 'extraordinary' rise in temperature of the 20th century is just astonishing. We would seem to be living in exceptional times from the climatic point of view, the like of which has not been experienced in the last 1,000, or even 2,000, years! For some climatologists, there is more to this than just a surprising observation: they are aware that the present epoch is part of a very

long falling off in temperature, irregular but continual, which has been going on since the Holocene Climatic Optimum (HCO) (see Chapter 9). The most significant fall, which affected the whole planet to a greater or lesser degree, occurred between 5 kyr and 4.8 kyr BP. This was followed by an alternating pattern of warmer and cooler episodes, the latter for example having been defined, for the Alps, by Furrer *et al.* (1987). The fourth warming since the HCO is precisely the Roman Warm Period (Figure 45), and the fifth corresponds to the MWP.

10.1.4 We are living in an 'exceptional' time!

The modern period (Figure 46) is therefore part of an alternating (and well-known) series of warmer and cooler periods, though the general trend has been one of gradual cooling for the last 5,000 years. If we take no account of this fact, and go beyond simple temperature estimates, we could suggest that **today's climate is truly 'outside of the normal'**; we would have no element of comparison, nor any past references to assess its nature ...

Let us therefore step back a little from these dubious approximations. Instead, let us consider a single example, that of sub-Saharan Africa. This is a representative one in this context, since no region evolves in isolation, all being influenced by general circulation. If the epoch in which we find ourselves were really exceptional, and especially from the point of view of heat, this would mean that the current drought in the Sahel was much worse, for a period of more than 1,000 years, than it is today! Which is absolutely not the case. Since the end of the so-called 'Sahara of Tchadian lakes' period, more than 5,000 years ago during the HCO, rainfall has gradually, step by step, been decreasing. Given that a cold period corresponds to a period of less rainfall (Chapters 9 and 11), the Sahel should now be experiencing, in what is claimed to be a 'hyper-warm' period, unusually abundant rain ... and all observations confirm that this is just not happening (Chapter 11). So, what is so exceptional about the time in which we live?

Since the IPCC is telling us that the modern period is exceptionally warm, this means (at the risk of stating the obvious) that it isn't cold, and certainly never *very* cold: it stands to reason! Well ... on 17 January 2004, the mercury fell to −28°C in Montreal, and as low as −17°C in New York (a 100-year record). In Massachusetts a renewed wave of cold on 25–27 January brought a minimum of −40°C, and 48 people died as a result. On 23 January 2004, there were snowstorms across Turkey, and Istanbul ground to a halt as temperatures fell as low as −10°C. In Romania, the city of Bucharest recorded −15°C, and in snowbound Athens it was −8°C. In Egypt there were seven successive cold snaps, accompanied by rain, strong winds, and sandstorms. On 13 February 2004, another wave of cold swept across Greece, bringing snow, and temperatures well below zero, with long-lasting frosts. Yet another cold spell began on 17 February. On 4 March 2004, the heaviest snowfall for a hundred years occurred in Korea, where records for cold weather were broken, and at the same time a cold snap in Bangladesh claimed 100 lives.

The winter of 2003 saw similar events: '*In January–March 2003, eastern North*

America experienced one of the longest and coldest winters in many years' (Khandekar, 2003), and there were record snowfalls of over 100 cm in 24 hours. The Gulf of Finland, between Scandinavia and Russia, froze over for the first time since 1947! In January, vast areas of central and eastern Europe experienced very cold episodes, and in Russia temperatures fell to −45°C. In Mongolia, for the third consecutive year, a dry summer was followed by a cold (or even particularly icy) winter, with effects devastating enough to oblige the state to call for international aid! In France at this time, although no temperature records were broken, records for electricity consumption were, during very cold spells. In January 2000, 20 cm of snow fell in Saudi Arabia, where it had not snowed for 50 years. Pieces of ice were observed in the Adriatic Sea (something which had not been seen for over half a century), and there were frosts in Kenya, with 75% of the tea crop lost ...

One could go on and on in this fashion, just picking out only cold spells from the abundant weather literature available (notably the *WMO Bulletin*). It is an easy thing to do, and might be construed as 'cheating'. It is also a practice quite common among partisans of global warming, but they choose only the warm episodes! What this brief glimpse does show, in the final analysis, is that we are not really living in such an 'exceptional' time, and that the modern epoch does not match the image which the IPCC would like to attribute to it!

10.1.5 A shoddy stick...

Finally, after all the criticisms and accusations of false methodology and errors in calculation, Mann and his co-authors corrected their papers in June 2004 (in the *Journal of Geophysical Research*), and in July 2004 (in *Nature*). After describing their errors, they still considered (2004) that '*none of these errors affect our previously published results*'! Persistent even in their error, they reached that surprising conclusion: and what is the good of a corrigendum if it is seen as serving no purpose? McKitrick and McIntyre are more categorical, stating: '*we have done the calculations and can assert categorically that the claim is false*'. They are also more sensible, considering that the corrigenda issued by Mann *et al.* are '*a clear admission that the disclosure of data and methods ... was materially inaccurate*'. This *affaire* still rumbles on, and the interested reader should consult www.uoguelph.ca/~rmkitri/research/fallupdate04/, where (s)he will discover that, after ten months of procrastination and prevarication, *Nature* recently turned down a new update/correction by McKitrick and McIntyre, sent to the journal in November 2003. Apparently the length of the article was the main bone of contention!

So the 'hockey stick' has had its day, and, as Corcoran (2004) puts it, '*is about to get swept away as a piece of junk science*'. However, this saga has had its uses, and in more than one way, because:

- It is symptomatic of the state of mind of the IPCC, whose scientific rigour and credibility seem rather tenuous: it appears to be preoccupied with using 'scientific reasoning' only to facilitate propaganda. It comes as no surprise that it was

Watson, true to form in his capacity as IPCC president (see Chapter 4), who proclaimed at the conference at The Hague in November 2000 that '*the Earth's surface temperature this century is clearly warmer than in any other century during the last 1,000 years*'!

- It highlights just how far some 'scientists' are prepared to go to gain recognition, to be seen as original, even to curry favour with some existing or hoped-for sponsor, or otherwise to seek benefits for themselves and their groups.
- It reveals the degree to which scientific journals (or those considered as such) can 'follow the fashion' and adopt the methods of the 'media types', or just deliberately adopt the *parti pris* of the IPCC. An editorial team can, by selecting the 'right' referee, launch onto the market any so-called 'scientific' idea, just as if it were some kind of washing powder or fizzy drink, and then shamelessly block any corrections, for the most feeble of reasons!

All this represents an obvious danger to science, and to its credibility. It must be pointed out, nevertheless, that there has been one beneficial effect: the keen reaction of responsible scientists, which shows that not everybody is always ready to swallow anything though there are evidently those who are only too willing to do so, out of naivety or self-interest!

10.2 THE RECENT EVOLUTION OF MEAN GLOBAL TEMPERATURE

The curve for the variation of the anomalies of global mean temperature from 1860 onwards (Figure 46) is regarded by the IPCC as the veritable standard curve as far as the anthropic greenhouse effect is concerned. This curve is established, or, more properly, 'reconstituted', from the data of approximately 7,000 weather stations scattered across the Earth's surface. Air temperature measurements for ocean areas (i.e., 71% of that surface) lack homogeneity, since they may be made on

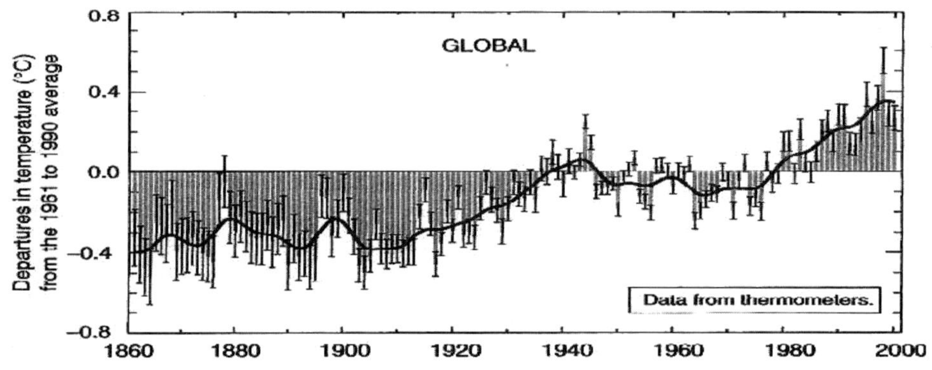

Figure 46. Variations of the anomalies of annual global mean surface temperature over the last 140 years (1860–2000).
From IPCC (2001).

islands, by automatic weather buoys, and also on moving platforms at various heights on ships. Most of the weather stations on large land masses are to be found in urban areas. Two data series are recognised by the IPCC: the Global Historical Climate Network (GHCN), compiled in the USA, and data from the Climatic Research Unit (CRU) of the University of East Anglia in the UK. After verification (with some rejected), and correction (especially for the 'urban effect'), data are applied, using the techniques of modelling, to base units (5° by 5°), within which is determined the thermal anomaly for each box of the surface area. From the combination of trends thus revealed is created the curve of the anomalies of 'annual global mean temperature'.

There have been numerous criticisms of this process of evaluating temperature, with reservations being expressed about the quality of the data (especially for ocean areas), the density of the observational network, regional disparities, and methods used (Daly, 2000). Particular reference has been made, for example, to '*uncertainties associated with combining land and marine instrumental records to produce regional-average series*' (Jones and Briffa, 1992). Another comment involves data from Europe for the period 1901–1999: '*94% of the temperature series and 25% of the precipitation series are labelled doubtful or suspect ... about 65% of the statistically detected inhomogeneities in the temperature series labelled doubtful or suspect in the period 1946–1999 can be attributed to observational changes ...*' (Wijngaard *et al.*, 2003). Other important divergences concern the evaluation of variations on the regional scale (Gray, 2003b). Moreover, disagreements appear between the two series of data (from GHCN and CRU), and criticisms are levelled at the methods of selection and treatment employed: '*... they disagree this rate of change by 42% over the most recent 30 years*'; since the two are based on the same measurements, at least one of the methods must be in error. However, we do not know which one it is (Courtney, 2003).

So, estimates involving the evolution of temperature can really be only approximations: this is inherent in this parameter, which is not strictly speaking a 'measurable' quantity. Let us look at the curve in Figure 46 more closely (as an 'accountancy document'), knowing that it has always been the tradition to try to determine some kind of mean hemispheric or global temperature. We shall ask the question: What is the real value of this curve from the climatic point of view? What does it really mean, or represent?

From 1860 onwards, temperatures rise, a manifestation of the gradual climb out of the LIA. Between 1920 and 1940, the climb is much steeper. Between 1930 and 1950, there is a relatively warm period which may be described as the 'Contemporary Climatic Optimum' (CCO). Thereafter, temperatures are lower through the 1960s, and, at the time, some envisaged another 'Little Ice Age' (cf. Chapter 2). In the 1970s, the curve levels out and again begins to climb as the 1980s begin: a veritable 'climatic upturn' (Leroux, 1996a). From then onwards, the trend is upwards, and the Recent Climatic Optimum (RCO) is established.

The qualifiers used to designate the evolution of temperature here have only a very relative value, given that the range of variation is only 0.6°C (±0.2°C) for the period from 1860–2000 (i.e., 0.4–0.8°C). This is a very narrow range of variation, and it pales into insignificance when we consider the fact that this period of one-and-a-half centuries includes the Industrial Revolution. To help us form an idea of this range, it is useful to consider that *this value (0.6°C), on a mean annual scale, represents the difference in temperature between Nice and Marseille*, two Mediterranean cities only 200 km apart (14.8°C compared with 14.2°C, according to values for 1931–1960). Such a negligible difference spread across 140 years (representing 0.0043°C per year!) hardly constitutes a 'climatic upheaval'!

10.2.1 The CO_2/temperature 'relationship'(?)

The link between greenhouse gases and temperature has not been demonstrated on the palaeoclimatic scale (Chapter 9). Is the present era so special that a claimed 'link' which has not been active for thousands of years was, a century and a half ago, switched on, and in an 'exceptional' fashion?

Let us compare the respective evolutions of mean annual global temperature and CO_2 levels. The variation in CO_2 levels (from 1960 to 2000) is fairly regular (Figure 32, p. 174, insert), displaying a continuous rise with no 'jolts' other than those which can be ascribed to seasonal variation. Such an evolution is, however, quite surprising in its way, if it is being attributed to human activities: its almost perfect regularity might suggest that those human activities also evolve in a regular and harmonious way, without accelerations or slackenings, in spite of the many vicissitudes of history and economic progress!

On the other hand, the secular global 'reconstituted' curve of temperature anomalies (Figure 46) is by no means regular, and the co-variation is far from being as close-fitting as it is with the variation in solar activity (Chapter 6):

- The marked warming of 1918–1940, of the same order of magnitude as that of recent decades, corresponded to only a slight rise in CO_2 levels, of only 7 ppmv (from 301 to 308 ppmv).
- Between 1940 and 1970, CO_2 levels rose by 18 ppmv (from 308 to 326 ppmv) but the temperature did not rise: on the contrary, it was supposed that a new 'Little Ice Age' might be on its way.
- Only the (presumed) rise in temperature towards the end of the century (RCO), from 1980 onwards, coincides with a rise in CO_2 levels (of more than 22 ppmv).

The greenhouse effect scenario therefore cannot be invoked to explain recent thermal evolution, since:

- *It does not tally with the warming from 1918 to 1940.*
- *Neither does it account,* a fortiori, *for the cooling from 1940 to 1970.*
- Therefore, *there is only one period, at the end of several thousand years analysed, during which some 'relationship' between CO_2 and temperature could be envisaged: the RCO, about thirty years long!*

Could this be just a happy chance? Or as one might say, an opportune 'happy ending'? Whatever we make of it, the only 'relationship' which we can immediately discern may be but a *co-variation*, a simultaneous variation which leaves still undefined the real nature of the supposed link between these parameters.

The cause of the evolution of temperature between 1860 and the period 1970–1980, with its conflicting trends, is not the greenhouse effect. This is moreover confirmed by the IPCC's own simulations (Figure 20, Chapter 6), which attribute most of the variation to natural forcing. And what is more, it is worth remembering that the 'cooling' which took place between 1960 and 1970, which, oddly, proceeded in parallel with rising CO_2 levels, **has never been explained**! Here we have a crucial question, but one that is constantly sidestepped, since to allow it would *ipso facto* bring down the IPCC's fragile house of cards!

We can only focus, therefore, on recent decades (the supposed RCO) to attempt any 'proof' of the IPCC's scenario! So the question remains as to whether the cause(s) of thermal evolution before the climatic 'switch' of the 1970s is/are responsible for the trend in the curve for the most recent decades. This question brings us back to another: what is the true climatic significance of the curve (Figure 46) of mean global temperature anomalies (reconstituted), and especially its last section? We shall reply to this question after analysis of the aerological dynamic (at the end of Chapter 13).

10.2.2 The urban greenhouse effect

The presumed global warming might well be merely an urban phenomenon, as is pollution. The presence of the built-up, paved area, and the existence in urban locations of activities, especially those concerned with industry and road traffic, which generate gases, interfere considerably with the way in which solar radiation is absorbed and returned. Locally, the result of all this is a rise in temperature, which can only be aggravated as the built area expands and the urban population increases. Weather stations once situated in rural locations have been progressively swallowed up as growing towns and/or their widening heat domes have absorbed them. The measurements made at these urban stations, which produce most land-based data, therefore largely reflect climatic evolution on a local scale. Also, the modelling of the climatic system is badly handicapped by the '*poor representation of the urban environment in global and regional climate models*' (Sheperd and Jin, 2004).

So the so-called 'global' curve of thermal evolution (Figure 46) may represent, at least partially, an 'urban greenhouse effect'. The comparison made by Daly (2003) of the evolution of temperature for the period 1822–2001 in Central Park, in the heart of New York, and at the West Point Military Academy, which lies to the north of the conurbation, is particularly revealing here. The two curves, which run almost together until the 1890s, thereafter drift progressively further apart. The curve for West Point proceeds practically unchanged, but the Central Park curve climbs away, eventually exceeding mean annual temperatures by 3°C (Figure 47).

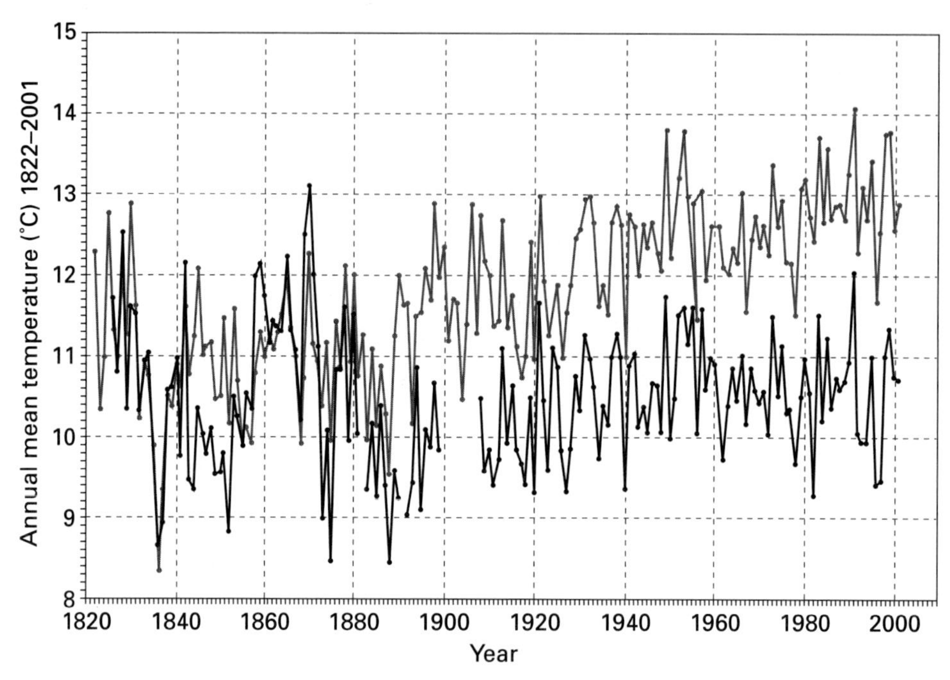

Figure 47. Annual mean temperature (°C), from 1822 to 2001, at: (*top*) Central Park in New York City; and (*bottom*) US Military Academy, West Point, NY.
From Daly (2003).

Goodridge (1996) gave a similar demonstration of this urban effect for California, by comparing the thermal evolution of urban areas of:

- more than a million inhabitants;
- between one million and 100,000 inhabitants; and
- less than 100,000 inhabitants.

Increases in temperature diminished in a very telling way as a function of the decreasing size of these towns and cities: for the largest, 3.14°F per century, and for the smallest, only 0.04°F per century (Figure 48). Goodridge concluded that '*the apparent "global warming" is in reality urban waste heat affecting only urban areas*'.

In France, as in other countries, the principal weather stations once situated outside the towns and cities have been incorporated by them as they have grown, and are now within their urban heat domes. Thermal evolution studies reveal a sustained rise in minimum temperatures (i.e., night-time temperatures), while maximum (daytime) temperatures show a more irregular pattern, with a much less definite trend (Leroux, 1997; Moisselin *et al.*, 2002). This differential evolution, which is another consequence of the urban heat effect, was of particular interest to Karl *et al.* (1988): '*Urbanization decreases the daily maxima in all seasons except winter, and the temperature range in all seasons; it increases the diurnal minima and the daily*

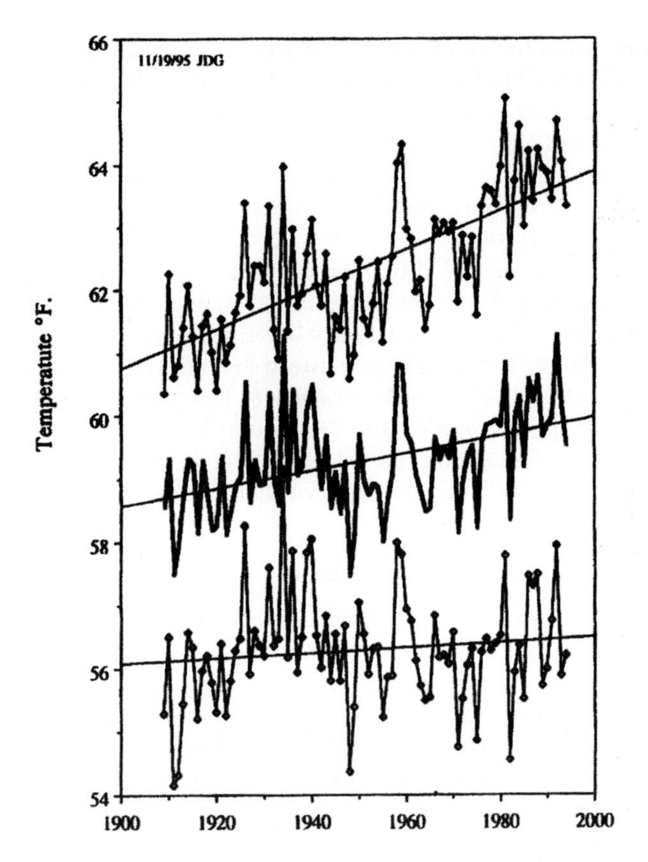

Figure 48. Temperature trends at 107 stations in California for the period 1909–1994. (*top*) Average of 29 stations in large counties of more than 1 million people. (*middle*) Average of 51 stations in mid-sized counties of 100,000 to 1,000,000 people. (*bottom*) Average of 27 stations in small counties of less than 100,000 people. (Note that temperatures are in °F.)
From Goodridge (1996).

means in all seasons'. This difference in the way in which night-time and daytime temperatures behave is considered by the IPCC as another obvious sign of the greenhouse effect, but it is in fact only a consequence of urbanisation, the effects of which '*are detectable even for small towns with populations under 10,000*' (Karl *et al.*, 1988).

A correction is therefore brought into the data to account for this urban thermal bias. However, the amount by which data are adjusted is also variable. For example, Karl and Jones (1989) considered that '*temperature data sets have an urban bias between +0.1°C and +0.4°C*' (from 1901–1984), a departure of the same, or even greater, magnitude than the trend in mean temperatures (+0.16°C) for the USA for the same period. If we compare this urban bias to the amplitude of the variation apparent in Figure 46 over 140 years (of the order of 0.6°C), the climatic significance of the global thermal curve is already considerably diminished. And even more so when we consider that the adjustment to be applied to urban temperatures is a fixed value. Figures 47 and 48 in fact show that the bias increases (with temperature), as a function (ignoring other factors such as the aerological dynamic) of the growth of the built-up area and the increase in its population and associated activities. This means that estimates of what temperatures are really doing are more and more unreliable. Is there any move afoot to apply an ongoing correction to observed values for urban temperatures?

The significance of the global thermal curve is even more diminished if regional thermal peculiarities are taken into account. A single example will suffice: Rupa Kumar and Hingane (1988) analysed variations in surface temperatures for major industrial cities in India for periods of between 86 and 112 years. Their astonishing conclusion: '*there was either a cooling tendency or cessation of warming after the late 1950s for most of the industrial cities*'. This example, based on the enormous and ever-growing cities of India, indicates yet again that the greenhouse effect, even on a local or only urban scale, is not the cause of thermal evolution.

10.2.3 Satellite and radiosonde measurements

The debate about the urban effect, a debate which continued through the 1990s, was still echoing when, from 2000 onwards, another controversy arose, showing the differences in opinion about trends in recent warming. The rising surface temperatures of the last decades seemed not to be confirmed by trends in the mean temperature of the troposphere. According to Christy and Spencer (2000), who were the first to analyse satellite data, microwave sounding units (MSUs), which had been carried by satellites since January 1979, revealed a warming tendency of only +0.085°C for the lower troposphere (i.e., +0.155°C for the northern hemisphere and +0.014°C for the southern hemisphere). This insignificant value is far below that derived from observations at the surface.

These satellite measurements give much better coverage than the data produced at the Earth's surface. They show good evidence of the effect of solar cycles (in this case, cycles Nos 22 and 23), and detect changes due to volcanic eruptions (El Chichón in 1982, and Pinatubo in 1991) or changes associated in time with El Niño, for example, the event of 1997–1998 (Figure 49). Their representativeness seems difficult to contest. They are in agreement with radiosonde (balloon) measurements analysed since 1958, which indicate, for example, a cooling of –0.07°C per decade for the USA, and of –0.02°C per decade for the UK, and confirm broadly that there has been a '*lack of a warming trend in the mid-troposphere*' (Parker *et al.*, 1997).

However, proponents of global warming immediately found fault with these satellite-based measurements: they certainly were particularly difficult to digest for some. It was even suggested that there might be some 'slowing' effect caused by the impact of the solar wind on the satellites, which would cause them to descend a little in their orbits so that their instruments would be 'seeing' a smaller area (Gaffen, 1998). The lesser value for the warming trend was therefore reduced to a mere artefact, which could be expressed thus: smaller surface observed = less radiation detected = lower temperature ... Further analyses were then carried out, with a

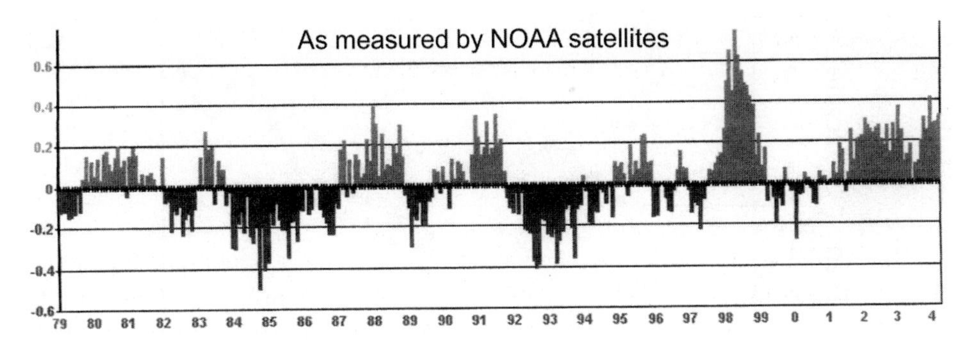

Figure 49. Global mean temperature anomalies (°C) of the lower troposphere, from January 1979 to March 2004. From Global Hydrology and Climate Center, University of Alabama, USA.
From Daly (2004).

view to 'nailing' these series of measurements: for example, by Vinnikov and Grody (2003), or Mears *et al.* (2003) who found '*a global trend of* 0.097 ± 0.020 *K decade*$^{-1}$*, ... in disagreement with the MSU analysis of Christy and Spencer, which shows significantly less* (*0.09 K decade*$^{-1}$) *warming*'.

So the great debate descended into quibbling, taking as its basic postulate that the values for the data at the surface (cf. Figure 46) were necessarily correct, which has by no means been successfully demonstrated given the lack of precision of the 'reference' curve. Anyway, questions involving differences of hundredths or even thousandths of a degree Celsius add little in the way of any real climatic significance to the 'official' curve of the IPCC, and neither do they provide proof that the so-called greenhouse effect scenario has a basis in fact! This controversy seems somewhat unnecessary, especially in the light of such statements as: '*we now have independent confirmation that satellite results are correct and that the climate is not warming*' (Singer, *The Week That Was*, 17 July, 2004).

Note the comparison by Douglass *et al.* (2004), for the last two decades, (Figure 50), of:

1 surface temperatures (ST), 'classic' IPCC curve (based on data from Jones *et al.* (2001, CRU);
2 satellite observations of the lower troposphere (after Christy *et al.*, 2000, MSU); and
3 near-surface temperatures (at 2 m) calculated mainly from observations by balloons (R2-2m, after Kanamitsu *et al.*, 2002, NCEP–NCAR data).

Thus, Douglass *et al.* (2004) showed the idea of 'global' warming to be a mere fiction: only the ST curve is almost entirely positive, while the MSU and R2–2m curves, which match each other quite well, are negative for the tropical zone (Figure 50). One can appreciate the agreement of the three curves for middle latitudes in the

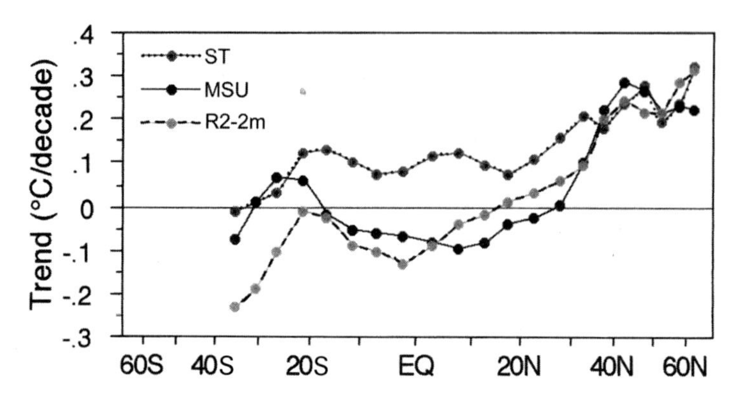

Figure 50. Temperature trend by latitude band, from 1979 to 1996. ST: surface temperature from CRU data; MSU: microwave satellite unit; R2–2m: NCEP–NCAR data.
From Douglass *et al.* (2004).

northern hemisphere, but high latitudes are unfortunately not included in the diagram. We shall, however, be visiting them later.

It is worth recalling here that *only* the presumed RCO of the last three decades could be used by the IPCC to proclaim a link between the greenhouse effect and so-called 'global warming'. Figure 50 shows very clearly that the notion of a global effect is but a chimera, and that **any actual global warming is not observed**. **So NOTHING remains of the IPCC's greenhouse effect scenario**. What Figure 50 does show, however, is that there are considerable disparities between latitudes, and that the agreement of the curves for middle latitudes in the northern hemisphere invites analysis of the aerological dynamic which is responsible for this kind of evolution (cf. below, and Chapters 12 and 13).

10.3 THERMAL EVOLUTION IS A REGIONAL EVOLUTION

Models, and the IPCC, try to accredit the notion of a global climate. From a strictly climatic point of view, everyone is well aware that such a notion is without basis in fact: **the climate may be defined on a local, or even on a regional scale, but certainly not on a global one**. Consequently, mean values for one hemisphere, and *a fortiori* global values, for temperatures *reconstituted* from observations, either made at the surface or involving satellites or balloons, have only a statistical and numerical significance. Obviously, they can have only a limited significance in terms of the climate, if indeed they have any significance at all. The mean values in question are in fact the result of the artificial sum of different weather behaviours, most notably as a function of latitude (Figure 50) and/or regions, or even of contrasting evolutions for which some 'mean' value represents, strictly speaking, nothing.

Climatic evolution occurs on a regional basis, and in diverse ways. For example, in the case of North America, Balling and Idso (1989) pointed out '*a widespread cooling in the major south-central portion of the US and general warming in the northeast and west*'. For the northern hemisphere, Morgan and Pocklington (1995) used two indices: first, the difference between the decade 1981–1990 and the preceding (warmest) decade, and second, the difference between temperature norms for 1931–1960 and 1961–1990. They concluded that '*there has been significant cooling in eastern North America, around the northern North Atlantic, and in northwestern Europe since the 1950s*', and they stated that '*warming as a whole has been marginal over the past 60 years*'. Litynski (2000) compared temperature norms for the periods 1931–1960 and 1961–1990, published in 1971 and 1996 by the WMO. The first period corresponds *grosso modo* to the CCO, and the second period to the so-called RCO, which includes the most marked presumed rise in temperature. This should be a very telling comparison, but it shows quite clearly that '*there is no planetary warming during the period 1961–1990*'. On the other hand, on the regional scale, warming and cooling episodes are apparent. Take an example for the northern hemisphere: the decline in temperature is of the order of: −0.40°C in North America; −0.35°C in northern Europe; −0.70°C in northern Asia; and as much as −1.1°C in the Nile Valley. Other regions have experienced warming: for example, the western side of North America (from Alaska to California), or, elsewhere, the Ukraine and southern Russia (Litynski, 2000).

Douglass *et al.* (2004) add: '*one of the most striking observations is that the values are geographically highly non-uniform*': tropical latitudes show no positive trends (Figure 50), yet middle latitudes do experience them, though they are only slight in the southern hemisphere and more obvious in the northern hemisphere. Again, from Douglass *et al.* (2004): '*in the northern hemisphere the highest trend-line values are localized in three areas*', areas which are located precisely within specific units of circulation (cf. Chapters 12 and 13).

So what really is the representative trend for temperatures: that of regions where there is a rise, or that of regions showing a decline? Let it be remembered that, on the one hand, a mean value based on contradictory data has little climatological significance; and on the other hand, **models have never predicted, nor have they revealed, such regional disparities, and they are, as yet, just as incapable of explaining them as they are at predicting them**.

Later, we shall go into more detail about regional disparities in certain aerological spaces (Chapters 12 and 13). Among these regional behaviours are those which are essentially local, being the consequence of general circulation and having only local consequences themselves. There are, however, other regions within which a modification in temperature is responsible for consequences far away, even on a planetary scale, through the intermediary of general circulation. These regions are in higher latitudes, and deserve our special attention here. It is also worth noting that studying past climates (Chapter 9) has shown us that climate change in tropical latitudes is generally of only limited amplitude, being essentially advected by circulation and driven by events in higher latitudes.

In Chapter 7 we mentioned the rise in temperature, by as much as 10°–12°C in each winter hemisphere (paradoxically), at high latitudes, as suggested by models.

We also looked into the physical reasons invoked by the models, reasons which might cause such a warming effect:

- Greater intensity of radiation?
- Increased greenhouse gas levels?
- The existence of a polar circulation cell?
- More rapid meridional exchanges?

We found no answer, and no known relationship with the greenhouse effect. The predicted climate change, which dates from a time before models, in fact from 100 years ago with the work of Arrhenius (1903), remains therefore very hypothetical. If the hypothesis has some basis in fact, as has been claimed for these last 20 years by the modellers, we ought already to be seeing very strong indications of the supposed thermal evolution at high latitudes. Let us examine what is really going on in the polar regions, areas of crucial importance in the mechanics of general circulation.

The pattern of thermal evolution actually observed at high latitudes is nothing like that predicted by the models. They did not predict it, are incapable of reconstructing it, and – assuredly – have not explained it, neither for the Antarctic, with its relatively simple evolution, nor for the Arctic, where things are more complex.

10.4 THERMAL AND DYNAMIC EVOLUTION IN THE ANTARCTIC

In the Antarctic there are few weather stations, and the existing ones tend to be of recent origin, dating from the 1960s (cf. Daly, *Tempreature in Antarctica*, 2004). Temperature data from the Antarctic show absolutely no general warming trend in that continent.

A study by Doran *et al.* (2002), concentrating on the period 1966–2000, reached this conclusion: '*our spatial analysis of Antarctic meteorological data demonstrates a net cooling on the Antarctic continent between 1966 and 2000, particularly during summer and autumn*'. Leaving aside the Peninsula (WAIS), the proportions (%) of the Antarctic which exhibit warming (+) or cooling (−), on either the mean seasonal or annual scales, are respectively:

Annual:	**+33.8**	**−65.9%**
Winter:	+56	−43%
Spring:	+49.4	−50.4%
Summer:	**+22.8**	**−76.3%**
Autumn:	**+0.3**	**−99.7%**

Vaughan *et al.* (2001) note that the general trend, established using data from all stations for the period 1959–1996, is '*+1.2°C per century*', though given the length of the period of the analysis this represents in reality a mean rise of the order of 0.4°C. What, though, is the real significance of such a mean value, established for the whole continent (a value which only serves to point up the inadequacy of such 'number-

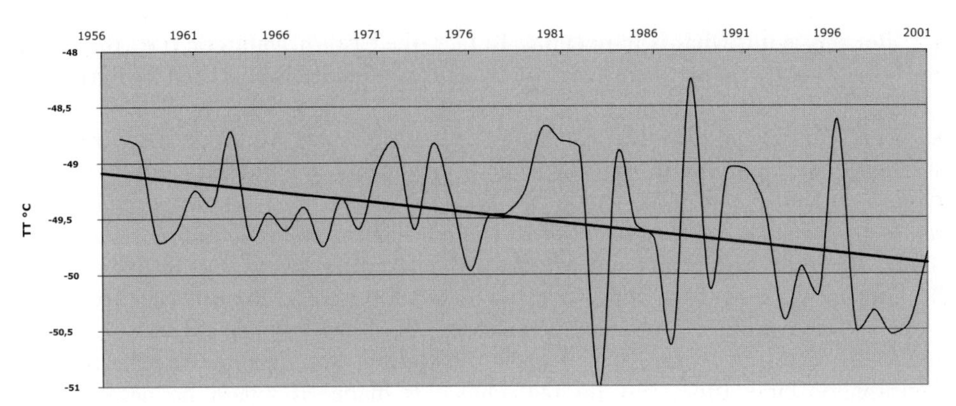

Figure 51. Temperature variation (annual mean, °C) at Amundsen–Scott South Pole station (2,835 m) from 1956 to 2001: linear trend.
Source: SCAR.

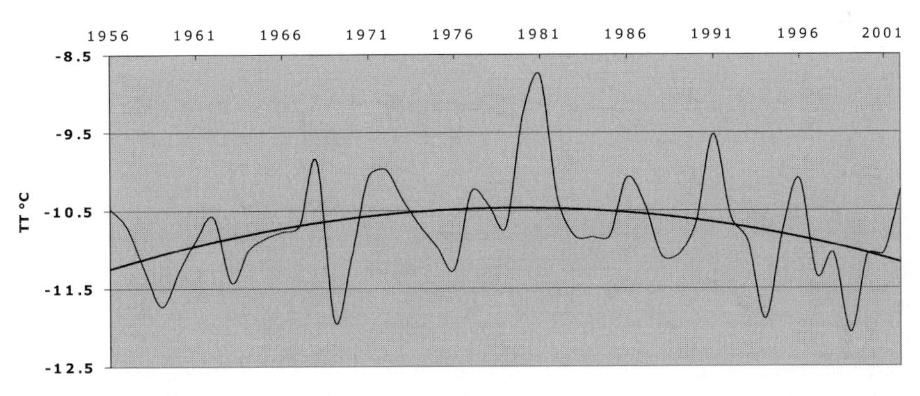

Figure 52. Temperature variation (annual mean, °C) at Dumont d'Urville (Adélie Land) from 1956 to 2002: polynomial trend.
Source: SCAR–WMO.

based' analyses), when we consider the dissymmetry in weather behaviour of different parts of the Antarctic, during the current era and, more importantly, in the past (Chapter 9)? At the Amundsen–Scott station at the South Pole itself, there has been a slightly negative trend since 1958 (Figure 51). On the coast, at the Dumont d'Urville station, this weak trend has also been recorded over the last two decades (Figure 52). However, the Antarctic Peninsula shows a warming trend, with higher figures recorded for its north-western part (+3.6°C over 32 years; confidence level: 90%), on Bellingshausen Island off the tip of the Peninsula. Temperatures on Faraday Island, a little further to the south, have risen by +5.6°C in the last 50 years (confidence level: 99%). This warming is also recorded further to the south and east, in the South Orkney Islands, with an increase of +2°C in 96 years (confidence level: 99%). How easy it is to claim that the Antarctic is warming up, if one relies on mean temperatures, and concentrates on figures from the Peninsula, which reinforce the argument for a rising trend!

But this is not what is happening. In fact, the eastern Antarctic (i.e., the major part of the southern continent), is cooling. This cooling is also detailed by Parkinson (2002), who observed, during the period 1979–1999, a general growth of the ice sheets around the continent of Antarctica. Yet the western Antarctic is experiencing warming, most notably in the north of the Peninsula. Considering this particular case, Vaughan *et al.* (2001), wrote: '*it may be tempting to cite anthropogenic greenhouse gases as the culprit, but to do so without offering a mechanism is superficial*'. There are many who would not trouble with such reservations, but in this case the authors are quite ready to admit that '*we do not know the mechanism that causes it*'. Over and above the hypothetical greenhouse effect, two further culprits are habitually put forward: a change in oceanic circulation, and/or some modification in atmospheric circulation; but no real choice is made (it would be necessary in either case to supply reasons for the choice), and the question has been asked: '*can global circulation models (GCMs) help determine the mechanism*?' The reply is unfortunately in the negative (until such time as the opposite case is proved), because models cannot do this. They are unable to reproduce regional changes, and cannot integrate the real mechanisms of meridional exchanges (Chapter 8).

Palaeoclimatology (Chapter 9), and the current dynamic of the circulation over the Antarctic and in its environs (Figures 37 and 38), can help us to understand these phenomena. The marked cooling observed in summer and autumn (essentially, during the period of insolation) increases the power of the MPHs which slide off the eastern side of the Antarctic, the most extensive and highest part of the continent. As a consequence, the volume of the cyclonic southerly flux returning along the leading edge of the Mobile Polar Highs (MPHs) is increased, and the general area of the Antarctic Peninsula is thereby warmed (Figure 38). This intensification is attested by Gillett and Thompson (2003), who made the observation that '*the high latitudes of the southern hemisphere are dominated by a strengthening of the circumpolar westerly flow ... from the surface to the stratosphere*'; this is a rapid circulation (cf. Figure 33) driven by strengthened MPHs.

The (geographical) near-equilibrium in winter and spring for regions experiencing either cooling or warming (cf. Doran *et al.*, 2003) requires a much more sophisticated dynamic analysis. The (apparent and relative) warming might well be the result of a greater intensity in returning warmer air reaching the Antarctic, itself a result of the greater winter energy of the MPHs ... Remember that a mean temperature value represents only the result of many intervening factors, and among the most important of these is the aerological dynamic, which brings alternate warming and cooling.

Evidence of this cooling is provided by the recent evolution of temperatures (Figure 53) at weather stations at Punta Arenas (Chile), Cape Town and Calvinia (South Africa), and Adelaide (Australia). These stations, which are coastal and windy (meaning that any urban effect will be much attenuated), receive the full force of the Antarctic MPHs. The three curves show the existence of the CCO, but any sign of the RCO has been replaced by a more or less regular cooling. Figure 50 confirms the trend observed around latitude 40°S during recent decades. The alternation of the warm cyclonic northerly flux and the cold anticyclonic

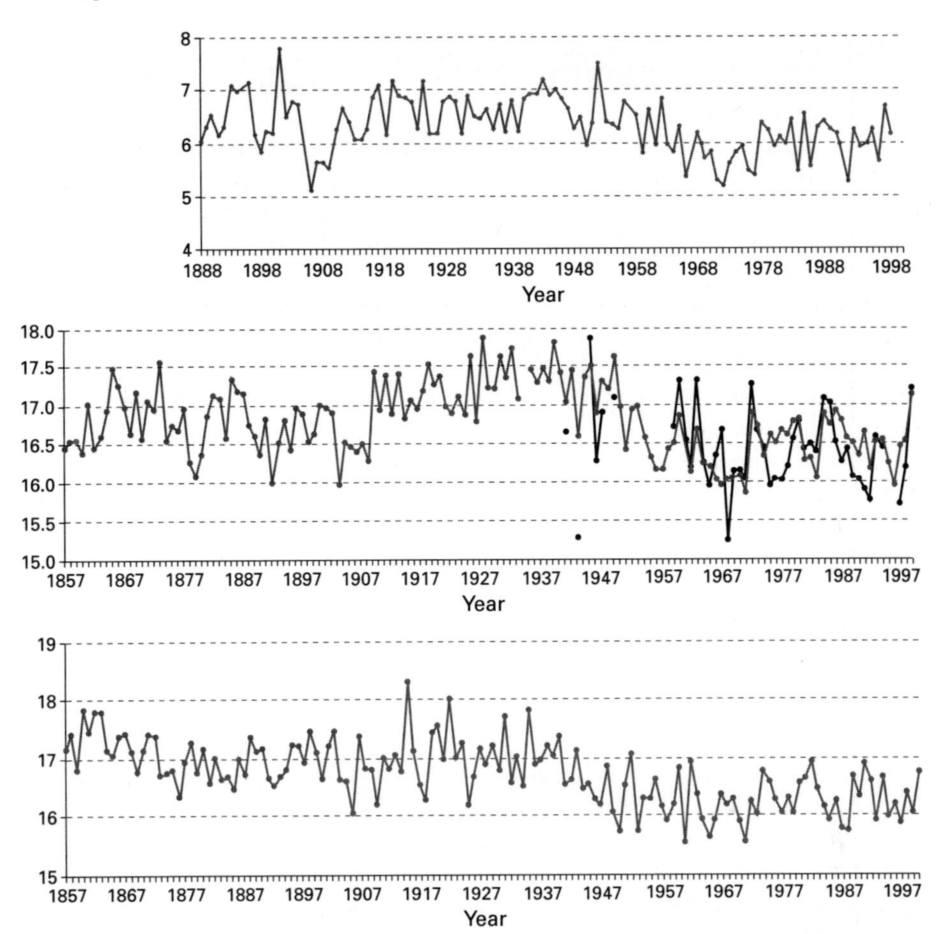

Figure 53. Annual mean temperature (°C), at: (*top*) Punta Arenas, Chile (1888–1998); (*middle*) Cape Town (1857–1997) and Calvinia (1946–1997), South Africa; (*bottom*) Adelaide, Australia (1857–1997).
From Daly (2003).

southerly flux, associated with the passage of MPHs, attenuates the mean value for cooling, but the overall continuous diminution of the temperature is a sign of the predominance of the cold air transported by Antarctic MPHs since the 1950s.

10.5 THERMAL AND DYNAMIC EVOLUTION IN THE ARCTIC

As has already been pointed out, the Arctic is much more complex, in both geographical and dynamical terms, than the Antarctic. Analyses of temperature are mostly concerned with the land areas bordering the Arctic Ocean – the Ocean itself has only recently revealed its thermal behaviour. The contributions made by

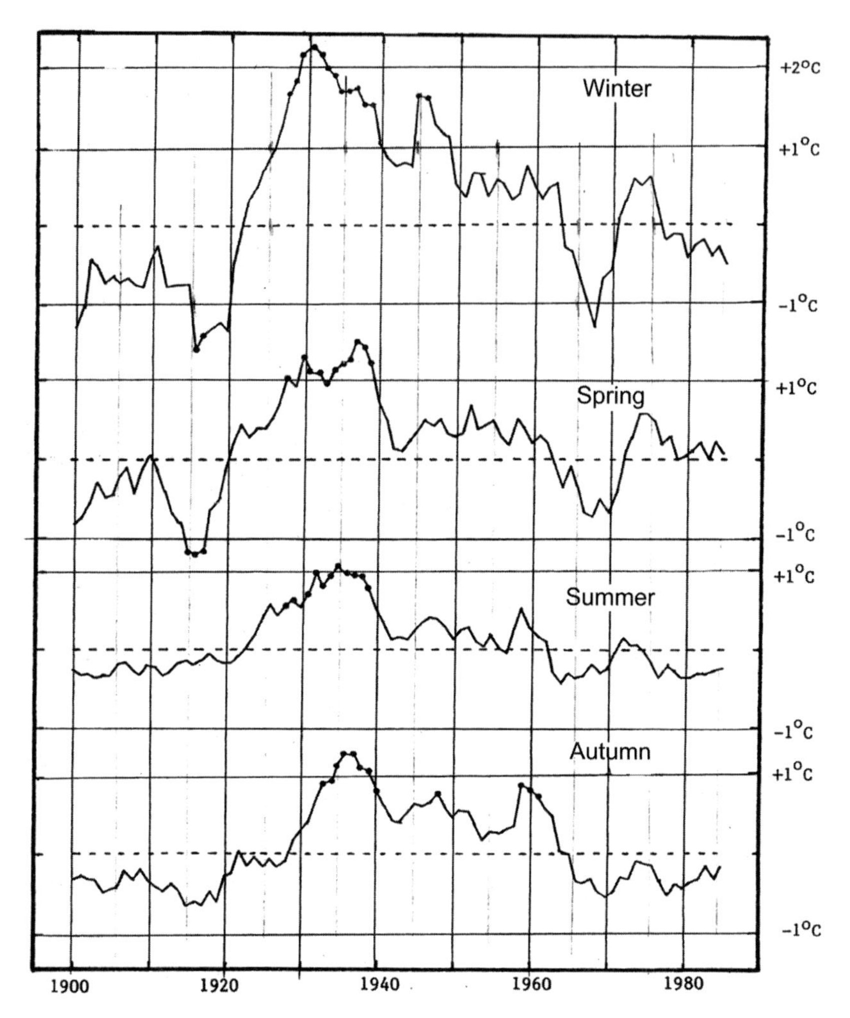

Figure 54. Five-year averages of seasonal temperature anomalies (°C) over the Arctic Atlantic, 1900–1987.
From Rogers (1989).

Rogers (1989) and Kahl *et al.* (1993) to our understanding of Arctic thermal phenomena are particularly useful.

10.5.1 Thermal evolution in the Arctic

Rogers (1989) analysed the seasonal variability of temperatures for the period 1900–1987 (Figure 54) in the Arctic part of the Atlantic, from eastern Canada to the Barents Sea. He showed that the highest temperatures, for all seasons, occurred in the period 1930–1940, and thereafter there was '*a steady decline in temperature in all seasons from the 1930s to the present*'. The 1960s were particularly cold in winter and spring. The mean winter temperature in the Arctic was then 2°C lower than it was in the 1930s, and 1°C lower in other seasons. Rogers also analysed pressures at the surface, and came up with the following facts: '*the largest pressure differences take place across a region extending from southern Greenland to Baffin Island ... each seasonal pressure time series is characterized by higher pressure in more recent times ... periods of more intense cyclonic activity are associated with higher values in the regional mean temperature, while higher pressure is associated with lower temperature*'. These are very important facts to bear in mind, and we shall come back to them later.

During the Cold War, the large volume of data collected in the Arctic was considered strategically sensitive, and the Arctic Ocean was at that time *Mare Incognita* as far as its meteorology was concerned. This was therefore quite useful for the modellers and for the IPCC, who could predict at their leisure and come to whatever conclusions they might, even if those conclusions seemed unlikely! In 1991 the great fund of data was finally available, having been gathered by radiosonde balloons released from floating platforms, and by sondes dropped from aircraft. The analysis of these data for the years 1940–1990 by Kahl *et al.* (1993) flatly contradicted the predictions of the models. The truth is that, in the lower atmospheric layers, the Arctic became markedly colder during a 40-year period, while there has been warming in the middle layers:

- For the whole of the ocean area, the cooling has been especially marked in winter (−2.44°C; confidence level: 0.95%) and in autumn (−4.14°C; confidence level: 0.99%); in spring and summer, the results were not significant.
- In the western part of the ocean (i.e., to the north of North America, precisely where the majority of MPHs originate, cooling has been as much as −4.40°C in winter (confidence level: 0.97%) and −4.99°C in autumn (confidence level: 0.96%); in spring and summer (the latter being negative), the results were again not significant.
- The cooling in the lower layers is confirmed (with reference to the aerological dynamic) above the western part of the ocean, by a marked warming (+3.74°C in winter; confidence level: 0.99%) in the 850–700-hPa layer (between 1,400 and 2,800 m). This warming in the middle layers is the manifestation of the intensification induced in meridional exchanges arriving from the south, above MPHs in the lower layers moving out of the Arctic: on the leading edge of these powerful MPHs, the returning cyclonic flux is accelerated.

So, contrary to all the much-vaunted predictions of the models, **the Arctic is not warming**. Far from it: cooling there is a certain fact. As Lenoir (2001) put it: '*the reception given to this news was derisory*' in the case of the IPCC, the partisans of the greenhouse effect, and even the journal *Nature*. Everybody should have been delighted finally to have access to such hugely important and as yet unworked data about an as yet little-known region: yet it was just as if the aforementioned article by Kahl *et al.* – exceptional in its novelty and accuracy – had never been

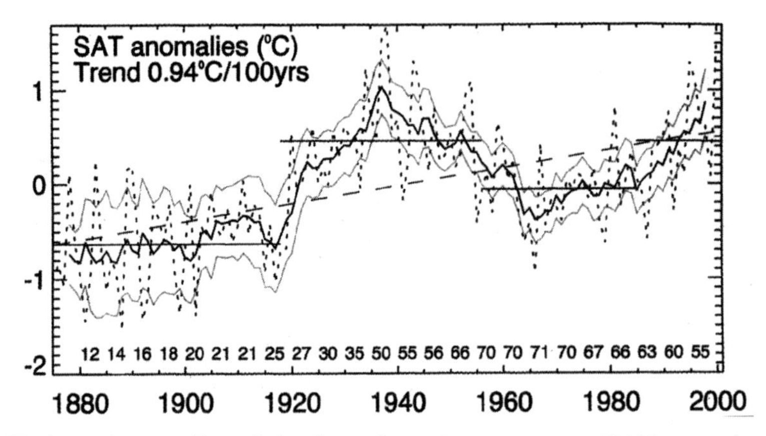

Figure 55. Annual anomalies of Arctic surface air temperature (SAT; annual mean °C). Numbers along the bottom indicate the number of stations used for averaging.
From Polyakov *et al.* (2002).

written. It was (and is still now) deliberately ignored, which is symptomatic of the IPCC's tendency to brush aside any 'problematical' arguments! Meanwhile, the IPCC and the modellers continued to churn out their pronouncements for the future in high latitudes, even though these ideas have been invalidated ever since 1993!

Rigor *et al.* (2000) used data from meteorological buoys, manned sea stations, and land-based weather units to confirm the cooling trend for the period 1979–1997, for the Beaufort Sea, eastern Siberia, and Alaska. On the annual scale, the eastern Arctic is experiencing warming, while the western area is getting slightly cooler. In autumn, Greenland, Iceland, and Siberia are warming up (by +2°C per decade), but the Beaufort Sea and Alaska are getting cooler (−1°C). In winter, the rise in temperature (+2°C per decade) involves the same regions, while for the Beaufort Sea, eastern Siberia, and into Alaska, the trend is −2°C per decade. In spring, there is warming almost everywhere, but in the western Arctic, the melting season is curtailed. No significant mean trend is observed in the summer.

In turn, Polyakov *et al.* (2002, 2003) '*conclude that the data do not support the hypothesized polar amplification of global warming*' predicted by the IPCC. Surface temperatures and pressures were examined for the period 1875–2000, north of latitude 62°N. The surface temperatures exhibited '*a stronger multidecadal variability in the polar region than in lower latitudes*'. Figure 55 shows two periods when temperatures were higher, the first being from 1930–1940 (the warmest years), and the second during recent decades, with, between these warmer periods, a marked cooling between 1940–1970.

The general mean across the Arctic is one of warming, by 1.2°C since 1875, which works out at 0.094°C per decade. It might seem surprising to calculate a trend for the whole of the period in question, since we do find in fact the same alternating tendencies (Figure 55) as are seen on the curves by Rogers (1989; Figure 54), with the exception of the final part of the curve, from the 1980s onwards. Polyakov *et al.*

(2003) recognise moreover that '*a trend calculated for the period 1920–present actually shows cooling*'; the trend would be more evident if the starting point were around 1940 (i.e., from the time of the thermal optimum). This cooling trend is more marked in winter and autumn. The regionally based chart for winter temperatures (in Polyakov *et al.*, 2002) shows, for the period 1961–1990, a strong concentration of higher temperature values stretching from eastern Canada to the Kara Sea (a sector swept by strong, warm returning southerly fluxes, cf. Section 10.5.2 and Chapter 12).

The curves presented by Douglass *et al.* (2004) and reproduced in Figure 50 go only as far as 60°N, but their charts showing trends in the northern hemisphere reveal, for the MSU and R2-2m aspects, an extended warming in the north-west Pacific, the north-east Pacific (west coast of the USA and Alaska), and the north-eastern Atlantic. However, a marked cooling is apparent in the Canadian Arctic, Greenland, and western Siberia.

The final part of the curve, from 1981–2001, was analysed by Comiso (2003) using clear sky satellite data. The mean temperature shows a generally positive trend, of 0.33°C ± 0.16°C per decade for the ice cap, 0.50°C ± 0.22°C per decade for Eurasia, and 1.06°C ± 0.22°C per decade for North America. The trend is slightly negative (−0.09°C ± 0.25°C per decade) for Greenland, and especially for its highest area. Seasonal trends are positive in spring, summer, and autumn, whilst they are generally negative in winter.

In summary, it may seem that the thermal behaviour of the Arctic is now relatively well understood, but thermal analyses still bring about some confusion, especially where mean values are employed, in both time and space. And if the Arctic, and notably its central part, constitutes what Kukla (2004) called '*a battleground of natural and man-made climate forcing*', this is because mean values have been used in an undiscerning way, and there is a lack of any perception of the overall dynamic of meridional exchanges. What we really need to do is to avoid mixing everything up, and establishing mean general trends which do not really correspond to what is happening. Above all, we must try to understand the dynamic of the thermal field, which is controlled by the aerological dynamic, since **the Arctic is by no means an isolated entity** which we could analyse without taking into account the rest of the northern hemisphere.

10.5.2 The aerological dynamic of the Arctic

Many different reasons for the way in which the Arctic climate behaves have been proposed, and the greenhouse effect figures in a good number of such propositions. Critics of the latter explanation are beginning to make themselves known, but their voices are still not much heard. Among other possible causes put forward, we may note: the LFO (Low-Frequency Oscillation); the NAM (Northern Annular Mode); the AO (Arctic Oscillation); the NAO (North Atlantic Oscillation); the effect of the oceanic thermohaline circulation in the north-eastern Atlantic; ... or even the El Niño–Southern Oscillation (ENSO) effect.

In fact, the cause and the mechanisms are still largely unknown, and certain researchers '*are now turning to regional models*' (Whitfield, 2003). But the way in

Figure 56. Dispersion of MPHs from the Arctic Ocean (bold arrows), and poleward return of cyclonic and warm airstreams (broken arrows), ahead of MPHs.
From Leroux (1996a).

which these models are constructed cannot supply the desired answer, since they do not integrate the mechanisms of circulation, and importantly, that circulation which involves transport to and from high latitudes.

Now, the concept of the MPH *can* provide an explanation of the mechanisms involved in thermal variations in the Arctic, because it offers a coherent schema of aerological organisation and function. Figure 56 shows the pattern of the dispersion of MPHs from the Arctic Ocean, and the way in which relief channels the cold air on its way to lower latitudes. For example, Arctic MPHs are channelled southwards mainly across North America, as a result of the presence of the mountains of Greenland, Ellesmere Island, and Baffin Island, with the American–Atlantic trajectory being favoured: one MPH follows this path on average every 2.3 days (Leroux, 1996a). In the northern section of this corridor, between the Rockies and Baffin

Island, mainly in winter when snow and ice lie all around, the leading edges of the MPHs propel only dry, cold continental air northwards. It is not surprising, then, that the temperature in northern Canada (the western Arctic) varies only slightly, or shows a negative trend. As they move further south, the MPHs divert air which is more moist, and warmer, northwards. This southern air, on its cyclonic path, is associated with areas of low pressure on the leading edges of the MPHs, with their low-pressure corridors and closed lows (or cyclones). This warmer air, like the cold air running in the opposite direction, is channelled by relief towards the pole in the lower layers, along favoured corridors.

As is the case in the Antarctic, the very dynamic of MPHs is a complicating factor when we try to determine the resultant values for temperature and pressure, since the colder and the more powerful an MPH is, the more intense and warm its returning cyclonic flux will be (Chapter 8). What is more, during the passage of an MPH system, with its alternating warmer and colder spells (Figure 24), it is not the extreme values for temperature associated with the low or the anticyclone, but the (shorter) duration of the warm sequence compared with the (longer) duration of the cold sequence which is the true criterion upon which we should base our investigation. However, the durations of these warm and cold phases are not considered when mean temperatures are worked out, and so thermal variation is only imperfectly factored in. This is because of the method by which mean temperatures are calculated ($Tm = Tx + Tn/2$), where only the maximum (Tx) and minimum (Tn) temperatures are used as a basis, and not – as should ideally be the case – 24-hourly measurements. Further complications arise with the lack of representativeness of mean temperatures, since some regions will be cooling while others are warming, as a function of the dynamic of each unit of circulation affecting the Arctic (Figure 26, Chapter 8; Chapters 12 and 13). Also, most of the data come from weather stations situated around the periphery of the Arctic Ocean, an area itself as yet imperfectly described (and when it is described, by, for example, Kahl *et al.* (1993) their results are ignored!). Any analysis based on mean values across the whole of the Arctic and its peripheral regions will not bring immediate enlightenment on the subject of their actual thermal evolution.

Serreze *et al.* (1993) made a synthesis of previous studies, and analysed the dynamic of anticyclones and cyclones in the Arctic basin north of 65°N, between 1952 and 1989:

- *In winter* – cyclones are most frequent in the eastern part of the Arctic, from eastern Canada to the Kara Sea, with the maximum number in the vicinity of Iceland and the Norwegian Sea, where depressions are deepest; they occur over the Baffin and Kara Seas, too, though pressures are not as low. Cyclones are markedly less frequent at the centre of the Arctic Ocean, and in Siberia and Alaska.
  – In the case of anticyclones, they are at their most frequent over the western Arctic, eastern Siberia, northern Alaska, and Greenland. Anticyclones '*associated with the Siberian, central Arctic Ocean and Alaska/Yukon frequency maxima tend to be strongest*' (Serreze *et al.*, 1993). In winter, the situation is

dominated over Asia by the existence of an enormous lenticular Anticyclonic Agglutination (AA), fed by MPHs crossing it and passing out through China. It is maintained by cold continental ground. This cell is known as the 'Siberian', or 'Sibero–Mongolian' anticyclone. The enormous size and robust nature of this high-pressure cell prevents cyclonic circulation from regaining the polar regions.

- *In summer* – the general dynamic resembles that of the winter, though there are differences in both strengths and distribution: the anticyclones are weaker, and the cyclones are less deep than in the winter; and, importantly, the 'Siberian' anticyclone has disappeared. Cyclonic activity is now more extensive, and '*cyclones are distributed more widely throughout the Arctic*'. Differences exist also in the pattern of distribution of high-pressure areas, and anticyclones are generally further apart, with a maximum still observed over Greenland, Siberia, and Alaska, whilst anticyclones are more numerous over the Beaufort Sea and the Canadian Archipelago. The highest pressures are encountered to the north of Alaska and Canada. Yet another difference is apparent between winter and summer: the numbers of anticylones and cyclones, which are '*least numerous, but strongest during the winter months*' (Serreze *et al.*, 1993). So the more intense cold airstream flows out through the intermediary of larger MPHs, which create deeper depressions and the expected stormy conditions of the winter.

MPHs, vectors of cold air, originate preferentially over the western part of the Arctic (to the north of both Canada and eastern Siberia), whilst returning cyclonic fluxes, formed further to the south within adjacent units of circulation (Figure 26), vectors of warm air, arrive preferentially in the eastern part of the Arctic: the Norwegian and Barents Seas are '*common entrance zones for systems migrating into the Arctic*'. This dynamic and climatic dissymmetry is fundamental in the Atlantic Arctic, or in the Pacific Arctic, as are their seasonal configurations. If these aspects are not carefully taken into account when calculating mean temperatures and pressures, the resulting conclusions will definitely be unrepresentative.

Another important aspect arising from the work of Serreze *et al.* (1993) is an analysis of the evolution of the seasonal dynamic. Generally speaking, whatever the season, anticyclones and cyclones have been on the increase during the period studied (1952–1989) (Figure 57). In the case of anticyclones, with their important role in the formation and characters of depressions, they show a more marked increase in spring, summer, and autumn. In winter, it is not easy to estimate the number of anticyclones in areas north of latitude 65°N, since the most energetic anticyclones soon spill over beyond this line of latitude. But the number of winter cyclones is also on the increase, which suggests that the anticyclones which engender them are also more frequent during the winter.

10.5.3 The intensification of meridional exchanges in the Arctic

So the number of MPHs (i.e., the volume of cold air exported by anticyclones) has been constantly increasing since the 1950s, in all seasons (Figure 57). This has caused

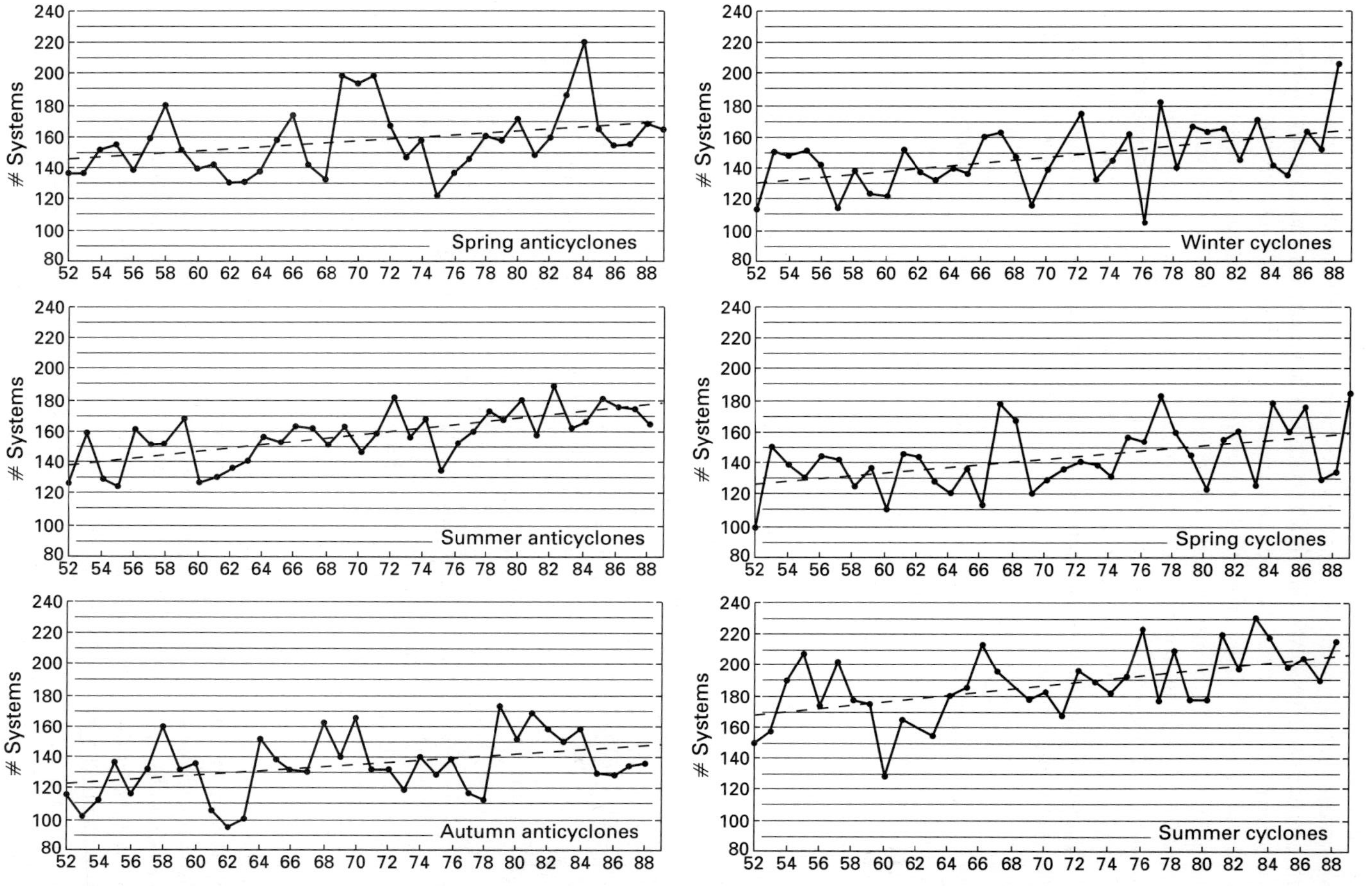

Figure 57. Evolution of the number of anticyclones (*left*) formed north of 65°N for spring, summer, and autumn, and of cyclones (*right*) observed north of 65°N for winter, spring, and summer, from 1952–1988 (in autumn the trend is not significant). Dashed line: linear regression.
After Serreze *et al.* (1993).

a similar increase in the number of associated cyclones (i.e., an increase in the volume of warm air returned towards the pole), with, as a result, an increase in stormy conditions. Serreze *et al.* (1997) also confirm that '*the number of cyclones penetrating into the Arctic from the North Atlantic has increased*'. This penetration has led to a rise in temperatures on the mean regional scale, together with lower pressures (cf. Rogers, 1989); or, as Polyakov *et al.* (2003) noted, '*a shift in the atmospheric pressure pattern from anticyclonic to cyclonic*', a remark which might also be appropriate on a geographical level.

This kind of evolution towards a rapid mode of circulation can be due only to cooling in the Arctic.

- Remember that the kind of warming (which has not been observed) proposed by the IPCC would in fact bring about a diminution in the outflow of cold air, less powerful MPHs, fewer anticyclones, fewer associated depressions, and a less stormy scenario, all associated with a slow mode of circulation (cf. Figure 35).
- The fact that the air exported by MPHs is colder during the winter is immediately apparent from mean temperature data involving the western Arctic, as a result of the marked thermal (and barometric) division within the Arctic. Additional evidence has come from sonde investigations in the western Arctic (Kahl *et al.*, 1993, 2000): while the lower layers are cooling, the middle layers are becoming warmer above the inversion which marks the top of the MPHs, and here is evidence of the more intense nature of the warm returning flux heading for the pole above those MPHs.
- On the other hand, during the summer, and because of the more balanced distribution of cyclones right across the Arctic basin, warming associated with incoming air from the south tends largely to mask this evolution in mean temperature. There is, however, no ambiguity in the curve showing the evolution in the numbers of MPHs (Figure 57). This is supported by Serreze *et al.* (1993) when they note that '*the trends in both cyclone and anticyclone numbers are largest during summer*', a fact also observed by Pommier (2004) for the period 1950–2001.

Temperature analyses can provide only an imperfect picture of the true dynamic, and it may even be a contradictory or distorted picture. Take as an example the curves of Figures 54 and 55: their general trend is the same, and only for recent decades do they show an accentuated warming (Figure 54). After the climatic optimum of the period 1930–1940, the temperature falls (as Wallen (1984) put it: '*for the Arctic and northern regions the cooling started as early as in the beginning of the 1940s*'), and activity associated with reinforced MPHs increases. Until the period 1970–1980, the outflow and ingress of cold and warm air respectively were on average quite well balanced in the Arctic (and in the means of temperature). From the 1980s onwards, the mean warming trend was evidence of the intensification of advections of warm air towards the pole, from areas all around the Arctic. But we are dealing here only with a mean 'warming', since, within the adjacent aerological units from which the warm air comes, the weather may show contrasting behaviours, and regional temperatures rise and fall (as we shall see later in Chapters 12 and 13).

The acceleration of the southerly cyclonic fluxes can be driven only by one other factor: the increasing frequency and/or strength of the MPHs (i.e., by the legacy of the cooling which began after the climatic optimum of the period 1930–1940. The incongruity inherent in the fact that the evolution of mean temperature rises and then falls (Figure 55), while trends in anticyclonic and cyclonic activity rise undiminished (Figure 57), shows that the temperature 'rise' during the period 1980–2000 is an artefact resulting from the use of means in the evaluation of **'the' temperature of the Arctic, which does not exist in the singular**, and is not uniform. On the contrary, it is very diverse (see Part Two conclusion after Chapter 13).

The concept of the MPH restores the logic of phenomena. Since the birth of MPHs is associated with the polar thermal deficit, a modification in their frequency, as noted here, with an undeniable and generalised increase in activity (Figure 57), must necessarily be the result of a widening of this deficit.

Thus, in the Arctic, just as in the Antarctic, the thermal deficit intensifies the outflow of cold air from high latitudes.

This dynamic evolution in the Arctic is a fundamental datum, not just for the Arctic itself, but for the whole of the northern hemisphere. This has been confirmed for the Arctic, mainly through the changes in the extent and thickness of sea ice (Chapter 14) at the outlets to adjacent aerological spaces. In fact, within those spaces develop the depressions which will complete their journeys in the Arctic. These spaces, such as the North Pacific and the North Atlantic, are experiencing what Flohn *et al.* (1990) have already called a definite '*rise in kinetic energy*'. Here, some regions are cooling, while others are warming up, as a function of their location *vis-à-vis* the trajectories of MPHs and their associated depressions. The consequences of the dynamical effects triggered at the pole by MPHs will have their repercussions by return within the Arctic. In Chapters 12 and 13, we shall examine these units of circulation, within which the temperature of the Arctic is determined, since they are the birthplaces of the lows (and warm fluxes) which will end their journeys in the Arctic basin.

10.6 CONCLUSION: THE GREENHOUSE EFFECT DOES NOT CONTROL THE EVOLUTION OF TEMPERATURE

The IPCC maintains that the evolution of temperature is controlled by the levels of concentration of emissive gases in the atmosphere. Various arguments are put forward in support of its stance: palaeoclimatological evidence (Chapter 9), the 'hockey stick' curve for the last thousand years (Figure 43), and the 'official' curve of global temperature anomalies since 1860 (Figure 46). What is the validity of these proofs, which are said to be 'irrefutable'?

- **Palaeoclimatology does not show any causal link between greenhouse gases and temperature**, but proves, on the contrary, that it is temperature (which is itself determined by other factors) which indirectly causes, with some delay, variations in the levels of greenhouse gas concentrations. Even if the concentration of these gases later intervenes, to enhance the greenhouse effect, this has absolutely no effect upon the initial process.

- *The 'hockey stick' curve* (Figure 43) hastily adopted by the IPCC – following a reversal in thinking and a denial of previous climatological research – has not been validated. It **deliberately distorts the last 1,000 years of climatic history**, a millennium which has nevertheless been well documented (Figure 42).
- A comparison of Figure 32, which shows the variation in CO_2 levels over the last 1,000 years, and Figure 42, which confirms the existence of the MWP and the LIA, reveals that there is no causal 'relationship' between the respective evolutions. The absence of any variation in CO_2 levels (in Figure 32) cannot really explain the important climatic changes experienced in the millennium in question: most notably the warm episode of the MWP and the very cold episode of the LIA, during which CO_2 levels show (apparently) no change.
- The highly dubious juxtaposition (Figure 43) of a long section of the curve, based on (unconfirmed) estimated temperatures, and a shorter section based on 'reconstituted' (and equally debatable) temperatures, does not demonstrate the so-called 'exceptional' nature of temperatures in recent times. Observations reveal, though, that this period is not as exceptional as the IPCC would have us believe!
- The 'official' curve for the evolution of mean global temperature ('reconstituted') for the period since 1860 (Figure 46), the climatic significance of which has not been analysed, is also very debatable. The reasons for this are many, chief among them being the reliability or otherwise of the data, and the urban influence question. This curve does not correspond to the evolution of greenhouse gas concentrations, and **the comparison does not even demonstrate a co-variation during the whole of the period analysed**:
  - from 1918 to 1940: a marked temperature rise, but only an insignificant rise in CO_2 levels;
  - from 1940 to 1970: a fall in temperature, but a marked rise in CO_2 levels; and
  - since the 1980s: a marked temperature rise (RCO) and a marked rise in CO_2 levels.
- It can therefore be seen that, in the course of the thousands of years mentioned in both the previous and the present chapters, there is only one short period, the RCO, about 20 years long, for which one might claim some hypothetical 'relationship' between changes in greenhouse gas levels and the evolution of temperatures (reconstituted)! And that 'relationship' might well be (unless better information becomes available; cf. Chapter 13) just a co-variation (i.e., a simple coincidence), **without any proven link to some causal mechanism**.
- The very reality of the RCO is seriously called into doubt (Figure 50): no evidence of global warming is apparent. So, **nothing remains of the IPCC's greenhouse effect scenario**!
- The notion of 'global temperature' is really only a mental construct, *since thermal evolution is both regional and diverse* in nature. An analysis of the

dynamic of temperature at high latitudes will show **how very wrong the predictions of the models are**, demonstrating that:

- warming in the western part of the Antarctic, held by the IPCC and the media to be proof of 'global warming', has absolutely nothing to do with the greenhouse effect. This regional warming is associated with the dynamical nature of MPHs (associated with the geographical factor), and is a result of cooling in the eastern Antarctic, which brings more intense volumes of warm air towards the south pole; and
- the mean 'warming' claimed for the Arctic also has no confirmed link with the greenhouse effect. This rise in temperature involves only certain well-defined aerological regions of the Arctic basin, towards which cyclonic air masses are channelled, heading for the pole, and intensified by the cooling of more frequent and more powerful MPHs.

- The climatic significance of the IPCC's temperature curve (Figure 46), not normally a subject for discussion, is considerably reduced, in both temporal and spatial terms. For neither tropical nor higher latitudes is the 'greenhouse effect' scenario confirmed for recent decades, and the same is true for the predictions of the models. The rise in temperature in polar regions is localised, and is not a feature of the whole of either the Arctic or the Antarctic; it is, however, associated with the dynamic of MPHs, **driven by cooling, which, through the intermediary of MPHs, intensifies meridional exchanges**.

Temperature is only one of the many parameters of climate. Remember that **no climatic parameter can evolve in isolation**. It is therefore necessary to examine the way in which weather evolves to provide (or deprive us of) precipitation (Chapter 11). **The way in which temperatures evolve at high latitudes is of enormous significance for the intensity of meridional exchanges**. It will also be of use to observe, within certain units of circulation (Chapters 12 and 13), the regional climatic consequences of the initial cooling of MPHs, those motors of that general circulation through which changes in the climate are transmitted.

11

The observational facts: Weather, rainfall, and drought

> *Drought, floods, heat and cold, snow ..., we've never seen the like!'. This is the unflagging leitmotiv which is being heard ever more frequently. Meteorology is truly the source of all our woes! ...*
>
> G. Dady. Direction de la Météorologie, Paris, *Bulletin d'Information*, No. 43, 1979.

As was mentioned in Chapters 3 and 7, the Intergovernmental Panel on Climate Change (IPCC) has forecast an intensification in the hydrological cycle (i.e., an increase in rainfall and especially in severe downpours). This prediction is based upon calculations using models, or, to put it another way, upon the supposed relationship within the models between rising temperature (T), evaporation (ev), and the resulting hypothetical rain (R), according to the very simple 'equation' 'T/ev/R', which may be further simplified to 'T/R'. They even envisage a proportion of 2–3% more precipitation per 1 additional degree Celsius. If the only thing that is fed into the model is the equation 'T/R', it is fairly obvious that R will increase if T increases! Amazing! Using the same equation, we can also 'predict' that the intensity, as well as the quantity, of rain will increase, another quite automatic result of the postulated increased temperature!

At the same time, however, it is considered that heat leads to aridity, from the moment that evaporation outweighs rainfall! So the equation can be schematically reworked as either 'T/ev < R' or 'T/ev > R', and the prediction then becomes: '*more rainfall in tropical regions and in high latitudes, but less in subtropical latitudes*'. Which may be more clearly expressed as: '*more rain in already rainy tropical regions, more rain in high latitudes which are supposed to be warming up, but less rain in already dry regions, especially in summer, when it must be hotter and therefore evaporation will be more intense*'!

Here, then, is a most practical and universal equation, which must – evidently – function in both directions, and also explain the absence of rain. Now, what has to happen to make rainfall diminish, or drought dominate? This same 'equation' suggests that if the temperature falls, then rainfall decreases (with 'mathematical' certainty, since if R increases, T increases): it's logical! This means, then, that cooling is synonymous with drought! In other words, the idea of an association between a heatwave and an episode of drought, a situation which has quite frequently been observed, is 'erroneous', and can only be a false impression, inappropriate in this instance ...

This is obviously rather simplistic; might one say numbing, in its simplicity? In fact, only those who do not understand the rudiments of climatology will believe such predictions! Once again, let it be said that there is no direct relationship between precipitable water and water which is actually precipitated, since pluviogenesis (i.e., the *ensemble* of processes which either promote or hinder precipitation), require other conditions which the models do not take into account. **Now, the conditions for pluviogenesis are closely associated with disturbances**, and the precipitation involved is of symbolic significance, because it is produced during situations of disequilibrium. The absence of rain, or drought, is similarly associated with certain dynamical conditions. Consequently, to predict rain or the lack of it by using magical equations, which hold that phenomena exist *in situ*, while not invoking the mechanisms of weather and without taking into account the linkage between general circulation and disturbances, is no more than an aberration, scientifically speaking. However, this is what the models do, and it would seem that they are now telling us of the likelihood of catastrophic weather to come: there will be extreme rainfall, floods, storms, cyclones, and with all this the threat of droughts and heatwaves! And yet, as the IPCC itself states (cf. Chapter 7), models cannot predict the weather of the future. They are certainly absolutely incapable of foretelling the way in which rainfall and/or droughts (a part of that weather) will evolve.

There is another considerable obstacle to the prediction of the weather: we simply do not fully understand its mechanisms, since no climatic parameter such as rainfall (or the lack of it), or temperature, can vary in isolation. All these parameters are components of the weather, and more especially of the disturbances which stamp its character. However, there are deep disagreements about explanation of these disturbances, and especially about the origin of disturbances in temperate latitudes, while much still remains to be discovered about tropical disturbances. So how can we 'predict' eventual consequences of the greenhouse effect – if we do not even know at what stage, and with what degree of intensity, it may intervene in the development of a perturbed system? Do we even know if it is capable of such intervention?

11.1 WEATHER

We have already mentioned the conceptual *impasse* in which meteorology finds itself. Obviously, it is out of the question to try to cover this subject exhaustively, which would involve 'revising' the whole spectrum of meteorology. But we definitely should discuss why the models and the IPCC, which pride themselves on the omniscience of their predictions, cannot claim that the greenhouse effect (i.e., humanity) is responsible for rain, and for fine weather ... responsible, in brief, for all the vagaries of the weather now and in the future!

11.1.1 Meteorological schools of thought, and concepts

Why is it necessary to discuss these basic points again? Firstly, because they are essential points, and secondly because the general public (and indeed others) are not aware of them. Everyone in fact seems to believe that all is known in the field of meteorology/climatology, and that it has reached such a state of perfection and maturity that it can no longer make mistakes. It can therefore only be a good thing to look at the current state of the discipline, and to point out its shortcomings. We shall review the formulae and/or false accounts of the weather, and the 'tunnel visions' and fixations which are the legacy of the various schools of thought. If we bear in mind how much remains to be done before we can even think about predicting the weather in the year 2100, we can then make the distinction between the possibilities for *real* prediction, and the results of mere crystal-gazing.

The climatological school

The climatological or 'statistical' school, which flourished in the late 19th and early 20th centuries, is concerned with the analysis of mean data. Its description of the weather is derived from the mean meteorological situation, particularly as shown on charts of mean pressure and resulting winds. This secular school of thought is still very much with us, through the names of so-called permanent 'centres of action'. These are said to organise (temperate westerly and tropical easterly) circulation, and the behaviour of disturbances. In the context of Europe, and the whole of the North Atlantic, there are constant references to 'the Azores Anticyclone' (responsible for fine weather) and to the 'Icelandic Low' (bringing bad weather). In the North Pacific, the statistical low is the 'Aleutian'. Elsewhere, these 'centres of action' bear names such as the Hawaiian Anticyclone, the Easter Island Anticyclone, the Saint-Helena Anticyclone, and the Mascarene Anticyclone.

The origin of anticyclones is supposed to involve air subsidence (cf. Chapter 8), but this suggestion is unable to explain annual migrations, or reinforcements ('inflations') or diminutions ('deflations') in pressure. Neither can it explain how tropical circulation is nourished (to cause, for example, trade wind accelerations), or the specific stratification observed in the trades. Also, anticyclonic cells are talked about as though they were living beings. In France, weather forecasters and the media go in for '*meteorological animism*', announcing the 'return' (bringing good weather) of the anticyclone, '*retreated to its native Azores*' (as if it had, by its absence, allowed bad weather to break through)!

The reliance on mean values leads people to believe that these 'centres of action' are somehow 'fixed', while, in reality, everything moves. Another result of these assumptions is the **incredible confusion between scales of phenomena: (statistical) entities as defined by mean values cannot (by definition) exist on the synoptic scale**. It is therefore not possible to explain weather, as it is happening (on synoptic scale), by referring to some (statistical) 'centre of action'. Physically, a system which is, hastily, considered for example as the 'Azores Anticyclone' can only be, on this synoptic scale, a Mobile Polar High (MPH), or an agglutination of MPHs!

The use of models, with its predilection for the use of mean data, really represents a strong swing back to the climatological school, with its baggage of mythical meteorological 'personages'. This school also proposed indices for the measurement of 'oscillations' (i.e., variations in the respective strengths of anticyclones and depressions). A case in point is that of the North Atlantic Oscillation (NAO), between the 'Azores Anticyclone' and the 'Icelandic Low', or the North Pacific Oscillation (NPO), between the 'Hawaiian Anticyclone' and the 'Aleutian Low'. These oscillations have in turn become real live (and mysterious) 'entities', but this still does not define the relationship between highs and lows, and the origins of these entities are still not known, and neither are the causes of their variations (cf. Chapters 12 and 13).

Remember, the MPH concept does away with any ambiguities of scale, with its message that 'anticyclonic cells' are nourished by MPHs, which gather to form Anticyclonic Agglutinations (AAs) (cf. Figure 25, Chapter 8). The concept also explains the true nature of the 'oscillations' (Chapters 12 and 13).

The frontological or Norwegian school

Founded in the 1920s in Norway, and also called the Bergen school, this way of thinking admits the existence of air masses of different densities, with cold, heavy air lifting lighter, warm air above it. The line along which these masses meet and interact is known as a front. Bjerknes and Solberg (1921) proposed their 'Norwegian' model of mid-latitude disturbances, with two fronts, centred on a depression at the heart of the disturbance. The frontological concept took on the character of a dogma, and synoptic meteorology outside the tropics became dominated by the idea of a 'Polar Front', with analysis of surface charts still now normally based on these principles.

However, many imperfections remained, chief among them the fact that the frontological schema did not specify the horizontal and vertical boundaries of the fronts. This might have seemed justifiable in the days when the concept was elaborated, since only a limited geographical area was involved, and little was known about features at altitude. But since the use of radiosonde methods from 1927 onwards and the satellite technology which flourished during the 1960s, our vision of phenomena has been much broadened. We can no longer therefore allow such a lack of accuracy, although fronts are still shown as discontinuous on surface charts, while no vertical structure is defined, with diagrams sometimes showing fronts climbing to the tropopause!

The Norwegian school focused upon the depression, at the centre of disturbed systems, but omitted to discuss the origin of the initial depression, and it is still unknown, or attributed to altitude (cf. the FASTEX experiment below). This focus is still very much with us, and manifests itself in our practice of looking for the lows which are 'responsible' for the weather, and giving them names like the Icelandic Low or the Aleutian Low, or even Ligurian or Cyprus Lows ... But these depressions cannot in fact be responsible for anything, because they are dynamical consequences, and not causal. The MPH which is really responsible for these depressions is never actually identified.

It is impossible to form a synthesis of the two schools already mentioned, as the first one works on the scale of statistics, with centres of action defined by means, and the second on the synoptic scale, or the scale of real time. This *impasse* has led to the belief that there exist two independent fields, the '*permanent field*' and the '*perturbed field*'. This dichotomy is, scientifically speaking, a considerable aberration, since it is quite obvious that there is but one field, that of reality, and **this real field is both mobile and perturbed**.

The concept of the MPH does away with this false distinction between fields, and keeps, from the 'mass' concept, the notion of density, with cold air lifting warm air above it along a front (which is cold, since the notion of a 'warm front' is nonsensical), and that front has defined boundaries. Moreover, the origin, trajectories and characteristics of depressions are attributed to MPHs.

The dynamical school

The dynamical school, also known as the kinematic or 'disturbances' school, invokes neither contrasts in density nor frontal discontinuities, but rather discontinuities in speed and the directions of fluxes. The construction of the kinematic field is based upon wind data, and the method of streamlines is applied when dealing with all high-altitude charts. The discovery by radiosondes of high-speed winds at altitude (jets) led this school to attribute a primary role to the upper levels of the atmosphere. The planetary wave, in the form of 'Rossby waves' (1939), undulations in the high-altitude westerly jet stream, became the causal atmospheric phenomenon in temperate regions. Phenomena in the lower layers, both anticyclones and depressions, and the weather associated with them (and in particular the Norwegian Low, the origin of which was now at last 'known'), were relegated to being mere consequences of the undulations of high-altitude jets.

The impossibility of control of phenomena from higher levels This school still dominates when high-altitude charts are compiled, but, since surface charts depend upon the 'mass' method, there is in any analysis of them no immediate liaison between the lower layers and the upper levels: they are understood using different concepts and different methods. In particular, the fronts drawn for surface charts are no longer there on charts for the lower layers, where the frontal surface is deployed. This incongruity cannot facilitate the depiction of the vertical structure of the fronts, or other aerological discontinuities. The dynamical school has also given the impression that it has solved the problem of the origin of the 'initial depression' located at the centre of disturbances in middle latitudes. We hear more and more often that '*updrafts are caused by the diverging part of the high-altitude wave*' (Figure 58), but this hypothesis is unexplained. Neither is any explanation given for the existence of fronts within the vortex; the fact that these fronts are not drawn into the motion of the eddy; or the presence of a meridional

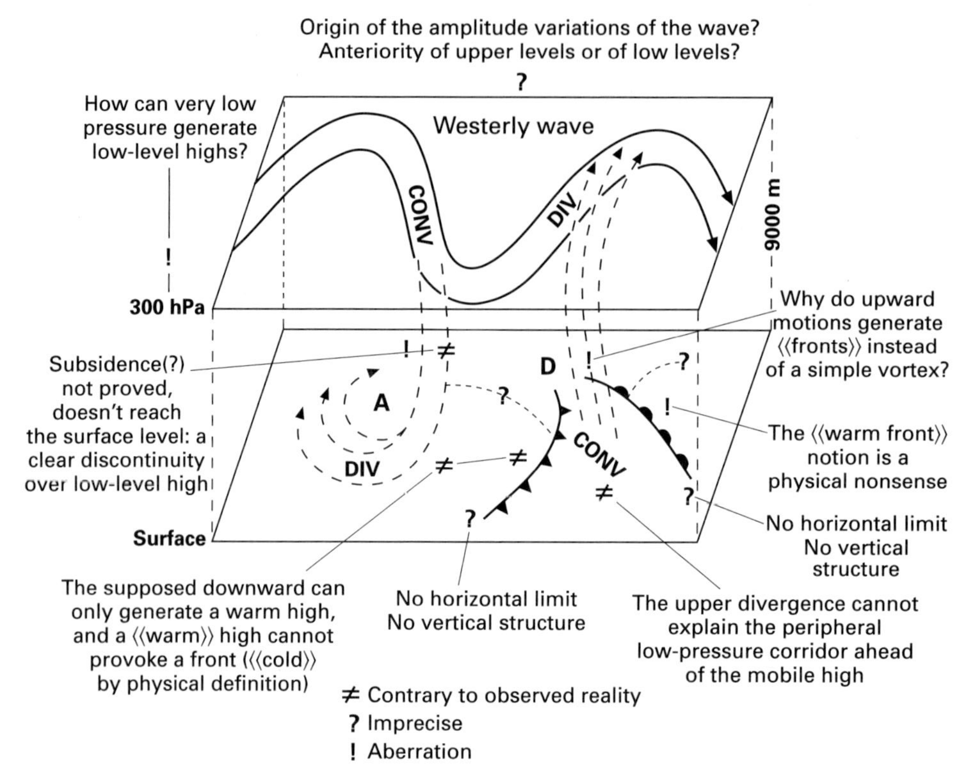

Figure 58. Scheme of the supposed connection between upper and lower levels of the troposphere.
From Leroux (2000).

low-pressure corridor (Leroux, 2000). The dynamical concept does not, however, dispense with the possibility that the high-altitude undulation is quite simply the consequence of lower-layer phenomena.

As for the formation of an anticyclone in the lower layers by subsident movements, this is even more hypothetical. It would indeed take a real 'physical miracle' for levels at which the air is rarefied to feed, with very rapid downward movements (a process never observed anywhere) into high-pressure areas, often at low temperatures. But according to theory, high-pressure areas can only be very warm, as a result of compression, and they are therefore utterly incapable of engendering a front!

Nevertheless, Malardel, of the *Centre National de Recherches Météorologiques* (CNRM) at Météo-France/CNRS, could still write in 2004, in explanation of the formation of a front in middle latitudes, that '*particles of air travelling within the disturbance ... came from a high-altitude level situated just below the tropopause ...*', and that '*the wind caused the particles to set off on a downward path, southwards ...*'; then, with the particles now having descended into the lower layers, '*this air is drier and colder than neighbouring air on the other side of the front*' (Malardel, 2004). So this extract from the 'hymn sheet' of the dynamical school states that the pressure of this subsident flux (supplied from an 'airless' level) goes from approximately 200 hPa to a value of more than 1015 hPa at the surface, but manages to remain cold in spite of tremendous compression! Of course, the CNRM article does not specify what factor might have caused this air to descend from above! Could there be a better example of a 'physical miracle'?

We might mention another verse from the same hymn sheet, which tells us that 'subtropical highs' (cf. Chapter 8) are similarly the result of subsident movements originating from levels above (the descending branches of Hadley cells), which this time cause considerable warming! What physical law is it, which makes descending air, according to its whims, become sometimes warm and sometimes cold?

The use of models has not brought with it any new concept, and indeed the modellers seem to pride themselves on doing without concepts. Nevertheless, the dynamical school constitutes '*the doctrinal corps of numerical forecasting*' (De Moor and Veyre, 1991). It comes therefore as no surprise, since there is (theoretically) no other concept within the models, that priority is always given to upper level phenomena, and especially the 'high-altitude jet', which is automatically considered to be responsible for almost all meteorological phenomena of the lower layers. Figure 58 aims to show schematically both the impossibility of lower layer phenomena being controlled by undulations in the high-altitude westerly wave, and the incoherence of the relationships claimed by the dynamical school, although these relationships are still in common use.

Dynamics of weather over North America Weather over North America, where the 'Rossby waves' are born, is said to be marshalled by a high-altitude wave 'anchored' over the Rocky Mountains following the principle that '*the stationary planetary waves in the atmosphere are forced by orography*' (Trenberth, 1990). This wave is said to form a crest lying in a north–south direction along the mountains, and a subsequent trough above the Mississippi valley. Thereafter, it rises towards the north-east as it approaches the eastern seaboard. If all this seems to 'fit' within North America, with its Rocky Mountains, how does it sit elsewhere, for example in Europe, or even over the Pacific, especially south of the equator (cf. Figure 26, Chapter 8)? Theory tells us that this high-level wave is capable of sending polar air southwards and that, as a function of its curvature, which is more pronounced in winter than in summer, it modulates the intensity of cold air in the lower layers, whose depth is, however, not more than 1,500 m ...

The wave therefore becomes the deciding factor in the weather and in its excesses, for example the rigours of winter: '*the winter of 1993–1994 was one of the coldest ever known in eastern Canada*' (according to *Environnement Canada*, 1994). The reason is immediately obvious: '*this winter, the high-pressure crest in the west and the barometric trough in the east were more marked than usual*' (Gergye, 1994). So, as is shown in Figure 59, it is necessary only for the (high-altitude) wave to be more pronounced, for a 'better' warm airstream across Alaska, and for cold air to descend

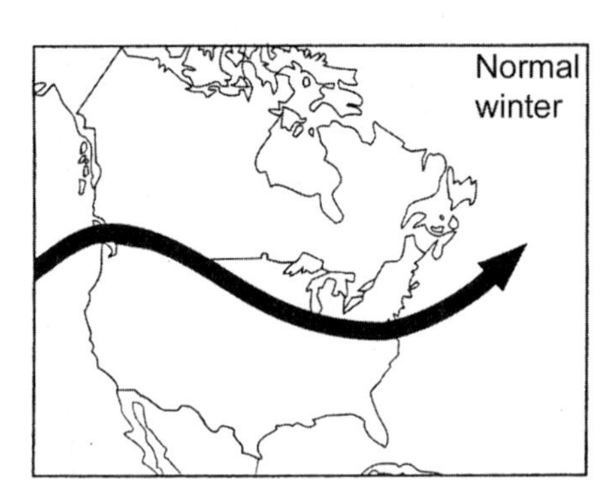

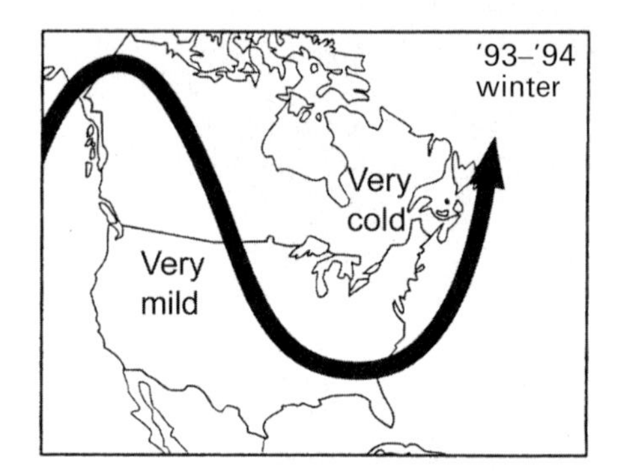

Figure 59. (*left*) Mean circulation in winter. (*right*) Circulation during the winter of 1993–1994.
From Gergye (1994).

'more' (in the lower layers) towards the east coast. But this document does not reveal just why the wave is more pronounced (has the relief suddenly become higher?), and, though it might seem banal, why the cold air has suddenly got colder! Observations of what is really happening, based on synoptic charts and satellite images, in no way confirm this simplistic and unlikely hypothesis (Figures 58 and 60, and Chapter 12).

Figure 60 shows, for example, a succession of five MPHs making their way across North America during an unremarkable week in terms of the weather: 18–24 June 2003. MPH1, already over the Atlantic, is losing its identity around 21 June as it is integrated into the AA known as the 'Azores–Bermuda AA', and its burden of air is fed into the trade circulation. The leading edge of MPH2 reaches the east coast on 18 June, moving slowly, and is then reinforced by MPH3, which was over the Great Lakes on the 18th. MPH3 merges with MPH2, and on 21 June divides into two, the northern part travelling faster and leaving Canada in the direction of the North Atlantic. Between the two sections of MPH3, a low-pressure area develops from 20 June onwards to the east of the Great Lakes, concentrating the northward flowing energy onto the leading edge of MPHs 2 and 3 (especially on the 21st). This depression dissipates on the 23rd. MPH4 moves rapidly northwards on its trajectory, crossing Canada in 3 days and merging into the northern part of MPH3. MPH5 appears on the 22nd and moves off slowly. This seven-day sequence shows real events, directly observed, not hypothetical theories. It raises the obvious question of how some high-altitude wave 'created' and 'anchored' by relief might be responsible for the birth (above a region where the wave does not exist (i.e., the polar region)) of these moving masses of cold anticyclonic air in the lower layers: the MPHs. A further question: how to explain the simultaneous presence of these MPHs, their nature, their trajectories (meridional or more zonal), the way in which they move well beyond the reach of the supposed wave, their speeds, possible divisions and/or mergings, the formation of low-pressure corridors, the deepening of depressions between two MPHs (especially along the east coast) ... and the diversity of weather associated with the passage of MPHs ... The question

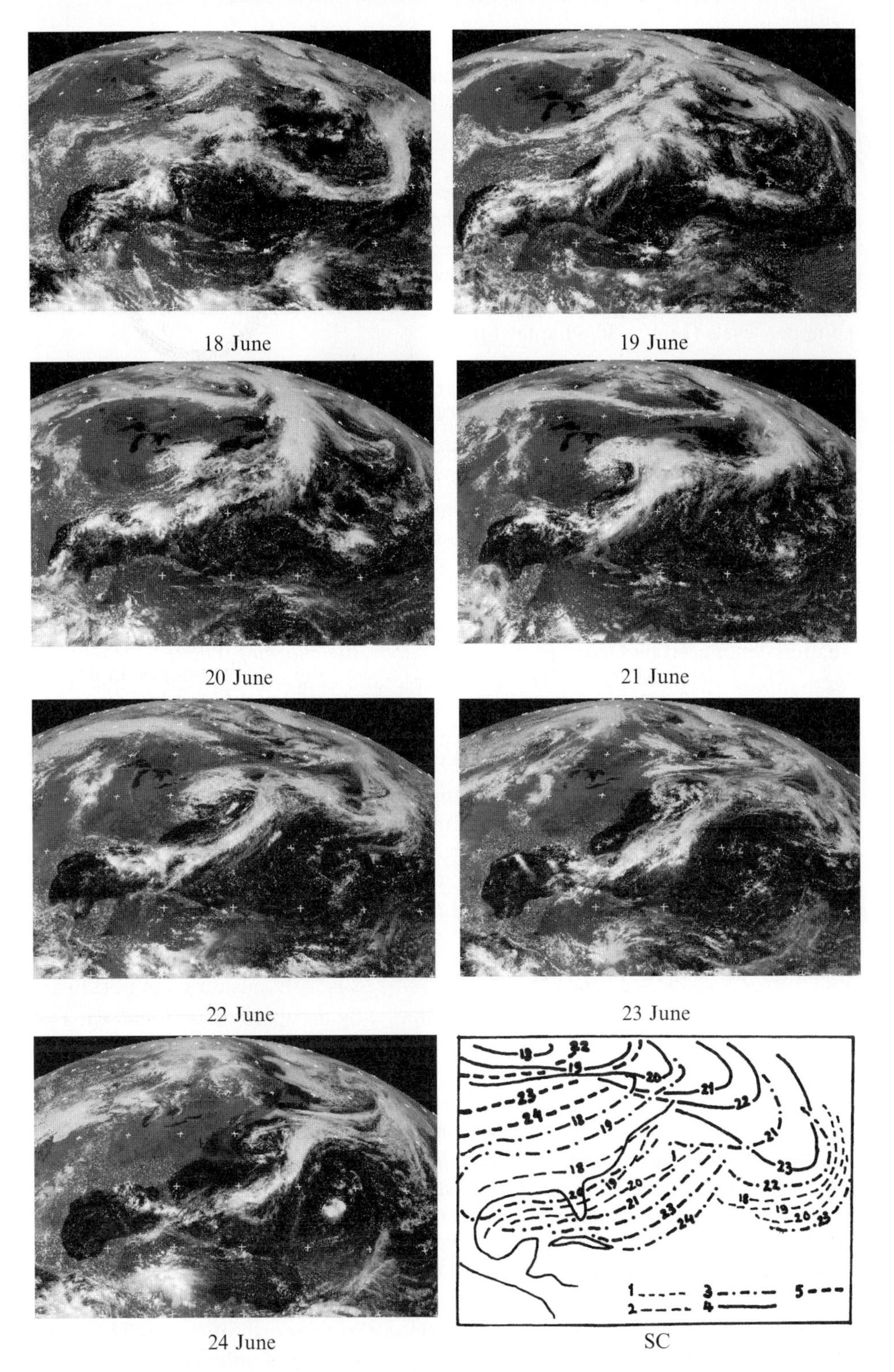

Figure 60. 18–24 June 2003 (*GOES E*, visible, 18 h NOAA). Over North America, during this week, five MPHs born in the western Arctic go southward, following two directions, the one (more meridional) toward the Gulf of Mexico, the Caribbean, and the Atlantic Ocean, the other (more zonal) crossing mainly through Canada and south of Greenland, towards the northern part of the Atlantic Ocean and Europe. Summary chart (SC): 1 2 3 – number of the MPH (MPH1, MPH2...); 22 23 – date of the position of the leading edge of an MPH.

remains very much an open one: wide open (cf. Figure 58, and the FASTEX experiment below).

The MPH concept diametrically inverts the priorities of the levels, and does away with the importance allotted to the upper layers, where the very tenuous nature of the air belies the role attributed to it by the theory. Instead, a primal role is allocated to phenomena in the lower layers, where the air is denser. Also, updrafts from the lower levels, engendered by MPHs, supply the rapid air currents of the upper levels (Figure 30, chapter 8).

11.1.2 The MPH: A major factor in extra-tropical weather

There is no reconciling the theories of the different schools of thought mentioned above, since each one follows its own internal logic. No synthesis can embrace them all, and for practical purposes they simply exist side by side. Even when satellite data began to be available, the necessary questioning of the principles on which these schools were founded did not happen. Their shortcomings are many, and their existence leads to irremediable contradictions and serious inconsistencies (Leroux, 1996a, 2000). A new concept is therefore greatly needed, in order to put an end to the situation denounced by Dady (1995) as '*the intellectual blocking, universal in meteorology*'. The concept of the MPH offers simple and coherent answers to questions which 'traditional' concepts leave unresolved.

MPHs, the driving force of general circulation (Chapter 8), are also the major player in the weather of middle and higher latitudes. Formed initially by cold air, MPHs retain the advantage of density along the length of their trajectories, allowing them to brush aside or lift other fluxes in their paths (Figure 24, Chapter 8). As this air is raised, a pressure deficit is created around the MPH, which as a result becomes surrounded by a low-pressure corridor, within which cyclonic circulation begins, deflecting the lifted air in the direction of the pole.

Dynamical link between the MPH and low, on all timescales

- The lifting of the surrounding air is facilitated by the speed of displacement and the thickness of the MPH (its leading edge being the most dynamic section) and by the qualities of the uplifted air: warm air has a natural tendency to rise, and moist air provides additional energy, from the liberation of latent heat carried in water vapour. This means that **the more energetic the MPH is, the stronger the convection will be**, the lower the peripheral pressure will become, and the more the attractive force acting upon the surrounding air will be intensified, leading to *a more intense and rapid cyclonic deflection of the surrounding air towards the pole* (Figure 61(a-1 and a-2)).
- As soon as the air deflected towards the north in the peripheral corridor has passed the MPH, the vorticity, which has until now been held in check along the leading edge of the MPH, is free to develop. To the north of the MPH (in the northern hemisphere) a closed low (cyclone) may deepen unhindered. *The depth of this closed depression depends* upon the volume, speed, and energy content of

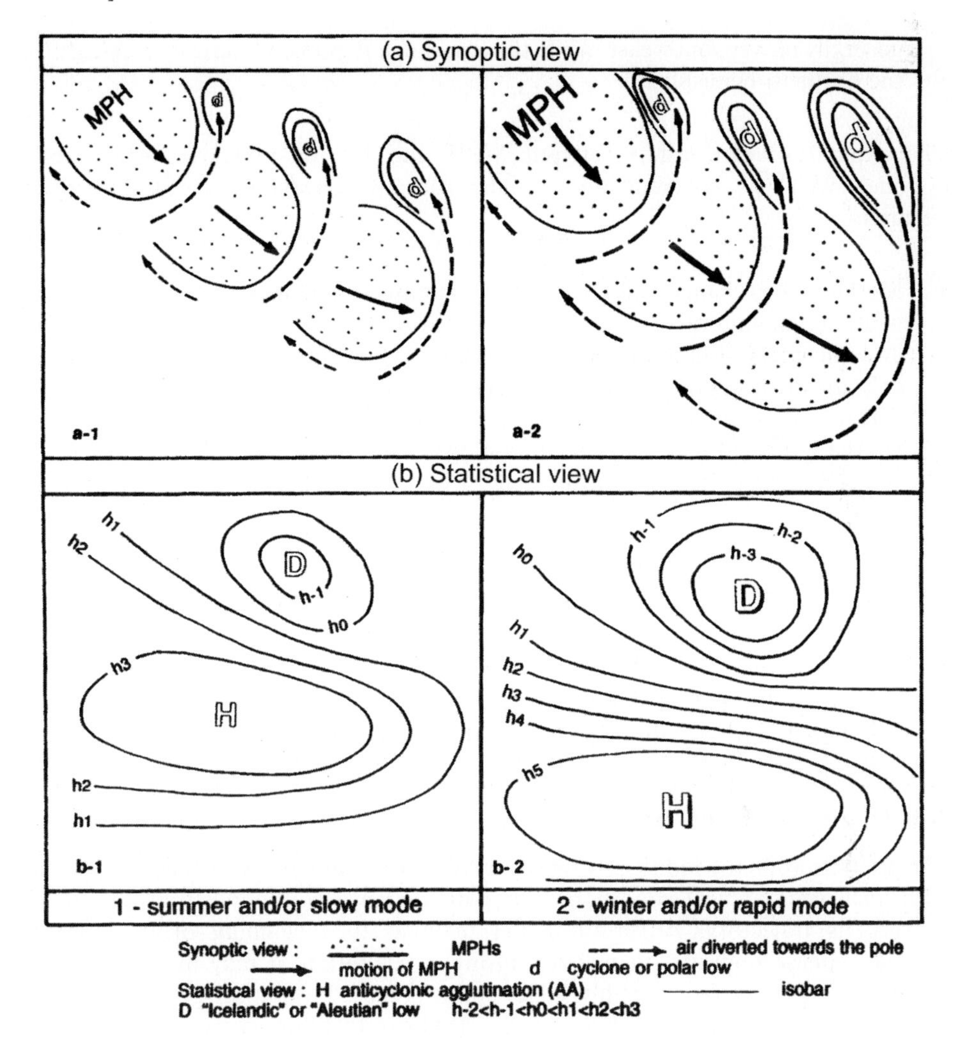

Figure 61. Dynamical link between the MPH, cyclonic circulation, and D (cyclone) on:

- All timescales: synoptic (a); seasonal (1 and 2); and statistical (b).
- Scales of phenomena: weak (1) MPH, or strong (2) MPH.
- General circulation modes: slow mode (1) – low polar thermal deficit; rapid mode (2) – high polar thermal deficit.

the cyclonic deflected air (i.e., for the most part, *upon the strength of the MPH* orchestrating this circulation (Figure 61(a-1 and a-2)).

- Since vorticity depends upon latitude, the poleward face of the depression is the more active, and the geostrophic force combines with the intensity of the deflected flux to progressively move the depression away from its parent MPH. So, **the more powerful an MPH is, the deeper its associated cyclone will be, and the faster the cyclone will distance itself from the moving MPH**. The MPH generally moves south-east, while the associated depression moves north-east (in the northern hemisphere).

The dynamical relationship between MPH and D (low-pressure corridor and cyclone), which is coherent on all timescales, is schematically represented in Figure 61.

- Meridional exchanges are particularly intense (rapid mode of circulation) in winter (contemporary) or during a cold period (palaeoclimatic): there is strong cyclonic circulation in the direction of the pole, and vigorous transportation of cold air in the opposite direction. The reverse is true when the polar thermal deficit is attenuated (slow mode of circulation).
- On the statistical scale used in climatology, a low mean value for pressure in an AA (H) corresponds to a mean, shallow depression (D), and conversely, a high mean value for pressure, more extensive and shifted further south (**H**), means that the associated mean Depression (**D**) is very deep and very extensive (Figure 61(b-1 and b-2)).
- Displacement towards the tropics weakens the MPH (warming, divergence), but at the same time brings its leading edge in contact with warmer surrounding air, with added moisture if the contact occurs over the ocean, enhancing the energy content of the deflected flux. In winter, the trajectories of MPHs are more meridional, moving more deeply into the tropical margins (Figure 61(a-2)).

Weather associated with MPHs

The confrontational conditions briefly mentioned above vary with the seasons and with local geographical conditions. They also change continuously as the MPH pursues its trajectory. A certain MPH may, at the beginning of its journey, provoke intense updrafts, then, later during its evolution, it may in turn nourish (at least along its 'warmed' trailing edge) updrafts caused by a more powerful and/or more recent MPH. The air carried by an MPH can therefore either be partially or totally dispersed, feeding into a cyclonic circulation of (more or less) 'warm' air; be integrated into another MPH of equivalent density (i.e., temperature); or be reinforced by the arrival of another MPH (in an AA of variable strength and duration). **These conditions are extremely changeable, and constantly 'in the melting pot'**. Even if the processes of weather are always identical, the weather associated with the dynamic of MPHs is always different in its intensity and its extent (Figure 62).

MPHs control the weather, both 'bad' and 'good' (Figures 24 and 62):

- In their 'warm' surroundings (peripheral corridor) and within the 'warm' closed low (cyclone) where updrafts (convection) are concentrated, instability prevails, with low pressures, often violent winds, and storms; the updrafts manifest themselves in cloud formations and precipitation (rain or snow, or sometimes hail, according to the season). After the passing of the (cold) front fringing the

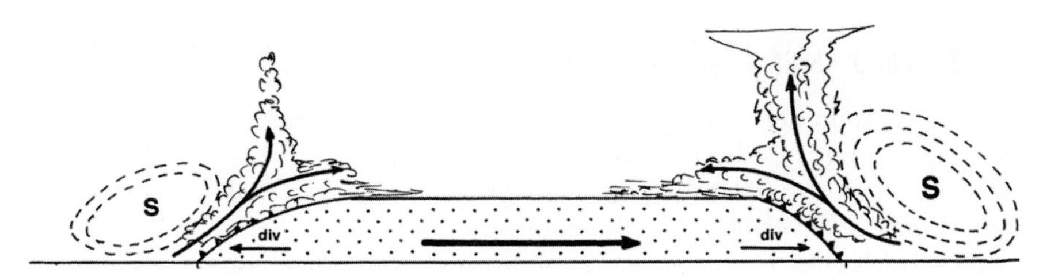

	per. corridor	front		MPH		front	periph. corridor
conv/div	convergence	**conv**	divergence	**DIV**	divergence	CONV	convergence
PP	falling	low	falling	anticyclonic stability		**rising**	**falling**
wind	South	S.E		**calm**	**North**	**storm**	**South, S.W**
TT	rising	cool	moderate cold	strong T range	**cold**	**falling**	**rising**
RH	high	high		**dry air**		**high**	**rising**
clouds	clearing sky	vertic.	stratified	**clear sky**	stratified	**vertical**	darkening sky
weather	fine, warm and windy	rain snow	drizzle	sunny, cold or warm radiation fog	rain	**shower** snow, hail	fine, warm and windy

Figure 62. Weather connected with an MPH (northern hemisphere) and the vertical structure of an MPH and associated clouds (see Figure 24, in Chapter 8).
Notes: per. corridor = peripheral low-pressure corridor; front = cold front; PP = surface pressure; TT = temperature; RH = relative humidity.

MPH, there is a period of relative calm, and the wind changes direction, veering from south to north.

- Within the 'cold' MPH, and especially towards the centre where all is calm, (anticyclonic) stability means that there is a lack of rain, radiation fog (freezing fog in winter), and good, sunny conditions to a varying degree (as a function of the density of the stratified cloud formations above the anticyclone).

The MPH concept does away with the notion of a 'polar front' which has been retailed for these past 80 years. The general physical principle stands: confrontations between air masses of different densities are constantly observed. But it is not necessarily a question of a confrontation between *polar air* and *tropical air* along an uninterrupted discontinuity. There are no *characteristic* air masses (e.g., arctic, polar, tropical, maritime, or continental). But there *are* 'mobile air masses' (fluxes), unceasingly evolving from a cold to a warm state, and/or from dry to moist and vice versa. The surface along which these encounters take place is not a continuous one. Each MPH (as long as it has the advantage of density) organises around itself a confrontational surface (only as a cold front) to a depth of the order of 1,500 m, and cloud formations develop well above the MPH if updrafts are intense. No 'front' ever *installs* itself permanently or seasonally, at any location whatsoever: **everything is mobile**. The peripheral front and the *cyclone* born of the MPH move with it, and evolve as the MPH evolves, finally fading away and disappearing when the MPH loses its vigour, its coherence, or its independent existence as it merges with another MPH or AA (Figure 25, Chapter 8).

11.1.3 The FASTEX (non-)experiment

What is the standing of the argument above within the meteorological community, in the face of more 'traditional' concepts? The recent FASTEX international experiment in the North Atlantic, little known to the general public, provides a good opportunity to investigate this fundamental question.

FASTEX (the 'Fronts and Atlantic Storms Track Experiment') was organised in January and February 1997 by several countries, both near and not so near to the North Atlantic: Canada, Denmark, France, Germany, Iceland, Ireland, the Netherlands, Norway, Portugal, Spain, Switzerland, the Ukraine, the UK, and USA (Joly *et al.*, 1997). The region studied (cf. NAO, Chapter 12), and the international nature of FASTEX, are of particular interest to anyone considering the way in which meteorologists deal nowadays with disturbances in high and middle latitudes.

The objectives and unexpected results of the experiment

One of the scientific objectives of the experiment was the observation of disturbances, and more particularly '*the entirety of the evolutionary cycle of a storm and the determination of the mechanisms which have contributed to its formation*' (Chalon and Joly, 1996). The existence of this objective also suggests the obvious fact that the origin of disturbances in the North Atlantic, and particularly those associated with storms, is still not understood! The aim of this part of the experiment was to show that, in line with the supporting theory put forward by Farrell (1989, 1994) '*involving organised precursors*' (Joly *et al.*, 1997), storms are engendered by '*a vortex at an altitude of about 10 km, far from the ground*', the so-called *precursor*. With a view to improving the quality of weather forecasting, the idea was to demonstrate the 'determining' influence of altitude upon phenomena in the lower layers, with special reference to depressions and storms whose evolution depends '*uniquely upon coupling with another anomaly within the flow of circulation in the upper atmosphere*' (Chalon and Joly, 1996). Such an idea, which is very 'traditional' in nature since it is derived from the dynamical school of thought, sees bad weather as '*the fruit of chance and opportunity*', arising from the fortuitous (but not necessarily fortunate!) encounters of hypothetical independent vortices, one in the lower layers and one in the upper layers (Joly, 1995).

In November 1997, eight months after the experiment, Arbogast and Joly (both of the CNRM, and the latter being in charge of the FASTEX project) were quick to unveil the fundamental conclusion of the FASTEX experiment, by means of an urgent communication to the Paris *Académie des Sciences*. **This conclusion was quite the opposite of what was expected, since it discounted the supposed role played by altitude**, and confirmed '*the very unexpected role of a precursor confined to the lower layers*' (Arbogast and Joly, 1997)!

The release mechanism of the storms in the North Atlantic was finally 'discovered': it was located in the lower layers, and was unambiguously, in the words of Arbogast and Joly themselves, '*the true release mechanism*' (1997, p. 230). As satellite

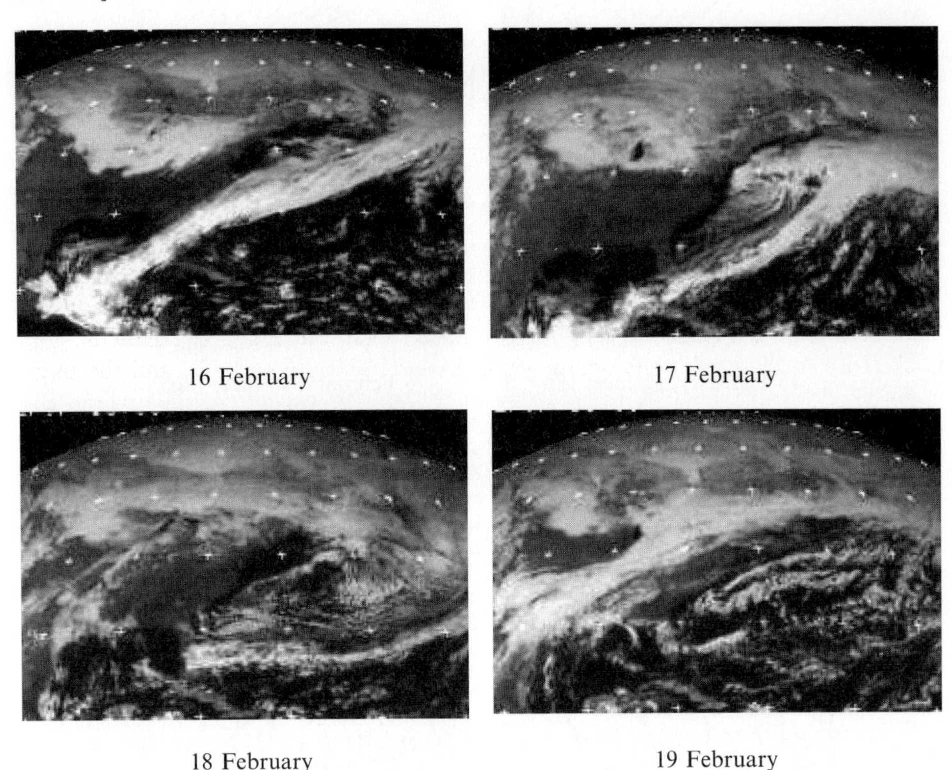

Figure 63. 16–19 February 1997 (*GOES E*, visible, 18 h). Displacement of MPHs, H1 and H2, and of the intermediate low-pressure corridor, eastwards, over North America, from the Arctic to the Atlantic.

images and the synoptic surface charts of *Environnement Canada* clearly showed, the 'new depression' – the 'precursor' – is located between two mobile anticyclones, MPH 'H1' (attaining 1038 hPa) and MPH 'H2' (1024 hPa). This depression deepened on the 16th between H1 and H2, and it owes its existence, its depth, and its mobility only to these two centres of high pressure, pursuing their course in the lower layers across North America, in the direction of the Atlantic (Figures 63 and 64).

The observational facts

MPH H1 is born on 14 February over the western Arctic, and by 15 February at 06h, its pressure has reached 1036 hPa (Figure 64). It moves south-eastwards, its western part running along the foot of the Rockies, which form an impassable barrier to cold air. On the 16th, the intermediate corridor on its leading edge shrinks, as it catches up with a preceding MPH, into which it merges, reaching the coastline on the 17th and moving out across the ocean on the 18th. The asymptote of confluence (the front) which precedes it is very obvious from the Gulf of Mexico to Greenland on the

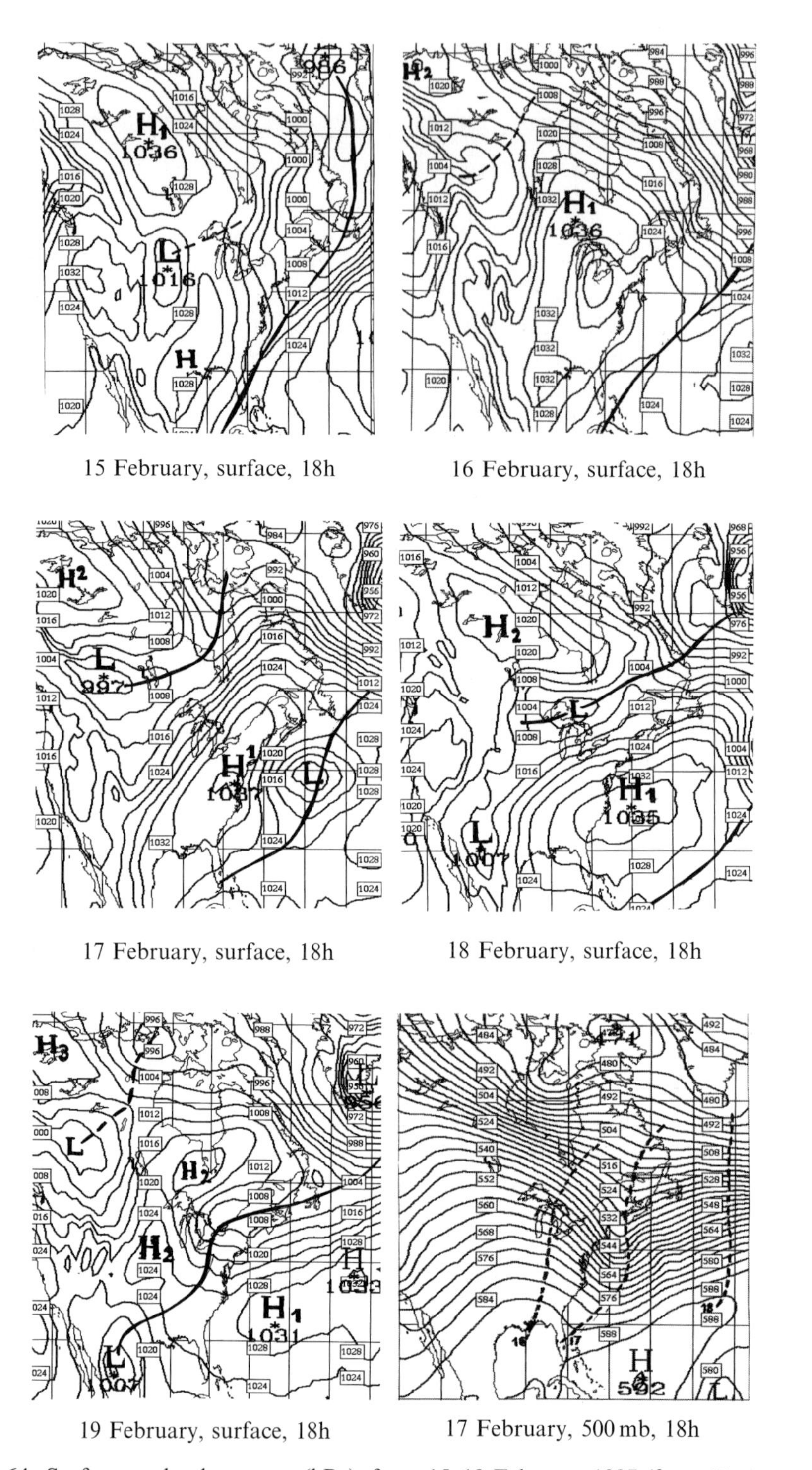

Figure 64. Surface sea-level pressure (hPa), from 15–19 February 1997 (from *Environnement Canada*). Bold line: asymptote of confluence (front). Also shown are the 17 February 1997 500-mb contours (from *Environnement Canada*); broken line – axis of talweg, on 16, 17 and 18 February.

images for the 16 and 17 February, and shifts towards the Atlantic over which H1 sits almost completely on the 18th. On the 19th, it becomes part of the North Atlantic Anticyclonic Agglutination. H2, a less powerful MPH, is born over the Arctic on 15 February, and follows the same path. The intermediate corridor between H1 and H2 deepens and grows, and moves with the MPHs. H2 splits into two on the 19th and reaches the ocean. Behind it, a new depression forms in front of MPH H3, which follows the same route. The topographical situation at the 500-mb level on 17 February 1997 shows that the high-altitude talweg is overhanging the front in the lower layers (showing the typical westward drift); the superimposition on the chart for 17 February of the positions of the axis on the 16th and the 18th indicates that the talweg is moving eastwards with the updrafts associated with the front. This configuration at altitude sheds no light whatsoever on how the upper layers might be involved in the renewal and trajectories of MPHs, and the deepening of intermediate depressions (Figure 58).

So the FASTEX experiment is (or could have been), in a very real sense, a scientific event! In fact, 50 years of hesitations and inertia might have been swept away, since the origin of the 'initial depression' which had been sought for more than a century had at last been found: it is created by the updraft powered by an MPH on its leading edge, and/or by the narrowing of the intermediate corridor between two MPHs. Unfortunately, the '*true release mechanism*' is ... a mobile anticyclone, a 'character' which does not figure in either the mythology of meteorology, or the 'toolbox' of the modellers! The existing dogma, its 100-year credentials unlikely to be challenged, will only allow, as a reference, a 'depression' ('fixed' in more ways than one, see above). Anticyclones, for their part, are still not taken into account, particularly when they move ...

A (mis-)interpretation unconnected with the synoptic reality

So **the authors did not rely on observational facts and the real dynamic**, but stuck to hoary old 'explanations':

- First, they invented a 'Great Lakes Depression', an unknown entity of indeterminate origin, which appeared *ex nihilo*, but might have been a '*survivor of a former depressionary system*' (sic!). One had only to consult surface charts to ascertain quite simply that there was no such thing, since the situation was anticyclonic on the 16th and 17th over the Great Lakes, and indeed was again anticyclonic on the 19th (Figure 64)!
- They then attributed to this (apparently surviving) depression a '*crucial role*', stating that '*because of its presence, the Great Lakes Depression logically induces a cyclonic circulation within the lower layers*' (sic)! Could there be a better example not only of spontaneous generation, but also of stating the obvious?
- Finally, since everything was on the move and at some distance from the Great Lakes, it was considered that '*the distant effect of this lower layer system ...*' etc. ... What is this 'distant effect'? Some sort of magic formula, perhaps, but lacking any physical basis whatever! The opportune and so miraculous appear-

ance of such a *deus ex machina* is obviously a 'mystery', showing that ideology precedes in this case the findings of direct observation, and the factual reality.

The FASTEX experiment might have demonstrated the validity of the MPH concept in the study of the genesis of depressions, but in spite of this the experiment became a non-event, and indeed it is as if it never happened ... Organised at great expense to further our knowledge of weather disturbances, it has as yet changed nothing, nor challenged any belief:

- The true origin of Atlantic disturbances, and in the wider view, of disturbances in middle latitudes, is still (and deliberately so) little understood.
- The (suspect) Norwegian theory still predominates in the preparation of synoptic surface charts.
- The (undemonstrated) dynamic concept, with its insistence on effects at altitude, still dominates the doctrines of the modellers.

Moreover, the NAO remains as mysterious as ever (cf. Chapter 12). And Météo-France has not changed its views by one iota, still talking about the origin of storms, in the face of what real observations formally teach, in terms such as: '*a string of high-altitude depressions*' (cf. Météo-France website). Given all this obscurantism, it comes as no surprise at all that storms like those which caused such havoc in France and western Europe in December 1999 (Lothar and Martin) were not properly predicted! Neither can we be astonished that an unpredicted Atlantic tempest caused such damage to the multi-hulled craft taking part in the 'Rum Race' on 13 November 2002! There is unfortunately no shortage of other examples.

11.1.4 Tropical cyclones

There exist major ambiguities as far as the processes that trigger disturbances in middle latitudes are concerned. In the tropics, too, the situation is no clearer. The concepts of the various aforementioned schools of thought have been extended to the tropical zone, where 'fronts' have been reported and 'centres of action' are still to be found. Specific named phenomena still appear, and the tropics are seen by some as a 'world apart', even though they are, aerologically, closely connected to the extra-tropical zones (Figure 30, Chapter 8). Disturbances in the tropics, which begin as extensions of MPH activity at tropical margins, progressively acquire their own separate identities, as a function (especially) of the amount of potential energy (water vapour) available, and the particular structural conditions within the troposphere above the tropics (Leroux, 1983, 1996a). For the moment, we shall confine ourselves to that most typical of tropical disturbances, the cyclone, which, although born at the heart of the tropical zone, most often ends its life within the temperate zone.

A memorable example is that of Hurricane Mitch (1998), still discussed in awed tones in Central America, where it claimed many victims and caused vast damage to property. At the time, the Miami Hurricane Center forecast that the storm was unlikely to cross the Central American isthmus, but instead would be following a 'mean' (i.e., statistically predicted) trajectory, moving off towards the north-east. However, '*observations based on satellite images showed early on that Mitch would definitely move onto the land*' (Barbier, 1999, 2004), because a vigorous MPH on the North American path had moved down towards the Gulf of Mexico, and was blocking any northward progress of the cyclone. Between 29 October and 3 November 1998, Mitch moved, as logic dictated, across Honduras, and then El Salvador, and on into Guatemala and the Mexican Yucatan, sowing its sad memories ... (Leroux, 2000).

The IPCC has predicted an increase in the intensity of tropical cyclones. In the words of Fellous (2003), echoing the findings of the models, '*the increase in sea surface temperatures in the tropical zones may bring about a recrudescence of cyclones, which may be triggered only beyond a threshold temperature of 27°C ...*'. Now here we have the famous 'T/R' equation again, this time applied to cyclogenesis: when T increases, there are more cyclones! Can the recipe be that simple? This 'sea surface temperature/cyclogenesis' relationship is yet another piece of meteorological baggage. It is a particularly tenacious one, carried for example by Météo-France, to whom a '*sea temperature above 26°C involving a depth of at least 50 m*' (Météo-France website, 2004) is an indispensable thermal condition. It elevates to the status of a physical 'law' a simple co-variation between marine temperature distribution and the location where cyclones originate (which location is determined by many other causes). Fortunately, this meteorological myth is beginning to lose currency: '*The popular belief that the region of cyclogenesis will expand with the 26°C SST isotherm is a fallacy*' (Henderson-Sellers *et al.*, 1998).

Tropical cyclogenesis conditions

To recap briefly, the formation of a tropical cyclone requires the simultaneous existence of several very necessary conditions (Leroux, 1983, 1996a, 2001):

- The preliminary existence of a depressionary field in the lower layers, with low pressure represented especially by the meteorological equator (ME) or the low-pressure corridor on the leading edge of an MPH moving into the tropical margins (cf. Chapter 12).
- The triggering of an updraft dictated by a dynamical process, usually the influence of a pulse line in the trade or monsoon circulation, which can be the starting point for a vortex.
- The formation or accentuation of a vortex, which is indispensable to the acceleration of the convergence (convection). Vorticity is a function of latitude (the geostrophic force), and so cyclones form only beyond latitudes 5°N and S.
- The self-maintenance of the updraft through the liberation of energy advected by maritime trade and monsoon fluxes, warm and moisture-laden, and the rapid renewal of this energy, vitally necessary if the convective movements are not to degenerate. Raised sea surface temperatures therefore represent only a co-variation of the parameters in the vicinity of the surface line of the vertical

meteorological equator (VME), the axis of confluence of the fluxes and the energy advected beneath inversions in the lower layers, and the confluence of ocean currents.

- A vertical structure favouring updrafts. Cyclogenesis is therefore not a feature of stratified aerological structures (the Trade Inversion (TI) or the inclined meteorological equator, IME), but it finds ideal conditions within the structure of the VME, the axis of confluence, of convection, and of the concentration of energy (water vapour).

These draconian conditions are not often all present at one time, and also, they must be maintained along the whole length of the cyclone's trajectory: these two facts combine, fortunately, to limit the number of tropical cyclones.

Recent evolution of tropical cyclones

How have these disturbances evolved in the various cyclone regions? A reply comes from a public study: '*there is no discernible global trend in tropical cyclone numbers, intensity, or location from historical data analyses*' (Henderson-Sellers *et al.*, 1998). Neither is there any link with the greenhouse effect scenario: '*there is no evidence to suggest any major changes in the area or global location of tropical cyclone genesis in greenhouse conditions*' (Henderson-Sellers *et al.*, 1998). Also, we do not see at what particular moment the greenhouse effect might intervene in the cyclogenetic processes described above!

Barbier (2004) confirms this last conclusion. He shows in fact the absence of the evolution of cyclones over the Atlantic (Figure 65, upper). The study by Smith (1999), centred on the east coast of the USA between the years 1900 and 1998, pointed out that the maximum frequency took place during the 1970s, and that there has since that decade been quite a marked decline in the number of hurricanes.

Barbier showed that there has been, on the other hand, a net increase across the north-eastern Pacific (Figure 65, lower). We cannot ascribe this increase to some intervention by the greenhouse effect, but rather to the increased number of MPHs affecting the region. Pacific cyclones (or *cordonazos*) develop and then track north-eastwards (towards the Californian peninsula), within the low-pressure corridor on the leading edge of MPHs moving down along the west coast of North America (Leroux, 2000; Chapter 13).

In circumstances such as those briefly mentioned here, where the real processes taking place in both tropical and temperate areas are so misunderstood, *how can we boldly presume to forecast the weather of the future,* and especially the multiplication of extreme events, all the way to the year 2100? How can we predict storms, if we have no idea how they form? How can we say what tropical cyclones are going to do, if we involve a mechanism which does not happen?

And is it possible, also, to predict rainfall, or the lack of it, while completely ignoring the indispensable physical mechanisms, which either promote or hinder it? Do greenhouse gases have any part to play in these processes? To find answers to

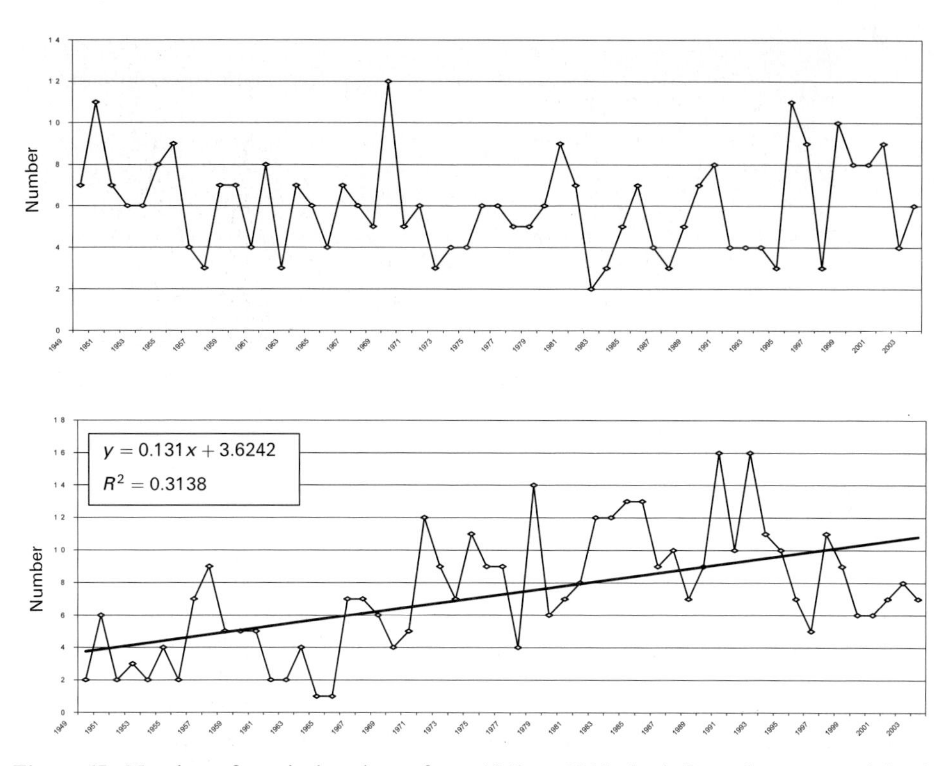

Figure 65. Number of tropical cyclones from 1949 to 2003. (*top*) Over the western Atlantic Ocean; (*bottom*) over the eastern North Pacific Ocean.
From Barbier (2004).

these questions, we shall examine some particularly eloquent French/European/ Mediterranean examples.

11.2 PRECIPITATION

There will be more rain: the models tell us this, so it's 'inevitable'! To extol the virtues of the 'T/R' equation, it is necessary only to cite the rainfall statistics for Scandinavia, or Alaska (Chapters 12 and 13), and to grasp at the least example of torrential rains or flooding, to be able to cry wolf! The media are not slow to do this ... but how much noise do they make about the many examples from other regions, such as sub-Saharan Africa, where rainfall has tailed away dramatically; what about the prolonged periods of drought which are occurring in many places? We should avoid relying upon the pluviometric outcomes predicted by models, whose results in this aspect are remarkable only because they are so far from the truth.

Notable regional disparities arise, even within units of circulation, with some regions being well watered, while next to them are others which receive less rain, in

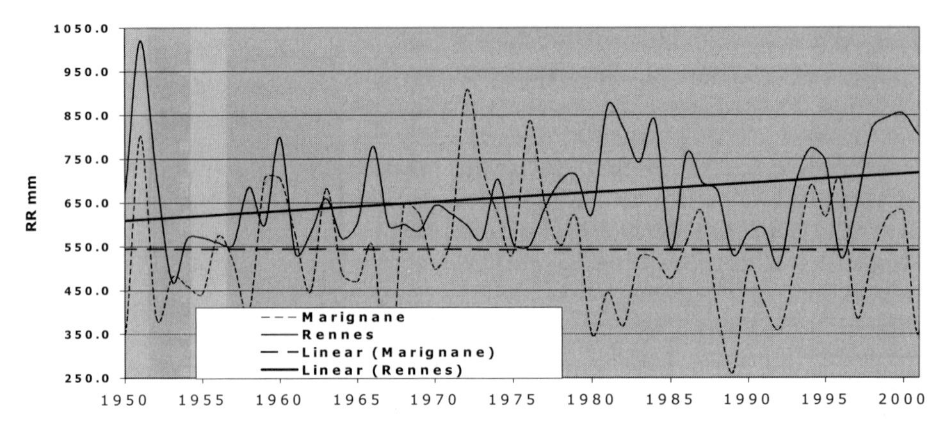

Figure 66. Mean annual rainfall from 1950–2001 in France, at Rennes, Brittany, and at Marseille-Marignane, Provence. Linear trend at Rennes (unbroken line) and at Marignane (broken line).
Source: Météo-France data.

spite of the fact that the regions are subject to the same general dynamic (Chapters 12 and 13). Also, at the heart of a country or large region, there may be differences.

Take the example of France, at the eastern side of the aerological Atlantic unit. Here, rainfall figures showed a gradual increase during the 20th century, the trend being insignificant for most of the country, except in areas bordering the Atlantic (Leroux, 1997) or along a single north–south band in western central France, from the Pyrenees to the Paris area (Moisselin *et al.*, 2002). However, in southern France there was a decrease, notably in Provence, in far-south regions, where drought was, by contrast, accentuated. Figure 66 clearly shows the increase at Rennes, while at Marseille-Marignane the trend is neutral for the whole of the period 1950–2001, with a slight downward trend seen after the 1970s.

So things are by no means as certain as the models proclaim! What is needed is to return to basics and to define the conditions which bring about rainfall, in order to be able to determine the validity of the schematic relationship used by the models, and of any possible hypothetical link between precipitation and the greenhouse effect.

11.2.1 Conditions for pluviogenesis

Precipitation is an uncertain and discontinuous phenomenon in both time and space. The process of pluviogenesis involves the necessary, simultaneous, and (variably) time-limited co-existence of precise conditions, to do with: precipitable potential, the triggering of updrafts, and the vertical aerological structure.

Precipitable potential

The existence of precipitable potential is necessary (i.e., there must already exist a quantity of water contained within a volume of air capable of being precipitated).

This potential varies with the temperature of the air, and the energy available for its acquisition (vaporisation heat) is conserved in the form of latent heat, which is liberated during a change of state (condensation). This energy lends support to the other factors involved in updrafts, *after* these have been triggered.

But the existence of a rich potential is not by itself sufficient: it is also necessary that it can be utilised. There is not, in fact, in any given location, any direct link between precipitable potential and water actually precipitated. Consider the area around the Red Sea, where moisture is present, and in quantity, yet it does not rain. The enormous potential of the Indian monsoon streams overhead in Somalia during summer, with the Atlantic monsoon above it (Leroux, 1983), to little effect. There are humid coastal deserts over which water will condense (as fog and dew), but it does not rain.

Water vapour transfer

The supply of precipitable potential must be both abundant and, above all, constant, or the associated phenomenon will degenerate. A phenomenon which uses only a source of supply in its vicinity will rapidly cease its activity, unless the precipitable potential is continuously renewed. This is what happens in the case of localised storms, which have a relatively limited lifetime. As soon as the meteorological phenomenon attains a higher scale, the advection of remote precipitable potential is indispensable to its survival. The transfer of water vapour across long distances, and the continuation of this supply, are evidence of the crucial importance of the factor responsible for this advection which takes place in the lower layers. In extra-tropical zones, this fundamental factor is the MPH, which is capable of tapping into the surplus precipitable potential of the tropical zone (Chapters 8 and 9). In the tropical zone, the transfers of precipitable potential are guaranteed by the maritime trade and monsoon circulations, the latter driving deeply into the interiors of continents. But the presence of a maritime trade, or of a monsoon, even if they are close to the point of saturation, does not necessarily mean that rain is on the way. These fluxes are only vectors, and the precipitable potential which they transport has to be set into action by a disturbance (e.g., a pulse line, squall line, VME, or cyclone).

The updraft factor

Because of the cooling it causes, updraft (convection) is an indispensable part of the processes of change of state in water and the consequent liberation of latent heat. The thermal factor (convection linked with continental warming) involves only phenomena of moderate amplitude, even though some local storms can become, briefly, quite violent. The dynamic factor is a multiple one, manifesting itself on all scales. It may be associated with relief (which causes not only rain, but also the *foehn* effect) and/or aerological phenomena such as confluence and convergence, or the intrusion of one air mass into another, with the difference in densities determining the slope of the front. The MPH is, these things considered, the most important mobile factor in upward air movements. The upraising (convection) of moist air allows the progressive liberation of latent heat, and it supplies the extra thermal energy which encourages the continuance of the updraft. In this manner, moist air is capable of maintaining an intense disturbance, and local storms, or organised storms (i.e., more extended weather systems) are more active in summer, and/or in the tropics. But the presence of water vapour is not in itself sufficient to create an updraft, and it cannot therefore be claimed (as modellers do) that '*(deep) convection will be triggered ... if there is an importation of water vapour due to strong evaporation or a considerable convergence of moisture*' (Coiffier, 1997). Some other factor has to have already provided the trigger for the updraft, and it is this triggering factor which it is most necessary to identify.

Structural conditions

Aerological structural conditions in the troposphere have to be favourable to convection (i.e., without shearing (fluxes in opposing directions), without subsidence (which causes heat and drying), and without stratification (fluxes of different origins and natures)). In other words, the troposphere must not prevent the vertical development of cloud formations. The tropical zone is characterised by a highly organised vertical structure, with particularly unproductive discontinuities and/or inversions (Leroux, 1983, 1996a).

These pluviogenetically favourable conditions, whose simultaneity and persistence are indispensable, are extremely variable, on both the synoptic and seasonal scales. They also vary with geographical conditions, the identity of the determining factor, and with structural conditions; all of which give the various disturbances their individual characters. **There is obviously a lot more to this than a simple 'T/R' relationship**.

Now, at what level might greenhouse gases intervene, given these conditions? Perhaps it could be a question of the quality of the precipitable potential, and it is at this level that models bring it in. But this is only one of the conditions, which cannot by itself determine the nature of rainfall ... if this were the case, it would never rain in our latitudes in winter, which is the season when precipitable potential is locally at its lowest level! On the other hand, winter is the season of the most intense long-distance air transfers, because of the strength of MPHs... and in certain regions it is the season when rain is at its most abundant (the 'T/R' relationship, applied literally, would here mean: 'cold = rain'!).

11.2.2 Extreme rainfall in southern France

Let us consider, as an example of extreme pluviometric activity, the heavy rainfall which periodically strikes southern France, and many other regions around the Mediterranean: activity of the kind that the IPCC tells us is on the increase. Torrential downpours are frequent around the Mediterranean. They are a feature of this type of climate. In 1996, Météo-France compiled an 'inventory of torrential rains' for the south-eastern Mediterranean and Corsica, for the period 1958–1994. 'Torrential' episodes are considered to be those where rainfall exceeds 190 mm in 24 hours. In the 37 years in question, 119 such events were recorded for the south-eastern area, and for Corsica, 25; with 72% of all these occurring in autumn. Episodes when values exceeded 400 mm occurred 22 times during those 37 years. Now, there seems to have been a noticeable increase in such extreme events during recent decades: events involving flooding, with often tragic consequences. The most dramatic cases in point:

Nîmes, 3 October 1988.
Narbonne, 5 August 1989.
Pézenas-Montpellier, 19 September 1989.
Châteauneuf-du-Pape, 30 July 1991.
Vaison-la-Romaine, 22 September 1992.
Rennes-les-Bains, 26 September 1992.
In the Languedoc and in Corsica, 22 and 23 September 1993.
In Corsica, 31 October 1993.
In the Alpes-Maritimes, 26 June 1994.
In Corsica, 4 November 1994.
In the Gard, 3 October 1995, and again on 13 October.
Puisserguier (Hérault), 28 January 1996.
In the Languedoc-Roussillon, 13 October 1996.
In the Hérault, 6 November 1997, and again on 20 December 1997.

... and, to cut a long story short, more recently on 8–9 September 2002; and then in Nîmes on 9 September 2003 (cf. Figure 28, Chapter 8); and also on 2 December 2003, in Marseille ... The list is by no means exhaustive. Comby (1998) analysed 160 events, and concluded that very intense rains were increasing in frequency during the period 1950–1997, according to 40–50% of weather stations in the southern part of the Rhône valley, beyond certain thresholds. The cumulative value of these torrential rains exceeded 400 mm, or even 500 mm, and heavy rainfall, normal in autumn, now tends to occur also in the winter.

This trend is, of course, immediately associated by some with 'climatic warming', and after each of these painful events, the media wastes no time in the shameless trumpeting of catastrophist scenarios to come. What truth is there in this? What link *can* exist between these rains and the greenhouse effect? The best way to reply to these questions is to analyse one of these situations.

Torrential rain and flooding in November 1999

On 12 and 13 November 1999, extremely violent storms raged across the Languedoc-Roussillon region, between the Rhône and the Pyrenees (Figures 67 and 68). This same region had experienced more than 300 mm of rain on 13 and 14 October 1996. In November 1999, the rain was particularly heavy. Over nearly 1,000 km^2, 400 mm were recorded, and over 500 km^2, more than 500 mm; one area of 100 km^2 received 600 mm, and the maximum figure was that recorded at Lézignan-Corbières (to the east of Carcassonne) which was 621.2 mm. According to Météo-France, '*50-year records have been largely broken*'. The heaviest downpours were concentrated

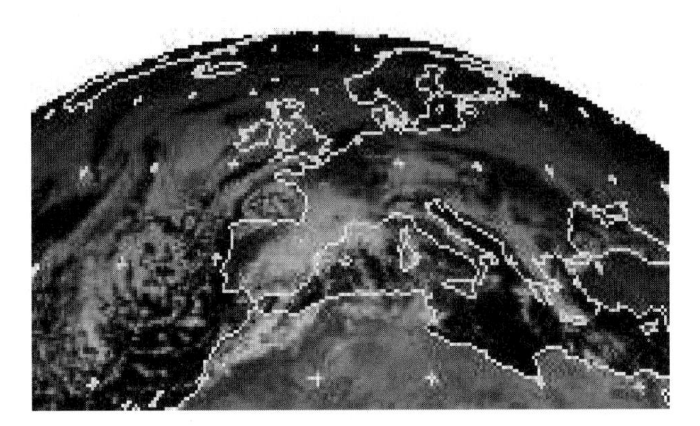

12 November

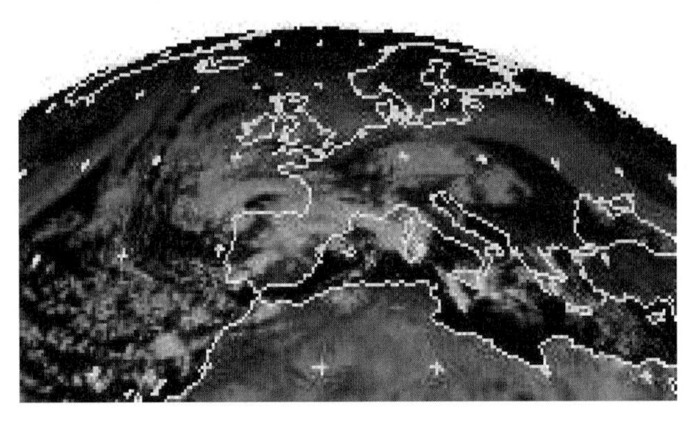

13 November

Figure 67. 12 and 13 November 1999 (*Meteosat*, visible, 12 h). An enormous meridian MPH, propagated rapidly southwards over the eastern Atlantic, invades the western Mediterranean basin, and provokes an intense northward advection of warm, moist Mediterranean air.

along an axis running from north to south (and especially along its southern part), across relatively flat land and therefore they were largely uninfluenced by relief. This axis was determined rather by the confluence of anticyclonic air coming in along the northern edge of the Pyrenees ('a' in Figure 68(b)), and warm cyclonic air from the Mediterranean. The considerable quantities of rain involved must have required a colossal volume of precipitable potential to be brought in from the south. It can be roughly estimated that about 50 billion cubic metres of saturated Mediterranean air passed above each square kilometre of the region of Lézignan-Corbières, where more than 600 mm fell, with a tally of 5,000 billion cubic metres for the 600+ mm region alone.

The essential question remains: **what dynamical factor can be capable of unleashing such formidable transfers of warm, moist air at such high speeds in the lower layers?**

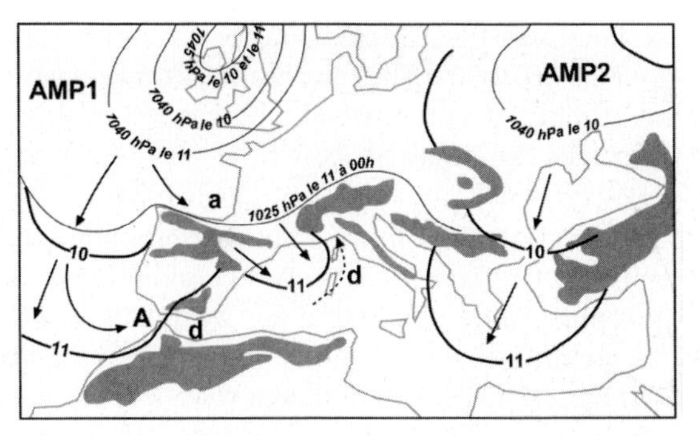

(a) 10 and 11 November

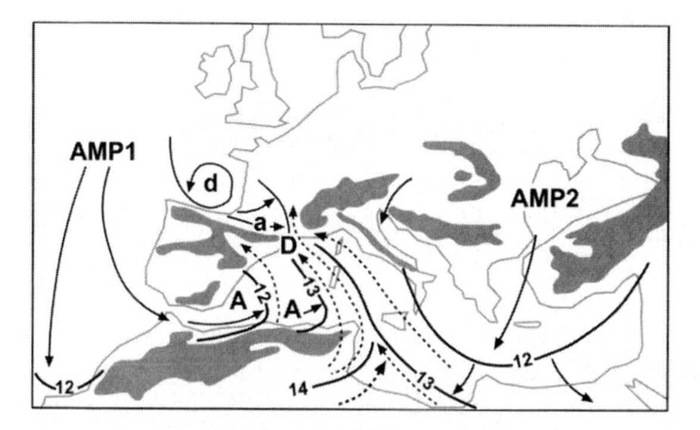

(b) 12 and 13 November

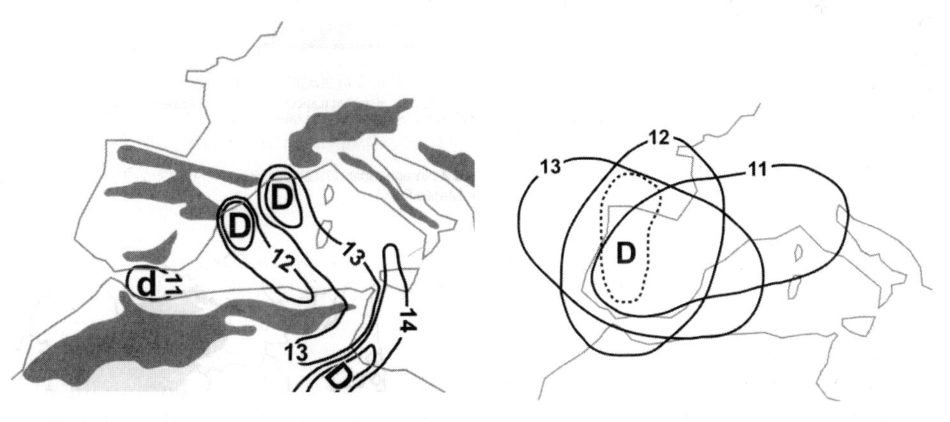

(c) D and corridor, surface, 11–14 November (d) 560-damgp contours, 11–13 November

Figure 68. Synoptic conditions inducing the torrential rains of 12 and 13 November 1999, over Languedoc-Roussillon, southern France. (a) Surface, 10 and 11 November 1999. (b) Surface, 12 and 13 November 1999. (c) Surface position of the low (D), and low-pressure peripheral corridor, from 11–14 November. (d) Position at altitude of the low (560 damgp) for 11–13 November, 12 h, and position of the 552 damgp on 12 November, 12 h (broken line).
Notes: AMP = MPH, A or a = high, D or d = low.
From *European Meteorological Bulletin* and satellite imagery from NOAA and *Meteosat*.

The 'official' explanations from Météo-France were of a very 'traditional' kind (in *La Météorologie,* No. 29, 2000). The four main factors suggested were: warm sea waters, the influence of relief, the presence of a depression at altitude (at the 500-hPa level), and the influence of the 'Balearic Low':

- Warm sea waters (at about 20.4°C on 12 November, not an excessive value) ... this more or less direct reference to supposed 'global warming' seems to have been a mere unsophisticated concession to 'weather fashion': here we were in early November, and the sea had certainly been warmer than this in the summer!
- The influence of relief is, as usual, ascribed a role: '*a mass of absolutely saturated warm air came up against the mountains*'. However, the heavy rain fell over lower lying ground, and the orientation of the north–south band bearing the most torrential downpours had nothing whatsoever to do with any mountains.
- Low pressures at altitude, or even a '*cold-air drop*' at the 500-hPa level ... but there was neither a cold drop nor a particularly marked depression, as can clearly be seen in Figure 68(d): the depression sitting over Spain was very far away from the zone of the strongest updrafts and the heaviest rain, and had nothing to do with them. The reference to the 500-hPa level appears here, yet again, only as a simple concession to the 60-year-old dogma which stresses the primal role of altitude.
- A '*Balearic Low*' ... a (secular) piece of nostalgia, focusing on 'the' depression! However, this depression was over the Alboran Sea on the 11th, and did not become 'Balearic' until the 12th–13th, being as mobile as the anticyclone ('**A**') which was instrumental in its formation (Figure 68(b,c)).

So these various hypotheses do not explain the origin of the depression and the low-pressure corridor at the surface, neither do they explain their mobility, nor the formidable advection of moist Mediterranean air moving in so fast in the lower layers. There is also little explanation of the precise localisation of the rain (especially its axial orientation), and even less explanation of its intensity ...

Evolution of the situation, 10–13 November 1999

On 9 and 10 November, two very energetic MPHs, brought together by an anti-cyclonic ridge of high pressure, moved across the North Atlantic and Europe:

- MPH1 came down rapidly, directly from the Arctic following a path to the east of Greenland. It attained a pressure of 1045 hPa on the 10th and 11th (Figure 68(a)), which is a very high value for the season and for an MPH on an oceanic course.
- MPH2, reaching 1040 hPa by the 10th, stretched across central Europe, towards the eastern basin of the Mediterranean (Figure 68(a)).

On 11 November, these two powerful MPHs moved on southwards. The Cantabrian Mountains divided MPH1 into two parts. Its northern part moved off eastwards into southern France, spilling out over the western Mediterranean. Its southern part moved on over the Atlantic, and invaded the Iberian Peninsula on its way to the Mediterranean, for the most part between the Atlas range and the Sierra Nevada. Its arrival over the Alboran Sea led to the formation of a shallow depression (Figure 68(c)), which began to move slowly north-eastwards across the Mediterranean coast of Spain. MPH2 extended itself across the eastern Mediterranean, in the direction of Africa, and westwards.

By 12 November (Figure 68(b)), the two MPHs were progressively spreading across the Mediterranean, and the formation of the intermediary low-pressure corridor was proceeding, as was the ongoing narrowing of this corridor, channelling the energetic precipitable potential into the Gulf of Lions.

- The southern apophysis (offshoot) '**A**' of MPH1 (which reached Tunisia by the end of the day) moved rapidly eastwards. On its leading edge, the vanguard of its air was being violently uplifted, and it slewed northwards, first across Spain and thence across southern France.
- MPH2, progressing southwards, stirred up along its trailing edge an intense south-easterly circulation across the central and western Mediterranean.
- The low-pressure corridor between MPH1 and MPH2 shrank progressively, and during the night of 12–13 November, it was drawn out between the Gulf of Gabes and the Gulf of Lions, advecting first African (Saharan) air, and then Mediterranean air. The northward movement of the Mediterranean precipitable potential was accelerated, to speeds of more than 100 kilometres an hour. A violent storm lashed the Mediterranean, and there were very heavy seas, accompanied by a rise in the level of the Gulf of Lions: three ships were driven aground.
- The northern apophysis 'a' of MPH1, which was less than 1,500 m deep, was diverted towards Aquitaine by the Cantabrian Mountains and the Pyrenees, and blocked the westward passage of the intense south-easterly flux associated with the low-pressure corridor, proceeding at speeds of over 100 kilometres an hour. This Mediterranean flux was forced to rise abruptly, and its energy was liberated, causing the torrential downpours all along the asymptote of confluence 'stuck' to the leading edge of the anticyclonic apophysis 'a' (Figure 68b).

On 13 November, MPH1 and MPH2 met over the northern part of the western Mediterranean basin, closing the low-pressure corridor and shutting off the advection in the lower layers. The rain eased, and finally stopped. The southern part of the Mediterranean basin remained host to low pressures on 14 November, south of the junction between the two MPHs, between Sicily and southern Tunisia (Figure 68(c)).

There is nothing really exceptional about the dynamic of this situation: all such paroxysmal events occurring in these southerly areas are the result of intervention by MPHs, either singly or severally. What distinguished the events of November 1999 was their sheer intensity, the result of the combined influence of two powerful MPHs: pressures of more than 1040 hPa are not very common, especially over the

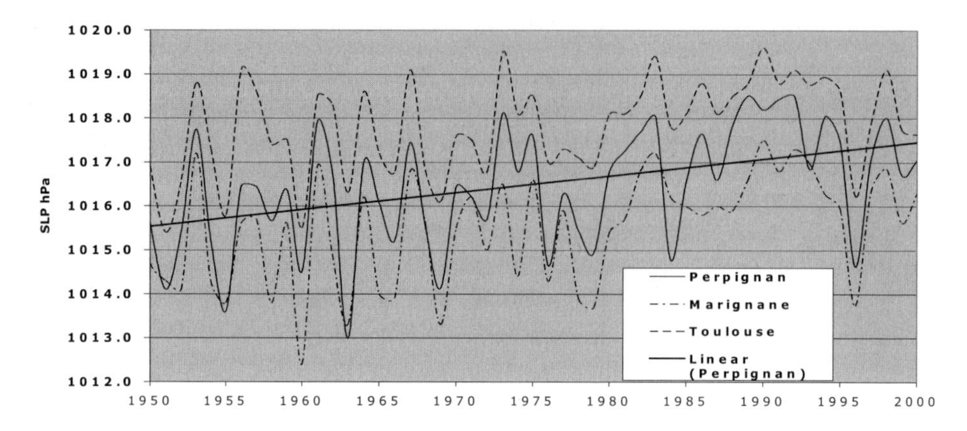

Figure 69. Annual mean sea level pressure (SLP) at Toulouse, Marseille-Marignane, and Perpignan (France) from 1950–2000; linear trend at Perpignan.
Source: Météo-France data.

ocean and at the end of summer. The progressive intrusion of these two MPHs over the Mediterranean basin ensured, as the intermediary corridor narrowed, enormous transports of precipitable potential towards the north.

The powerful nature of these MPHs is of course in absolute contradiction to the 'global warming' scenario (cf. Chapter 10) ...

- ... as are, consequently, the rising pressures recorded right across this southern part of France. Figure 69 shows, for Toulouse, Perpignan, and Marseille-Marignane, and in spite of much variability, a continuous and pronounced trend towards higher pressures since the decade 1960–1970; this increase is of the order of 2 hPa, on the scale of annual mean values.
- ... as is also the decrease in rainfall in these Mediterranean regions, where, although total rainfall figures are not much affected (cf. Figure 66), the greater instability of the weather has brought about a greater concentration in rainfall patterns: events are now more intense, or even very severe, and there are (anticyclonic) periods when no rain falls, and these last longer.

11.2.3 The dynamic of weather in the Mediterranean

The recent climatic evolution observed in southern France reflects, to differing degrees, what is happening over the whole of the Mediterranean basin.

North winds and south winds

The Mediterranean basin is surrounded by mountains, some of which are very high, notably to the north. These mountains vigorously channel the air advected by MPHs: the *cierzo*, *mistral*, and *tramontane* blow between the Pyrenees and the

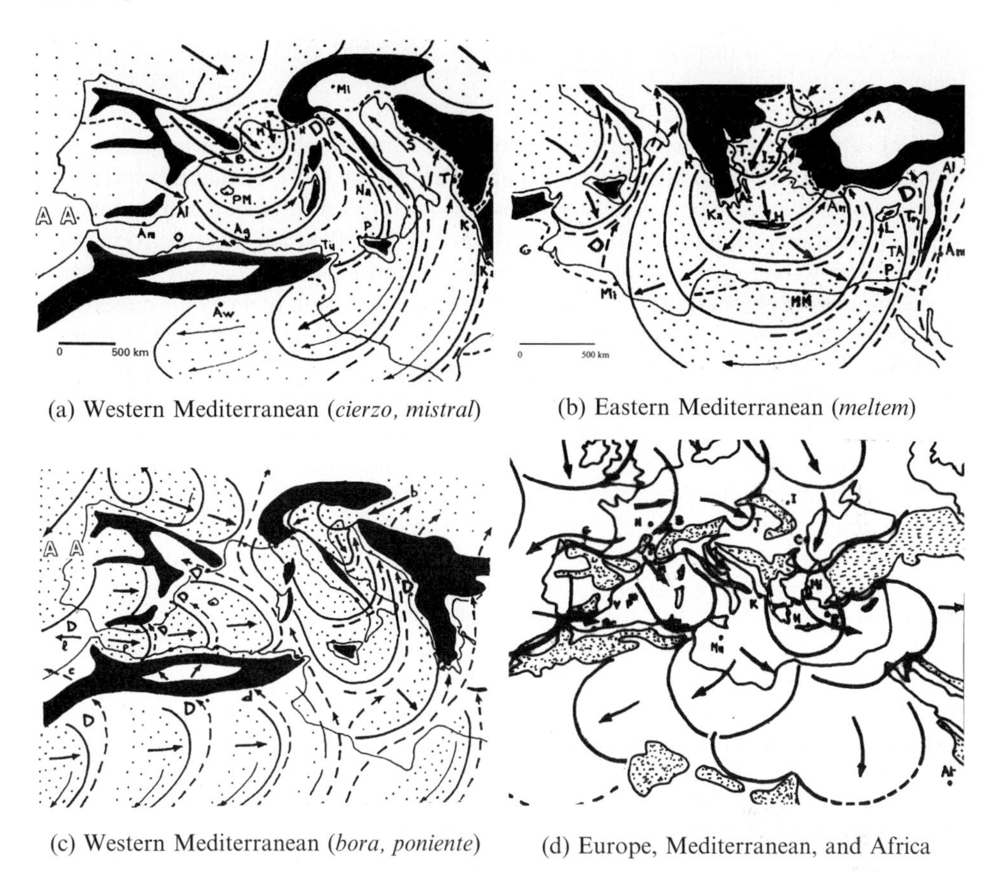

(a) Western Mediterranean (*cierzo, mistral*) (b) Eastern Mediterranean (*meltem*)

(c) Western Mediterranean (*bora, poniente*) (d) Europe, Mediterranean, and Africa

Figure 70. Interference between MPHs and relief; north winds and south winds over the Mediterranean area.

Notes: D = Low; bold arrow = displacement of MPH; broken arrow = cyclonic wind from south.

Alps (Figure 28, Chapter 8; Figure 70(a)); the *bora* between the Alps and the Dinaric Alps (Figure 70(c)); the *meltem* (or *Etesian winds*) between the Balkans and Anatolia (Figure 70(b)), and the *poniente* between the Sierra Nevada and the Atlas range (Figure 70(a,c)). These fluxes, accelerated through defiles, are responsible for winter cold snaps, hard frosts, sudden and violent winds, unexpected snow on low-lying land, and sometimes disastrous visitations of severe storms and disturbances. Winds from the north are either cold or cool, intense, anticyclonic, and divergent.

Reconstitution of a reduced size MPH at the exit of these 'corridors of penetration' causes a proportionally intense northward deviation, on their leading edges, of the Mediterranean flux (or sometimes, an African flux). The southerly fluxes known as the *sirocco*, *livas*, *khamsin*, and *sharav*, are cyclonic in nature, and are warm, turbulent, and moist. They rapidly take up water from the sea, and are involved in violent confrontations with cold air masses and/or relief. If they originate in the

Sahara, they also transport large amounts of dust, and even clouds of migrating locusts (Figure 70(c)).

Polar air flows constantly into the Mediterranean area. The flow is more vigorous and winds are accelerated in winter; in summer, they are attenuated, and slowed. It is during the summer period that MPHs merge most frequently, spreading a relatively thin anticyclonic covering across the Mediterranean basin, with, above it, a pronounced inversion at an altitude of about 1,000 m. Beneath this (unproductive) inversion, water vapour concentrates – as does pollution – but it cannot be utilised, so the summer is characterised by the absence of rain, though real dry air conditions are only found normally in the interior of continents, while coastal areas remain quite humid. At this time, precipitable potential is at a maximum, and much of it is exported southwards from the eastern basin (to the east of the Atlas range) by the trade wind, which is destined to become the *harmattan*, blowing across the Sahara (Figure 70(d)). If a more vigorous MPH intervenes, this exported air changes direction abruptly, and now passes northwards. The updrafts associated with this reversal briefly open the barrier formed by the inversion (Figure 68). Towards the end of summer, and in autumn, the most 'explosive' conditions come together, and the strength of MPHs is restored – the precipitable potential is still considerable. Hence, catastrophic situations like the one analysed above. Other examples: the torrential rain which fell on the Spanish Levante (Gil Olcina *et al.*, 1983; Gil Olcina and Olcina Cantos, 1997; Fabre, 2001); or the dramatic flooding which occurred on 10 November 2001, in Algiers (Bab-el-Oued); or the violent storms which characterise southern Tunisia (Leroux, 2001).

The recent evolution: cooling and surface pressure rise

Generally, the recent evolution of the weather in the Mediterranean basin has seen an increase in paroxysmal phenomena, accompanied by floods, with adverse consequences to humans and to property. At the same time, however, there have been more periods of drought, which have begun to cause particular concern since the 1990s: for example, in Spain (Gil Olcina and Morales Gil, 2001), in Italy (Conte and Palmieri, 1990), in Greece (Nalbantis *et al.*, 1993; Nastos, 1993), and in Algeria (Djellouli and Daget, 1993).

One might 'associate' this pluviometric evolution (following the IPCC's precept to the letter) with 'global warming', according to the equation 'T/R' for torrential rain, also supposing that, for droughts, 'heat = aridity' (thereby jumping to hasty or erroneous conclusions) ... but this would assume that the Mediterranean is really getting warmer. Now, even though its temperature rose until the decade 1940–1950, it has since undergone '*a marked diminution*' (Maheras, 1989), by as much as 1°C in the course of the last 30 years in the central and eastern Mediterranean (Kutiel, 1999). Note that Litinski (2000) found evidence for this cooling in the eastern basin for the periods 1931–1960 and 1961–1990, with a fall at Alexandria, for example, of 1.1°C.

This cooling has manifested itself in recent weather events (cf. Chapter 10): in January 2000, northern Italy experienced its '*coldest winter for a century*', and pieces

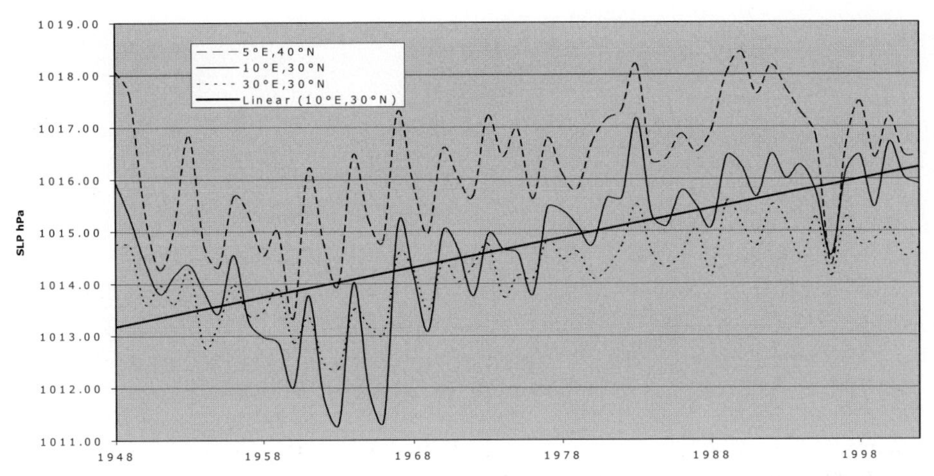

Figure 71. Mean annual pressure evolution over the Mediterranean area from 1948–2002. Western Mediterranean basin: 5°E/40°N (near the Balearic Islands). Eastern Mediterranean basin: 10°E/30°N (near Ghadames, Algeria–Tunisia–Libya) and 30°E/30°N (near Cairo, Egypt). Linear trend at 10°E/30°N.
From Pommier (2004) (CDC/NCEP–NCAR data).

of ice were seen floating in the Adriatic Sea (something which had not been observed for 50 years); at the same time, the Middle East was gripped by cold, and snow fell in Damascus and Jerusalem, and in Saudi Arabia. More recently still, in January 2004, the temperature in Istanbul fell to −10°C, and the city was brought to a halt by snow. In Athens, there was snow, and the temperature was as low as −8°C, and in Egypt seven successive cold snaps occurred, bringing rain and sandstorms. Also, falls of dust lifted from the Sahara are on the increase in Europe, evidence of more meridional trajectories taken by MPHs (Figure 70(c)). The 'idyllic Mediterranean' seems rather hard to pin down, and the same might be said for the simplistic relationships and predictions of the models.

More evidence of this cooling can be seen in the continual rise since the 1970s, across the whole of the Mediterranean, of atmospheric pressure at the surface (Makrogiannis and Sashamanoglou, 1990; Leroux, 1994b; Maheras *et al.*, 2000; Kutiel and Paz, 2000). Higher pressures (Figure 71) have been general throughout the Mediterranean basin and beyond, regularly since 1960–1970. The eastern basin has experienced the most noticeable rise, lying as it does in the path of MPHs which have come down from central Europe, mostly via the Aegean Sea (Figure 70(d)). The linear trend seen for pressure in the vicinity of Ghadames (Figure 71) echoes the increase in the advection of air by MPHs across northern Africa. Aubert (1994) has drawn attention to rising pressures in central Europe, and Omar Haroun (1997) has done the same for northern Africa as far as the Sudan.

Figure 72 uses the example of Kerkyra, in Greece, to show that the rise is mainly a winter phenomenon, especially in the eastern basin where winter values can reach 4 hPa on the mean scale. The maximum values were recorded in the 1990s, a period

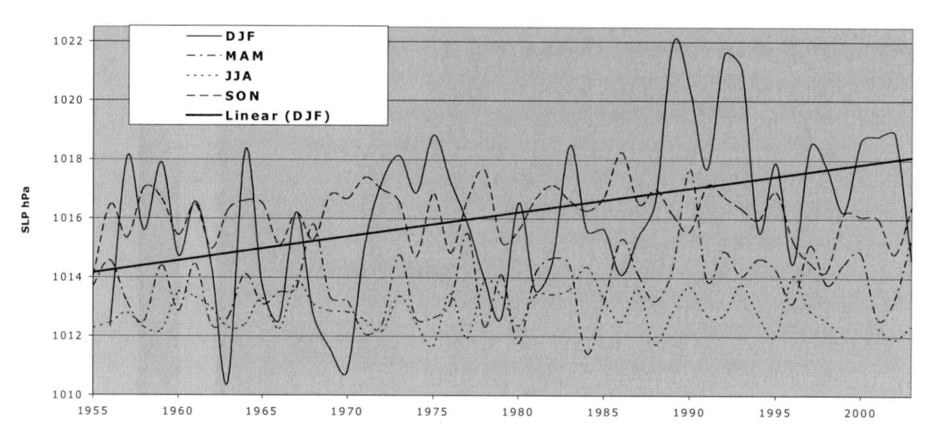

Figure 72. Seasonal mean sea level pressure (SLP, hPa) at Kerkyra, (Greece), 39°37′N, 19°55′E, from 1955–2003. Linear trend in winter: DJF.
From Greek Meteorological Service data.

when heavier rainfall became of particular concern, as has been stated above. Since then, pressures have been measurably lower, but they still remain relatively high. Cooling, and higher pressures, are both indications of the intensification of imports of air by MPHs which are initially both colder and more powerful (Chapter 10). A sign of this is the formation of long-lived AAs in winter and/or summer, and prolonged droughts (as long as a vigorous MPH has not managed to dislocate the agglutination, before reinforcing it). Periods of severe pollution also result from these conditions, above extensive conurbations, with the urban and industrial pollution concentrated in the lower layers below the level of the inversion. These agglutinations are also conducive to higher temperatures during the daytime, since pressure, insolation, and water vapour (and even the pollution itself) considerably encourage warming in the lower layers: thermal contrasts are therefore much more marked.

Consequently, the evolution of the weather over the Mediterranean basin has absolutely nothing to do with the greenhouse effect. Cooling and higher pressures, which have actually been observed, run counter to the predictions, based on hypotheses, of the IPCC.

To try to assess the future climate of the area around the Mediterranean without taking real weather mechanisms into account, and without an appreciation of their actual evolution, is a matter of pure divination ... and this is also the case with predictions involving drought.

11.3 DROUGHT

The term 'drought' means different things to meteorologists, hydrologists, soil experts, agronomists ... but here, we confine ourselves to its meteorological sense (i.e., the absence of precipitation). The equation 'T/R' suggests that there is more rain all the while T increases; but there are limits: heat, by causing enhanced evaporation, ends up signifying drought, which runs counter to the initial equation. We need only look at the ravaged, sun-baked landscapes of the Sahel, where the Great Drought has held sway these last 30 years (Leroux, 1995, 2001), to have this fact dramatically reinforced in our minds. It is often forgotten that, although the heaviest rains fall in warm tropical areas, and rainfall values there are the highest, these areas also may exhibit the lowest values; *ipso facto*, heat cannot be said to be conducive either to rain or to the lack of rain. Other factors, as well as temperature, intervene (i.e., one, or several, of the conditions necessary for pluviogenesis may or may not be present (see above)). So some unthought-out association between heat and aridity, or heatwaves, or droughts, is yet another lamentable simplification, *and does nothing to explain the causes of drought.* But this association is still put forward, whenever there is a drought. The drought in western Europe in the summer of 2003, was 'obviously', nay, 'very obviously', symptomatic of 'global warming' ...

11.3.1 Heat and drought in the summer of 2003 in western Europe

The 'dog days', or veritable heatwave, of summer 2003, when little rain fell, were a remarkable event for western Europe: sadly, very many people succumbed to the heat. The media were not slow to exploit the scorching weather, and its dramatic consequences, as had happened at the time of the Dust Bowl tragedy, and in the summer of 1988 (see Chapter 2). There was no shortage of peremptory declarations about the future of the world's weather, even though the phenomenon was limited to a relatively small area. It could in no way have been described as 'global'. The partisans of the 'global warming' scenario hastened (automatically) to attribute the heatwave, drought, and forest fires near the Mediterranean to the greenhouse effect (i.e., to the human factor). But ... a year before, they had stated that the cool, disappointing summer of 2002, with its rain and flooding (notably in Germany in August) was *also* a confirmation of the predictions of the IPCC's models! Everything seems to be fair game, and any scruples in this field have worn rather thin!

'Official explanations'...

In reply to the question: '*What mechanisms lie at the origin of this heatwave*?', Météo-France offered '*the presence of the Azores anticyclone ... with an extended ridge ..., and above it, a layer of unusually warm air from the south*'.

- Bringing in the 'Azores anticyclone' immediately puts us back into the realms of magic: more meteorological animism (see above)! Well may the man in the street or the media cherish their 'favourite anticyclone', but for so-called specialists to perpetuate this kind of reprehensible old confusion gives pause for thought about the state of official 'meteorological science'.
- The anticyclone in question has its '*extended ridge*', which seems to have appeared out of nowhere: we are told nothing about the origin of this

fabulous ridge, which reaches as far as about 60°N, an amazingly high latitude for a *subtropical* anticyclone called 'of the Azores'!

- The anticyclone in question has *above it* (indicating its discoid nature, which is correct) '*a mass of very warm, very dry air from the south, coming in through the Pyrenees ... both near the surface and at altitude*' (cf. Météo-France website). Has the Iberian relief had no blocking effect upon this air, even near the surface? And this flux was supposed to be heading northwards, into areas of high pressure ... a highly unlikely event! But it was warm, and so the air was coming from the south, where it's warm! There can be no argument against this! Except that it is very naive ... We even heard Bocrie of Météo-France saying it again, one year on, on the *France 2* TV channel's programme *Alerte Canicule* ('Heatwave Alert'), on 9 July 2004: '*a mass of warm air from the tropics settled over France*'! This was really absurd, for how could such a flux have physically established itself, and maintained itself, for such a long period? It would also have had to flow counter to an anticyclonic circulation connected with the supposed Azores anticyclone! Then again, what is the significance of the (never demonstrated) presence of warm air advected to high altitude, since we know that **warm air cannot reach the surface because of the inversion located above the anticyclone settled in the lower layers**? And, more importantly, we know that *it was near the surface, within the anticyclone itself, that it was warm.*
- Consider the replies given by Météo-France to the questions: '*why did this cap of warm air move so far to the north, and why did it remain fixed there for so long?*'. They said: '*There is no ready answer ... many mysteries remain*'! This is, to say the least, quite alarming. For Météo-France to rely blandly upon *mysteries* as arguments reflects poorly upon their scientific competence. This public service agency should not be indulging in pointless academic discussion, but ought to show its ability to predict meteorological phenomena and warn of their possible disastrous consequences. We are wide of the mark!

The meteorological conditions of a heat wave

In 2003, from June onwards, the meteorological situation was dominated, to a lesser or greater degree, by the presence of a vast area of high pressure, covering Europe, the Mediterranean, and the eastern side of the Atlantic. The great stability brought about by the anticyclonic conditions – calm air, or very slight breezes, and the absence of upward air movements – encouraged the warming of air in the lower layers. The conduction of heat, and infrared absorption, are in fact much enhanced when pressures are high and the air cannot rise; the layers nearest to the ground become (for the same amount of solar energy received) relatively very much warmer. The heating leads to a considerable diminution in relative humidity (i.e., the air becomes much drier). This drying effect is exaggerated by the fact that water vapour from the Atlantic and the Mediterranean cannot reach the interior of the anticyclonic area (which considerably reduces the natural greenhouse effect associated with water vapour, and allows in the daytime more energy to reach the surface). Now that the moisture content is absent, maximum insolation occurs, and the cumulative heating effect gradually assumes the proportions of a 'heatwave', especially in urban areas, which are less 'ventilated', and become warmer, drier, and more polluted. The effect of the so-called urban heat dome is reinforced, and more so as towns continue to grow. At the same time, the anticyclonic weather, which is limited to the lower layers, and the absence of horizontal and vertical air movements, concentrate pollution at lower levels. This pollution, however, is fortunately less severe during the summer (reduced activity), the inversion layer being located at an altitude of about 1,000 m, while the increased insolation accelerates photodissociation (the production of ozone, facilitating rising temperatures in the lower layers).

Heat, drought and pollution are consequences of the higher pressure. And not vice versa. Remember that the greenhouse effect scenario would envisage the opposite situation, where pollution is a cause of rising temperatures, and this would lead to lower pressures, since the air would then rise, except under the anticyclonic conditions which are precisely *the key* to this situation! So the high pressures certainly have to be explained, but certainly *not* by invoking some childish and imaginary transportation of warm air coming from somewhere like the Sahara, as simple minds might assume, for this warm air would, if it could actually reach ground level, cause pressure to fall!

The dynamics of weather over Europe during the 'dog days'

What really happened, then, during the summer of 2003, or, to be more precise, during the first two weeks of August, when the weather was at its hottest? To find the

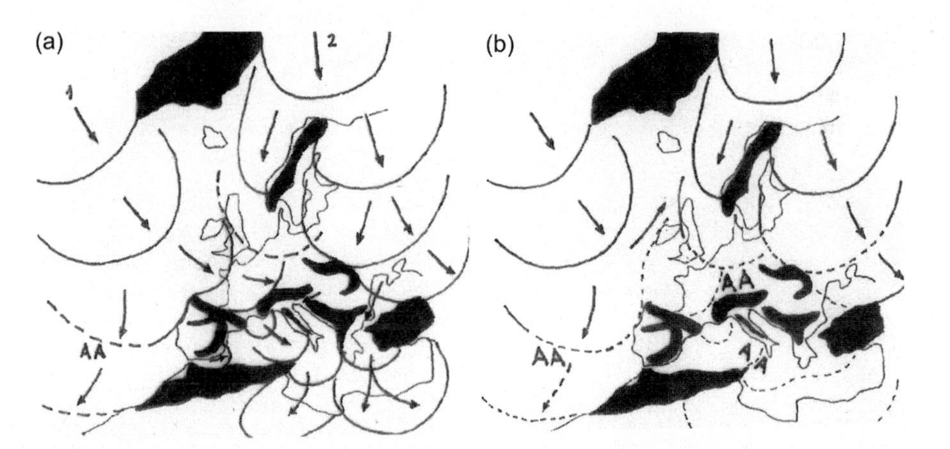

Figure 73. Dynamics of weather over Europe, the north-eastern Atlantic, and the Mediterranean. (a) Normal situation. (b) Situation with an extended AA (winter and/or summer).

Western Europe is the meeting area of MPHs, coming from the Arctic, and following two paths: one, the more frequent, on the western side, the other on the eastern side of Greenland. Paths of MPHs: 1 – west of Greenland, America, and the Atlantic; 2 – east of Greenland, Scandinavia, and Russia.

Only the leading edge of MPHs are represented, where 'bad weather' is concentrated. Broken lines show the overlapping of MPHs. The relief impassable to MPHs is schematised.

Note: AA = anticyclonic agglutination.

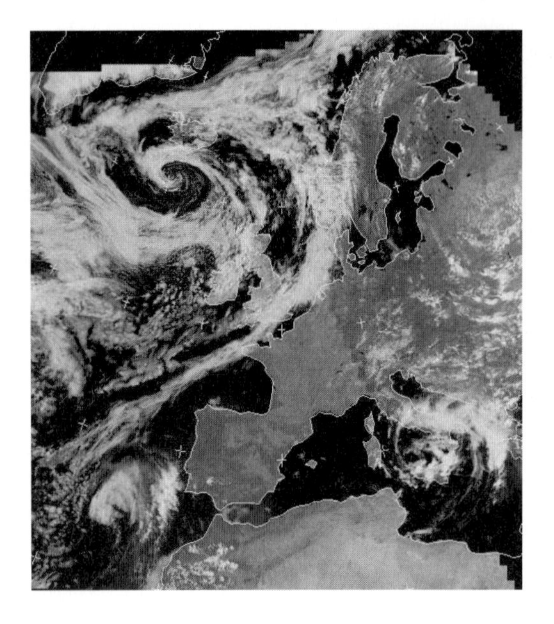

1 August

2 August

3 August

4 August

Figure 74. 1–15 August 2003 (NOAA, visible, 12 h 15 h, *SATMOS*). Meeting over Europe of 'American' and 'Scandinavian' MPHs, provoking high surface pressures and the connected 'dog days' during the first two weeks of August 2003.

5 August

6 August

7 August

8 August

9 August

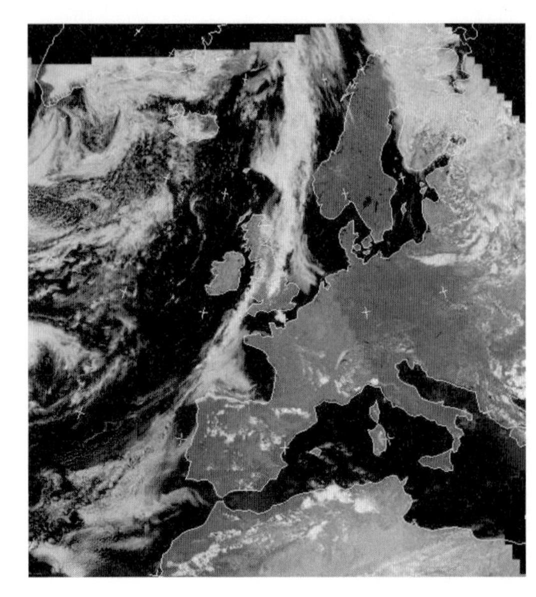

10 August

11 August

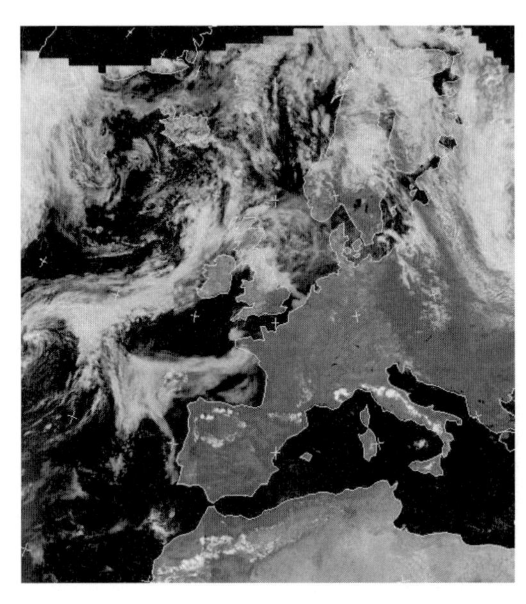

12 August

Figure 74. (*cont.*)

13 August

14 August

15 August

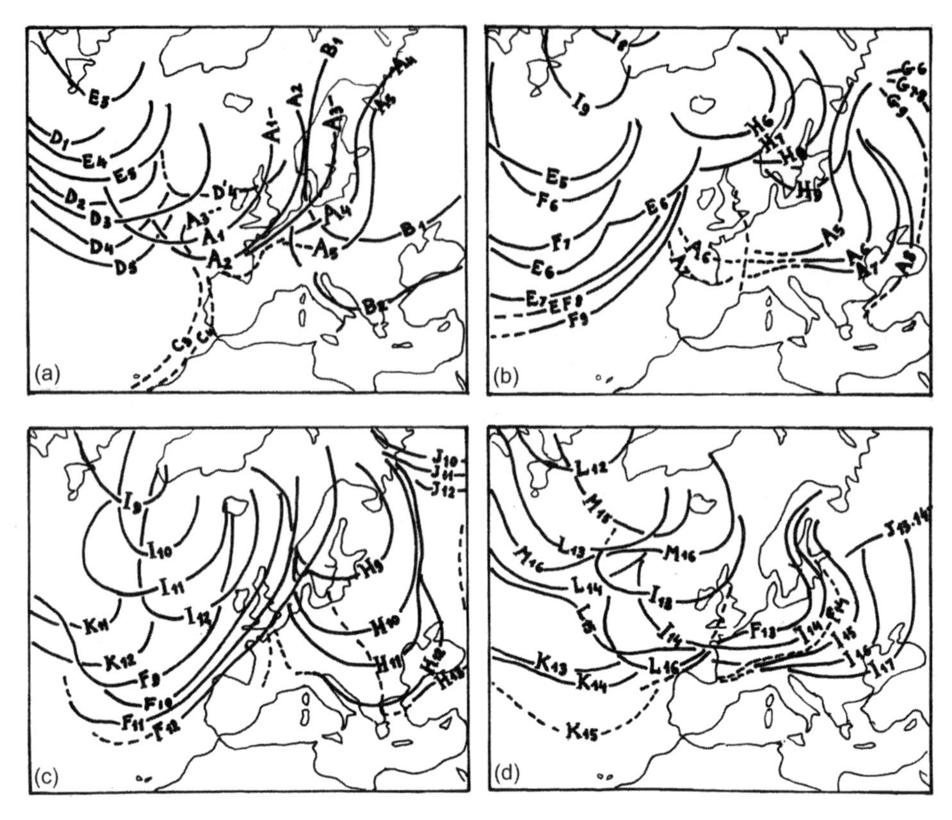

Figure 75. Surface evolution of MPHs from 1–17 August 2003, over the North Atlantic and western Europe. (a) 1–5 August, surface. (b) 5–9 August, surface. (c) 9–12 August, surface. (d) 12–17 August, surface.
From *European Meteorological Bulletin* and satellite images.
Notes: A, B, C . . . = leading edge of an MPH.

answer to this question, one has only to study satellite images (Figure 74) and analyse the synoptic charts (Figure 75). MPHs are approaching western Europe along two trajectories (Figure 73(a)): one from the Arctic via America and the Atlantic, and the other directly to the east of Greenland along the Scandinavian trajectory. As these MPHs meet and merge, and are slowed by the presence of relief, an AA is formed (Figure 25), as usual, over the ocean, its mean position being designated as 'of the Azores', but it also extended over the mainland of Europe and the Mediterranean. The unusually long duration of such an agglutination of high-pressure areas, and its very size, were a result of the continuous and regular feeding of this anticyclonic 'cap' by MPHs in the lower layers.

From 1 to 17 August, 12 MPHs, which we shall label A to L (Figures 74 and 75), exhibiting fairly high pressures for the season (1020–1025 hPa), made their contributions to the AA: seven of them followed the American–Atlantic trajectory, and five of them the Scandinavian trajectory. The distinction between these two trajectories is not always a very precise one, since certain 'American' MPHs are reinforced along their way by cold air from Greenland: for example, MPH D on 4 August (D4, Figure 75(a)). Scandinavian MPHs are colder, since Scandinavia is experiencing lower temperatures at the time than are normal for the rest of Europe (Pielke Sr. and Chase, 2004), and they form a barrier to the MPHs coming in from the west: for example, the enormous MPH B blocks the eastward path of MPH A from 1 August (Figures 74 and 75(a)). Another example is that of MPH H, hindering the progress of MPH F from 9–10 August, the latter having merged with MPH E on 8 August (Figures 74 and 75(c)). MPHs of various origins are coming together over Europe, which sits beneath the meeting point of the two trajectories, and their encounter maintains pressures on land as well as at sea. The agglutination is particularly homogeneous from 3 to 12 August, and it is then that the highest temperatures of this very stable anticyclonic period are reached. MPH I, which rounded the south of Greenland on 9 August, is considerably reinforced on its direct southward trajectory, and reaches Britain on 13 August, covering northern Europe from France to Denmark by 14 August. This MPH is moving mainly eastwards on 15 August (Figure 74), and on the 17th it extends from the Pyrenees to the Black Sea (Figure 75(d)). Now, the temperature drops by more than 10°C, and there is precipitation. The heatwave has come to an end, to be replaced by a relatively cool period, and a cool autumn.

Consequently, and indisputably, if we rely upon the observation of actual phenomena, instead of resorting too readily to 'magic' recipes and/or baseless statistical links, we can see that:

- the cause of this heatwave was the **presence of an AA**;
- it was absolutely not the result of air coming up from the south, neither at altitude nor at the surface; and
- it was, on the contrary, caused by the concentration and rapid warming of **anticyclonic air, coming from the north in the lower layers** (i.e., from the Arctic) – the vehicle being the MPH.

There is nothing exceptional about this type of situation, since it occurs regularly at lower latitudes (30°, even 40°) during this season in the eastern Atlantic and over the Mediterranean; the eastern Atlantic AA is the result of it, as are other agglutinations on the eastern sides of oceans (Figure 25). There is nothing really exceptional here. The peculiarity of this particular heatwave was the unusual extension of this anticyclonic dome, in both space and time. But barometric situations like this, with similar temperatures, are not actually all that rare. The summers of 1998, 1995, 1994, 1985, 1983, 1976, 1964, 1947, 1921 (which saw only a quarter of the usual rainfall), 1901, 1900 etc., were equally warm, or even warmer, locally ... In France in 2003, 70 records – out of 180 – were broken, but all-time records remained unchallenged. The national record is still held by Toulouse, where, on 8 August 1923, a temperature of 44°C was recorded, involving at that time a slight urban effect. The (drier) summer of 1976 is remembered everywhere in France as the symbolic 'drought summer'. The summer of 2003 created new national records in Portugal,

Germany, and Switzerland. In Great Britain, the summer of 2003 is still outranked by 1976 and 1995, when the very hot spells lasted longer.

The key to the heatwave: high pressure

What needs to happen for such an anticyclonic agglutination to form? The passage of MPHs must be slowed, or stopped, by 'fixed' relief (Figure 25, Chapter 8), or 'mobile' relief (i.e., by a denser MPH) (Figures 73, 74, 75).

- Usually, in summer, MPHs on mainly American trajectories move freely into Europe (unless relief intervenes). In France, for example, the northern edge of the Alps marks the limit of this translation. To the north, MPHs enter central Europe, and to the south the fact that they are blocked by the mountains of the Alps and the Cantabrian–Pyrenees chain means that southern France sits beneath an AA – an extension of the AA over the ocean stretching out over the Mediterranean.
- In winter, Europe, and especially central Europe, frequently lies beneath a robust agglutination, which increases in strength towards the east, into Asia. Atlantic MPHs then overlap, over western Europe (cf. below).

It was this type of situation, typically a winter one, which happened (but in the summer) in 2003, as a result of an unusually high frequency of MPHs on the Scandinavian trajectory. Figure 76 shows the mean evolution of pressure at the 'centre' of France: notice that mean summer pressures have been higher since the 1970s, of the order of +2 hPa on the scale of mean annual values (Pommier, 2003). Such a rise in pressure is observed across almost the whole of Europe, with the same 1970s 'shift' (Chapter 12). Bern, Switzerland, shows a very marked increase: more than 4 hPa on the mean annual scale (Leroux, 1996a, 2000). However, there is not a word about this essential parameter in the article by Schär *et al.* (2004), for whom the higher temperatures during these summer heatwaves in Europe derive from '*increased atmospheric greenhouse concentrations*'! It is that easy!

Let it be remembered that **this rise in pressure has nothing to do with warming**, on the contrary, but has much to do with the greater energy of MPHs, which has already been mentioned (Chapter 10), and with the constant increase in their frequency over France in summer (Figure 77). It is worth pointing out here the regularity of the summertime trend towards greater numbers of MPHs over France, echoing the curves of Figure 57, which show the increase in the numbers of MPHs over the Arctic (in Chapter 10).

In conclusion, such a heatwave is not some chance phenomenon, nor is it isolated in time. **It cannot in any way be associated with the greenhouse effect** (on the contrary). **The primary cause of the heatwave, and the drought it brings, is an aerological one**.

However, in urban areas, because of the peculiar climatic conditions pertaining there, emissive gases enter the equation as an aggravating factor. This local factor can only worsen with time, as cities continue to spread, and urban activities intensify:

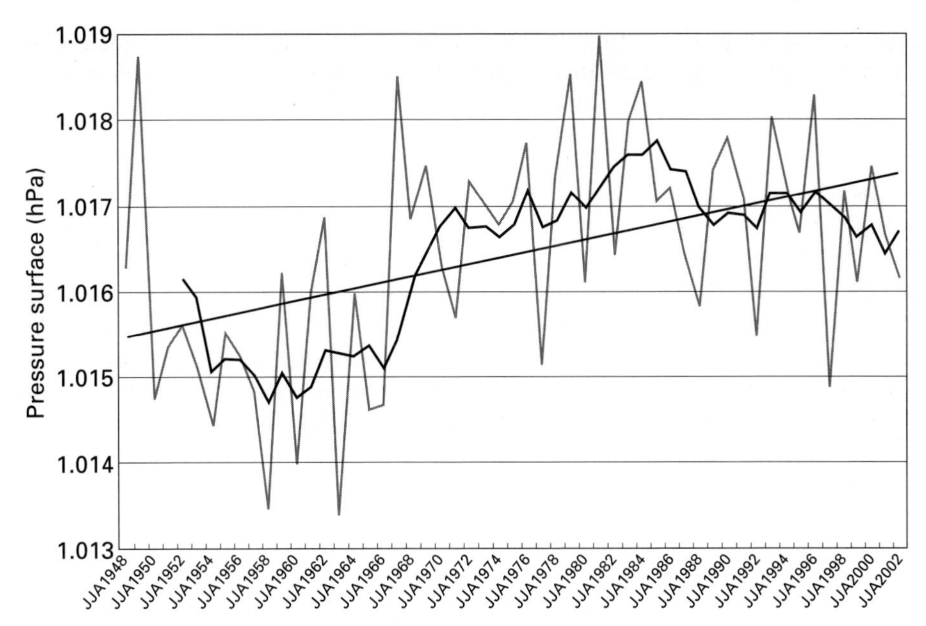

Figure 76. Evolution of mean pressure over the 'central point' of France, near 2.5°E, 45°N (i.e., south of Clermont-Ferrand, during June–July–August, for the period 1948–2002).
From Pommier (2003), (CDC/NCEP–NCAR data).

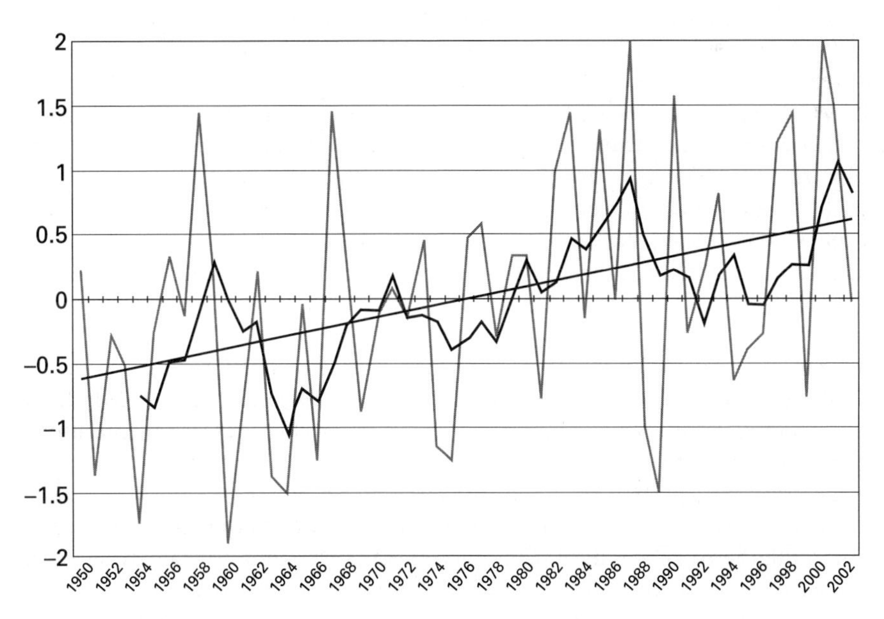

Figure 77. Annual frequency (standard deviation) of MPHs reaching French territory, in June–July–August, from 1950–2002.
From Pommier (2003) (CDC/NCEP–NCAR data).

this parameter ought to command more attention (particularly in the estimation of 'records'), since, leaving aside any possible climate changes, the effects of very hot spells will be (increasingly) amplified everywhere in built-up areas.

The failure of prevision

'*No operational model was able to predict even three months in advance the arrival of last summer's heatwave*'. This was the incontrovertible conclusion about the seasonal prediction capabilities of models reached by André and Rogel, of the *Centre Européen de Recherche et de Formation Avancée en Calcul Scientifique* (CERFACS), and by Dequé and Planton of CNRM of Météo-France, in a *Compte Rendu* for the Académie des Sciences (in *Journal du CNRS*, No. 172, May 2004). It seems, then, that '*climatologists thought that a sea surface temperature anomaly recorded between the months of April and July 2003 in the North Atlantic ... might be considered as one of the precursors of this heatwave*', but it had to be said that '*even if the models had 'known' the exact temperature of the ocean surface, they would have been no better at predicting the event*' (*Journal du CNRS*)! This was nothing unexpected, of course, since what have sea surface temperatures got to do with this 'affair', and how could they be seen as a 'precursor' of a prolonged increase in pressure? The same official article from the CNRS goes on to say, as if more evidence were needed of the disarray in forecasting: '*the causes of the heatwave cannot be reduced to this 'simple' anomaly ... complex phenomena, involving interactions between the ocean and the atmosphere, as yet little understood by climatologists and not easily worked by models, are probably in play here*'! 'Complex phenomena ..., little understood ..., not easily worked ..., probably ...', need we say more? At this level of 'scientific knowledge' or incompetence, perhaps it still remains only to consider 'the captain's age', or the 'colour of his braces'!

Such an admission of ignorance (presented to the *Académie des Sciences*!) is worrying in its *naïveté*, because it shows without any ambiguity at all why our modellers did not predict the heatwave three months in advance, one month in advance, or even a week in advance, or even one day in advance: because **the modellers do not know what real physical mechanisms lie at the origin of the heatwave of summer 2003**!

So the heatwave was not foreseen, had not been explained either at that time, nor since August 2003. But this did not stop some 'brass necks' from unashamedly announcing afterwards, with much media support, a 'Heatwave Alert' plan! Given their lack of understanding of actual meteorological conditions, such a plan is a senseless masquerade. If the causes of heatwaves are not yet properly understood, the only thing the so-called 'alert' can tell you is that 'it's hot!' (which anyone could work out anyway!), but it absolutely cannot tell you that 'it is going to be hot!'. If a similar heatwave occurred during this summer 2004, it would have still take us all by surprise, because (and do we need to say it?) no heatwave or lack of one has yet been predicted for that season! So what can be said about predictions for the year 2100?

The poverty of 'official climate science'...

This heatwave, and more especially the way in which the authorities and the media dealt with it, is **symptomatic of the current state of thinking on the subject of weather or climate**. The meteorologists, taking no risks, usually say that a heatwave '*conforms to the predictions of the models*', but they are unable, as we have stressed, accurately to describe the relevant mechanisms ... In fact, how could greenhouse gases be the explanation for an abrupt rise in atmospheric pressure, and how could they direct the formation and continuing existence of an AA?

Interviewed in his official capacity, the President of Météo-France claimed that '*the exceptional meteorological phenomena, which we are seeing more and more often, are early manifestations of climate change ... heatwave events could be five times more frequent than they are today*'. **But can he tell us why? No**. He is not a meteorologist and his agency cannot provide him with any coherent explanation at all, and he adds that this is '*more a question of conviction than of certainty*'! This may come as quite a surprise! What does science really require: 'conviction', or proof, and reasoned explanations?

Jones, director of the CRU, is more affirmative when discussing this heatwave: '*The 2003 extreme may be directly attributed, not to natural variability, but to global warming caused by human actions*' (in *The Independent*, 2003), but his references are only statistical ones; he does not venture to explain how humankind has caused the heatwave.

These Pavlovian, political responses from unswerving and 'automatic' partisans of 'global warming' are not based upon any scientific reasoning. Moreover, the only 'conviction' they share is that they should not kill the goose that lays the golden eggs (or, in their case, substantial funding opportunities)!

I did not hesitate to write to the principal relevant journals, particularly *Le Monde*, and self-styled 'scientific' magazines, in France, **challenging the apostles of 'global warming' to show how they related the heatwave to the greenhouse effect**. My attempt at correction was not published, because of what a journalist called '*the minority view within the community represented by your explanation*' (though other equally unscientific reasons were given)! Météo-France may say all manner of things 'officially', and does not fight shy of 'explanations' either unfounded or 'magical'; the media, however, prefer to go along with the current, especially if that current is a catastrophist one, and sells papers (Chapter 4). The 'scientific debate' holds little interest for the media, and the preoccupation of newspaper and television journalists, whether they are dealing with the climate or with news items, seems in the main to entertain rather than to inform!

The tendency of models to schematise everything, and to use simplistic 'recipes', masks the fact that the absence of precipitation is multi-faceted, and subject to various aerological factors. A distinction especially needs to be made between middle latitudes and the tropics. We shall examine below only a few salient aspects, for example winter drought, the Dust Bowl, and the drought in the Sahel.

11.3.2 Winter drought and limited mountain snowfall

Since drought is not necessarily associated with heat, AAs can frequently form in winter. For example, the climate of Siberia in winter is dominated by the presence of the 'Sibero–Mongolian' thermal anticyclone, supplied constantly by MPHs from the Arctic. Its density is maintained by the coldness of the continental land mass, and by the blocking presence of high Asian mountains. The harsh Siberian winter is characterised by clear skies, sunshine, very low temperatures, a lack of rain, and strong, biting winds. Central Europe is influenced by the presence of the western part of this anticyclonic 'cap' over the Touran Depression (where the Caspian and Aral Seas are situated). It has a considerable slowing effect upon eastward-moving MPHs. An offshoot of high-pressure areas is quite often established over central Europe in winter, and excess cold air streams down towards the eastern basin of the Mediterranean (Figure 70(b,d)).

Winter continental AAs

Western Europe, a region where the trajectories of MPHs meet (Figure 73), also experiences AAs of variable duration in winter. Let us find examples from the winters of 1988–1993, when there were periods of remarkable anticyclonic stability lasting:

- in 1988–1989, for 77 days in three periods, the longest being 49 days;
- in 1989–1990, for 86 days in four periods, the longest being 39 days;
- in 1990–1991, for 52 days in four periods, the longest being 26 days;
- in 1991–1992, for 111 days in five periods, the longest being 30 days; and
- in 1992–1993, for 105 days in five periods, the longest being 39 days.

The longest lived agglutinations maintain an anticyclonic eastern barrier in the lower layers (over central Europe and Asia), and the frequency of powerful MPHs is heightened. For instance, the AA of the winter of 1988–1989 was continuously supplied for 49 days by 30 MPHs, arriving with an (above average) frequency of one MPH every 1.6 days, with pressures of between 1025 and 1045 hPa (Leroux, 1996a, 2000).

The lack of rainfall is generally less unwelcome in the winter, with fewer clouds, and with the anticyclonic character of the weather even providing some fine, sunny spells and, paradoxically, 'warm' daytime temperatures for winter days. However, such 'fine' weather is not welcomed by everybody, since in mountain areas it also means a lack of snow, and winter sports resorts can be disadvantaged. Such is the case in the Alps (which, because of their east–west alignment, block the southward movement of MPHs), when this mountain mass sits within an immense area of high pressure in the lower layers. Disturbances occur only around the periphery of this AA, over the Atlantic, the Mediterranean, and more often over Scandinavia. As a result, there is an inverse relationship concerning the evolution of glaciers, noted by Reynaud (2003), in the Alps and in Scandinavia: when the supply of water to Alpine glaciers diminished (during the period 1967–1997), glaciers in Scandinavia were, conversely, 'in rude health', and vice versa (cf. Figure 138 in Chapter 14).

Since the AA is particularly stable in mid-winter, precipitation (rain and snow) is more likely to occur in the Alps during spring and autumn, when anticyclonic conditions are less dominant. Mid-winter (January–February) tends to be dry and sunny, with a very wide diurnal temperature range: nights are frosty, but the temperature soon picks up during the day, with the Sun shining upon the remaining snowfields.

The lack of snow in winter is certainly a cause of concern for the *montagnards* (highlanders) and those responsible for the winter tourism trade, even generating conferences such as the one held in June 2000 at Chamonix-Mont-Blanc, on the theme of 'Climatic changes and their implications for the mountain environment'. Naturally, its 'explanatory' aspect was ultra-simplistic (warming will melt the snow), as were the predictions: '*the snowline will increase in altitude by about 150 m for each degree of warming ... and snow conditions will become mediocre and less certain*' (Béniston, 2000). But for what reasons should they become '*mediocre and less certain*'? Delegates were content to parrot the same old pronouncements of the IPCC, but nobody, not even those representing Météo-France, mentioned the sequence of meteorological conditions responsible for a possible shortage of snow!

Rising pressure over Europe

These conditions are nevertheless quite simple, and Figure 78 will serve as an example to provide a quick and to-the-point answer to this question. The curve representing surface pressures to the north of the Alps shows a continuous, marked rise since 1960–1966 of the order of 4 hPa, a considerable value on the mean scale (and a rise already observed at Berne in Switzerland). The changing pressures (for the communication of which I am indebted to Thieme) at the

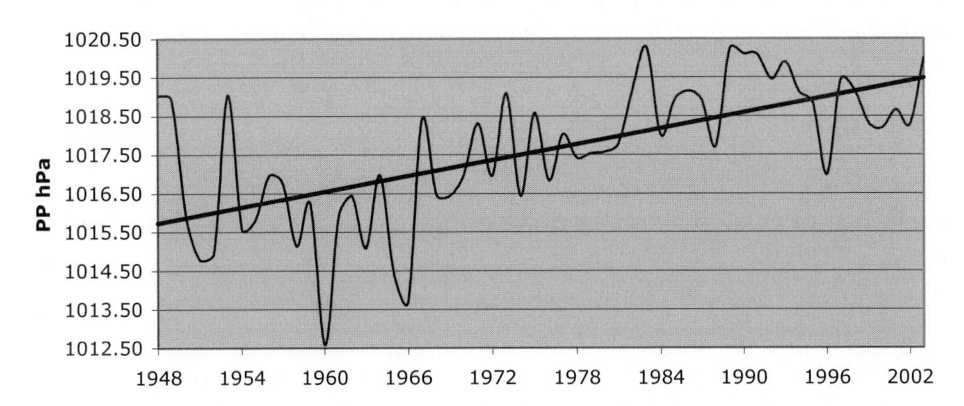

Figure 78. Evolution of surface pressure (mean annual values) at 47.5°N, 10°E, near Lake Constance, Southern Bavaria, Germany.
From Vigouroux (2004) (CDC/NCEP-NCAR data).

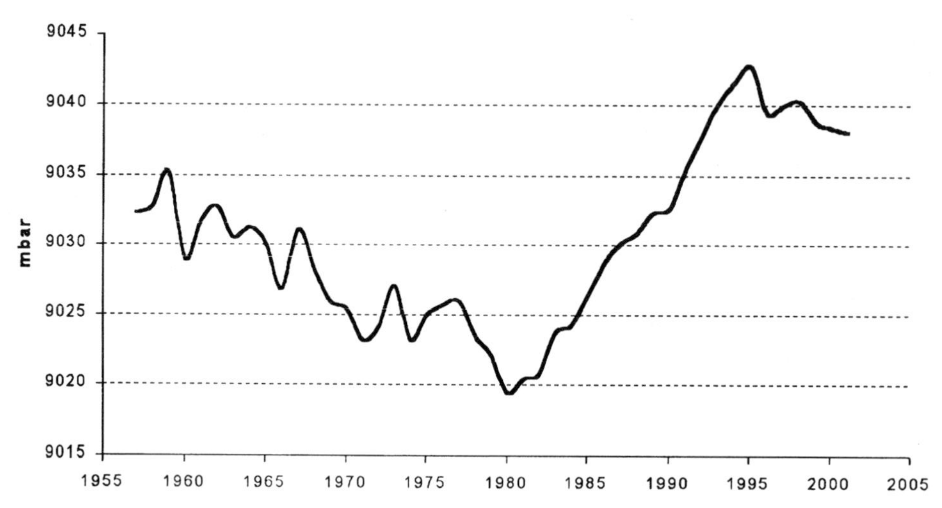

Figure 79. Evolution of surface pressure (mean annual values), Hohenpeissenberg, Bavaria (977 m above sea level), south of Munich, from 1957–2002.
From *Deutscher Wetterdienst*.

Hohenpeissenberg Meteorological Observatory (Figure 79) show, in turn, a rapid and spectacular rise to the north of the Alpine arc of the order of 2 hPa, after the 'trough' of 1970–1980. Because of the altitude of this weather station, at nearly 1,000 m, the curve reveals the progressive increase in the depth of MPHs.

This very obvious rise in atmospheric pressure to the north of the Alps (which has also been occurring to the south, in a more attenuated fashion) is evidence of the increasing frequency of winter AAs. Vigouroux (2004) compared 'highs and lows in snowfall in the French Alps'. He made the observation that, during a winter with less snow, such as that of 2001–2002, the number and strength of MPHs were greater, and agglutinations were more frequent, the situation being reversed when snow was more abundant, as in 2002–2003. Then, AAs were less frequent, and MPHs were able to approach the foot of the massifs and cause updrafts.

The abundance or scarcity of snow in the mountains is therefore certainly not dependent on the greenhouse effect!

This rise in pressure simply cannot appear as part of a 'warming scenario', and in fact the opposite is true. It may be necessary once again to underline the fact that warming (a cause) lowers pressure, but high pressure (a cause) facilitates the diurnal rise in temperature, by improving conductivity.

Again we find the stamp of strengthened MPHs associated with recent cooling in the Arctic.

11.3.3 The Dust Bowl and droughts in America

North America has also known its periods of drought, in both winter and summer (Redmond, 2002). Heatwaves and droughts are frequent occurrences in the USA, especially on the Great Plains of the Midwest. Thousands of people have fallen victim to the most dramatic episodes, in 1934–1936 (generally held to be the worst of all, cf. Chapter 2), 1952–1955, 1980, 1988, 1995 (Kunkel *et al.*, 1996, 1999; Kalkstein *et al.*, 1996), 1999 (Palecki *et al.*, 2001). With even more victims, the heatwave and drought of mid-July 1995 was a harbinger of the severe conditions of August 2003 in Europe. The greatest numbers of fatalities were recorded in cities such as Chicago, where (as happened in Paris) the urban heat island considerably aggravated the effects of the general meteorological situation. This latter event was essentially marked by '*a massive ridge of high pressure stalling over the Midwest*' (Changnon *et al.*, 1996) which moved slowly eastwards between 11–16 July 1995, on its path towards the Atlantic. The end of the hot period was defined by the passage of a '*weak cold front*' from 14–17 July (Kunkel *et al.*, 1996).

High pressure and the heatwave of July 1999 in the Midwest

The heatwave of 29–30 July 1999, in the Midwest, was similarly brought about by an anticyclonic situation, and preceded by an initial hot period lasting five days, from 22–26 July. The leading edge of the MPH responsible for this was visible over the Atlantic on 26 July, extending from Louisiana to Newfoundland (Figure 80). On the 26th, a front driven by a new MPH crossed the Midwest, accompanied by rain on 26–27 July, and afterwards, a '*surface high moved over the area*' (Palecki *et al.*, 2001). From the 27th onwards, the front, still visible over the Labrador Peninsula in Canada, weakens and disappears as the two MPHs merge, while what remains of the front reaches the Atlantic on 29 July (Figure 80). Nearly the whole of North America is now covered by an immense anticyclonic cap, practically cloudless, on 27–30 July.

It is then that the high pressure, anticyclonic stability, and strong insolation cause the temperature to rise rapidly in the layers near the ground, a phenomenon exacerbated in urban areas. During the morning of 31 July, '*a strong cold front began to move across the Midwest from north-west to south-east*' (Palecki *et al.*, 2001), ending the heatwave (and exactly matching what happened in France in 2003), as a third, cooler MPH arrived (Figure 80 – 31 July).

North America lies beneath a favoured pathway for MPHs: here, they are still robust, since they are not far from their birthplace, and so this continent is particularly likely to be covered by an AA. The MPHs overlap and are reinforced before they move out over the Atlantic, and they may sometimes cover the whole of North America (Figure 36, Chapter 9). If MPHs are very powerful, and move only slowly, anticyclonic stability is likely to result for variable amounts of time. But here in this aspect, the principal thinking seems to favour phenomena at altitude, with special reference to an '*upper level ridge*' (Kunkel *et al.*, 1996). Phenomena in the lower layers, and particularly MPHs, seem to have gone unnoticed (cf. Figure 58).

How can sea surface temperatures cause continental highs?

Even then, it seems that the supposed link is not seen to be a completely satisfactory

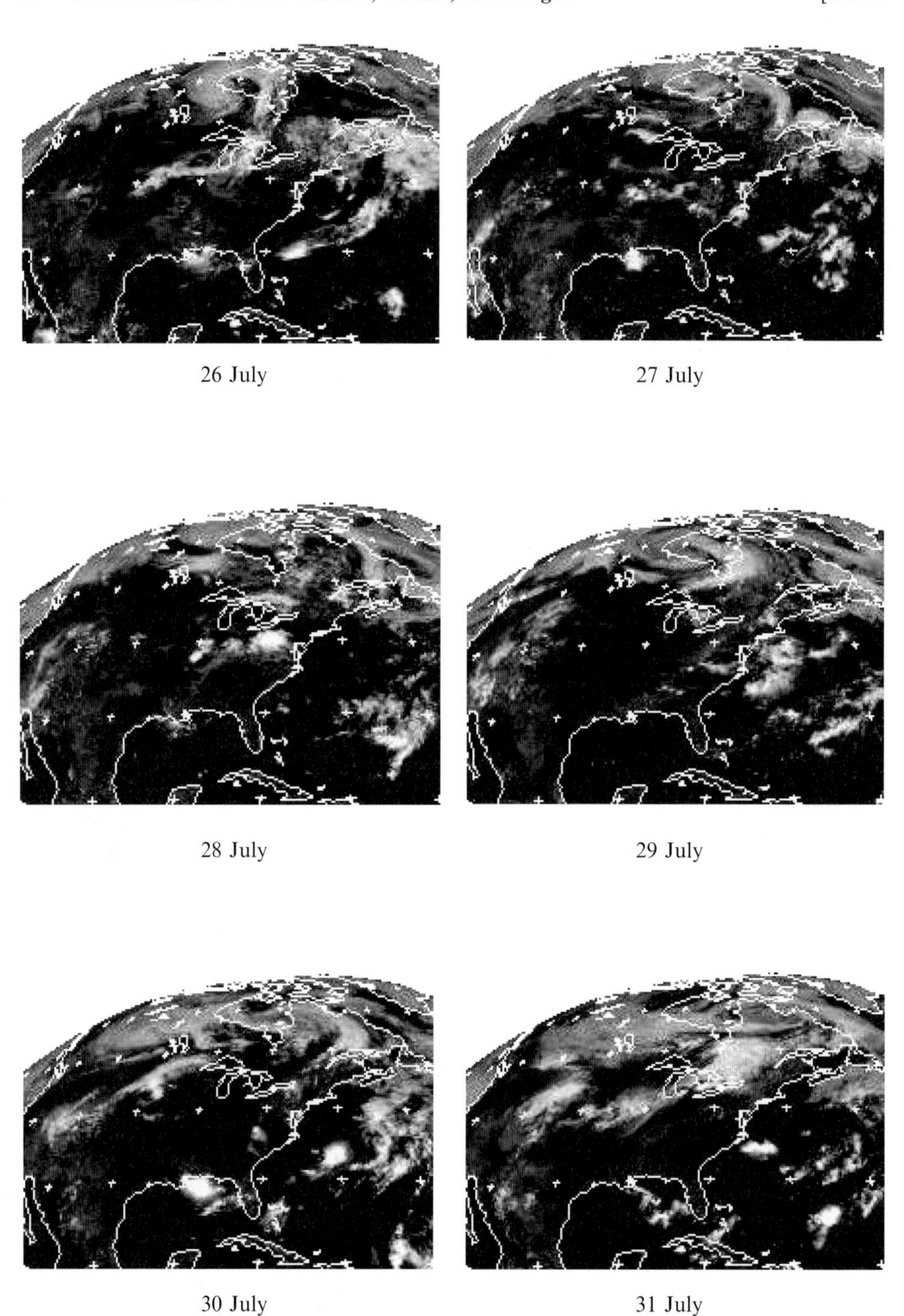

Figure 80. 26–31 July 1999 (visible, *GOES 08*/NOAA, 18 h). Slow displacement and merging of MPHs over North America responsible for the Midwest heatwave.

one, since 'causes' are sought elsewhere (Trenberth *et al.*, 2004); in particular, in remote connections with sea surface temperatures in the Atlantic and the Pacific. McCabe *et al.* (2004) and Schubert *et al.* (2004) considered, especially with reference to the Dust Bowl, that '*the drought was caused by anomalous tropical sea surface temperatures*'. To quote from Schubert *et al.* (2004): '*the study found (that) cooler than normal tropical Pacific Ocean surface temperatures combined with warmer tropical Atlantic Ocean temperatures to create conditions in the atmosphere that turned America's breadbasket into a dust bowl from 1931 to 1939*', and '*other major US droughts of the 1900s suggest a cool tropical Pacific was a common factor*'. However, there is no explanation given of just how sea surface temperatures (especially temperatures of seas lying beyond the barrier of the Rockies!) can cause the highs, which – oddly – are not mentioned! How, for example, could very distant SSTs dictate short-lived synoptic conditions such as local pressures, the motions and mergers of MPHs, or even the local 'dusters'? Also, how could sea surface temperatures 'encourage' a powerful MPH to form within the Arctic and then slowly move across the Great Plains?

All that is happening here in fact, is that we (or rather the modeller) are falling back one way or another on the good old, indispensable equation 'T/R', with statements like '*these changes in sea surface temperatures ... reduced the normal supply of moisture from the Gulf of Mexico and inhibited rainfall throughout the Great Plains*' (Schubert *et al.*, 2004)!

All this does nothing to answer the problem posed, since the presence of an anticyclonic cap hinders the advection of precipitable potential. In any case, the supposed 'sea surface temperature solution' adds nothing to our understanding, since, even if we assume that the influence attributed to sea surface temperatures exists, where do these sea temperature anomalies themselves come from? They are quite insignificant in amplitude, in comparison with the offending phenomena and with the powerful MPHs and AAs. There is no reply to this question, a question which was not asked anyway in the articles mentioned above.

These studies are symptomatic of the skewed approach built into the models. Not surprisingly, they provide the results which are expected of them: if one 'enters' data sets for sea temperatures, one obtains – without fail – statistical 'relationships' with sea surface temperatures (it's elementary!). Moreover, these data ought themselves to be reliable, which does not seem to have been the case, according to Hurrell and Trenberth (1999), who report that '*the sea surface temperature data sets all contain problems of one sort or another*'. However one looks at it, these relationships can only be co-variations; we simply do not know (and *it would be quite a coup if somebody managed to explain it!)* **how sea surface temperatures can determine the pressure field over the Great Plains**. Sea surface temperatures are becoming a 'universal panacea', so we read that '*tropical sea surface temperature anomalies were found to contribute to recent prolonged drought conditions over much of the northern middle latitudes, to drought in the Great Plains, and to drought conditions in the African Sahel region ...*' (Schubert *et al.*, 2004).

We might even say on reading such things that sea surface temperatures are being seen as the 'Jack of all trades': all you need is a data set of them and you can straight away provide the solution to all those unresolved questions! In spite of the fundamental differences in the conditions for pluviogenesis in middle latitudes and/or to the south of the Sahara, sea surface temperatures are here being held responsible for the heatwave in Europe, for the Dust Bowl effect, and for the drought in the Sahel; looking at other publications, we can add to this (infinite!) list the rains in South Africa and Ethiopia, the Indian monsoon, the floods in China, and more! This puerile approach is something of a bad joke, and we might ask in these circumstances just *where meteorology is going*!

The existence of high-pressure areas is the cause of drought conditions and hot spells. So it would not be too great a task, if we are following the reasoning of real meteorology, to 'enter' a data set of surface pressure values for North America, and to seek thereafter the reasons for the way in which pressures evolve. The way in which it is done, however, is to grope blindly, since, unfortunately, '*the mechanisms are not well understood and cannot yet be used to help predict the likelihood of droughts*' (McCabe *et al.*, 2004). The last straw! But it is inevitable, if we do not refer to a schema of circulation which lays down the order of linked events, after first defining precisely the general (real) meteorological framework and conditions which are responsible for the absence of rainfall (or for its abundance).

11.3.4 The Great Sahel drought

In middle latitudes, the process that leads to long-term droughts involves the formation of an AA by MPHs. In the tropics, conditions for pluviogenesis exhibit variations, especially as far as the necessary factors for the transfer of water vapour and the causes of convection are concerned. There are also differences in structural conditions, mainly when discontinuities are unproductive, as is the case with the TI, and the IME.

The tragic and alarming lack of summer rainfall to the south of the Sahara since 1970 (Figure 81) is still going on, with only brief local remissions. To have actually lived through this period (Leroux, 1970, 1983, 1995, 1996a, 2001), and to have seen the dried-up oxbow lakes, the fall in river levels, and the disappearance of the small interdunal lakes, known locally as *niayes*, in the region of Dakar; to have watched Lake Chad shrinking, and the interruption of the flow of the Niger, actually divided briefly into two sections by the Gao sill; and to have watched the withering of crops and the exodus of destitute rural populations: is to be amazed that some people can imagine that this drought is only some kind of '*artefact*' (Chappell and Agnew, 2004)!

The attempts at explanation of the Sahel drought

There have been many attempts at an explanation of the sub-Saharan drought, and faddish and short-lived suggestions have come and gone, or have blended into each other.

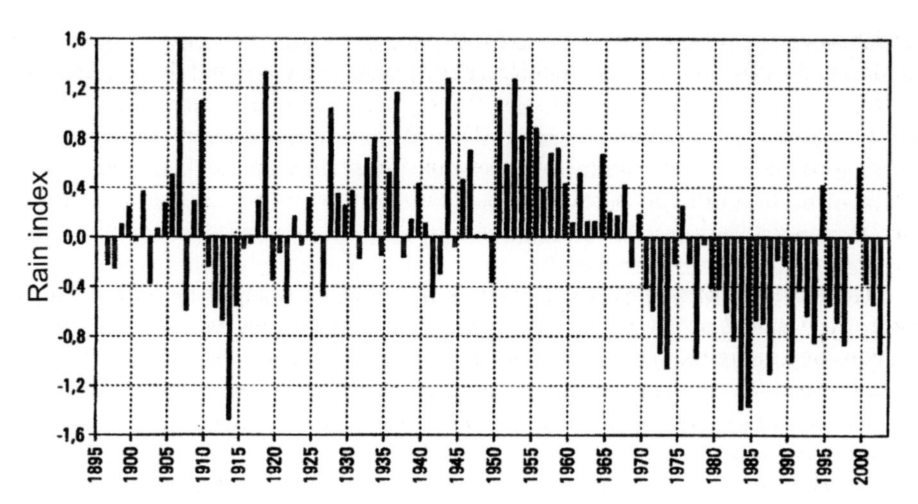

Figure 81. Annual index of rainfall in the Sahel from 1896–2002.
After Mahé and L'Hôte (2004).

- The 'albedo variation' hypothesis – associated principally with overgrazing – may be summed up thus: '*increase in the albedo of the surface: energy deficit facilitates the descent of air masses, which inhibits precipitation*' (Sadourny, 1992). A wholly theoretical vision, and the reverse of reality, since denuded ground, on the contrary, is conducive to strong convection! According to this fable, the people of the Sahel are held to have brought the drought upon themselves! Now, however, thanks to the IPCC, they carry this burden of responsibility no longer, because the enhanced greenhouse effect seems to have taken up the baton!
- The decline in rainfall is caused by changes in the vegetation, since it is re-evaporation above the dense vegetation growth which supplies the monsoon flux. However, the presence of precipitable potential is only one of the conditions necessary for pluviogenesis, and coastal regions such as Senegal have themselves experienced much reduced rainfall; the Cape Verde Islands have recorded a deficit even more pronounced than that of the nearby mainland.
- The variation in rainfall is also 'explained' (as already mentioned) by involving sea surface temperatures. First of all, the sea surface temperatures studied were those near Africa, in the Gulf of Guinea, which seemed a logical assumption since the available water originates essentially in the Atlantic. The results, however, were contradictory, with warm sea surface temperatures being associated sometimes with a lack of rainfall, and sometimes with an excess of rain. Cooler sea surface temperatures also gave such ambiguous and paradoxical results! 'Explanations' such as these are based on nothing more than mere co-variations, and the positive or negative outcomes really depend on the sea sectors chosen. Sea surface temperatures further afield were then examined, for example in the North Atlantic, in spite of the fact that the maritime trade,

beginning its journey towards the Caribbean, has almost no opportunity to play a role in providing rain to the mainland. The net was cast still wider, even over the Indian Ocean (Gianini *et al.*, 2003). Such supposed links are really the product of an 'exercise in statistics', and there is no attempt to specify the type of physical relationship which might exist between SSTs and the factors involved in rainfall in the Sahel!

- Other possible causes have been put forward: variations in the area covered by the monsoon; variations in the speed and/or amplitude of undulations in the African Easterly Jet in the middle layers or the Tropical Easterly Jet in the upper layers; the hypothetical (and nowadays unfashionable!) 'Walker' cells; the El Niño-Southern Oscillation effect (ENSO); and, of course, the inevitable 'greenhouse effect'. Higher temperatures can only in fact – theoretically – improve the quality of the precipitable potential (cf. 'T/R'), and deepen the thermal depressions that draw in the monsoon fluxes towards the interior of the continent. These circumstances, if all other necessary conditions are adhered to, could only really improve rainfall, as is the case with climates in the past, when each warm period has been a wet period. But monsoon fluxes are only vectors of water vapour: they are not 'rainmakers' (i.e., disturbances able to cause rain). So it is the pluviometric efficiency of these disturbances to which we should turn our attention.

The causes set out above mainly involve statistical co-variations, while not taking into account the real physical mechanisms governing rainfall in the Sahel (Leroux, 1983, 2001; Sagna, 1988, 2004). In particular, when the 'monsoon' is mentioned, it is always in a very theoretical light. No mention, however, is made of the actual dynamic of the monsoon flux, of its vertical structure and its interaction with the energy of the opposite hemisphere (cf. Figure 27, Chapter 8), or of the way in which the advected precipitable potential is utilised. The southern Sahara, as far as about latitude 12°N, lies in summer underneath the unproductive structure of the IME, and within this structure the moist monsoon flux of the lower layers moves beneath a dry trade wind blowing in the opposite direction (Chapter 8). The VME structure to the south of the IME provides abundant precipitation, regularly and without much storm activity. The IME structure is activated only by squall lines, which are mobile disturbances travelling from east to west, briefly suspending the conditions which hinder rainfall. Precipitation is therefore very limited in its duration, and is accompanied by violent storms of great intensity, though these are infrequent.

In the Sahel and northern Sudanian region, the IME structure and, therefore, the squall lines, are responsible for nearly all summer rain. The pluviometric efficiency of these mobile disturbances, with their north–south alignment, varies with their greater or lesser proximity to that fundamental pluviogenic structure, the VME, which does not trespass in its summer northward migration beyond latitude 12°N (approximately). The northern sections of the squall lines give much more modest rain than their southern parts, which are more fully developed and larger in extent. But are these structural and dynamic meteorological parameters of rainfall in the Sahel incorporated in the models?

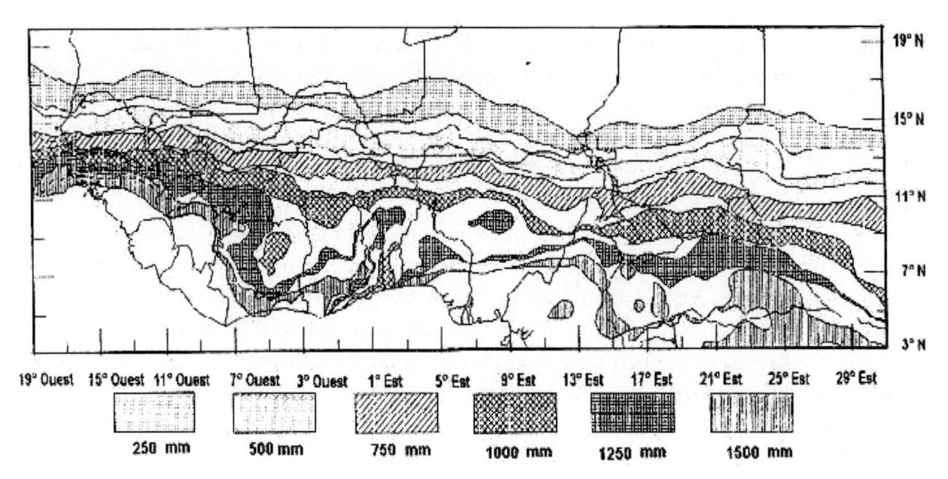

Figure 82. Southward shift of isohyets for 250, 500, 1,000, 1,250 and 1,500 mm over Western and Central Africa, from the period 1951–1969 to the period 1970–1989.
From Morel (1995).

The southward shift of pluviogenic structures

It took about 20 years for us to realise that the deterioration in rainfall to the south of the Sahara was without doubt a spatially organised event. Figure 82 shows the remarkable **overall southward shift of isohyets over sub-Saharan Africa**, from the Atlantic to Ethiopia. The pattern is particularly clear to the north, but more complicated to the south because of relief. Such a general southward movement rules out those 'explanations' which rely on local factors, or factors of limited amplitude: such a shift must form part of a vast dynamic framework, bigger than Africa itself. It can only be the result of a general southward migration of pluviogenic structures. This migration is therefore a part of general circulation, and **belongs to a scenario of 'rapid circulation**' (Figures 33 and 34, Chapter 9) within which the ME, or to be more precise the VME with its east–west orientation, and behind it the IME and the squall lines, are migrating *en bloc* towards the south. Remember that a slow scenario, associated with warming, would cause the opposite effect (i.e., a general northward shift (Figure 35)), and that all the warm periods fitting in with this scenario have been benign, both to the south of the Sahara, and everywhere else in the tropical margins. This is one thing that palaeoclimatology teaches us (Chapter 9): that tropical pluviogenic structures move northwards in warm periods (as happened during the Holocene Climatic Optimum (HCO)), and move southwards during periods of thermal deficit which are accentuated at high northern latitudes (as at the time of the Last Glacial Maximum (LGM), when the desert extended for about 1,000 km further south).

The Great Sahel drought, which has lasted for more than three decades now, is part of a cold scenario and a rapid circulation mode (Figure 33), and it has therefore absolutely nothing to do with a 'greenhouse effect' scenario.

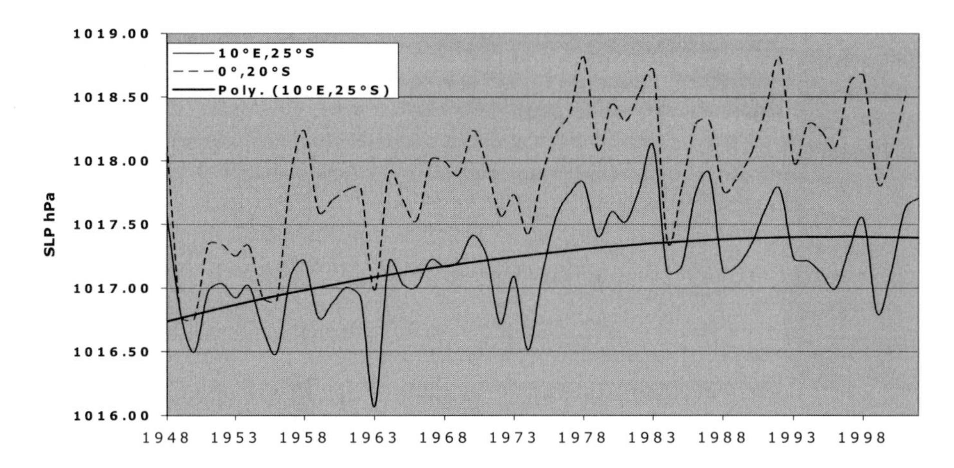

Figure 83. Annual mean sea level pressure (hPa) at 10°E/25°S (off Lüderitz, Southern Namibia), and at 0°/20°S (between St. Helena and Northern Namibia) from 1948–2002; polynomial trend at 10°E/25°S.
From Pommier (2004) (CDC/NCEP-NCAR data).

Instead of seeking out hypothetical links which are remote and impossible, it makes more sense to dwell upon realities. How can we quickly summarise the dynamic of the structure of the ME, its migration being responsible for the prolonged Sahel drought? Remember that the position of the ME depends upon the opposing dynamics of the two meteorological hemispheres, and that the Atlantic monsoon which blows across the northern part of tropical Africa comes up from the southern hemisphere, fed by the maritime trade which has its origins off southern Africa and Namibia (cf. Figures 26 and 27, Chapter 8).

The pressures reached within the so-called 'St. Helena' AA, at the source of the maritime trade which becomes the monsoon, are evidence of this initial dynamic. Figure 83 shows that the mean pressure off the coast of Namibia was rising until 1983, and thereafter levelled off. Of most interest, however, is the comparison between the irregularities in this curve and the rainfall of the Sahel (Figure 81). After the positive peak in pressure values in 1970, the first visitation of drought corresponded, from 1971 until 1974, to a very obvious fall in pressure off southern Africa. The definite improvement in rainfall thereafter corresponded to a pronounced rise in pressure, until 1983, with an abrupt reversal in 1984. 1983 and 1984 certainly saw a second very severe dry episode. The slight fall in pressure after this was accompanied by a further deterioration in rainfall figures to the south of the Sahara. So, modest pressures in the southern AA give evidence of a less than energetic trade–monsoon stream, while high pressures in the AA give evidence of a more active dynamic, with enhanced advection of precipitable potential, without prejudging the necessary conditions for the utilisation over Africa of this potential by pluviogenic factors.

However, these correspondences are not perfect, since the northern meteoro-

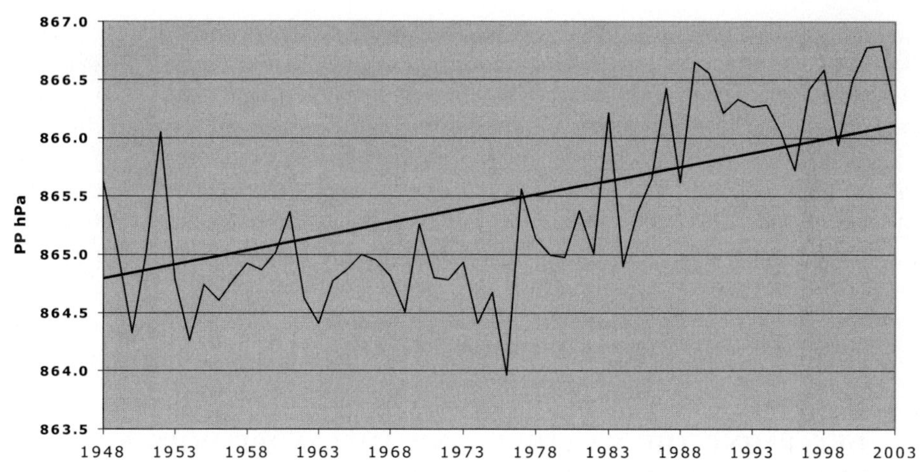

Figure 84. Annual mean station-level pressure (hPa) at Tamanrasset (Algeria, 22° 47′N, 05° 31′E, at 1,378 m); linear trend shown.
From the Algerian Meteorological Service/WMO.

logical hemisphere intervenes in its turn. Figure 84 shows the evolution of pressures to the south of the Sahara, at Tamanrasset (Algeria) during the same period: note the net rise in pressures since the 1970s, when the Sahel drought began. This rise, which was observed from the Atlantic across to Ethiopia and beyond over the Red Sea, was studied, for the ocean and coastal areas by Sagna (2004) and Amraoui (2004), for Mauritania by Nouaceur (1999), and for the Sudan by Omar Haroun (1997). It is this general rise in pressure (following on from that recorded for the Mediterranean, Figures 71 and 72) which prevents the IME structure from moving further to the north, and which, contrary to all hopes, spells the continuation of these rainless times, reinforced by the simultaneous stagnation or slight fall in pressures to the south (Figure 83). Such relationships still need to be analysed in detail beyond that which is presented here, on different scales of time and space, since the Sahel is very extensive in longitude. But these relationships have a demonstrated physical reality, and an analysis of them goes beyond the level of the simple statistical co-variation which is usually enough for the models.

It is worth mentioning again that, if some link were to be established, for example with SSTs in the southern Atlantic, it would only be further proof of a general co-variation of parameters within an already recognised general dynamic process. However, to envisage a link between SSTs in the Indian Ocean and rainfall in the Sahel, in a logically reasoned way, it would first of all be necessary to determine what physical connection there might be between, on the one hand, the sea surface temperatures which are in one unit of circulation, and on the other hand, the rains which fall in a different one (cf. Figure 26).

The same would be true in the case of the Indian Ocean sea surface temperatures if we wish to evaluate, via pressures and/or wind speeds, the dynamism of the Indian monsoon, the quality of the precipitable potential advected in the direction of Asia, and perhaps also its pluviometric efficiency. For it is not enough to state, as the IPCC does, that there exists (IPCC, 2001) '*a regional meridional temperature gradient extending from the tropical Indian Ocean north to mid-latitude Asia*'. The Indian monsoon cannot be taken for some kind of large-scale, regional 'offshore breeze', with its intensity determined only by the temperature gradient according to the formula '*the stronger this meridional temperature gradient, the stronger the monsoon*' (IPCC, 2001). The Indian, and Chinese, monsoons are actually supplied with energy by the trade circulation from the opposite hemisphere, and therefore, initially, by anticyclonic agglutinations formed by southern MPHs.

11.4 CONCLUSION: THE GREENHOUSE EFFECT DOES NOT DETERMINE THE WEATHER, AND THEREFORE DOES NOT CONTROL RAINFALL OR DROUGHTS

Those few examples which have been briefly analysed above show clearly that the mechanisms behind the weather are still far from being well understood, and that they are considerably more complex than the parameterisations within the climatic models. Meteorology, or whatever it is that meteorology has become nowadays, is stuck in a conceptual *impasse*, and tangled up in the skeins of thought of the various schools, whether it is high latitudes, mid-latitudes or the tropics which are involved.

So meteorology fails to explain the mechanisms by which temperate disturbances are formed, and can throw even less light upon the origins of storms and floods, or the trajectories of tropical cyclones. Neither does meteorology explain how drought conditions come about in both temperate and tropical latitudes, why snow sometimes does not fall, or why heatwaves occur. It also gives no insight into particular regional weather behaviours, for example, those in the Mediterranean basin.

In summary, meteorology/climatology is not in a position to claim an ability to predict the future, since it is now unable to understand the present weather.

The use of climatic models has not furthered our understanding of the weather, of the mechanisms of pluviogenesis, or of situations where rain is absent. It is content to employ extremely simplistic 'equations', and to establish very remote and undetermined 'relationships' which are in fact simply statistical co-variations. These latter are rightly labelled 'teleconnections', meaning that they are (very) far-reaching (and far-fetched?), and definitely very hypothetical connections, relying as they do upon illusory and very distant 'causes' of phenomena, even though no physical *rapport* has been discovered. All this, instead of an attentive examination of the immediate and visible causes of actual phenomena. Modelling has, moreover, reduced pluviogenic 'mechanisms' to a few elementary 'formulae', and thereby has 'dictated' a supposed link between greenhouse gases, temperatures, and rainfall (or the lack of it). *That link is by no means proven.*

The greenhouse effect does not determine the weather, and therefore has nothing at all to do with the evolution of rainfall or drought conditions.

Models have dissociated precipitation or the lack of precipitation (effects) from the weather (cause), a dissociation which represents a considerable scientific aberration. As a result, *these models can in no way tell us what the weather will be like in the future*, and certainly not what it will be like in the year 2100. The IPCC recognises this is the case of small-scale phenomena '*such as thunderstorms, tornadoes, hail, and lightning, ... not simulated in climate models*' (IPCC, 2001). For other phenomena, the same honesty is not in evidence.

So we are presented with suggestions and predictions of a probable catastrophic decline in our weather, with extremes of rainfall, floods, storms, cyclones, droughts, scorching heat. Why? Because tomorrow's 'climatic drama', so beloved of the media, who exaggerate it in their fondness for the sensational, has to be constantly reviewed and exploited, with fears about the greenhouse effect scenario constituting the major argument of the IPCC and its partisans. Such, then, is the task of Group II of the IPCC: to concoct alarmist scenarios, and to imagine maximum disaster!

The fictional situations and wild imaginings of the IPCC, attributed of course to the greenhouse effect and therefore to humankind, and derived from 'models', are wrongly seen as predictions. **They are the product of false science**.

Conversely, real facts are ignored to such an extent that **the actual evolution of weather and climate is going completely unnoticed**! Observations show, in fact, and especially in Europe and the Mediterranean basin, a generalised rise in pressure, in the context of a rapid mode of circulation associated with cooling in the western Arctic (Chapter 10), whence the greater vigour of MPHs and an acceleration in meridional exchanges. This evolution is, for example, observed within the aerological units of the North Atlantic and the North Pacific.

The observational facts: Climate and aerological units

Mainstream science never sets out to throw light upon new phenomena: those which do not fit into the (preformed and inflexible) frame furnished by the paradigm can often go unperceived ...

Scientists do not normally aim to put forward new theories, and they are often intolerant of those put forward by others ...

In fact, normal scientific research is directed towards the articulation of phenomena and theories already within the paradigm.

T.S. Kuhn. *La Structure des Révolutions Scientifiques*, Flammarion, Paris, 1983.

As has already been stated, **the notion of a 'global climate' is but a fiction**. Climate cannot be defined on the planetary scale. Moreover, the notion of 'general circulation' is essentially only a mental construct. In the middle and upper layers, away from the meteorological equator (ME), which is the fundamental boundary between the two meteorological hemispheres, no discontinuity is present to interrupt circulation as it becomes ever more simplified with altitude in each hemisphere. However, it is a completely different story in the lower layers, where circulation is divided as a result of the interference of the geographical (relief) and dynamical (Mobile Polar High (MPH)) factors into six major aerological units (Figure 26, Chapter 8). The perceptions of 'climate change' on a global scale, of the homogeneity of the atmosphere, or of some planet-wide uniformity of phenomena involving mysterious 'teleconnections', all of which are proposed by models and the Intergovernmental Panel on Climate Change (IPCC), are mere flights of fancy which have nothing to do with meteorological reality and which are therefore in error.

Circulation within the lower layers is divided into individualised units, three in each hemisphere. Each one extends from the polar regions to the ME, whose position fluctuates constantly with the passing seasons. These aerological spaces are more or less defined by relief, as continuous mountain chains form barriers impassable to cold air, while discontinuous ranges, or the absence of relief, permit communication and interaction with neighbouring units.

Within each aerological space, circulation begins at the pole, in the form of MPHs, which, as they travel through high and middle latitudes, constitute the major factor in the behaviour of the weather, and cause warm air to be sent in the direction of the pole throughout their journeys. As they progressively lose their energy, new MPHs overhaul and merge with those preceding them, leading to the formation of an Anticyclonic Agglutination (AA), and the revictualling of the trade circulation. Within their lower strata, and beneath the Trade Inversion (TI), the trade winds accumulate both the perceptible and the latent heat of the tropics, transferring them more or less directly to the pole, to the benefit of new MPHs arriving to contribute to the circulation of the aerological unit (Chapter 8).

Consequently:

- There is no hiatus in the circulation within the lower layers, from the pole towards the ME and back towards the pole – unless relief, such as the east–west chain of the Zagros–Himalaya–Tibetan mountains, interrupts interzonal meridional exchanges, deflecting them towards their margins.
- Schematically, circulation describes a gigantic 'figure 8' within each unit, and this is true not only of the air of the lower layers, but also of the surface water of the oceans, transporting cold towards the equator, and warmth towards the pole; above, advected air sinks and is recycled within new MPHs.
- Pronounced climatic regional differences exist, depending upon which trajectories associated MPHs follow, the various regions being situated near the departure points of MPHs, along their main trajectories, or away from these beneath an AA, within a trade flux, or on one of the preferred paths of the returning cyclonic circulation. The extent of these climatic regions varies according to the season, especially in the tropical zone if the trade flux is continued as a monsoon.
- Within each unit, in spite of regional climatic differences, the dynamic is determined by these same MPHs, and therefore, **all the climatic parameters co-vary** because they depend for their characteristics upon the intensity of the initial impetus, originally defined near the pole. **Consequently, if one of the parameters varies, so do all the others, and it is therefore not possible to treat one particular element in isolation without taking into account all the others**.

The co-variation of the parameters within an aerological space has been noted – often, a very long time ago – but it is still not explained, because the cause (i.e., the dynamic of the MPH), has not been identified. As a result, it has been translated into an 'oscillation', a term as enigmatic now as it was when it was coined. It has recently become fashionable to invoke multiple oscillations, but this brings us no nearer to being able to explain phenomena.

In the end, the 'oscillation' became responsible for the evolution of the weather, and is now yet another 'character' on the meteorological stage, a recognised 'factor' ranking as one of the fundamental causes of climatic variations. For the IPCC, it has even assumed the status of an independent 'cause' of climate *change* (cf. Chapter 6).

It will now be useful to our argument to examine circulation, the variations in its intensity, and its climatic consequences in two aerological units: the North Atlantic (Chapter 12) and the North Pacific (Chapter 13), with a view to demystifying the 'mystery' still surrounding some of these 'oscillations'.

12

The North Atlantic aerological unit

If we are too preoccupied with certain principles within a science, we find it harder to accept other ideas within that same science, and a new method ...

Luc de Clapiers, Marquis de Vauvenargues,
Introduction à la Connaissance de l'Esprit Humain, Paris (1746).

This unit was the first to be properly studied, and it provided the first hypotheses concerning mechanisms of circulation. But as the recent FASTEX experiment showed (Chapter 11), it still harbours many 'mysteries', not least of which are those involving the origin of the disturbances occurring there, the mechanism of its 'oscillation', and the evolution of its climates.

12.1 THE DYNAMIC OF CIRCULATION IN THE NORTH ATLANTIC SPACE

12.1.1 Description of the North Atlantic aerological unit

The North Atlantic aerological space has a clearly defined western boundary, but is more open to the east (Figures 85 and 86). In the west, the Rocky Mountains form a barrier which is impassable to the cold air of MPHs, which means that the western edge of North America belongs aerologically to another unit, that of the North Pacific. Warm Pacific air can however rise above this mountain chain, and, having crossed it, create a *foehn* effect (or *chinook*) as it descends towards the Great Plains. South of the Mexican Sierra Madre mountains, the relief is lower, and the Atlantic space can spill over in winter across the Central American Isthmus into the near Pacific, with the Trade Discontinuity (TD) as the prolongation over the ocean of the Mexican mountains to the north of it. On the eastern side, north of the Cantabrian-Pyrenees and the Alps, the North Atlantic space is open, and Western Europe is a

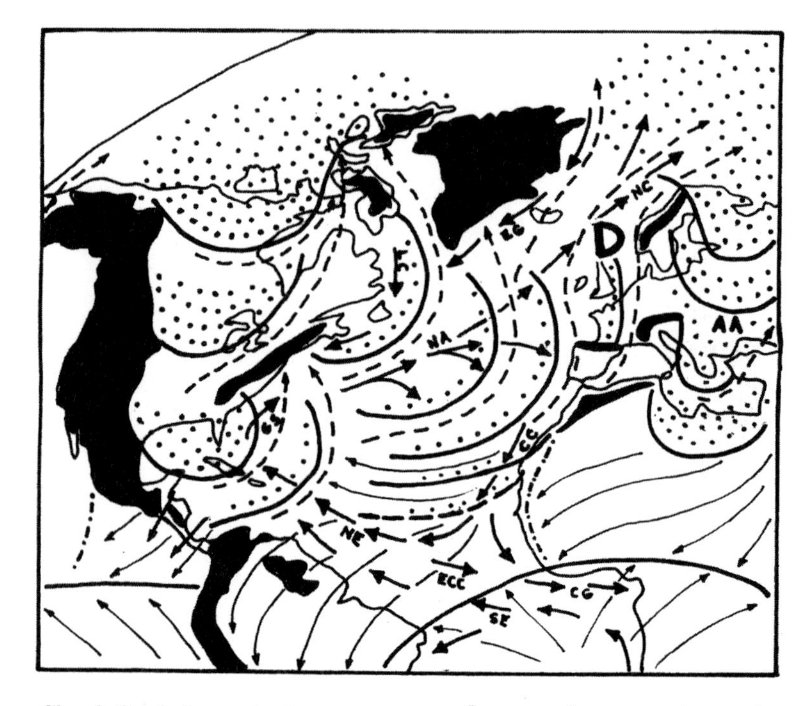

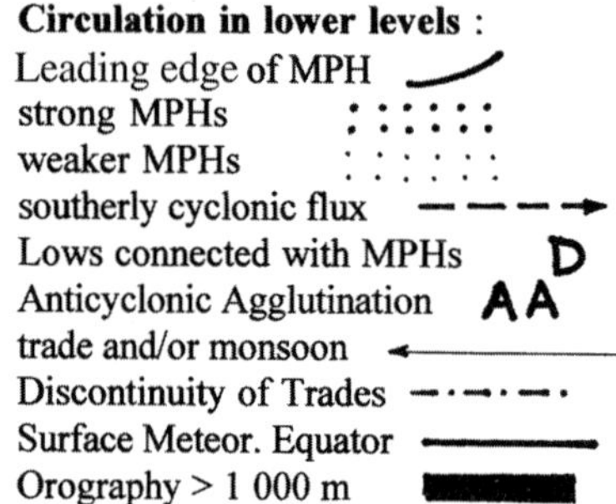

Ocean surface currents :
EG : Eastern-Greenland Current
LC : Labrador Current
NC : Norway Current
NA : North Atlantic Drift
GS : Gulf Stream
CC : Canaries Current
NE : North Equatorial Current
ECC : Equat. Counter-Current
CG : Current of Guinea
SE : South Equatorial Current

Figure 85. The North Atlantic aerological unit, at the time of a high polar thermal deficit: winter and/or cold period, rapid circulation mode.

region of interactions with northern and Central Europe (Figure 73, Chapter 11). The mountains of the Iberian Peninsula (the elevated plateau of the Meseta and the high Sierras), and the Atlas range, especially in Morocco, form a 'leaky' barrier, channelling the fluxes towards the western basin of the Mediterranean (Figure 70). South of the Atlas Mountains, a TD runs along the west coast of Africa, mainly in the winter. This is the season during which MPHs can pass to the south of the Atlas on their way eastwards into North Africa, gradually transforming themselves into the continental trade flux, or *harmattan* (Figure 70(c)). The aerological space is more extensive in the winter, when the maritime trade extends itself across South America

Figure 86. The North Atlantic aerological unit, at the time of a weaker polar thermal deficit: summer and/or warm period, slow circulation mode. For information see the key of Figure 85.

as the Amazon monsoon. The southern edge of this unit is not clearly defined, as the ME shifts southwards in northern winter, and northwards in the summer.

MPHs originate preferentially in the Arctic to the north of Canada, and they follow a path between the Rockies and the highlands of Ellesmere and Baffin Islands. This American trajectory has two main branches, one of which runs directly south between the Rockies and the Appalachians towards the Gulf of Mexico, while the other, more zonal, reaches the Atlantic between southern Greenland and the northern end of the Appalachians (cf. Figure 60). MPHs on the southward path, more frequently taken in the winter, bring about a rise in pressure, forming the (statistical) 'Bermuda' AA. These southbound MPHs are the cause of the Atlantic aerological space trespassing into the Pacific by way of the Isthmus, and of the associated *northers* (or *nortes*) – strong north winds which can be cool or even cold. These storm-bringing winds can cause violent disturbances along the leading edge of MPHs, and severe frosts can follow (Barbier, 2004). MPHs on the zonal path, more frequently taken in the summer, cross the Atlantic heading roughly south-east, in the general direction of Spain and Morocco. As MPHs slow down and encounter relief, they form the (statistical) 'Azores' AA. In winter, this trajectory lies further to the south, as does the AA, and in summer it is shifted northwards. The two AAs, the 'Bermuda' and the 'Azores', artificially individualised by mean pressure values, are merged, and the Atlantic anticyclonic 'subtropical' cell thus reaches further to the west in winter.

Other MPHs penetrate this unit to the east of Greenland, coming directly off the *inlandsis* (mainly in summer), or down from the Arctic. Most of these MPHs, on the Scandinavian trajectory, head for central Europe, and beyond, into Asia or the eastern Mediterranean. Some of them, on a more directly southward trajectory, move down over western Europe, in both winter and summer (cf. Figure 74), though they are most frequent in the winter. It is these more meridional MPHs which provoke severe weather events in western Europe, with cold snaps and/or frosts, snow in low-lying areas, and torrential rain (Figure 67).

Figure 87 (cf. Figure 21, Chapter 7) shows one of these descents of cold air, which occurred in October 2003. MPHs had begun to follow this direct trajectory on 17 October, and pulses of cold air succeeded each other until 27 October. One of these cold pulses took the form of two lobes, easily seen on the image taken on the 22nd. The western lobe reached Scotland on 21 October. By the 22nd, it had encountered the Cantabrian mountains and the Pyrenees, which affected its flow (Figure 87 – 22 October). A modest amount of this cold air continued its southerly progress into the Atlantic alongside the relief of Galicia and Portugal (Figure 87 – 23 October). Most of it, however, was channelled eastwards along the northern edge of the east–west mountain mass, to rejoin, over Europe, the cold air of the eastern lobe, coming down from the direction of Scandinavia and Germany, which it had crossed on the 22nd and 23rd respectively. On 24 October, western Europe lay beneath the cold air, which proceeded into the Mediterranean on the 25th via the corridor between the Alps and the Pyrenees, though much of it hastened out across the Atlantic in an east–west flow (Figure 87 – 25 October), along the northern edge of the Alps, encouraged by the anticyclonic rotation. Then it was 'bounced' by the Pyrenees and the Cantabrian mountains, this time in the opposite direction (i.e., running counter to the normal trajectory followed by MPHs).

It was this situation which was responsible for setting new records for cold in October, during the night of 24–25 October 2003. For example, in Switzerland, at La Chaux-de-Fonds: −16.1°C (vs. −8.8°C); at La Brévine: −23.2°C (vs. −14.7°C); at Fahy: −6.5°C (vs. −4.9°C); in France, at Millau: −4.1°C (vs. −2.9°C); at Tarbes: −3.0°C (vs. −2.2°C); and at Brest: −1.4°C (vs. −0.8°C). This early visitation of winter cold, coming after the heatwave of August 2003, was symptomatic of the great irregularity of weather, and underlined the fact, if such a reminder were needed, that a hot summer is not necessarily synonymous with continuous warming!

Figures 85 and 86, which for graphical convenience cannot show such rapid descents by MPHs, nevertheless represent schematically more than three-quarters of the meteorological situations in the context of which MPHs follow the American–Atlantic trajectory. Slightly fewer than one-quarter of MPHs drive more directly into western Europe by way of the Norwegian Sea, and/or indirectly by westward extension through northern and central Europe, mainly in winter (when MPHs are at their most powerful, and the continental AA slows the eastward motion of MPHs).

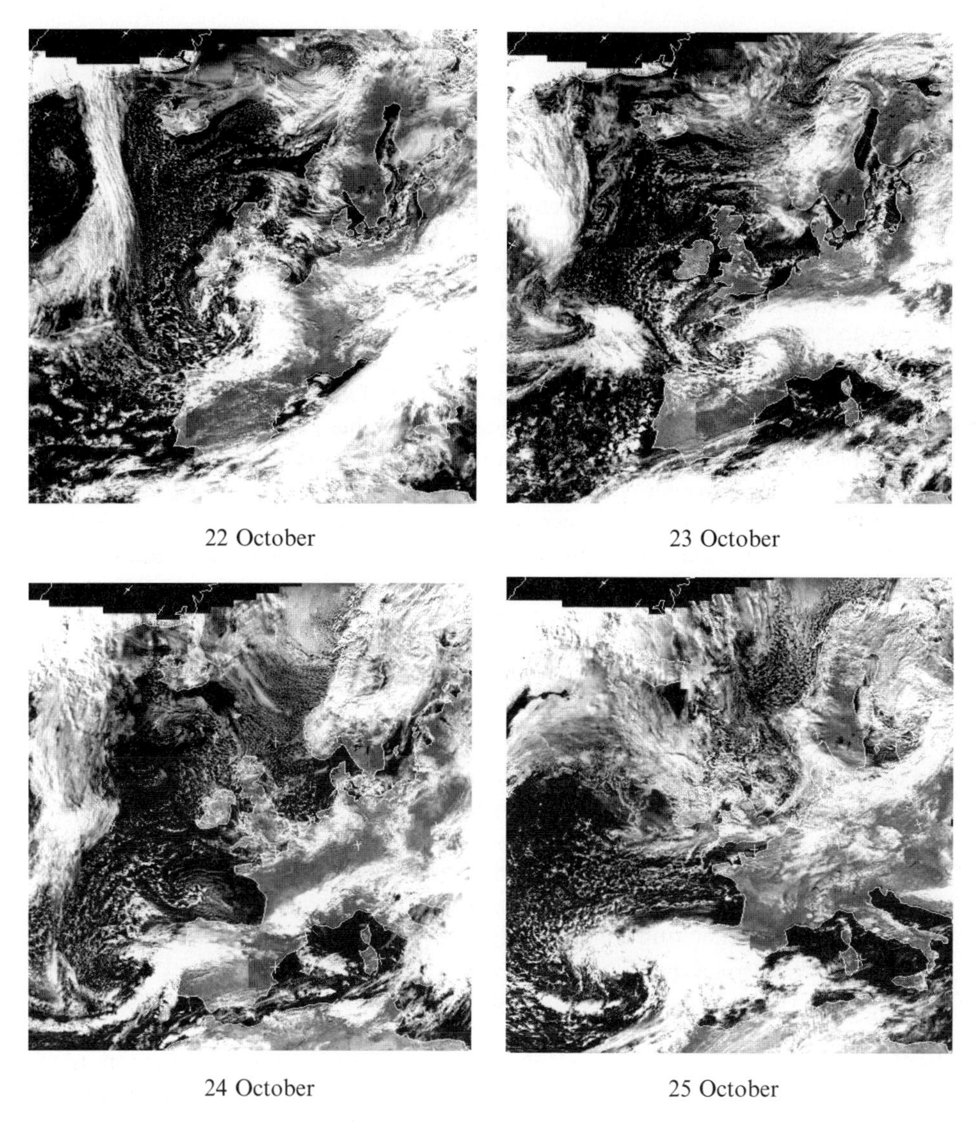

Figure 87. 22–25 October 2003 (visible, NOAA, 14h). Invasion of cold air across western Europe, with MPHs following the Greenland–Scandinavian path.

Oceanic surface circulation is determined by air circulation (wind-driven currents). Surface waters are driven eastwards off the American coasts, impelled by MPHs exerting maximum pressure upon the oceans. The North Atlantic Drift current, a prolongation of the Gulf Stream, splits into two branches: the northern branch heading for the Norwegian Sea (as the Norwegian Current) and thence into the Barents Sea, and the southern branch (the Canaries Current) moving southwards. One section of this current heads away towards the Gulf of Guinea (as the Equatorial Counter-Current and thereafter the Guinea Current), while the greater part becomes the North Equatorial Current, flowing towards the Caribbean and the Gulf of Mexico. The North Equatorial Current is reinforced by a branch of the South Equatorial Current, originating in the southern Atlantic (an extension of the Benguela Current). A smaller, southern fraction of the South Equatorial Current forms the Brazil Current. Dense (cold) currents flow along the coasts of Greenland: the East Greenland Current, coming out of the glacial Arctic Ocean, and the Labrador Current. Their density draws them down below the surface, and as they sink they give way to the North Atlantic Drift system.

In this immense aerological space there is a great variety of climates: polar, tropical, and even equatorial. In spite of this very wide diversity, co-variation is general, which means that **no sector considered in isolation can have its own independent evolution: if we study one location, we are obliged to take into account all other locations within the ensemble**.

Even though this interdependence of phenomena remains to be explained, it has long been hinted at, and expressed using the term 'North Atlantic Oscillation', or 'NAO'.

12.1.2 The North Atlantic Oscillation

The weather in the North Atlantic and in Europe is traditionally associated with the NAO, '*one of the most dominant and oldest known world weather patterns*'. Recognition of this phenomenon dates back certainly as far as the voyages of the ancient Vikings, who noticed that '*when the winter in Denmark was severe ... the winter in Greenland in its manner was mild, and conversely*' (in Stephenson *et al.*, 2003).

The NAO (Walker and Bliss, 1932) is defined thus: a difference in surface pressure between two 'centres of action' (i.e., the so-called 'Icelandic' Low and the 'Azores' Anticyclone). In other words (those of Hurrell *et al.*, 2003), the '*NAO refers to a redistribution of atmospheric mass between the Arctic and the subtropical Atlantic*'. The NAO is also called, variously, the Arctic Oscillation (AO) (Thompson and Wallace, 1998) and/or the Northern Annular Mode (NAM) (Thompson *et al.*, 2003).

This oscillation, or mode, '*dictates climate variability from the eastern seaboard of the USA to Siberia and from the Arctic to the subtropical Atlantic, especially during boreal winter*' (Hurrell *et al.*, 2003). However, in spite of the interest shown in this phenomenon across the centuries, '*unanswered questions remain regarding the climatic processes that govern NAO variability*' (Hurrell *et al.*, 2003). In fact, the usual 'official' concepts take no account of the mechanisms involved in this 'North Atlantic pendulum', neither do they seek the reasons why its modes alternate, a fact which is as yet unexplained. As Wanner (1999) put it: '*How and why does the NAO see-saw between one mode and the other? ... In spite of all the studies* (really, all?: ML) *... this remains an open question, and the "flip-flop" mechanism is quite mysterious*'.

The NAO still remains a 'mystery'!

Remember that the FASTEX experiment, with its profusion of observational methods and its interesting results concerning the pertinence of the MPH model (cf. Chapter 11), did not change this situation. The search is still on for the mechanisms which marshal the weather in the North Atlantic, and consequently, the causes of the evolution of the NAO remain unknown. Among those suggested we find (and are we surprised?) those irrepressible sea surface temperatures which thereby add another string to their bow! For example, '*on longer timescales, recent modelling evidence suggests that the NAO responds to slow changes in global ocean temperatures, with changes in the equatorial regions playing a central role*' (Hurrell *et al.*, 2003). We can underline that sea surface temperatures are here only one parameter in the unit's co-variation! Links with the stratosphere, and especially with the intensity of the polar vortex, are suggested: '*a stronger polar vortex was associated with positive NAO-like anomalies in the troposphere, and vice versa*' (Thompson *et al.*, 2003). This is a most interesting relationship, supporting the idea of the role of MPHs. And of course, we have the inevitable question of '*how the NAO might change in response to increasing concentrations of greenhouse gases*'. This question is, by the way, considered to be '*one of the most urgent challenges*' (Hurrell *et al.*, 2003) – need we say it?

A recent publication (Hurrell *et al.*, 2003) presents the current thinking on this question. The articles brought together in this volume deal with: mean pressures; 'centres of action', statistically defined; mean wind vectors; 'westerlies', involving the statistically based notion of a flux between the two aforementioned mean 'centres of action'; and high-altitude undulations or 'stationary waves', which are '*forced primarily by the continent–ocean heating contrasts and the presence of the Rocky and the Himalayan mountain ranges*' (Hurrell *et al.*, 2003) (i.e., involving fixed elements). Now, the Himalayas might seem rather a long way from the Atlantic, and not only do they extend in an east–west direction, but they are also at about latitude 30°N, well to the south of the supposed 'waves'; and what is more, these 'waves' are labelled '*stationary*', while everything is in fact moving! Changes in circulation are discussed, but what is essentially dealt with are changes in mean pressures within the 'centres of action'. Not one of the articles in this monograph, devoted to the NAO, refers to meteorological phenomena, on the synoptic scale, which have actually been observed in this aerological space!

Invented by the climatological school in the early 20th century (Chapter 11), the NAO has been incorporated into the models, doubtless because of the intensely statistical (mean-value) nature of its context. In all this, there is no questioning of the real significance of the mean values involved; in other words, *the question:* **What does a mean pressure value (high or low) really represent and, more importantly, what physical reality lies behind these 'means'? is never asked**. So the anticyclone and the depression, mere abstractions, are allowed to become 'real' and essential characters in their own right!

Synoptic and statistical: correspondences between scales...

Statistics, then, ignore the synoptic scale of things. As we have stated, a 'centre of action', as defined through mean pressure values, can have no actual existence in the reality of weather. Talking about the influence of the Azores Anticyclone or the Icelandic Low can only represent **a considerable confusion of scales of phenomena, and a scientific aberration**.

What we have to do, then, is to rethink this and bring back the idea of movement, which gives the weather its character; we need a radical re-appraisal of the scale of phenomena. Figures 85 and 86, which represent (schematically) the synoptic scale, should be read (by statisticians) with reference to Figure 61, which deals with correspondences between scales. For example, what is known as the 'Azores Anticyclone' is, in reality, and at a given time, either an MPH or an AA formed by the overlapping of successive MPHs. It is therefore quite clear that 'the anticyclone' is not 'inflating' or 'deflating' of its own accord, but varies in extent as a function of the intensity of the importation of air by MPHs in the lower layers. The 'westerlies' which are supposed to be blowing between the anticyclone and the low do not exist either, in reality, but the mean resultant westerly flow is the product of both south-westerly air currents to the fore of the MPHs, and the general eastward shift of phenomena. Similarly, the 'Icelandic Low' does not exist as such, but the mean pressures involved represent the mean depths of all the depressions formed to the fore of all the MPHs which have crossed the space in question during the period for which mean values are to be established. The size of the 'low', whether it is 'expanding' or 'shrinking', is really only a representation of the greater or lesser area swept by the individual depressions (Figure 61).

The variation of the NAO is studied by means of an index based on the difference in pressure between the 'anticyclone' and the 'depression'; in practice, it is the difference as observed at two weather stations, located for example in Iceland, or the Azores, or in Lisbon (Portugal), or Gibraltar. The NAO is said to be in 'positive' mode when the pressure is simultaneously high within the 'anticyclone' and low within the 'depression', and in 'negative' mode when the difference in pressure is slight. These positive and negative modes set up co-variations, but do not explain them, the cause in both cases (the dynamic of MPHs) not being identified by the traditional theories. We must recapitulate that it is the energy of MPHs which determines the rate at which the northward cyclonic transfer of warm air occurs, particularly on the leading edges of MPHs, and that this same factor determines the depths of the peripheral low-pressure corridor and closed low (cyclone). Moreover, the energy of the MPHs themselves depends upon the polar thermal deficit. When Hurrell *et al.* (2003) write that '*positive values of the index indicate stronger than average westerlies*', all that this statistical outcome really means is that, on the synoptic scale, air moving up from the south-west to the fore of MPHs is blowing more strongly northwards, and that MPHs are more robust and more rapid, causing meridional transfers to be more intense (Figure 61, b-2 = NAO, and a-2 = NAD).

12.1.3 The North Atlantic Dynamic

The NAO index is based upon data for mean pressures from weather stations. This reference to fixed stations, while all weather is in motion, distorts reality, since a given station may register the passage of a phenomenon, but not necessarily its real strength (e.g., the centre of an MPH may or may not pass across the station in question). In order to get round this problem, Pommier (2002) preferred to examine the intensity of the North Atlantic Dynamic (NAD). Using the NCEP–NCAR database, he established an index for the NAD (the NADi) based upon essentially dynamical criteria.

- First of all, three specific indices (for frequency, strength, and size) are determined for MPHs on the American trajectory: from these indices is calculated an index for the dynamism of American MPHs.
- The same analysis is carried out for MPHs on the Greenland–Scandinavian trajectory, and, using the same criteria, an index for the dynamism of these MPHs is established.
- The mean value of these two indices provides the index for the dynamic of MPHs within the North Atlantic space.
- Next, an index is determined for the dynamic of the lows associated with the MPHs, based on the frequency and depth of these lows, and the latitudes at which they begin their journeys and at which they fill.
- The combination of the two indices, the anticyclonic and the depressionary, determines the index for the aerological dynamic of the North Atlantic: the NADi (Figure 88).

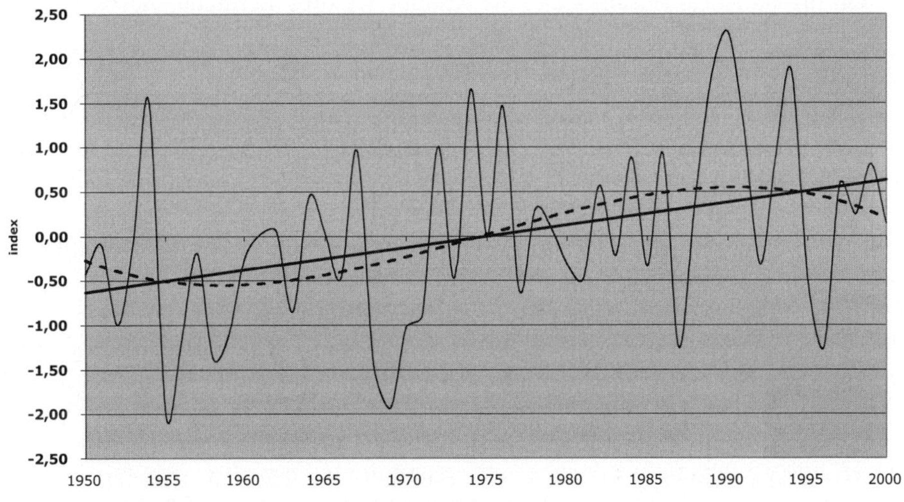

Figure 88. The NADi from 1950–2000; linear and polynomial trends.
From Pommier (2002).

This index gives a clearer picture of the intensity of meridional exchanges in the North Atlantic space. A negative index reflects an attenuated dynamic, and a positive index a more robust dynamic. For the period in question, the 1970s represent a 'turning point', after which the NADi was in a positive mode which would reach its culmination in the 1990s (Pommier, 2002).

12.1.4 Variations in the intensity of the aerological dynamic of the Atlantic

The mechanisms and climatic consequences listed below are easily verified for the North Atlantic space, on various scales; synoptic, seasonal (Figures 85 and 86), statistical, (cf. Figure 61), and even palaeoclimatic (Leroux, 1996a; Chapter 9).

The positive or high phase of the NAD (or of the NAO, on the mean scale)

The positive phase is illustrated diagrammatically in Figure 85, which, with inevitable differences in intensities, reflects the winter season and/or a cold period or even a glacial period, but with more striking modifications in the latter case (cf. Figure 41, Chapter 9). **The positive phase corresponds to a rapid mode of general circulation** (Figure 33, Chapter 8) (i.e., when there is a (more or less) greater thermal deficit over the Arctic). The equivalents on the statistical scale are given in Figure 61(a-2 and b-2).

- The Arctic is colder, and especially so in its western part, to the north of Canada, where most MPHs are born.
- MPHs are initially stronger and more extensive.
- Their motion is more rapid, and their more meridional trajectories (Figures 89 and 90) extend further into the tropical zone (which is itself shifted southwards).
- On the statistical (mean) scale, the Atlantic (Azores–Bermuda) AA is stronger, bigger and more meridional (Figure 61(a-2)).
- The increased strength of MPHs increases the probability of the formation of AAs not centred over the ocean (i.e., over North America and Europe), and they are more frequent and longer lasting in both winter and summer.
- The synoptic lows associated with the MPHs (low-pressure corridors and closed lows) are deeper, and they sweep over a larger area, reaching more northerly latitudes to penetrate further into the Arctic. There will therefore be a greater distance now between an MPH, pushing further to the south, and its closed low, moving further to the north.
- Statistically, the 'Icelandic' Low is deeper and bigger (Figure 61(a-2)).
- Consequently, on the synoptic scale, the difference in pressure between MPHs and their associated synoptic lows is wider (cf. NAD) (i.e., on the mean scale (cf. the NAO), the pressure differential is greater between the 'Azores' AA and the so-called Icelandic Low).
- *Meridional exchanges are intensified*, especially the warm air transfers from the south-west towards the north, and the cold air transfers from the north-west towards the south.

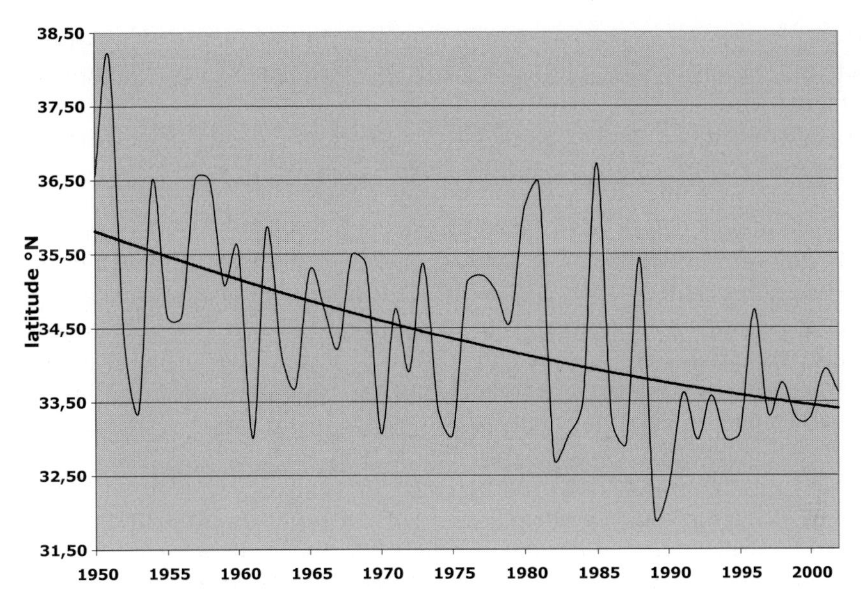

Figure 89. Minimal latitude reached by American MPHs in the North Atlantic aerological unit, from 1950–2002.
From Pommier (2004a) (CDC/NCEP–NCAR data).

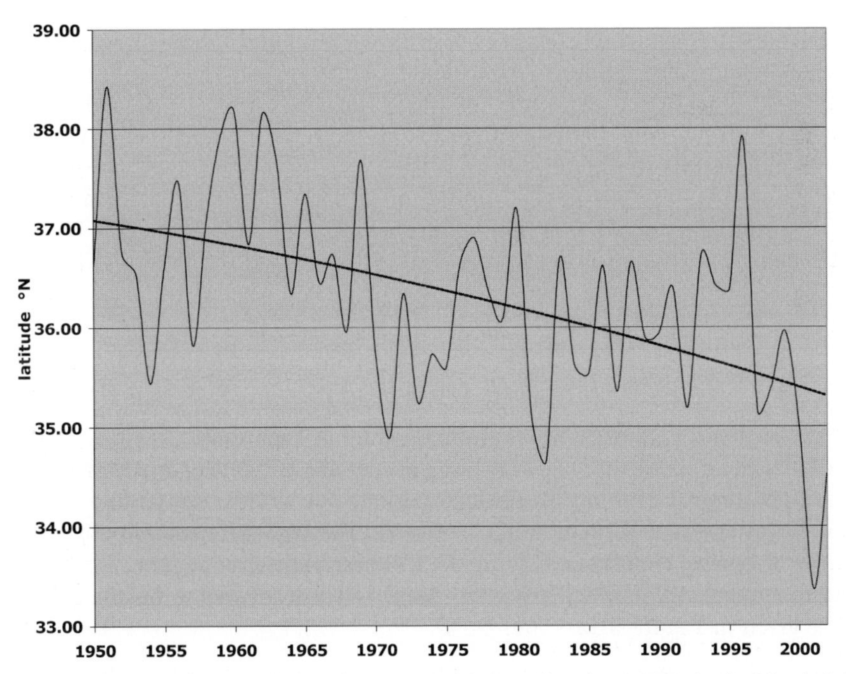

Figure 90. Minimal latitude reached by Greenland–Scandinavian MPHs in the North Atlantic aerological unit, from 1950–2002.
From Pommier (2004a) (CDC/NCEP–NCAR data).

- There is a general acceleration in circulation, in the lower layers (both within and around MPHs) and at altitude. General circulation assumes a rapid mode (Figure 33).
- This acceleration involves ocean surface waters, and wind-driven currents now flow more rapidly, while upwellings show a corresponding increase in intensity. Moreover, the cold adds vigour to density currents at high latitudes.
- The weather is more violent: MPHs are more powerful and move more rapidly, instigating more dynamic convection at their leading edges, facilitated by greater temperature contrasts between the MPHs and the south-west flux, and by an enhanced supply of energy from this same southerly flux. The weather is stormier, across a larger area.
- Dynamical contrasts are accentuated between the two sides of the Atlantic, and as a function of the climatic regions determined by the trajectories of the MPHs.
- Climatic contrasts are accentuated, especially in their thermal aspect, as a result of the alternation of fluxes, warm from the south and cold from the north. Contrasts in the pluviometric aspect vary with particular pluviogenic conditions.
- A *mean* temperature established for the whole of the aerological unit, with its different thermal behaviours, which may even be the opposite of each other, would therefore be completely meaningless as far as the real climate is concerned.
- The weather is more irregular, mainly due to the formation of AAs, particularly over the land, bringing anticyclonic stability which briefly raises temperatures (especially in summer, cf. Chapter 11 – heatwaves) and inhibits precipitation.
- Interactions with neighbouring aerological units are only modest in the case of the North Pacific unit (by reason of relief), but more robust with northern and central Europe, with Scandinavian MPHs and the continental AA possibly forming a barrier to the passage of MPHs.
- The North Atlantic aerological unit is at its largest, especially in the tropical zone, at the expense of the South Atlantic aerological unit; the surface line of the ME shifts to a greater or lesser degree southwards, and the IME and the VME, which belong to the southern unit, move with this general translation, principally in Brazil and western Africa.

The negative or low phase of the NAD (or of the NAO, on the mean scale)

The negative phase is illustrated diagrammatically in Figure 86, which, with inevitable differences in intensities, reflects the summer season and/or a warm period, a climatic optimum, or even an interglacial period, but with more striking modifications in the latter case (cf. Figure 35, Chapter 8). **The negative phase corresponds to a slow mode of general circulation** (Figure 35, Chapter 8) (i.e., when there is a (more or less) attenuated thermal deficit over the Arctic). The equivalents on the statistical scale are given in Figure 61(a-1 and b-1).

- The Arctic is less cold, and especially so in its western part, to the north of Canada, where most MPHs are born.
- MPHs are initially weaker and less extensive, but relatively more frequent.

- Their motion is slowed, and their less meridional trajectories (Figures 89 and 90, at the beginning of the curves) trespass very little into the tropical zone (which is itself shifted northwards).
- On the statistical (mean) scale, the Atlantic ('Azores') AA, its western offshoot for the most part 'amputated', is less strong, smaller, and less meridional (Figure 61(a-2)).
- The decreased strength of MPHs lowers the probability of the formation of AAs over continents (North America and Europe), and agglutinations are less frequent and of shorter duration, especially in winter.
- The synoptic lows associated with the MPHs (low-pressure corridors and closed lows) are less deep, they cover a smaller area, and are maintained at temperate latitudes, reaching more northerly latitudes less easily, especially the Arctic Ocean. The MPH and its corresponding low now stay relatively nearer to each other along the trajectory.
- Statistically, the 'Icelandic' Low is more shallow and smaller (Figure 61(a-2)).
- Consequently, the difference in pressure between MPHs and their associated synoptic lows is smaller (cf. NAD), and on the mean scale (cf. NAO) the pressure differential is weaker between the 'Azores' AA and the so-called Icelandic Low.
- **Meridional exchanges are less intense**, especially the warm air transfers from the south-west towards the north, and the cold air transfers from the north-west towards the south.
- There is a general slowing of circulation, in the lower layers (both within and around MPHs) and at altitude. General circulation assumes a slow mode (Figure 35).
- This deceleration involves ocean surface waters, and wind-driven currents now flow less rapidly, while upwellings show a corresponding decrease in intensity. Moreover, the relative warming diminishes the vigour of density currents at high latitudes.
- The weather is more clement: MPHs are less powerful and move more slowly, instigating less dynamic convection at their leading edges, while temperature contrasts between MPHs and the south-west flux are narrowed, and storminess is reduced.
- Dynamical contrasts are attenuated between the two sides of the Atlantic, and as a function of the climatic regions determined by the trajectories of the MPHs.
- Climatic contrasts are therefore narrowed, especially in their thermal aspect, as a result of the alternation of fluxes, now cooler from the south and less cold from the north. Contrasts in the pluviometric aspect vary with particular pluviogenic conditions.
- A *mean* temperature established for the whole of the aerological unit, with its still different thermal behaviours (although there is less contrast), is now a little more 'representative', although this still does not completely represent reality, which retains its contrasts.
- The weather is more regular, and disturbances to the fore of MPHs are better able to move eastwards, because the formation of AAs is less vigorous. This is a

result of weakening of the barrier formed by Scandinavian MPHs, which are now less frequent and less strong, and by the continental AA, also less frequent.

- The North Atlantic aerological unit now covers a smaller area, and is particularly reduced in the direction of the tropical zone. The surface line of the ME, and its associated vertical structures, shift generally northwards.

12.2 RECENT CLIMATIC EVOLUTION IN THE NORTH ATLANTIC SPACE

The North Atlantic aerological space exhibits various kinds of climatic evolution, their common characteristic being that they are all determined by the same original cause (i.e., by variations in the strengths of MPHs coming from the Arctic). What is the nature of the changes in this space during recent decades, changes which are reflected principally in the rising NADi?

12.2.1 Evolution in the Arctic

We have already mentioned thermal evolution in the Arctic (Chapter 10), because of its great importance, especially in its role as the motor of circulation in the northern hemisphere, and because of the resulting consequences for the weather (Chapter 11). Remember that the evolution of temperature in the Arctic, and notably that of the central area of the Arctic as studied by Kahl *et al.* (1993) and Rigor *et al.* (2000), **is characterised by cooling, mainly in the western part, north of Canada**. This cooling is not immediately apparent in the observational data from weather stations on the periphery of the Arctic, a fact welcomed by partisans of the scenario preferred by the IPCC.

Let us examine, for example, temperature data from the Resolute meteorological station located in the Northwest Territories of Canada (Devon Island), between Ellesmere Island and Baffin Land.

- At first sight, judging from mean annual temperatures (Figure 91), it is easy to reach the conclusion that the temperature has risen since the 1970s. Therefore 'the Arctic is getting warmer': by about 1°C. And here the argument ends – at least, this is often the case.
- On re-examining the data, and taking into consideration seasonal temperatures (Figure 92), we see that, in spite of the difference in scale of this figure (as compared with Figure 91), the recent rise is much less obvious. It seems not to appear in the summer (June–August), is only very slight in spring (March–May) and autumn (September–November), and for the winter months (December–February) it appears, and weakly at that, only towards the end of the curve.
- A third look at the data, at the temperatures for winter months (Figure 93), reveals that the weak rise during the winter occurs only for the month of

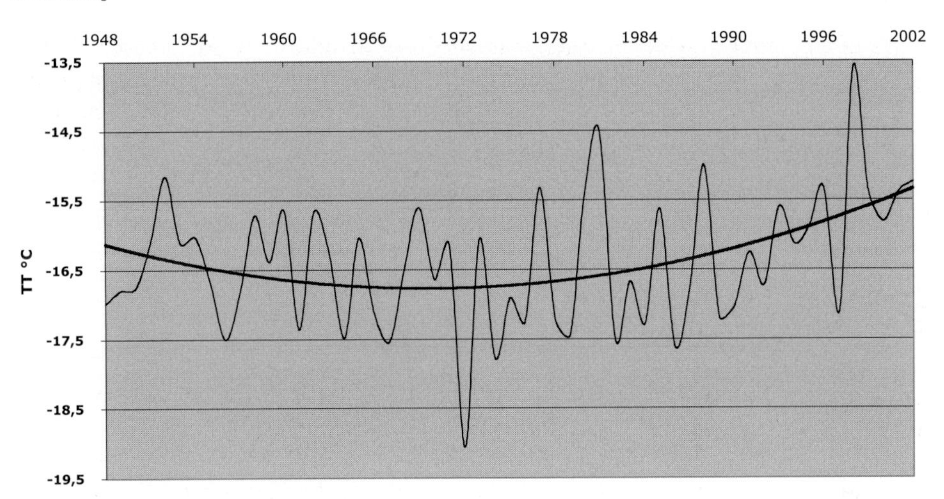

Figure 91. Mean annual temperature (°C) and polynomial trend at Resolute, Canada, 74°43′N, 94°59′W, from 1948–2002.
From *Environnement Canada* data.

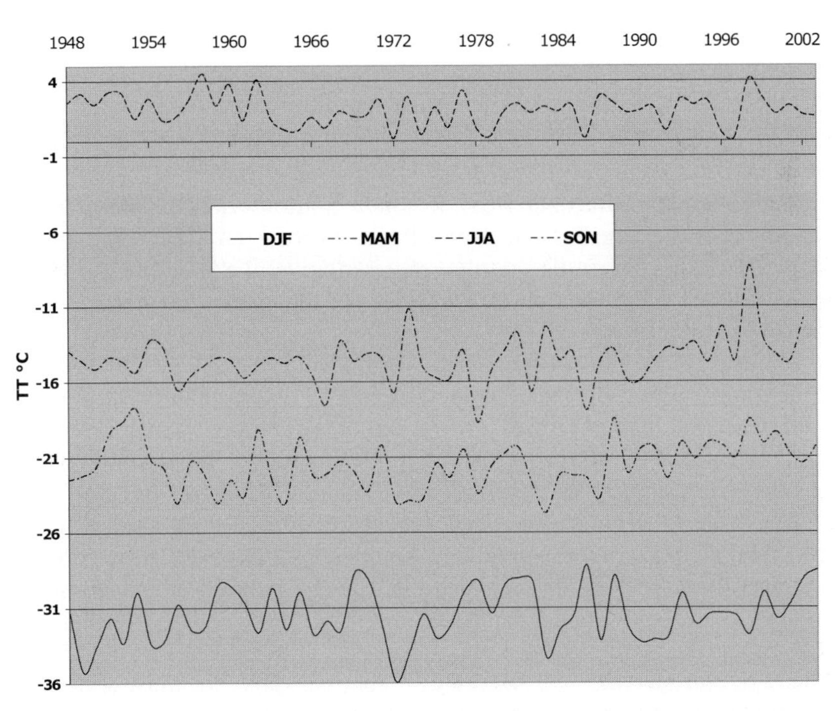

Figure 92. Mean seasonal temperature (°C): winter (DJF), spring (MAM), summer (JJA), autumn (SON) at Resolute, Canada, 74°43′N, 94°59′W, from 1948–2002.
From *Environnement Canada* data.

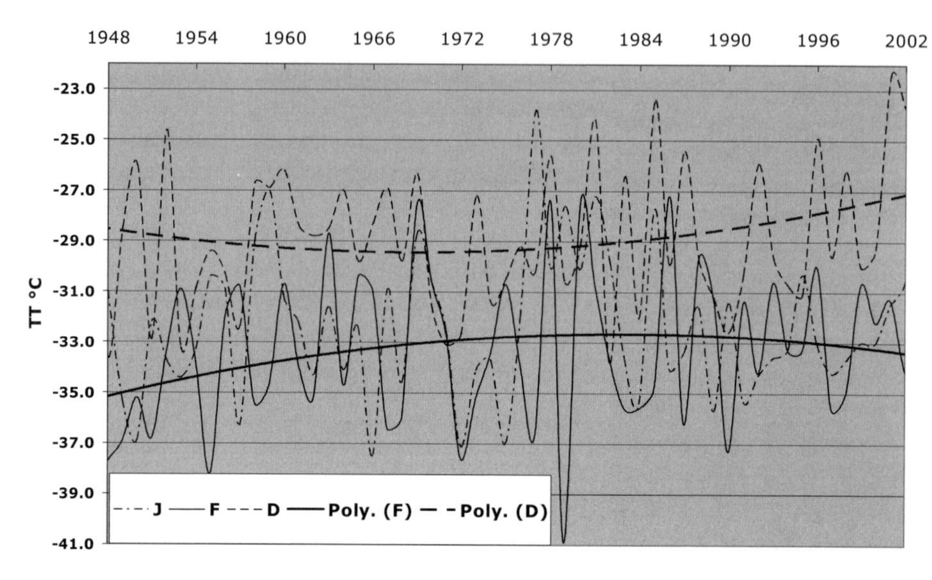

Figure 93. Mean monthly temperature (°C), winter months: December (D), January (J), February (F), and polynomial trends in December (*top*) and in February (*bottom*), at Resolute, Canada: 74°43′N, 94°59′W, from 1948–2002.
from *Environnement Canada* data.

December, while in January and February it is absent. The conclusion this time is that 'the Arctic is not getting warmer'!

Are the mean temperature data from Resolute Station really representative? Are those from Inuvik (Canada), or from Point Barrow (Alaska), in comparable locations and producing similar results, representative? To base one's conclusions on Figure 91 alone is to look only at the surface of the matter: Figures 92 and 93 do not send the same message.

In addition to the previous remarks on this subject (Chapter 10), it is worth reminding ourselves that the mean temperature datum gives an imperfect result, that of the alternation of different fluxes passing across the station, warmer ones from the south and colder ones from the north. Figure 94 eloquently illustrates the establishment of the mean annual temperature at Godthaab-Nuuk (on the west coast of Greenland), where the temperature fell between 1970 and 2000. An indication of strong MPH dynamism (mainly involving American MPHs) is represented here by a fall in temperature (passage of cold air). Conversely, reduced dynamism in MPHs corresponds to relatively milder temperatures. Note also (Figure 94) that the two trends intersect during the 1970s.

When we consider the real meaning of thermal data, we must still take into account the variations in dynamical conditions at the observation sites. For example, Litinsky and Genest (2002) inform us that the whole of northern and eastern Canada has experienced cooling between 1961 and 1990; however, there

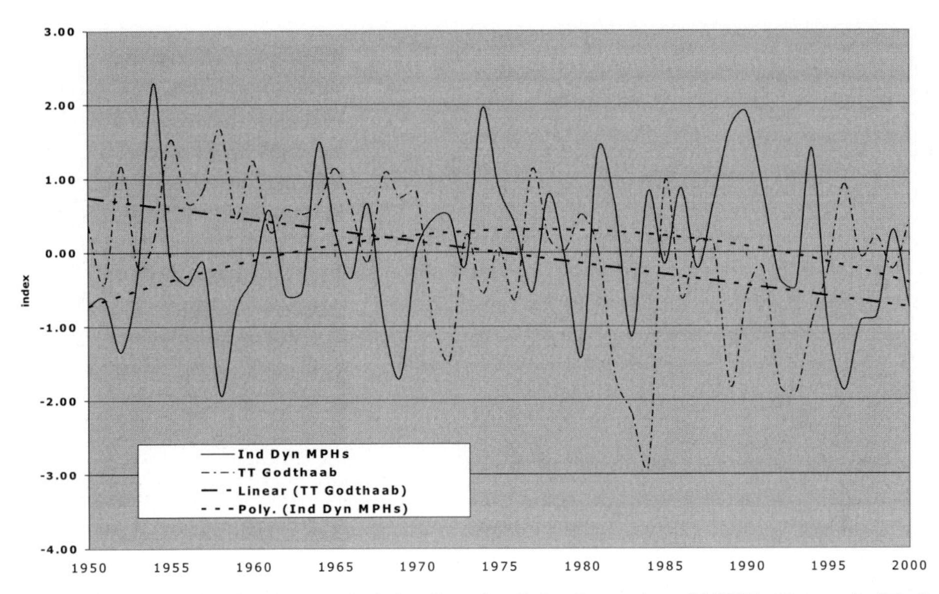

Figure 94. Evolution in the North Atlantic unit of the dynamics of MPHs (*Dynamical Index of MPHs)*, with polynomial trend, compared with the evolution of mean annual temperature (standard deviation) at Godthaab (Nuuk) on the western coast of Greenland (*Danmarks Meteorologiske Institut* data), with linear trend, from 1950–2000.
From Pommier (2002).

has been a slight warming at certain weather stations in the east, during the last decade (cf. also the temperature at Godthaab, Figure 94). What does this slight rise indicate? It could be that the warming is real; or perhaps the preceding cooling has itself continued, while causing warming at certain stations which are located where intense return airflows from the south are present. Once more, remember that **the Arctic is not an isolated aerological entity**, and certainly does not possess its own independent unit of circulation, unconnected to the three units lying to the south of it. So temperatures in the Arctic cannot be studied without at the same time examining the wider aerological dynamic (cf. Chapter 10). **Reality is all-embracing**, and the validity of any reasoned argument cannot be founded on a few (sometimes deliberately) selected aspects, superficial in nature: it must be rigorously founded on an appreciation of how *everything* works together. With this in mind, let us now examine, concentrating on salient facts, the different climatic behaviours encountered in the North Atlantic aerological space, to test the reality of this coherence.

12.2.2 North America (to the east of the Rocky Mountains)

Temperature is decreasing

The fall in temperature in the Arctic has been echoed in Canada, and in eastern Canada records for low temperatures have been broken again and again, particularly since the 1990s (cf. NADi, Figure 88). As *Environnement Canada* reported in 1993,

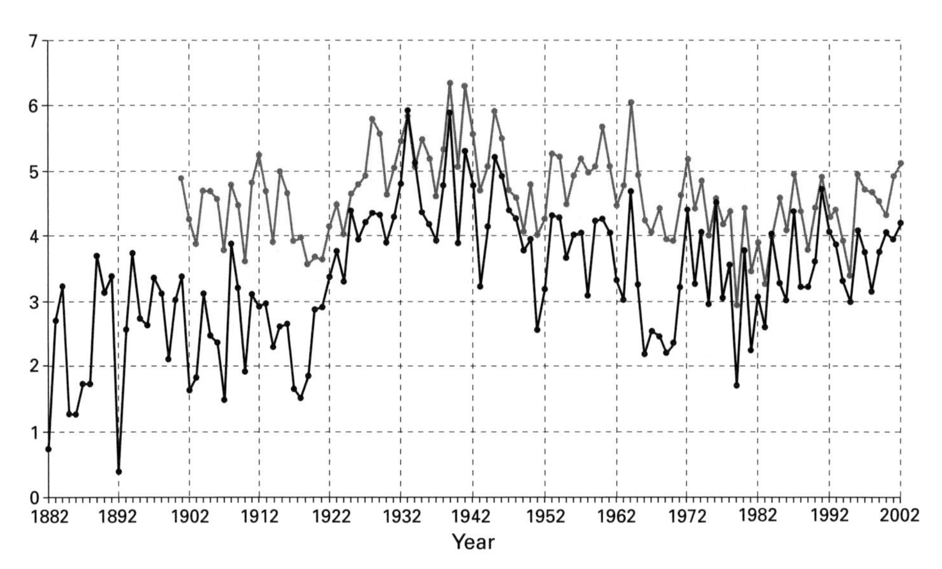

Figure 95. Annual mean temperature (°C) in Iceland, from 1882–2002. (*top*, grey line) At Reykjavik (in the south); (*bottom*, dark line) at Akureyri (in the north).
From Daly (2004).

under the heading 'Record Cold in Eastern Canada', there were '*new absolute minima of temperature*' (*Perspectives Climatiques*, vol. 15, No. 06). In Quebec, the winter of 1993–1994 was also the most severe ever known (WMO, 1995). In the winter of 1994, there was '*much-above normal ice cover*' on the Great Lakes. The Lakes were almost entirely frozen over on 7 March 1994, this drastic situation (surpassed only during the winter of 1979) being associated with '*anomalously strong anticyclonic circulation . . . that brought frequent air masses of Arctic and polar origin*' (Assel *et al.*, 1996). In January 1996, the temperature again fell to −52°C in Saskatchewan (*Environnement Canada*, 1996). In January 1998, a most dramatic 'ice storm' swept across eastern Canada and the north-eastern USA. This was dubbed the '*ice storm of the century*' (Abley, 1998; Martin, 1999). In turn, 2004 saw intense periods of cold, especially in March, though temperatures were also very low in May, and the summer unusually cold. Record low temperatures were recorded, the temperature falling to 0°C on 20 August; it even snowed in the centre of Winnipeg, a 'first' for August in this city! This tendency towards more extreme cold has been described by Morgan *et al.* (1993), Gachon (1994), and Litynski (1994, 2000). The same low temperatures were experienced in Greenland: on the west coast at Godthaab (Figure 94), and also on the east coast, according to observations made at Angmassalik. The same trend was observed in Iceland (Figure 95), in the south at Reykjavik, and in the north at Akureyri, with temperatures lower than those of the optimum of the period 1930–1960 (a thermal evolution identical to that observed in Figures 55 and 56, Chapter 10).

Mean thermal data from the USA (Figure 96) also reveal this evolution in mean temperatures: the whole of the central and eastern USA, down to the Gulf of

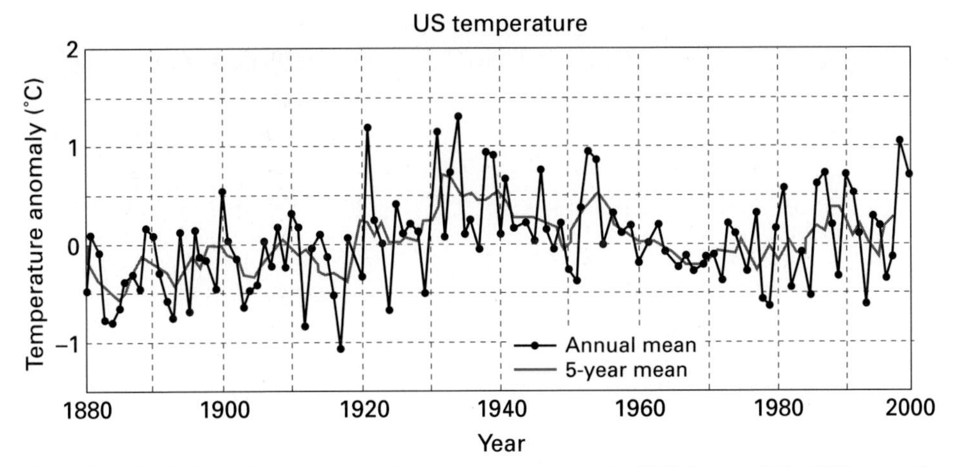

Figure 96. Evolution of mean annual temperature anomaly (°C) from 1880–2000 over the contiguous USA (urbanisation-adjusted).
From NASA-GISS.

Mexico, has seen a definite, continual cooling trend (Balling and Idso, 1989; Kukla, 1989; Erhardt, 1990; Agee, 1991; Litynski, 2000; Daly, 2004: in *What the stations say*). The curve for the whole of the territory (Figure 96) plays down the cooling in the central regions, while warming predominates west of the Rockies (cf. Chapter 13, North Pacific unit). To the east, and especially east of the Appalachians, intense returning airflows from the south mask the cooling which accompanies the passage of MPHs (Karl *et al.*, 1996). At the same time, waves of cold, brought about by vast MPHs with raised pressures, have been far more severe since the 1970s, much more so than in the 1950s; such severe spells were almost absent for a period of 40 years (Michaels, 1992). The hardest frosts have been experienced since the 1970s (Rogers and Röhli, 1991), with devastating effects upon the Florida citrus plantations. More recently, in 2003 and 2004, new records have been set for cold and snowfalls, most memorably in the east (cf. Chapter 10). Among explanations proffered for the cooling in the eastern central region has been the (not unexpected) model-based one, of course: '*It could be a direct response to anthropogenic climate forcing*' (Robinson *et al.*, 2002)(!) ... it is so obvious!

Increasing pressure and storminess

As the temperature has fallen, so the surface pressure has risen. In Canada, for example, the rise in pressure is more marked in autumn and winter in the south-eastern part of the country (Gachon, 1994). In the USA, the Great Plains have also experienced an increase, culminating in the 1990s (Figure 97). This rise, recorded as far down as the Gulf of Mexico, and beyond, is (as stated) not well borne out by mean observational data, because of the alternating rise and fall in pressure associated with the passage of MPHs.

The intensification of the MPH dynamic (Figure 88), which is reflected, however

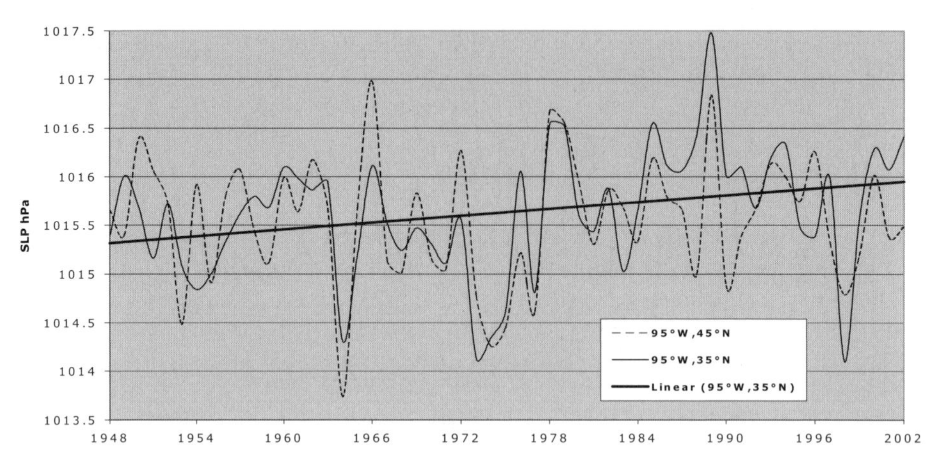

Figure 97. Mean annual surface pressure at 95°W, 45°N (near Minneapolis) and at 95°W, 35°N (between Oklahoma City and Little Rock), from 1948–2002. Linear trend at 95°W, 35°N.
From CDC/NCEP-NCAR data.

imperfectly, by the rise in pressure (Figure 97), explains the fact that '*the climate of the USA has become more extreme in recent decades*' and '*indices increased rather abruptly during the 1970s*' (Karl *et al.*, 1996).

Symptomatic of this increase in the strength (frequency and/or size) of MPHs are:

- A mean increase in precipitation, albeit not uniform, quite pronounced in the east, where it is associated with air returning from the south-west on the leading edge of MPHs. These '*increases of total precipitation are strongly affected by increases in both frequency and intensity of heavy and extreme precipitation events*' (Karl and Knight, 1998). Conversely, the diagonal strip east of the Rockies from Montana to Texas has seen less rain (Karl *et al.*, 1996).
- A continuous increase in the frequency of violent storms, accompanied first by rain and then by snow (blizzards), mainly on the eastern side of the country. 'Storms of the century' and 'blizzards of the century' succeed each other, examples being the events of 12–14 March 1993 (Forbes *et al.*, 1993; Kocin *et al.*, 1995), 17 January 1994, 7 January, and 6 February 1996, whence the popular name for the winter of 1995–1996: '*the incredible winter*'. New records for snowfall were established, to be broken again on 10 January 1998, 21–24 January 2000, and also in late December 2002, when snowstorms of great ferocity hit the central USA, with even Texas and Louisiana not escaping them. To all this should be added the winters of 2003 and 2004. Another sign of this increased severity in the weather is the continued increase in deep depressions (on the leading edges of MPHs) over the Great Lakes basin (Angel and Isard, 1998; Kunkel *et al.*, 1999). One of these depressions became remarkably

intense on 14 September 1996, and a satellite image of this feature earned it the nickname '*Hurricane Huron*' (Miner *et al.*, 2000).

- The much greater frequency of tornadoes during the period 1953–1995 (WMO, 1998). There is an ongoing increase (Johnson, 2002) in the numbers of tornadoes, born of the stormy 'supercells' which develop on the leading edges of MPHs, as cold air and moist, very energetic air (from the Gulf, and destabilised over the land) come into turbulent contact. This increase is in part due to more extensive recording of tornado occurrences (meaning that they may not be increasing as much as is thought), but the margin of error is unlikely to be great. Certainly, tornadoes (numbering about 1,000 per year, though in 1998 there were more than 1,400) are becoming considerably more frequent.

12.2.3 The Gulf of Mexico – Central America and the Caribbean

Here, the passage of air towards the tropical zone can be a troubled affair. There is of course no calming effect upon phenomena having travelled through an AA. Instead, there is a direct encounter between air transported by MPHs and the warm, unstable air advected by the maritime trade, with its great store of energy, beneath a high, modest inversion (Figures 85 and 86). This contrast leads to the formation of very deep lows on the leading edges of MPHs, and these (originally) tropical lows may become hurricanes, and/or move north-eastwards to the fore of MPHs, to become the blizzards of winter.

Cyclogenesis in the Atlantic

These MPHs are also responsible for the (statistically) preferred trajectory of hurricanes, which pull away to the north-east as they enter the Gulf of Mexico, or more often, just before they reach it: the arrival of a quite powerful MPH blocks their route and imposes this diversion upon them. In this way, MPHs (well deserving their name of 'anti-cyclone' in this context) can protect North America from these violent disturbances, turning them away along their leading edges and all the while integrating them into the dynamic of the temperate zone. So eastern coastal areas have benefited from the increase in the frequency/strength of MPHs, as Smith (1999) and Landsea *et al.* (1999) noted while confirming a fall in the number of hurricanes reaching the continent during the period 1944–1996.

However, the timing of an encounter between an MPH and a cyclone is not always well 'co-ordinated'. Take the example of Hurricane Charley, on 13 August 2004. This hurricane was born on 8 August within the VME structure to the north of Guyana, to be drawn in by the low-pressure corridor on the leading edge of an MPH which had crossed North America. Its intensity having been increased in this corridor, it went on to cause deaths and damage to property, in Florida. Sadly, a difference of only a few hours would have meant that Hurricane Charley would have been diverted towards the Atlantic.

So the circumstances in which these encounters take place are not always favourable (e.g., Hurricane Frances was approaching Florida on 2–3 September 2004, just as an MPH was moving off the USA into the Atlantic. Frances therefore continued its path to the south of the MPH, along its trailing edge, straight into Florida, this time not protected by an anticyclonic air mass.

A similar situation occurred in the case of hurricanes Ivan and Jeanne (Figure 98). Ivan began as a tropical depression on 28–29 August between Dakar (Senegal) and the Cape Verde Islands. It gradually developed, moving westwards within the structure of the VME, which lay at about 10°N, becoming a cyclone as it neared South America. It was moving off the northern coast of South America by 5–6 September. Its centre crossed the Caribbean to the south of Haiti and Cuba, and it passed over Cuba on 13 September. At the same time, an MPH, preceded by anticyclonic air of modest activity, was coming down more rapidly from Canada, to the west. By now Ivan had entered the Gulf of Mexico, and the two systems met on 16 September. Ivan, weakened, and now over Alabama, was forced back across Georgia and into the Atlantic, to be integrated into the low-pressure corridor to the fore of the MPH.

Hurricane Jeanne began its life to the north of South America, on 6–7 September, within the structure of the VME, again at about 10°N. This tropical depression followed fairly closely behind Ivan, and was therefore not well supplied with energy. It appears in Figure 98, and remained quite poorly organised until 19 September, when, with Ivan having been neutralised by the oncoming MPH, Jeanne adopted the classic circular form, a sign of reactivation, but was drawn on 20 September towards the low-pressure corridor of the same MPH. It then seemed that the hurricane would suffer the same fate as Ivan, and be integrated into the low-pressure corridor on 21 September. However, the MPH was tracking too rapidly eastwards, and this prevented Jeanne (by now once again a depression) from moving north-eastwards with the cyclonic flux on the MPH's leading edge. On 22 and 23 September Jeanne, profiting from the concentration of energy advected by the trade to the south of the MPH (forming a barrier), resumed, with increased vorticity, its westward passage as a hurricane towards Florida. An earlier, weak MPH blocked its path on 26 September, and an increase in its strength on the 27th forced Jeanne to move away to the north-east, off the land, and to integrate itself into the temperate disturbance.

It also happens occasionally that MPHs prevent hurricanes from being diverted northwards. A case in point is Hurricane Mitch, originating in the Caribbean on 23 October 1998, to the fore of an MPH. This MPH later barred its passage northwards and forced the hurricane to move across the Central American Isthmus, holding it to the south of the Cordillera until 2 November 1998 (cf. Chapter 11).

Figure 98 shows, as does the example of 'Mitch', that cyclones do not follow 'capricious' trajectories: they are, on the contrary, totally directed, obeying firstly the logic of the tropical east–west path, and thereafter the logic imposed upon them by MPHs. The reactivation of hurricane Jeanne between latitudes 25°N and 30°N confirms the fact that the relationship 'T/cyclogenesis' (cf. Chapter 11) cannot properly explain the dynamic of cyclones. Predicting cyclogenetic activity must, in fact, involve an appreciation of the latitude of the VME and of the part played by MPHs, the determinants of the eventual trajectories. It must be re-emphasised that a mere equation will not suffice!

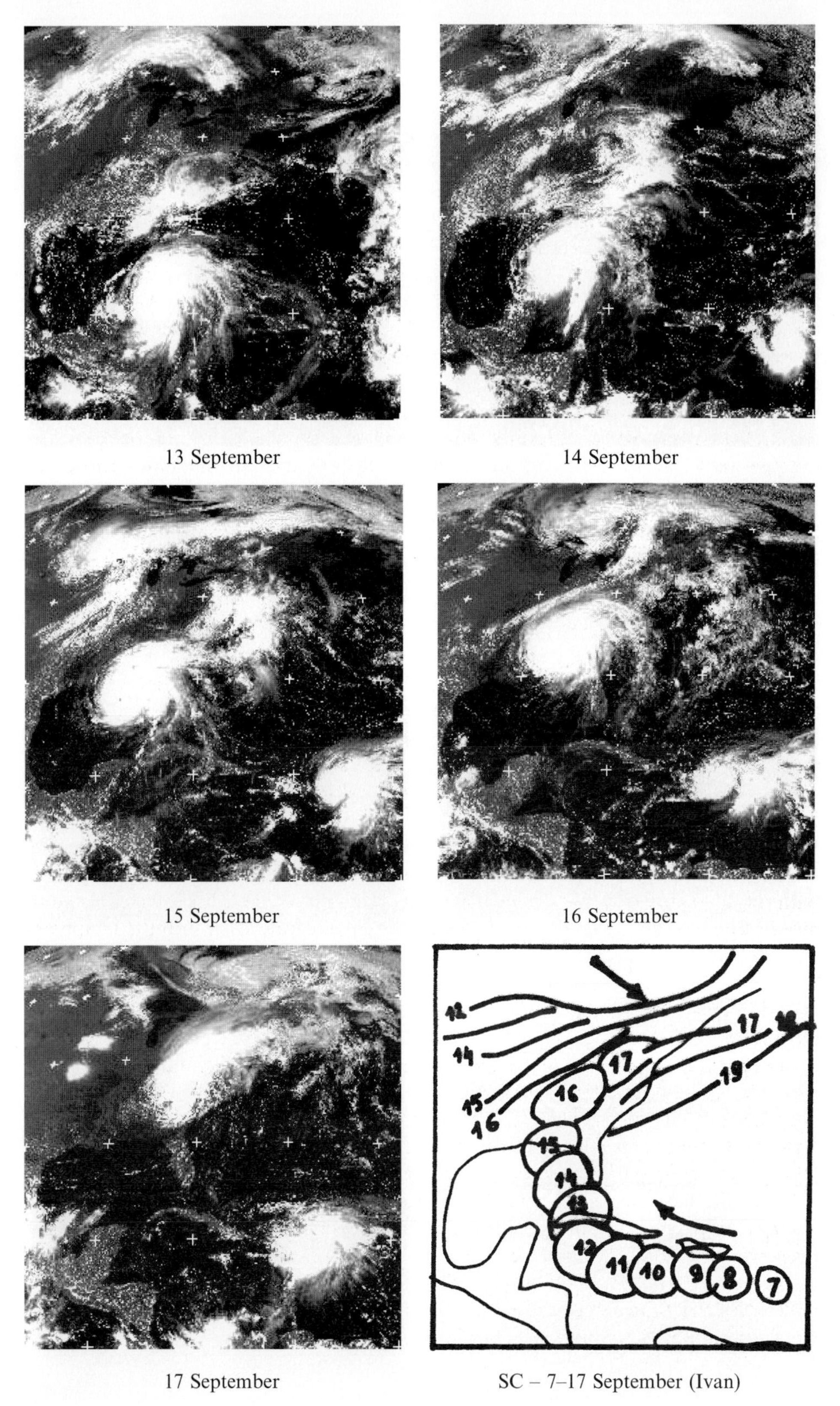

Figure 98. Paths of hurricanes Ivan and Jeanne, from 13–28 September 2004 (*GOES 12*, visible, 18h, NOAA). (*Continued overleaf.*)

Note: SC = summary chart.

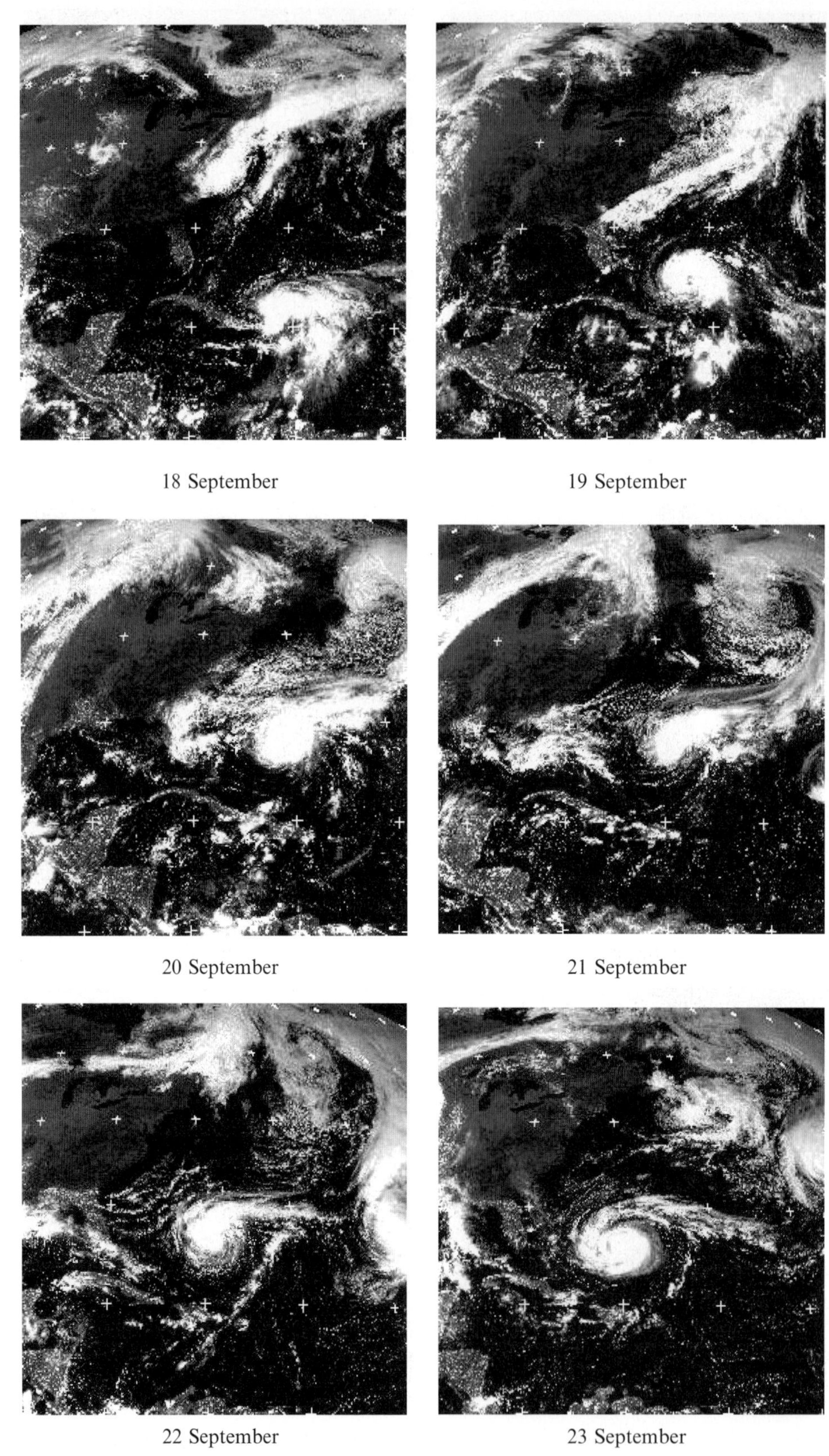

Figure 98. (*cont.*)

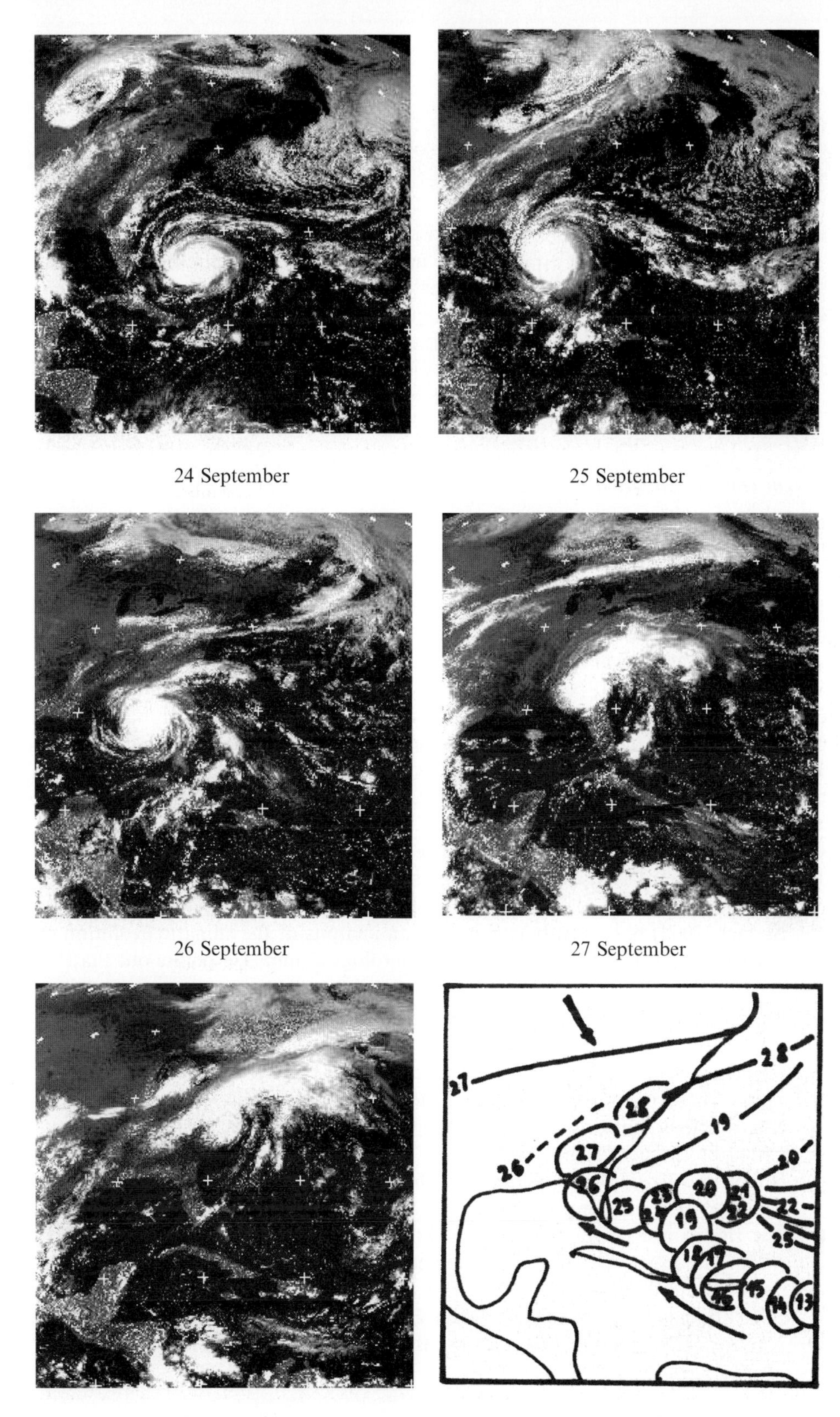

24 September

25 September

26 September

27 September

28 September

SC – 13–28 September (Jeanne)

Cyclogenesis, like all other phenomena, must be seen in the context of general circulation.

- So, a rapid mode (Figure 33), or positive NAO/NAD index, means that the VME is nearer to the equator and initial vorticity is diminished, while the frequency and strength of MPHs offer better protection to the continent.
- On the other hand, a slow mode (Figure 35), or negative NAO/NAD index, means that the VME is further from the equator in summer and initial vorticity is enhanced, increasing the probability of the formation of a vortex within this structure; MPHs, weaker now, offer less protection to the land (cf. Chapter 13, and the 'link' with El Niño).

Recent climatic evolution

Intervention by MPHs is obviously much more vigorous during the winter, when the influx of polar air initiates the *northers* or *nortes*, and strong winds, storms, and frosts descend upon the tropical plantations with dramatic effect (Hastenrath, 1991; Rogers and Röhli, 1991; Garcia, 1996; Barbier, 2004). These invasions of cold air cross the Isthmus towards the equatorial Pacific as the mountain-gap winds known as *tehuantepecers*. The most powerful MPHs can reach northern Venezuela (at around 10°N) and their leading edges join forces with the high relief (at the extremity of the Andean chain), to lift the moisture-laden maritime flux and bring about, in the middle of the dry season, the '*invernio de las chicharras*'. Figure 116 (Chapter 13) sets out the circumstances favourable to this type of weather, with the leading edge of an Atlantic MPH blocked in this fashion towards the south by the mountains of South America. Such a situation is characterised by abundant and sometimes torrential rains (e.g., 791 mm fell in two days on 15 and 16 December 1999, in the state of Vargas (Venezuela), an event which has been called '*the worst catastrophe that country has ever known*' (Barbier, 2004).

Barbier (2004) has recently carried out an analysis of the climatic evolution of this particular sector of the North Atlantic aerological unit. He points out that:

- There has been a general rise in surface pressure (Figure 99), an extension of that observed in North America. It is at its greatest over the Caribbean Sea, where the increase is of the order of 1 hPa, a remarkable value for the tropical environment, where pressures vary very little on the mean annual scale. Similarly remarkable is the consistent nature of the increase; this rise in pressure, transmitted by the *nortes* and by the trade flux, reaches the opposite shores of the Isthmus through lower lying areas of the mountains, and is also observed in the Pacific Ocean, off Central America.
- The rise in pressure is closely associated with the latitude reached by MPHs within this space, MPHs having managed to reach further southwards throughout the last 50 years (Figure 89). For example, MPHs averaging 1,020 hPa or 1,025 hPa have been reaching almost 2° further to the south in latitude (from 27°N to 25°N approximately) during the period 1950–2000 (Barbier, 2004).

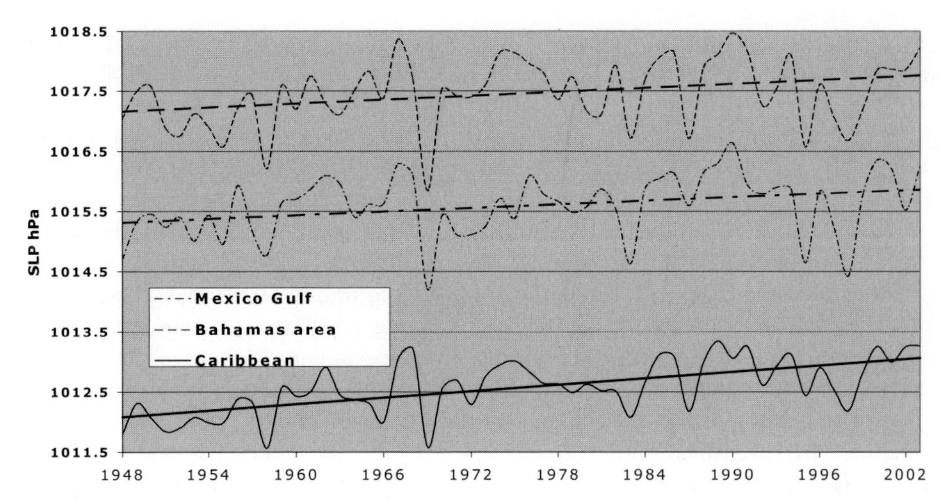

Figure 99. Surface pressure evolution over the Gulf of Mexico (*top*), the Bahamas area (*middle*), and the Caribbean Sea (*bottom*), and linear trends, from 1948–2003.
From Barbier (2004) (CDC/NCEP–NCAR data).

- The evolution of temperatures in the Gulf of Mexico, and their evolution in the Caribbean Sea, differ – a turning point having been reached in the 1970s:
  - in the Gulf of Mexico: before the turning point, temperature was falling, and it rose afterwards; and
  - in the Caribbean, before this point, temperature was rising, and fell afterwards, especially during the 1980s.

  This differentiation is symptomatic of the gradual southward progression of MPHs, with warming associated with the southerly fluxes diverted northwards by them; while the fall in temperature is associated with the arrival of the MPHs themselves, more belatedly in the Caribbean.
- The evolution of precipitation is, however, more simple, showing a general and quite marked diminution for the whole of the Isthmus during the period 1960–1995.
- Cyclonic activity shows no particularly significant trend (Barbier, 2004; cf. Figure 65 (top), Chapter 11).

12.2.4 The central and south-east Atlantic

MPHs normally move across the ocean from north-west to south-east, and they gradually overlap to form an AA, the so-called (statistical) 'Azores' AA. This means that the cooling observed on the American continent extends across most of the Atlantic Ocean, from Greenland as far as Europe, and down to the south into the tropical zone, in respect of both air and water. Deser and Blackmon (1993) observed, in winter, '*warming from 1920 until 1950, followed by cooling from 1950 to the present*', and a coincidence between '*colder than normal marine temperatures and winds stronger than normal*', as far away as the coastal waters off West Africa, and most notably in the area of the Canaries and the Cape Verde Islands; sea surface temperatures being cooler from the coast of Africa across to the Caribbean (Kaplan *et al.*, 1997; Visbeck *et al.*, 2003). Sea surface temperature charts, however, show no sign of this propagation of cold, because of the interference between the dynamic associated with MPHs themselves, and that of the currents driven by them. The greater strength of MPHs is reflected by an increase in wind speeds, involving both the statistical westerlies and the trade to the south of the AA. The driving force exerted upon the ocean surface by (weightier) MPHs and accelerated winds increases the intensity of sea surface circulation, especially in the case of the Canaries Current and its associated upwelling (Amraoui, 2004), a situation which happens in cold periods, according to the evidence of palaeoclimatology (Chapter 9). Off North America, the intensified Gulf Stream system, like the energetic, warm airflows returning from the south-west, to the fore of MPHs near the east coast, has a warming effect on sea surface temperatures towards Europe. However, the axis of the main MPH path (and the associated cooling) cuts through this warm transversal band (which recommences near to Europe): a statistical artifice which, of course, does not mean that the North Atlantic Drift is actually interrupted.

The reinforcement of the Anticyclonic Agglutination

Along the main trajectory taken by MPHs, the rise in pressure is general, and more marked as one goes eastwards, with the progressive reinforcement of the agglutination. Within the AA itself, mean pressure also increases, this time towards the south, as the intermediary low-pressure corridors progressively disappear between MPHs (cf. Figure 25, Chapter 8). Figure 100 shows this evolution between Gibraltar, where lows to the fore of MPHs still arrive regularly (and mainly in winter), and where pressures have stabilised or fallen, and the Canaries, further to the south, which more shallow low-pressure corridors find it harder to reach; here, pressures have risen by a value of the order of about 1 hPa over the last 50 years or so. This increase in pressure is even greater further to the south, in latitudes which low-pressure corridors do not reach; within this same period the rise has been of the order of 1.5 hPa at Nouadhibou (Mauritania), and in the Cape Verde Islands; at Dakar, the value is 2 hPa (Figure 101). This rise in pressure inhibits precipitation to the north of the AA, and so, on the Atlantic coast of Morocco, there is less rain, and chronic drought is a growing problem. Amraoui (2000) has detailed this diminishing trend in rainfall since the 1970s–1980s: rainfall in Morocco is showing the same trend as in the tropical regions of the Sahel, as the AA is now more extensive both to the north and to the south. Occasionally, however, powerful MPHs drive on far to the south within the AA, opening up a deep low-pressure corridor there, which is the seat of intense and violent rainstorms, leading to sometimes dramatic flooding. This is what happened in Morocco during the winter of 2002–2003, and most memorably in November 2002, when 91 mm of rain fell at Kénitra on the 15th and 16th, and

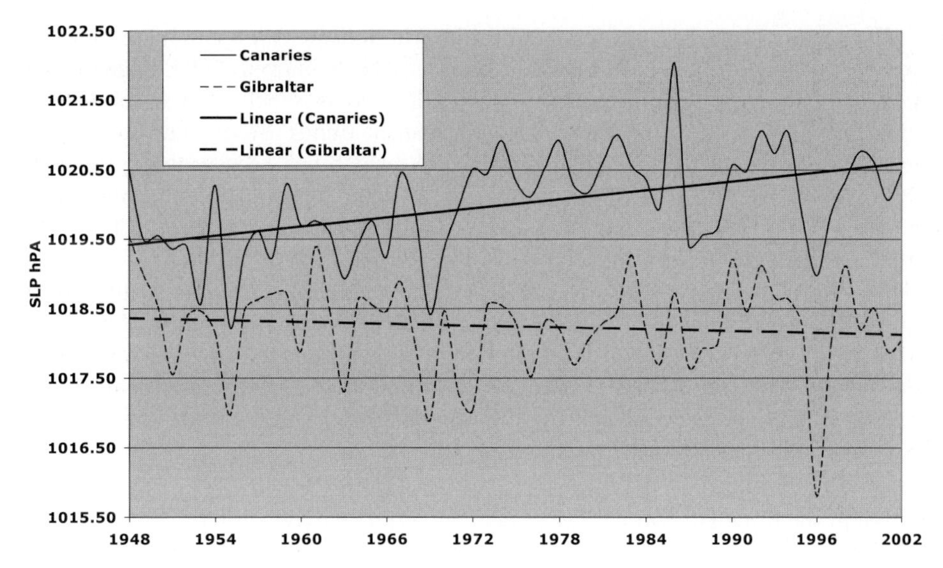

Figure 100. Evolution of surface pressure (hPa) with linear trends, from 1948–2002, at 20°W/30°N, near the Canary Islands (from CDC/NCEP–NCAR data), and at Gibraltar, 36°09′N, 05°21′W (data from www.cru.uea.ac.uk).

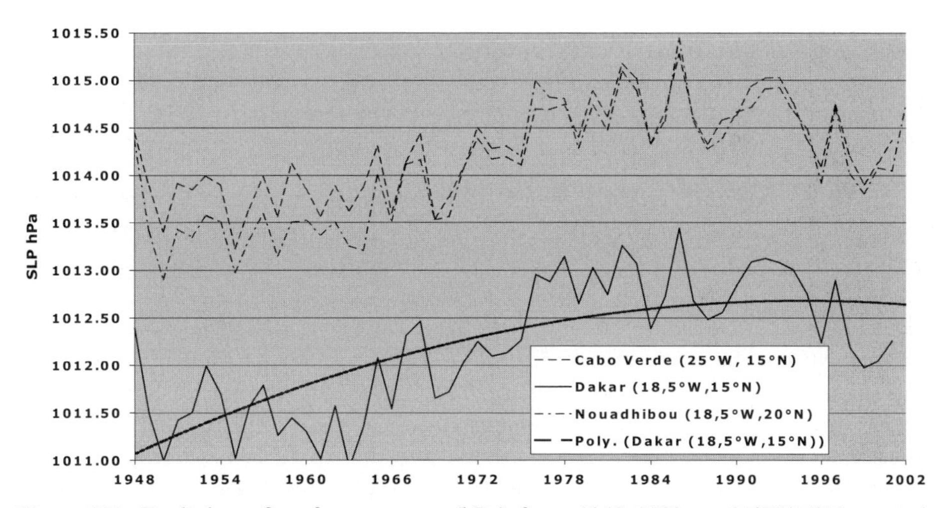

Figure 101. Evolution of surface pressure (hPa) from 1948–2002, at 25°W/15°N, near the Cape Verde Islands, at 18.5°W/20°N, near Nouadhibou (Mauritania), and at 18.5°W/15°N, near Dakar (Senegal), with polynomial trend.
From LCRE (CDC/NCEP-NCAR data).

84 mm at Casablanca. Soon afterwards, on 24 and 25 November, 112 mm were recorded at Rabat and Larache (Amraoui, 2004). Similar episodes of intense rainfall were also seen in the Canary Islands, where torrential rain caused unusually severe floods.

The rise in pressure is accompanied to the south of the AA by an increase in the speed of the maritime trade (Ndong and Dione, 1994; Nouaceur, 1999; Sagna, 2001, 2004). The frequency of wind-blown sand and dust events is increasing in Mauritania (Nouaceur, 1999); and in northern Senegal, ancient dunes are on the move again. The extended area of the so-called 'Azores' AA, which has been growing since the 1970s (Figures 100 and 101), shifts pluviogenic structures proportionally southwards (cf. Chapter 11), and the rainfall of the 'Atlantic Sahel', and especially the Cape Verde archipelago, is considerably affected (Sagna, 2004; Figures 81 and 82, Chapter 8). This same rise in pressure has occurred throughout the Sahel (Figure 84), as far east as Ethiopia, transmitted by MPHs crossing the Mediterranean basin, from the west or from the north, and subsequently by the continental trade issuing from the MPHs (Figure 70, Chapter 11). The progress made by the northerly fluxes is made easier by the simultaneous weakness of the Atlantic monsoon coming up from the south (cf. Figure 83).

Rainfall in the Brazilian *Nordeste* is associated with the arrival of the structure of the IME (Figure 85), which is active here because of the superposition of the moist southern maritime trade upon the equally moist Amazon monsoon from the north. A general southward shift, driven from the north, can only be advantageous (e.g., at Fortaleza, the rainy season was generally 'wetter between 1963 and 1979 than between 1948 and 1963' (Servain and Seva, 1987)). The situation of the Sahel, north of the equator, is therefore the reverse of what is seen in the *Nordeste*, south of the equator, where heavy rain can fall, while the Sahel on the other hand lacks rain (and vice versa). An identical relationship may also be seen in the case of so-called El Niño episodes, which are also connected with a southward shift of pluviogenic structures (cf. Chapter 13).

12.2.5 The north-east Atlantic

This sector, from the coast of Canada and Greenland across to western Europe, the Norwegian Sea and beyond, lies to the north of the favoured trajectory of the great majority of MPHs. It is a region of disturbances, visited by the northern (and the most active) parts of low-pressure corridors, and the closed lows (cyclones) at the extremities of these corridors. It is also, statistically, the home of the so-called 'Icelandic' Low. Nearly always dominated by low pressure, this part of the ocean is not protected from the visitations of MPHs coming down from Greenland or directly from the Arctic via the Scandinavian trajectory (Figure 87). This region of the North Atlantic has undergone a unique climatic evolution since the 1970s (Figure 88), which has been well documented: the curves for climatic parameters presented by the *Annual Bulletin on the Climate*, WMO Region VI (Europe and Middle West), in the rubric 'Climate in the region in the 20th century', have been established by various weather services over long periods, often more than a century.

Thermal evolution shows contrasts, but warming dominates

- In the vicinity of America and Greenland, cooling dominates, as observations at Godthaab (Figure 94) and in Iceland (Figure 95) confirm. This lowering of the temperature is also recorded for the ocean, mainly from south of Greenland down to the south of Iceland (Kaplan *et al.*, 1997). The low temperatures may be attributed to the passage of the cold air of MPHs, to descents of cold air from Greenland, and to density currents in the ocean.
- In western Europe, northern Europe, and the Norwegian Sea, however, temperatures are gradually on the increase. Figure 102, which shows this rise for Jan Mayen Island, halfway between Greenland and Scandinavia, is illustrative of this more general increase, which has been going on since the 1970s. In 1994, Reynaud drew attention to this continuous rise, more marked in winter, for the area between Iceland and Svalbard, and over the whole of Scandinavia. On the mean annual scale, this warming was as much as 1.5–2°C between 1962 and 1992, and in the winter it could be of the order of 2.5–3°C. The seasonal nature of this rise confirms the dynamical origin of this warming, which is advected by depressionary fluxes to the fore of MPHs during the season when they are at their most powerful. The southerly cyclonic fluxes drive on far into the north-east, beyond the Svalbard group of islands and the Barents Sea, and their effects are present even in temperature data from Franz Josef Land (80°N/58°E, Russia) in the middle of the Arctic Ocean, just 10° from the North Pole (cf. Daly website, whose rubric *What the stations say* offers a great number of temperature curves from this region).

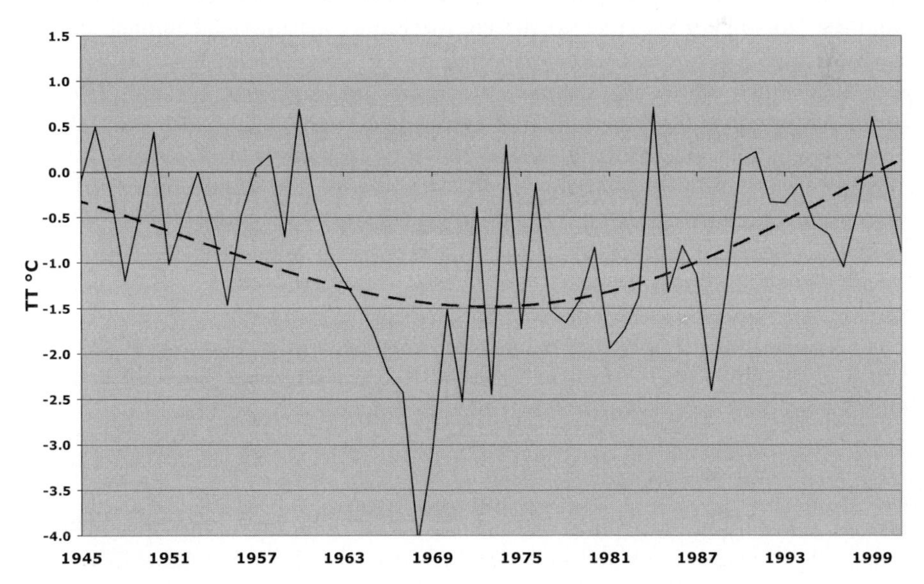

Figure 102. Evolution of mean annual temperature (°C) from 1950–2002, and polynomial trend at Jan Mayen Island, 70°56′N/8°40′W.
Data from the Norwegian Meteorological Institute, DNMI..

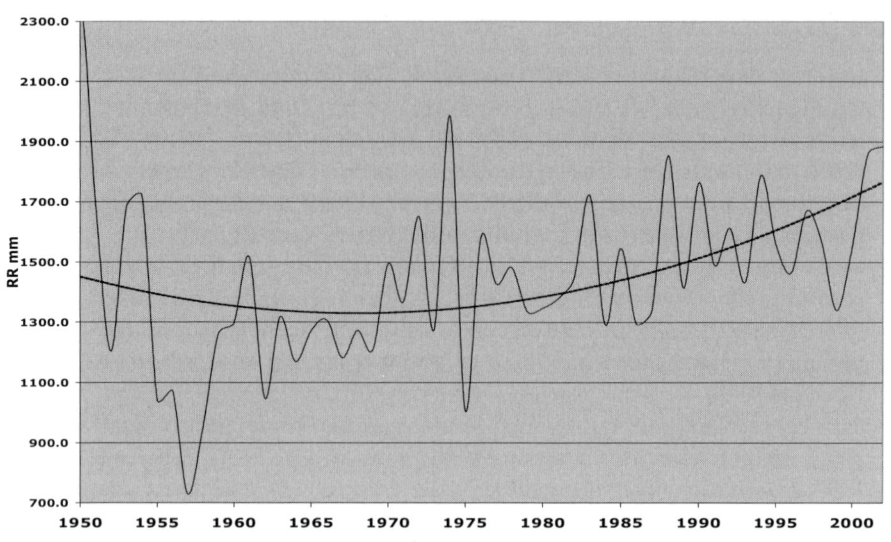

Figure 103. Evolution of annual rainfall (mm) from 1950–2002 at Dalatangi (Iceland), and polynomial trend.
From Icelandic Meteorological Office data.

Precipitation is increasing

A continuous increase in precipitation has been observed for the whole of the region around the Norwegian Sea, the North Sea, the Baltic, the English Channel, and the ocean near the Atlantic coastline, especially in France (Figure 66, Chapter 11). Figure 103, which shows the increase in rainfall at Dalatangi, Iceland, after the 1970s, is illustrative of this general pluviometric trend. This extra rainfall has caused Greenland's glaciers to gain considerably in mass, and the same thing is happening in Iceland and Scandinavia (WMO, 1998). This gain, confirmed by a survey of seven Norwegian and Swedish glaciers for the period 1967–1997, by Reynaud (2003), is very rarely (or perhaps never) mentioned by the media, although the thinning of the nearby ice sheets is discussed, always in exaggerated terms.

This increase in precipitation is associated with the intensification of southerly fluxes, and thereby with the greater availability of precipitable potential and with deeper depressions moving further northwards right into the Arctic.

Surface pressure is decreasing

The more pronounced deepening of lows associated with more vigorous MPHs is bringing about a continuous fall in pressure. Moreover, the most marked decreases are recorded in winter (Reynaud, 1994): during the period 1962–1992, on the mean annual scale, falls of as much as 5 hPa are recorded, and in winter this can even surpass 9 hPa, a considerable value. Figure 104, from observations on Jan Mayen

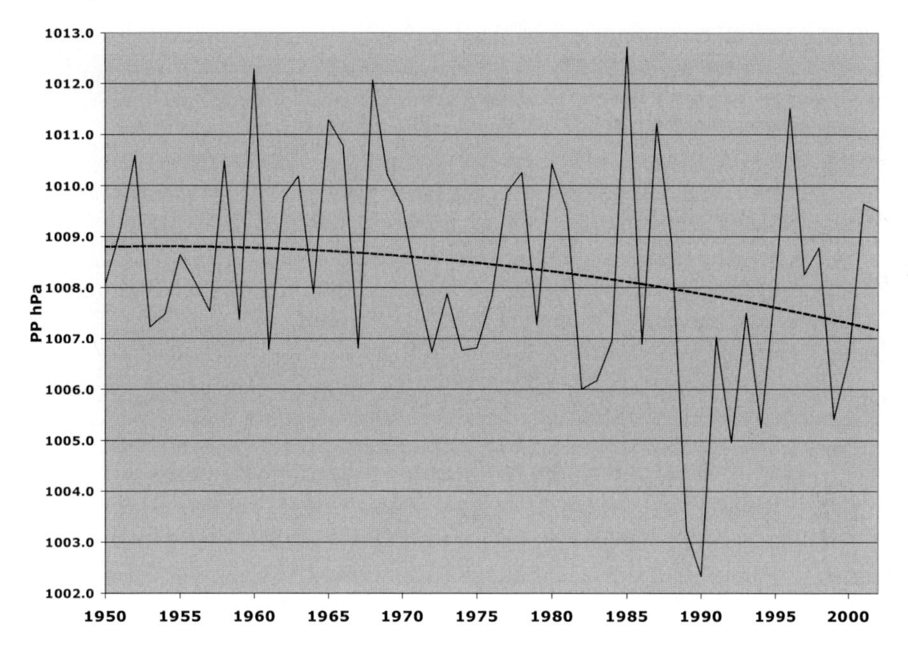

Figure 104. Evolution of mean annual sea level pressure from 1950–2002, and polynomial trend at Jan Mayen Island, 70°56′N/8°40′W.
Data from the Norwegian Meteorological Institute, DNMI.

Island, is illustrative of this progressive fall in surface pressure for the Norwegian Sea and its environs.

Statistically, the 'Icelandic' Low (an average of synoptic lows) is therefore deepening and increasing in extent. This spatial spread of lower pressures is progressive: indeed, in France, where pressures are generally rising (cf. Figures 69 and 76, Chapter 11), the northern part of the country is belatedly being affected by this spread. To take as examples Lille and Caen: average pressures in these cities have shown a slow decrease since the 1990s. In Brest, pressures have fallen more markedly, and in all three places figures for rainfall have shown an increase: sometimes through dramatic events such as those which have sometimes brought flooding to the Somme, Normandy, and Brittany. Further to the south, on the MPH trajectory, pressures have continued to rise (e.g., at Nantes and later at Bordeaux (cf. Météo-France data)), and the same is true even further south, in Spain, at the foot of the Cantabrian mountains, at La Coruña and Bilbao (WMO, 2000), and in southern France (Figure 69).

Stormy conditions are on the increase

The increasing strength of MPHs causes strong winds to be more frequent and storms to be more intense (Visbeck *et al.*, 2003). The greater thermal contrast between polar air and air from the south encourages more vigorous updrafts

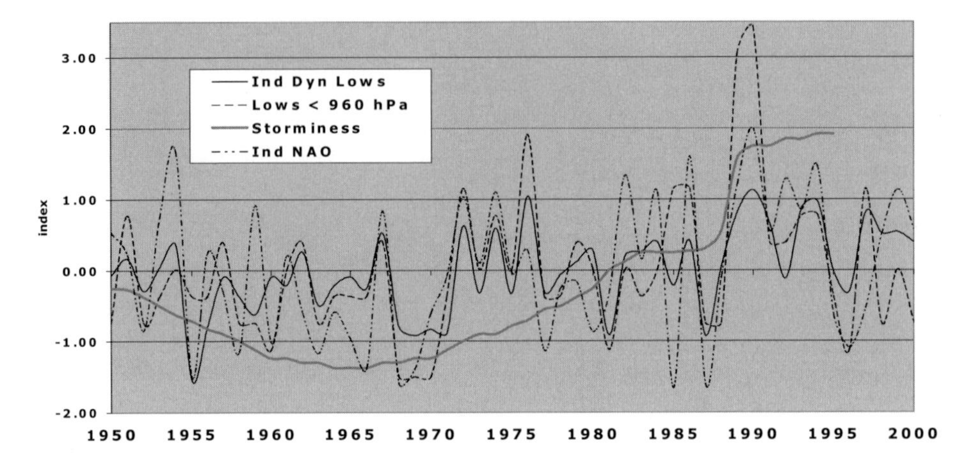

Figure 105. Evolution of the index of dynamics of lows, and of the index of dynamics of lows with pressure below 960 hPa, from 1950–2000 (Pommier, 2004a); of the index of storminess in the north-east Atlantic from 1950–1995 (WASA Project, 1998); and of the NAO index from 1950–2000 (Hurrell *et al.*, 2003).

(convection) enhanced by the greater potential energy of the cyclonic flux originating at latitudes further to the south. This increased convection has the effect of further deepening low-pressure corridors and closed lows. In this fashion, 'cyclones' with pressures below 950 hPa, associated with very severe winter storms, showed a remarkable increase in numbers between 1956 and 2000: in the 1990s, their numbers almost tripled (WMO, 2000). These storms have often spilled over into countries bordering the north-east Atlantic.

The results of the European WASA Project (1998), set up to observe pressures and wind strengths, reveal unequivocally that '*the main conclusion is that the climatology of storms and waves for most of the north-east Atlantic and the North Sea has really become more severe during recent decades, but the present intensity seems comparable to that of the beginning of the century*'. At the beginning of the 20th century, weather had certainly been violent, with the climatic optimum of the period 1930–1960 having seen a return to more clement weather. Figure 105 shows the agreement in evolutions for the last 50 years of the indices for lows in the North Atlantic space, the NAO index, and the index for storminess in the north-east Atlantic. These last values, for storminess in the 1990s, are in fact the highest for the whole century. At the same time, as a result of increasing wind strengths, the height of waves has progressively increased (Kushnir *et al.*, 1997; Gulev and Hasse, 1999). Storms on the Atlantic coasts of both France and Britain have become increasingly frequent and intense (Lemasson and Regnaud, 1997). In Brittany, there has been '*a greater frequency of strong winds and storms since the 1970s*' (Audran, 1998), with the warm and moist *suroît* winds from the south-west having intensified on the leading edges of MPHs. This has led to more frequent pluviogenic situations, and more rain has caused repeated flooding, with higher temperatures and more violent storms.

The intensification of air circulation, affecting ocean circulation with the arrival of MPHs over the ocean from the American continent, and responsible for the greater vigour of the Gulf Stream system and the North Atlantic Drift, has caused temperatures along the coasts of western Europe and Scandinavia to be higher (Kaplan *et al.*, 1997). The Norwegian current brings this warming to the Arctic Ocean and the Barents Sea. The ice sheets, warmed from above by the air and from below by the water, are thinning and retreating, a localised diminution attributable to aerological dynamics.

12.3 CONCLUSION: THE GREENHOUSE EFFECT DOES NOT CONTROL THE CLIMATIC EVOLUTION OF THE NORTH ATLANTIC AEROLOGICAL UNIT

The dynamic of weather and climate in the North Atlantic aerological unit obeys definite rules. MPHs from the Arctic organise meridional transportation of air and energy within this space, thereby marshalling the weather and its evolution, directly in high and middle latitudes, and in a more or less indirect fashion in tropical latitudes. Climatic consequences are very diverse: pressures may increase or decrease, some regions may become cooler, while others become warmer, and some may have torrential rain while others suffer from a lack of it. But, above and beyond these differences, **everything is dynamically linked**. Several lessons may be drawn from the behaviour of this aerological unit (and indeed, from the behaviours of other units).

12.3.1 Rigorously organised phenomena

The most important point here is that **meteorological phenomena are strictly organised**. This cannot be stated too often, since it contradicts the supposition that weather has a 'chaotic' character, and 'cannot' therefore be understood or predicted. The **general co-variation of phenomena** means that we cannot undiscerningly pick out one isolated element and/or area in this space without taking into account the *ensemble* of elements and/or areas. Neither can we 'explain' some characteristic, or a particular change, without first locating it in its correct place in the chain of mechanisms, and taking into account the polar origins of the evolution.

For example, we cannot study the evolution of the Arctic without taking note of the presence of intense fluxes, bringing warm, moist air from the south and returning northward, especially between Greenland and Scandinavia. These cause localised warming. Also, we must take into account the deep lows (with more numerous cyclones and localised falling pressures) which have spread in this area to the very centre of the Arctic (Maslanick *et al.*, 1996), while in the western Arctic it is, conversely, the number of anticyclones which is increasing (Walsh *et al.*, 1996).

Neither can we attempt an analysis of the evolution of rainfall in the Sahel without considering its place within the framework of general circulation, especially within the lower layers. Similarly, it is impossible to 'turn off' one isolated sea current, without immediately turning off all the others, and also all winds, and in the bargain doing away with all the MPHs! Moreover, before seeking out hypothetical and really remote causes in the Pacific Ocean, or even the Indian Ocean, it makes better sense to consider the actual physical relationships inside the aerological and oceanic unit, respecting the logic of phenomena within this absolutely coherent whole. This coherence means that we cannot claim that the weather is 'out of order' without having first specified how things work, and having defined the 'order' (which is surprisingly coherent!): for **it is not the nature of phenomena which changes, but only their intensity**!

12.3.2 The absurdity of a 'mean climate' within an aerological unit

Climatic contrasts have widened in the course of the last few decades across the width of the Atlantic Ocean, and even between the regions within this space. Consequently, there can be no *representative* average value, for any parameter, for this unit. In particular, a 'mean temperature' is, climatically, completely meaningless in this context. It would in fact be absurd to suggest that the Atlantic space is getting warmer, or colder, by relying on some arithmetical reckoning covering the whole area to establish a supremacy of warmth or cold, since there is no generalised thermal behaviour. We can work within a context displaying a relatively homogeneous evolution, using direct observations, to get a clearer idea of matters. For example, we can estimate that the western Arctic is cooling, as are the regions on the western side of the Atlantic (especially the central part of North America), by *n* degrees or tenths of a degree, and that the north-eastern sector is warming by *n* degrees or tenths of a degree ... but **a mean thermal (or pluviometric) picture, established for the whole of the unit, represents absolutely nothing in climatic terms**.

We can see, then, in this aerological space, a good example of what the IPCC is doing on a global scale, averaging out temperature values for regions with different climatic characters and evolutions. So a **mean global temperature** (the average of six aerological units) **makes no more sense than some hypothetical mean temperature for the North Atlantic** (which is only one of those units).

So a claim, like that made by the European Environment Agency (EEA, 2004), that '*Europe is warming faster than the global average*' has no meaning in climatic terms, because:

- It presupposes that the whole planet is warming up, which has yet to be proved: since, as already stated, **the mean official thermal curve proves nothing** (Figure 46, Chapter 10), no more than some artificial 'Atlantic average'. An even more artificial 'global average' cannot be a reliable reference.
- It also presupposes that the evolution of temperature is '*caused by human activities*', and that '*the temperature in Europe ... is projected to climb by a further 2–6.3°C this century as emissions of greenhouse gases continue building up*' (EEA, 2004). Again, just as hypothetical; let us say it again – *such a link has never been demonstrated.*
- The EEA's 'predictions' are in fact only *extrapolations*. One part of Europe is

getting warmer, a fact confirmed by observations. But the observational data deal with the past and the present: there is (obviously!) nothing to suggest that they represent the future. **Real prediction involves knowing not only what has happened, but how it happened.**

- Indeed, the EEA **does not know the real reason why some of Europe is experiencing warming, since it takes no account (like the models) of the dynamics of the aerological space of which Europe forms a part.** It does not appreciate that the warming of the north-eastern sector of the North Atlantic space is mainly due to the intensification of returning airflows from the south, on the leading edges of more vigorous MPHs, and to other reasons discussed above (Chapter 10), for example the urban effect, or the increase of pressure in the lower layers.

What is more, this localised warming, itself connected with localised cooling elsewhere, is certainly not a result of the greenhouse effect: **the future climate of Europe will not be shaped by changing rates of greenhouse gases, but by the evolution of the dynamics of MPHs.**

12.3.3 The 1970s were a 'climatic turning point'

The evidence of the NAD index (Figure 88) and the NAO index (Figure 105) is that '*since 1974 the positive mode has been preponderant*' (Wanner, 1999). **The 1970s represent a true climate turning point**, subsequent to which circulation within the Atlantic space entered a 'rapid' scenario. At the same time, intensified MPHs (cf. NADi, Figure 88) drove more and more deeply into the tropical zone (Figures 89 and 90). Their increasingly meridional trajectories now allowed them to seek out and export towards higher latitudes a greater quantity of energy from the tropics, warming certain areas in the space in question, and especially favouring the development of disturbances.

This kind of evolution, **associated with cooling in the western Arctic** and more vigorous MPHs, is quite the opposite of the scenario predicted by the IPCC for high latitudes, and indeed runs absolutely counter to the 'global warming' scenario. Observation therefore gives the lie to the predictions of the models. *Recent climatic evolution owes nothing whatsoever to the greenhouse effect.*

The weather is becoming increasingly violent and irregular. Wind speeds are greater, violent winds are on the increase, and snowstorms and heavy rain are more frequent; there are more tornadoes in North America, and 'mini-tornadoes' are now appearing in areas where they were previously rare or unknown (most notably in France). Thermal contrasts are greater, with cold snaps and heatwaves succeeding each other without transition; updrafts are more intense, and storms more severe, with abundant rain and resulting dramatic flooding, though droughts also last longer. Many meteorological agencies are deploring this deterioration in the weather, sometimes catastrophic in its consequences.

This evolution towards more violent weather since the 1970s is an observed fact.

- **It forms part of a cold scenario of general circulation** (*positive indices of the NAD and the NAO*), *and therefore has nothing to do with the greenhouse effect.*
- *A warm scenario* (*negative indices of the NAD and the NAO*), *such as that predicted by the models and the IPCC, could result only in clement weather* (cf. Chapter 11) (*i.e., the opposite of what is now happening*).

12.3.4 The 'hijacking' of 'bad weather' by the IPCC...

Why do the supporters of the IPCC try to claim a monopoly for the greenhouse effect when allotting responsibility for the evolution of the weather? Perhaps because it's an easy option, or because it's fashionable, or because they lack imagination; or is it a Pavlovian reflex? Mere ignorance of the real causes? Or a lack of scientific rigour dictated by self-interest?

The dramatic events associated with the evolution of the weather are shamelessly exploited. It was the EEA (2004), for example, which stated that '*the summer floods of 2002 and last year's summer heatwave are recent examples of how destructive extreme weather can be*'. Of course, no demonstration was offered as to how such dissimilar events as floods and heatwaves (or droughts) can be ascribed to the greenhouse effect, but if one is playing upon people's fears, then everything is grist to the mill!

In a kind of frenzied, copycat attempt to 'hijack' the drama of recent violent events, a link will always be found with changing amounts of greenhouse gases: in 2003, Hurrell *et al.* considered this to be '*one of the most urgent challenges*'. So, the Laboratoire de Météorologie Dynamique (LMD) can suppose that '*the well-known "NAO" could, under the influence of an enhanced greenhouse effect, ensure an ever more positive index*' (Le Treut, 2000), '*creating favourable conditions for new generations of storms ... this is one of our strongest research hypotheses*' (sic). A 'strong hypothesis'! even though the NAO is not a result of the greenhouse effect, and a warm scenario, as is widely known, corresponds unambiguously with a negative index of the NAO! This statement, issued in the wake of the storms known as Lothar and Martin, does not offer much hope of our being able to forecast Atlantic storms in the days to come!

This attribution of phenomena to the greenhouse effect is increasingly becoming an automatic, 'knee-jerk' reaction, a kind of conditioned reflex. Take, for example, the remark from the Hadley Centre in December 2003, that '*the average number of storms (per station) shows a significant increase in the UK winter period*'. Interviewed by the BBC on 9 December, Geoff Jenkins of the Hadley Centre '*insisted that it could not be explained other than via man-made climate change*' (Beran, 2004). So there could be no other explanation! A claim like this emanating from such a prominent centre for meteorological research seems unthinkable!

Similarly, the dramatic flash flood which tore through the village of Boscastle in Cornwall on 17 August 2004 was immediately blamed on the greenhouse effect by the press, and by others. In one of many such reports, the *NZ Herald*, beneath the headline '*More summer storms point to global warming*', did not hesitate to rush in with: '*Increasingly severe summer rainstorms like the one which led to the flood which devastated Boscastle, Cornwall on Monday are beginning to suggest the influence of global warming, a leading flood scientist said yesterday*' (18 August 2004, McCarthy). Quick and convenient news, without much fear of contradiction, since a 'scientist' has given it the 'green light'! And nobody has had to work too hard, since that so-called scientist has not troubled himself to try and analyse the meteorological situation! So we end up with utter falsehood.

Figure 106 shows how the torrential rain which fell on Cornwall was caused by a rapid and vigorous incoming MPH on the Scandinavian trajectory, which checked and then repelled an 'American' MPH. Violent updrafts were produced to the fore of this MPH, over the ocean on 17 August, along a north-west–south-east axis (the confluence between the two MPHs). Its leading edge was located over south-west England at the time (Figure 106). The depression associated with this same MPH caused torrential rain over Scotland on the 18th. On 18 and 19 August, this powerful MPH was stretched out across the ocean, and its leading edge, to the north of the Pyrenees and the Cantabrian range, was pushing eastwards over Europe, sowing storms, heavy downpours, hail, gusty winds, and even tornadoes, with much activity along the border between France and Belgium (Figure 106 – 18 August).

What relationship exists between this situation and the greenhouse effect? None! The reverse might even be the case: for such a powerful MPH to arrive on the Scandinavian trajectory, and in August, is very far removed from a 'warming' scenario! The 'scientist' in question, like so many of his peers in similar circumstances, has come up with the expected, modish answer. He may have confirmed his reputation with this interview, but his knowledge of the subject is lacking.

The dramatic parade of tropical storms across the Caribbean and Florida in the summer of 2004 (Figure 98) has left the same (unfounded) 'explanations' in its wake. These echoed throughout the press. For example, in *Le Monde*, September 19, 2004, we read: '*... climatic warming. Nothing is absolutely certain, but some of the experts are wondering ...*'. Similarly, Jean Jouzel, a glaciological chemist, confided during a television interview that, although there was no certainty, this '*conforms to what the IPCC predicts*'. And Stephen Byers MP, Co-chair of the International Task Force on Climate Change, was reported to have said that '*the hurricanes of this season are actually a result of human-made global warming*' (article by Soon and Baliunas, 2004). However, according to the records of the National Hurricane Center, Hurricane Ivan ('The Terrible'), which for two and a half days was classed as a 'Category Five' on the Saffir–Simpson scale, was equalled in ferocity by Hurricane Dog, in 1950. Dog was itself outclassed by Hurricane Allen (1980), which was a Category Five for a record three days.

The modellers, the IPCC, the EEA, etc. ... in brief, all the unquestioning partisans of the greenhouse effect scenario, have commented on an evolution which we can all observe, and they are obviously right so to do. This is not a difficult exercise, since the facts speak for themselves. They automatically agree

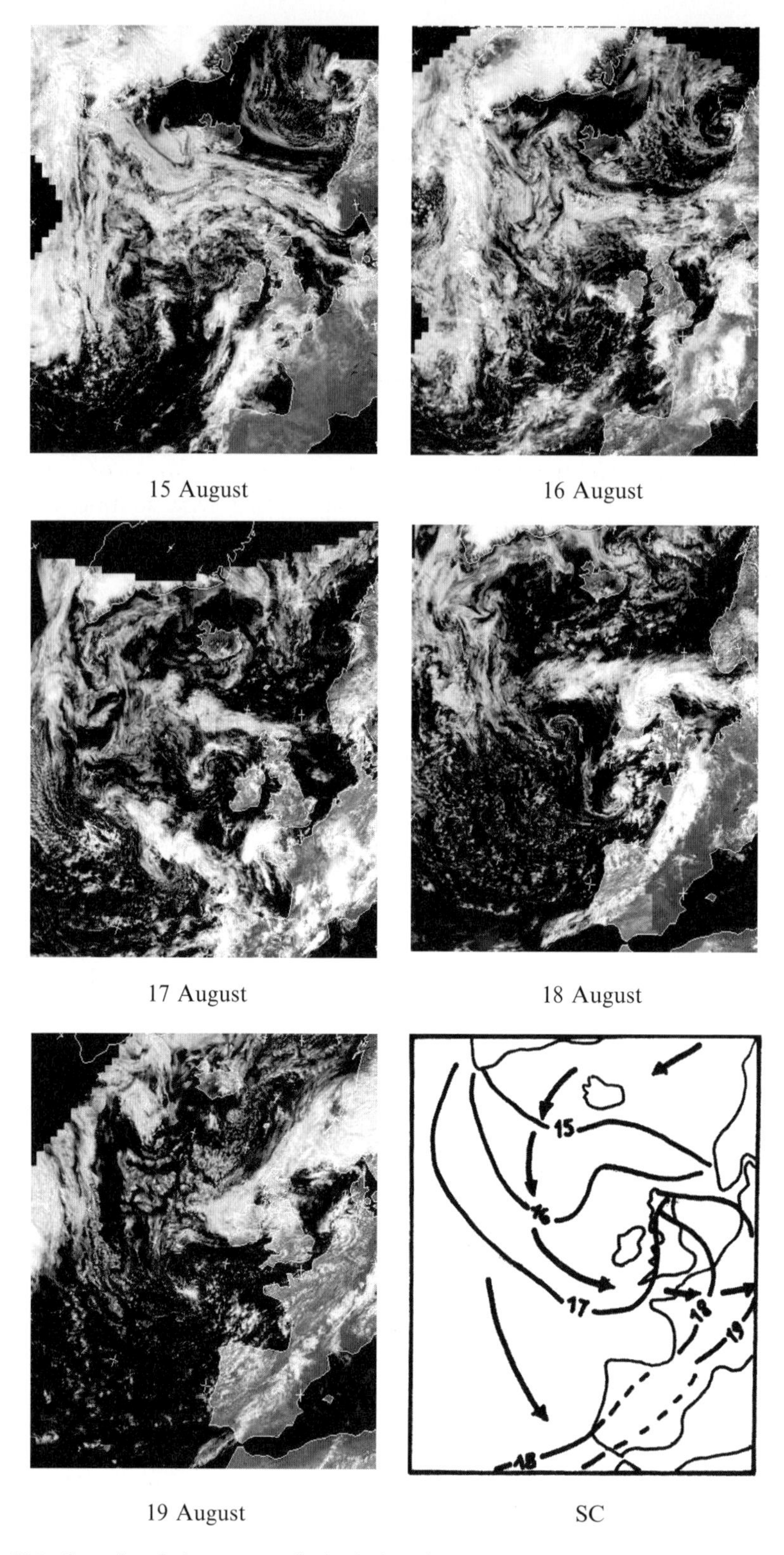

Figure 106. Genesis of the meteorological situation responsible for the dramatic floods in Cornwall and Scotland, 15–19 August 2004 (visible, NOAA, 14h).
Note: SC = summary chart.

to attribute this evolution, however, to a 'ready-made' cause, but it must be said that it is a **relationship which they (and especially the modellers) are incapable of proving**!

What are we to make, above and beyond the purely scientific aspect, of this 'hijacking of catastrophe'? If it helps us to face up to the consequences of violent events, all well and good; scientific truth could go out of the window, to be replaced by the 'non-science' of the popular press. Few seem very concerned about this aspect.

But how can we even think about predicting the evolution of meteorological phenomena if we brush aside the real physical mechanisms involved? How are we to put in place proper preventive measures, if we are not aiming at the right target, but wasting time and considerable amounts of money on completely useless projects? In 1994, *Environnement Canada*, which until that time had gone along with the 'warming' scenario, made a lucid comment on this paradox: '*It would be premature to develop national responses in anticipation of global warming, when so many regions of the North Atlantic, north-western Europe, and North America indicate, not warming, but a cooling trend*' (CGCP, Canadian Global Change Programme, 1994). Many of the policymakers of this North Atlantic space, and indeed in other areas, would benefit from dwelling upon this reflection! We shall return to this point.

13

The North Pacific aerological unit

El Niño is not in control of the climate. If climatology were less preoccupied with statistics, and more preoccupied with dynamics, this conclusion would have been reached long ago. What a pity!

Yves Lenoir. *Climat de Panique* (eds) Favre, Lausanne, 2001.

This is another unit of circulation which deserves our attention, because it has also been analysed in detail, and also because a modest local phenomenon, El Niño, has acquired an unexpected notoriety. Indeed, it has been thought by some to be controlling the entire climate; at least until 'global warming' came on the scene!

We do not intend to repeat in this present chapter all the remarks made during the analysis of the North Atlantic space, given the many similarities between the dynamical behaviours of these neighbouring aerological units (cf. Chapter 12).

13.1 THE DYNAMIC OF CIRCULATION IN THE NORTH PACIFIC SPACE

13.1.1 Description of the North Pacific aerological unit

The aerological space of the North Pacific is vast, and clearly defined (Figures 107 and 108). To the east, it is separated by the barrier of the Rockies, in the lower layers, from the North Atlantic unit, except in the south, at the latitude of the Central American Isthmus. To the west, the high mountains of eastern Siberia and China direct the paths of Arctic Mobile Polar Highs (MPHs) into the Pacific. In particular, the Verkhoyansk Mountains near the Bering Sea and the Yablonovy–Stanovoi ranges near the Sea of Okhotsk steer MPHs which have crossed the Central Siberian Plateau. MPHs coming from the direction of the Turan Depression (between the Caspian Sea and Lake Balkash), and which have come through the

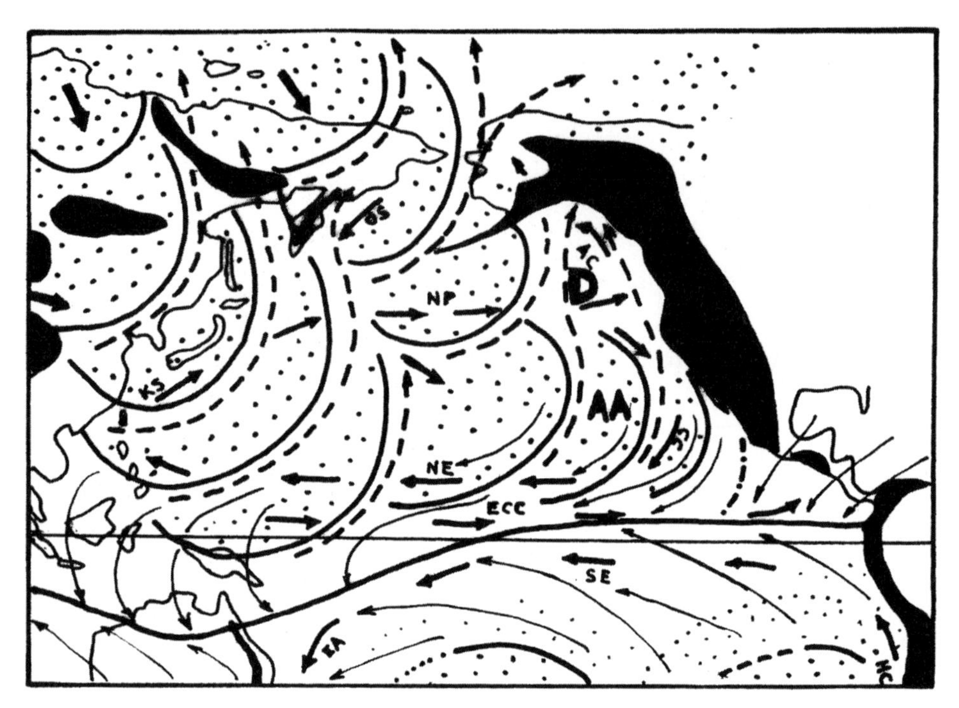

Figure 107. North Pacific aerological unit, during a high polar thermal deficit: winter and/or cold period, rapid circulation mode. For the circulation in lower levels see Figures 85 and 86.
Notes: Ocean surface currents: **OS** – Oya-Shivo Current; **KS** – Kuro-Shivo Current; **NP** – North Pacific Drift; **AC** – Alaska Current; **CC** – California Current; **NE** – North Equatorial Current; **ECC** – Equatorial Counter-Current; **SE** – South Equatorial Current; **HC** – Humboldt Current; **Cc** – Columbia Current; **EA** – East Australia Current.

Dzungarian corridor (between the Tien Shan and the Altai Mountains) arrive in China via the Gobi Desert. The convergence of these paths over eastern Asia means that MPHs will be vigorous, especially in winter (Figure 109).

Within the enormous North Pacific space, the extent of which is too great for homogeneity, two main trajectories are observed (Favre and Gershunov, 2004). The eastern trajectory is followed by MPHs coming down from the Arctic through the Bering Sea (cf. Figure 40, Chapter 9). These move towards the coast of North America, usually reaching it at the latitudes of California. The western trajectory is followed by the 'Asian' MPHs, which have crossed eastern Siberia to the south of the Verkhoyansk Mountains and through China, and which head for the centre of the aerological unit. These two trajectories meet at the centre of the North Pacific, and certain MPHs, reinforced by the encounter, can cross almost the whole space from west to east.

- The western trajectory, which is more meridional in direction in winter (Figure 107), is associated with high pressures which, statistically, form during this season the Anticyclonic Agglutination (AA) known as the 'Philippines AA'. The effect of this is to shift the western part of the aerological space markedly

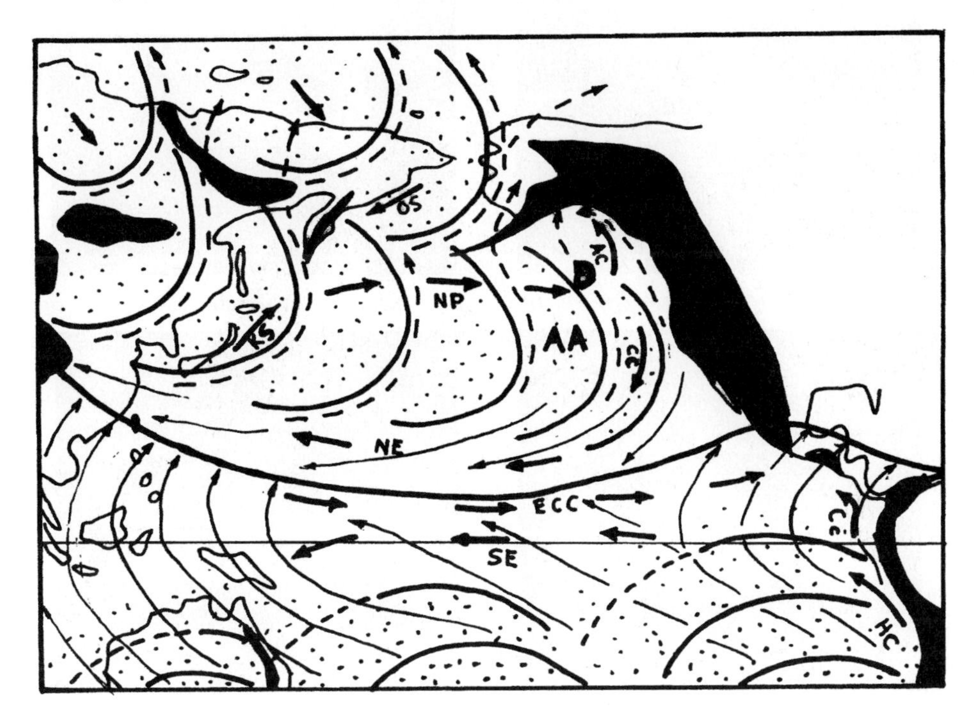

Figure 108. North Pacific aerological unit, during a weaker polar thermal deficit: summer and/or warm period, slow circulation mode. For the circulation in lower levels see Figures 85 and 86.

Notes: Ocean surface currents: **OS** – Oya-Shivo Current; **KS** – Kuro-Shivo Current; **NP** – North Pacific Drift; **AC** – Alaska Current; **CC** – California Current; **NE** – North Equatorial Current; **ECC** – Equatorial Counter-Current; **SE** – South Equatorial Current; **HC** – Humboldt Current; **Cc** – Columbia Current; **EA** – East Australia Current.

southwards, towards Indochina and Indonesia, the Indian Ocean, and Australia. This encroachment of the northern trade wind – incorrectly called the 'winter monsoon', since it is genetically a trade flux all the while it blows from the north-east and does not cross the equator – is drawn towards Australia by a thermal low. So the northern trade crosses the geographical equator, to become the Australian monsoon, blowing mainly from the north-west.

- On the eastern side of the Pacific, MPHs approaching North America cause a strong cyclonic flow from the south towards Alaska. This flow is powerfully channelled northwards, between the leading edge of the MPHs and the steep and continuous relief of the Rockies. The encounter between the cold air of the MPHs and the relief creates the (statistical) 'Hawaiian' or 'Californian' AA, which feeds into the trade circulation. Here, anticyclonic impetus is relatively weaker than in the west in winter, and the surface meteorological equator (ME) is not shifted as far southwards. It stays near the geographical equator, another factor being the force of the southern trade, well supplied by MPHs diverted northwards *en masse* by the Andes. However, the ME is displaced further (than shown on Figure 107) to the south in February and March (Barbier, 2004), in

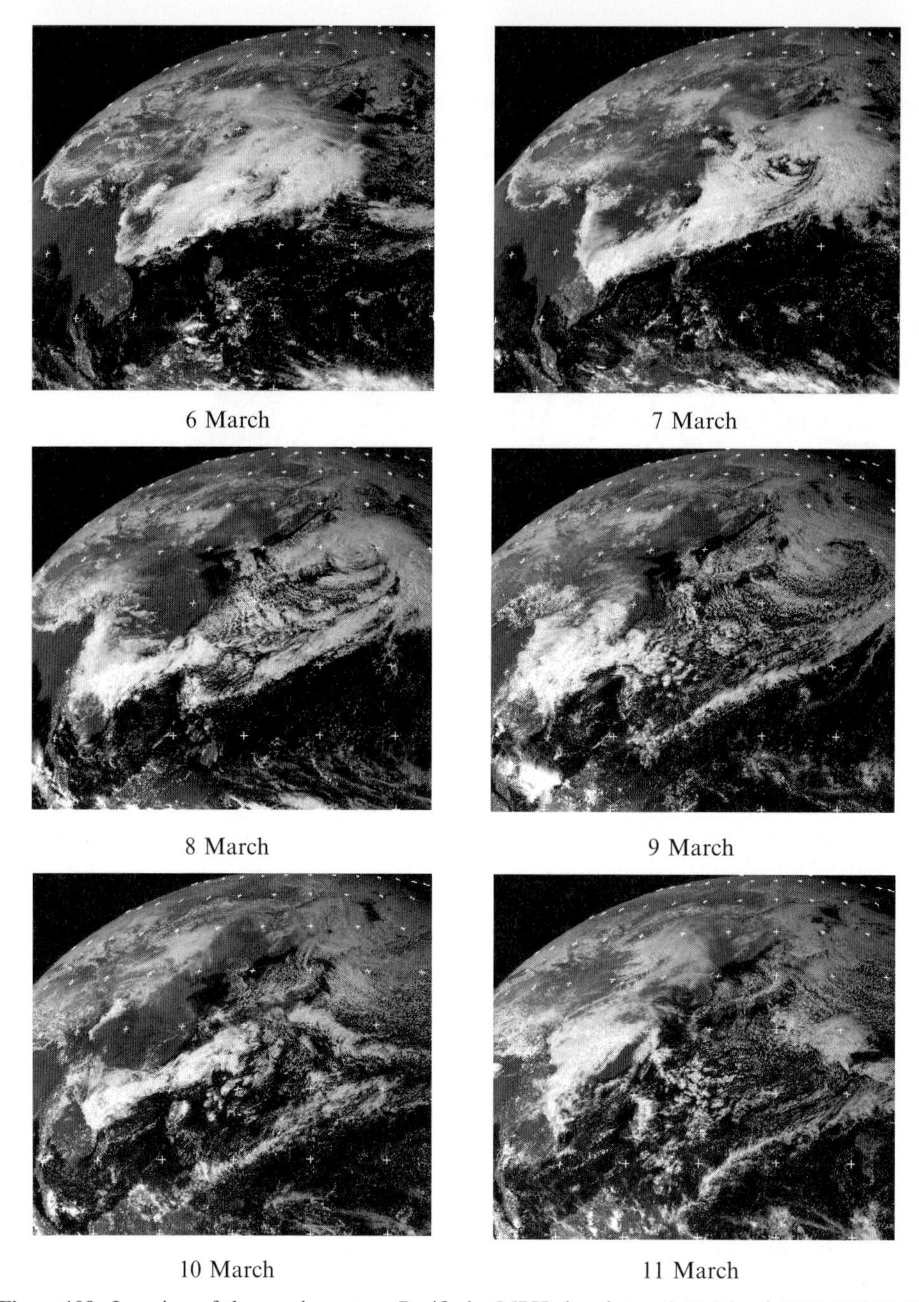

Figure 109. Invasion of the north-western Pacific by MPHs in winter. 6–11 March 2003 (*GMS 05* (JMA), visible, 06h). One MPH, visible in the top-right corner of the picture (6 March), coming from north-eastern Siberia, is running south-eastward and disappears from the picture, *en route* to the eastern side of the Pacific Ocean (8 March). At the same time, cold air, coming from the west and following the two other entrances (Siberian Plateau and Mongolian Plateau), converges over China. The line of southern high reliefs is well marked by clouds, the cold air passing round them eastward and then southward, right to the end of the modest Annamese range. The new, enormous MPH spills off the continent (6 and 7 March) and spreads out over the Pacific Ocean, provoking an intense cyclonic flow on its leading edge, from Vietnam to the Sea of Okhotsk (7 March), and invades the western side of the ocean during the following days. Axes of divergence of cold air over the ocean (8 and 9 March), fan-shaped, from cyclonic northward to anticyclonic southward, are noticeable. On 10 and 11 March, an MPH, which has already reached the Philippines on the 9th, is now covering the South China Sea and the western side of the Pacific Ocean. The air of the MPH which supplies the north-east trade, which is over Indonesia (i.e., the equator), becomes the Australian monsoon (the deviation of the airstream is clearly apparent over Borneo on the 11th). This MPH is at once followed and strengthened by new cold air advected across Asia.

the middle layers (vertical meteorological equator, VME), and as a function of the strength of the northern impetus, especially during an El Niño episode, which corresponds to a temporary, sometimes prolonged extension of the North Pacific aerological unit.

In winter (Figure 107), a particularly strong and extensive AA sits over Siberia. MPHs bearing cold air from the Arctic constantly cross the area, maintaining their low temperatures over the freezing continent. The presence of this cold anticyclone in the lower layers, the vast continental surface, and the barriers presented by relief, all hinder the return of warmer air from the south towards the Arctic. This is the reason why mean temperature data from the Arctic do not show any 'warming' here (which could only be a statistical artifice, as stated already). On the contrary, and as satellite measurements show, there has been cooling from Novaya Zemlya to the Bering Sea in the course of recent decades (Comiso and Parkinson, 2004). These observations confirm the cooling to which Litynski (2000) drew attention, based upon normals from 1931–1960 and 1961–1990. During this season MPH trajectories are shifted further south, as are AAs, and in summer they migrate northwards. The two AAs, the 'Philippines' and the 'Californian', artificially distinguished by statistical pressure values, are merged in reality, and the so-called 'subtropical' Pacific anticyclonic cell consequently extends further west during the winter, at the same time as it also stretches further southwards.

In summer (Figure 108), the aerological space is considerably reduced in size by the northward shift of the ME. This is a notable phenomenon in the east, due to the impetus of the Panamanian–Mexican monsoon, and even more obvious in the west because of the strength of the Chinese monsoon, itself drawn by a thermal low over southern China. Here, MPHs make direct contact with the northern maritime trade and the Chinese monsoon flux, which are both warm and moisture-laden. The strong thermal contrast and the abundant energy advected from the tropics are responsible for plentiful rain, although they may also bring extremely damaging torrential downpours and floods. Cyclones (or *typhoons*, in this part of the world) may start their careers in the mid-Pacific, within the structure of the VME, or even on the western Pacific side within low-pressure corridors to the fore of powerful MPHs. As MPHs move, they usually divert the typhoons north-eastwards (on their leading edges), often protecting the continent of Asia in this way (as happens also in the case of North America). In this area, typhoons can form in any season, because MPHs are so strong, although it is the summer (from May to November) which tends to bring together the optimum conditions for cyclogenesis.

Aerological circulation determines water circulation at the ocean surface (wind-driven currents). Surface waters are driven eastwards off the coasts of Asia, pushed by MPHs here exerting their maximum pressure upon the ocean. The North Pacific Drift, a prolongation of the Kuro-Shivo Current (the Pacific equivalent of the Gulf Stream), crosses the North Pacific and splits into two in the east: the northern branch goes off towards the Gulf of Alaska (as the Alaska Current), and thence into the Bering Sea, while the southern branch becomes the California Current, flowing southwards. A weak offshoot of this current veers towards the Central American

Isthmus, but most of it goes into the North Equatorial Current, flowing in the direction of the Philippines and on towards the Asian continent. Along these coasts, it flows northwards to become the Kuro-Shivo or Japan Current. The South Equatorial Current, originating in the South Pacific (a prolongation of the Humboldt or Peru Current), forms the East Australian Current. Between the two Equatorial Currents, North and South, impelled by trade winds and flowing westwards, there is a returning current, the Equatorial Counter-Current (ECC), flowing in the opposite direction. This compensating current is warm, flowing beneath the ME where updrafts cancel out the surface winds and pressure is reduced. The ECC encounters the Central American Isthmus, where it is turned both northwards and southwards. The southward flow forms the modest coastal current called 'del Niño'. A dense, cold current, the Oya-Shivo, flows beside the Kamchatka Peninsula and sinks rapidly, while to the east of the Bering Strait, a weak current from the south, a remnant of the Alaska Current, finds its way into the glacial Arctic Ocean.

In this immense aerological space there is a great variety of climates: polar, tropical, and even equatorial. In spite of this very wide diversity, co-variation is general, which means, as in the case of the North Atlantic, that **no sector considered in isolation can have its own independent evolution: if we study one location, we are obliged to take into account all other locations within the ensemble**. However, in the Pacific space, the existence of the two main individualised MPH trajectories makes matters more complex. Given that the *primum mobile* of the circulation of this aerological space (like that of the North Atlantic) lies within the Arctic, a 'statistical' relationship between the phenomena of these two units, adjacent but distinctly separated by the Rockies, is not, in physical terms, unthinkable, if it is based upon the common departure point of the units' MPHs. Such a relationship, however, is only established to the second degree.

Even though this interdependence of phenomena remains to be explained, it had certainly already been suspected, and pre-existing Atlantic phenomena now have their counterparts in this unit (e.g., the North Pacific Oscillation (NPO), itself later joined by other 'oscillations' such as the Pacific Decadal Oscillation (PDO)).

13.1.2 The North Pacific Oscillation and the Pacific Decadal Oscillation

Weather in the North Pacific has traditionally been associated with the NPO. This oscillation is the Pacific component of the Northern Oscillation, the other being the North Atlantic Oscillation (NAO) (Walker and Bliss, 1932). The NPO is defined on the basis of the difference in surface pressure between the so-called 'Aleutian' Low and the so-called 'Hawaiian' Anticyclone; an index is determined on the basis of these two 'centres of action', defined statistically from pressure values. The NPO is evaluated using an index calculated as a function of the mean gradient between the *anticyclone* and the *depression*. The NPO is said to be in 'positive' mode when the pressure is simultaneously high within the 'anticyclone' and low within the 'depression', and in 'negative' mode when the difference in pressure is slight. These positive and negative modes exhibit co-variations, but (as is the case with the NAO) they do not explain them, the cause in both cases (the dynamic of MPHs) not being identified by the traditional theories.

Remember, it is the energy of MPHs which determines the rate at which the northward cyclonic transfer of warm air, particularly on the leading edges of MPHs, occurs, and this same factor determines the depth of the peripheral low-pressure corridor and closed low (cyclone). The energy of the MPHs themselves depends upon the polar thermal deficit. The NPO, like its Atlantic counterpart, is established statistically, and must be converted to the synoptic scale (as in Figure 61). The 'Hawaiian' Anticyclone is in reality an AA, and the 'Aleutian' Low is the mean of all the depressions associated with all the MPHs crossing the North Pacific. These depressions are firmly channelled northwards, and considerably deepened by updrafts provoked by the Rockies. In this sector, unlike the situation in the Norwegian Sea, the Rockies close off the northern outlet to the Arctic. Access to the Arctic Ocean is therefore more difficult, and any advected warm air can make relatively little headway into the Arctic beyond the 70°N line of latitude.

The NPO, which is essentially based on the pressure field, is complemented by the PDO (Hare, 1996), arising from the variability of surface temperatures in the North Pacific. Given that '*interdecadal fluctuations in the dominant pattern of North Pacific sea level pressure (SLP) have closely paralleled those in the leading North Pacific sea surface temperature pattern*', the idea was to consider the '*North Pacific sea surface temperature as an index for the state of the PDO*' (Mantua *et al.*, 1997). So, '*when the PDO is positive, temperatures of the central North Pacific are cool and the eastern tropical Pacific Ocean are warm. When the PDO is negative, the central North Pacific is warm and the eastern tropical Pacific is cool*' (Mantua *et al.*, 1997; Hunt and Tsonis, 2000). The region of reference here is the American west coast, and the PDO is said to be in its 'warm' (positive) phase when this eastern flank of the ocean is warm (and conversely, in its 'cool' phase, the PDO is negative). In dynamical terms, a positive phase corresponds to a situation when MPHs are more powerful, bringing more intense flows of returning warm air from the south across the reference region. Consequently, and in reality, by broadening the perspective to embrace the whole of the North Pacific, a positive phase would correspond instead to a cool scenario, as in the western and central Pacific. As is the case with the NPO, the '*causes for the PDO are not currently known. Likewise, the potential predictability for this climate oscillation are not known*' (Mantua, 2004).

The variability of atmospheric circulation is also described by the North Pacific Index (NPi) (Trenberth and Hurrell, 1994), which measures the mean pressure at sea level in the zone between 30°N and 65°N/160°E and 140°W, a sector lying mostly (but not completely) to the north of MPH trajectories. This index therefore informs us of the deepening of the 'Aleutian Low', but '*it does not take fully into account the evolution of meridional exchanges, because, within this zone, synoptically, move MPHs and their associated depressions*' (Favre and Gershunov, 2004).

13.1.3 Variations in the intensity of the aerological dynamic of the Pacific

The NPO and PDO indices vary in intensity, as does the NAO. For example (Figure 110), on the scale of a century, the period known as the Contemporary Climatic Optimum (CCO), centred around the 1960s, saw a negative phase of the NPO, bracketed by two essentially positive phases, one in the early 20th century and one in recent decades. The PDO has been in its positive phase since the end of the 1970s. It came to maximum during the 1980s, with a short relapse around 1990, and has remained in the positive phase for a decade now (Mantua, 2004). Since it is that phase which is in progress at present in the North Pacific, let us examine the essential characteristics of the positive NPO, though briefly, since it closely resembles the NAO (see above).

The positive or high phase of the NPO/PDO

The positive phase is illustrated diagrammatically in Figure 107, which, with inevitable differences in intensities, reflects the winter season and/or a *cold period* or even a glacial period, but with more striking modifications in this last case (cf. Figure 41, Chapter 9). **The positive phase corresponds to a rapid mode of general circulation** (Figure 33, Chapter 8) (i.e., when there is a greater thermal deficit over the Arctic). The equivalents on the statistical scale are given in Figure 61(a-2 and b-2).

- The Arctic is colder, and especially so to the north of the Pacific and over eastern Siberia, across which stretches the powerful continental AA known as the 'Siberian cell'.
- MPHs are more frequent, as shown in Figure 111(top) for the whole North Pacific. It should be remembered, though, that frequency is only one of the aspects of variability, which also manifests itself through the strengths (in terms of pressure), speed of movement, and/or area covered by MPHs.
- Their motion is more rapid, and their more meridional trajectories extend further into the tropical zone (Figure 111, bottom), especially in the west near the outlet of the powerful Asian MPHs (Figure 109), reinforcing simultaneously the 'Philippines' AA and the supply to the Australian monsoon.

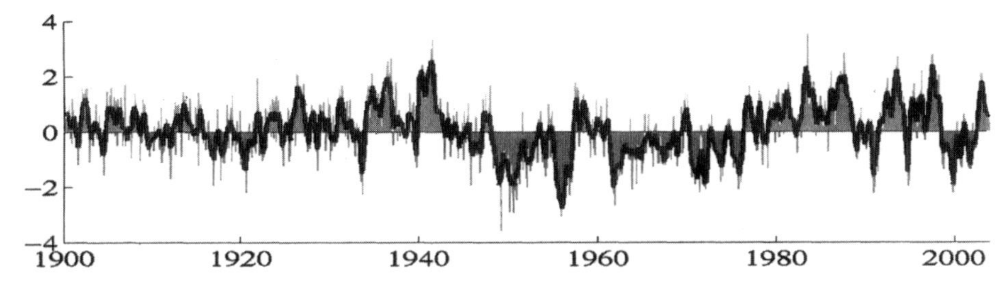

Figure 110. Secular evolution of the North Pacific (inter-)Decadal Oscillation (PDO), from 1900–2003.
From Mantua (2004).

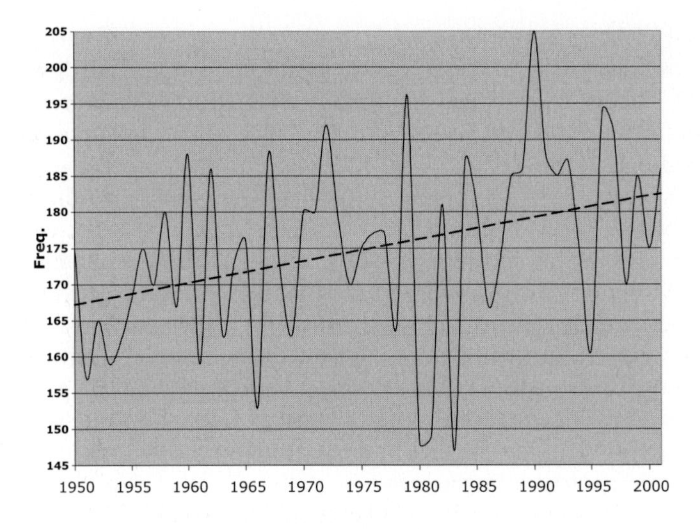

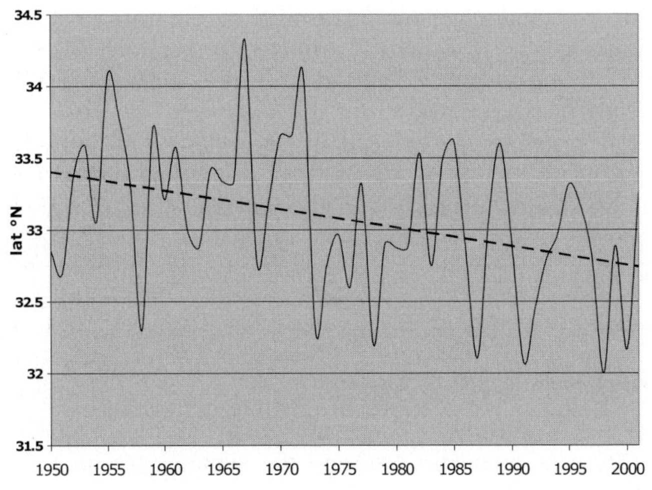

Figure 111. (*top*) Annual frequency of MPHs over the North Pacific aerological unit from 1950–2001. (*bottom*) Mean latitude reached by MPHs' centres over the North Pacific aerological unit from 1950–2001.
From Favre (2004).

- On the statistical (mean) scale, the Pacific ('Hawaii–Philippines') AA is stronger, bigger and more meridional (Figure 61(a-2)).
- The synoptic lows associated with the MPHs (low-pressure corridors and closed lows) are deeper, and they sweep over a larger area, reaching more northerly latitudes, but their progress towards the Arctic is hindered in the east by the high relief of the Rockies, and blocked to the north by the highlands of Alaska, which create strong updrafts. The statistical 'Aleutian' Low is deeper and bigger (Figure 61(a-2)).

- **Meridional exchanges are intensified** (both the warm air transfers from the south-west towards the north, and the cold air transfers from the north-west towards the south). There is a general acceleration in circulation, in the lower layers (both within and around MPHs) and at altitude. *General circulation assumes a rapid mode* (Figure 33, Chapter 9).
- This acceleration involves ocean surface waters, and wind-driven currents (e.g., the North Pacific Drift and its extensions) now flow more rapidly, while upwellings (notably along the coast of California) show a similar increase in intensity. Moreover, the cold adds vigour to the Oya-Shivo density current.
- The weather is more violent and more irregular: MPHs are more powerful and move more rapidly, instigating more dynamic convection at their leading edges, facilitated by greater temperature contrasts between the MPHs and the south-west flux, and by an enhanced supply of energy from this same southerly flux, now coming from latitudes further to the south. The weather associated with this flux is stormier, across a larger area.
- Dynamical contrasts are accentuated between the two sides of the Pacific, and as a function of the climatic regions determined by the trajectories of the MPHs. Climatic contrasts are accentuated. A *mean* temperature established for the whole of the aerological unit would therefore be completely meaningless as far as the real climate is concerned.
- The North Pacific aerological unit is at its largest, especially in the tropical zone, at the expense of the South Pacific aerological unit; the surface line of the ME shifts well southwards, especially in the western part of the unit, towards Australia and the Indian Ocean. During this phase, the El Niño situation can develop.

The negative or low phase of the NPO/PDO

The negative phase is illustrated diagrammatically in Figure 108, which, with necessary differences in intensities, reflects the summer season and/or *a warm period*, a climatic optimum, or even an interglacial period, but with more striking modifications in this last case. **The negative phase corresponds to a slow mode of general circulation** (Figure 35, Chapter 8) (i.e., when there is an attenuated thermal deficit over the Arctic. The equivalents on the statistical scale are given in Figure 61(a-1 and b-1). Such a mode prevailed at the time of the CCO (cf. Figure 110).

The characteristics of this kind of phase are roughly the opposite of what is observed during a positive phase, the dominant characteristic being the reduced frequency and strength of MPHs. We need not repeat here, for the Pacific, the details given to describe the nature of the negative phase of the NAO (see Chapter 12).

It must be stressed that, statistically, and in the medium term, especially since the 1970s, a positive NPO and a positive NAO (and their negative counterparts) are synchronous, since they are subject to the same impulsion from the direction of the

Arctic. This means that we can, for example, establish teleconnections between the Pacific and North America (even east of the Rockies), as suggested by the Pacific North American (PNA) pattern (Wallace and Gutzler, 1981). A 'link' made between, for example, sea surface temperatures in the North Pacific and meteorological phenomena occurring in the Great Plains, is thus only a **co-variation of the second degree**, where the first degree would be the direct relationship with the Arctic (and MPHs) in each of these units ... On the synoptic scale however, such synchronism (direct or indirect) is not automatic, and delays may operate in the short term between the respective dynamics of these neighbouring units.

13.2 RECENT CLIMATIC EVOLUTION ON THE EASTERN SIDE OF THE NORTH PACIFIC

The North Pacific space is too vast to be considered as a whole, and the dynamic of MPHs suggests that we study the eastern and western parts separately. Here we shall look in detail only at the eastern side of the North Pacific, the sector which is the better documented.

13.2.1 Indices of anticyclonic and depressionary activity

Favre (2002) aimed to calculate an index for the activity of MPHs and depressions, with a view to describing the nature of this eastern side of the Pacific and avoiding the shortcomings of previous definitions, with their static visions of phenomena, while everything is in fact on the move. More regrettably, these definitions had nothing to say about the origin of these phenomena. Indices of the intensity of anticyclonic activity (H Index) and depressionary activity (L Index) in winter (January, February, March), are calculated, between longitudes 180° and 110°W, as follows: H Index $= (f + pmx + latmn)/3$, and L Index $= (f - pmin - latmn)/3$, where $f =$ standardised frequency anomalies, $pmx =$ maximum pressure observed for MPHs, $latmn =$ minimum latitude reached (by H or L), and $pmin$ = minimum pressure observed in the depressions. Figure 112(top) shows that the evolution of the H index is the inverse of that of the NPO/PDO indices, given that for the whole of the period (1950–2001) anticyclonic activity has tended to increase in the western part of the North Pacific, the reverse of what is happening in the eastern part (Favre and Gershunov, 2004). However, this diminution in the east, reflected in the H index, is only relative: generally, the frequency of MPHs is increasing regularly in the North Pacific (Figure 111, top). The L index (Figure 112, bottom) varies inversely with that of anticyclonic activity (H index), showing – in more than 75% of cases – the meridian changes in fields of pressure, temperature, wind, and precipitation in this eastern part. So the (relative) reduction in anticyclonic activity, and/or the correlated increase in the depressionary index, since the mid-1970s, show the extension of the (statistical) 'Aleutian' Low, off and along the west coast of North America. This is

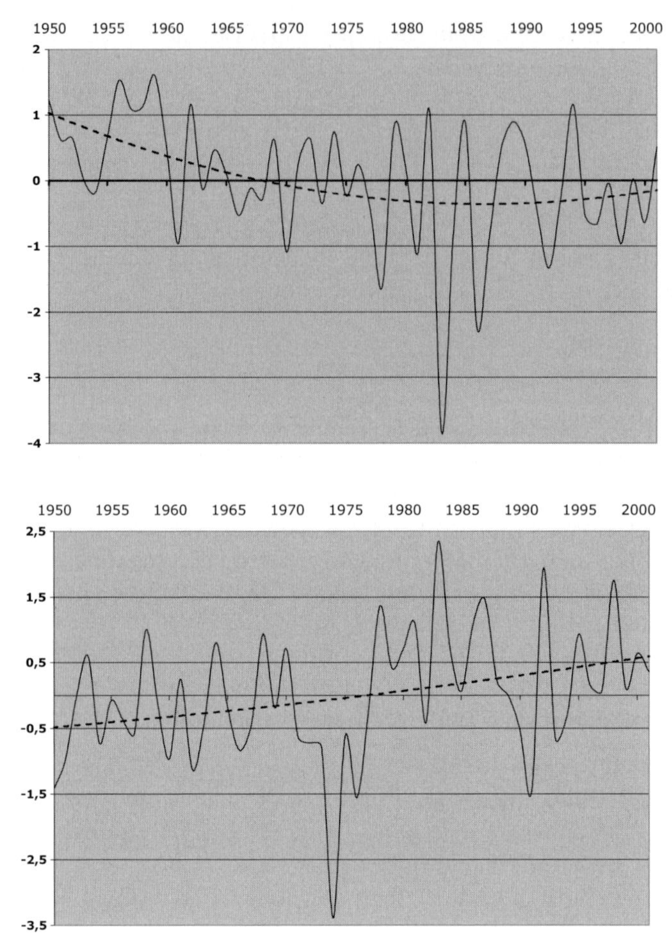

Figure 112. (*top*) H index = mean winter north-east Pacific index of MPH activity. (*bottom*) L index = mean winter north-east Pacific index of activity of lows, from 1950–2001.
From Favre (2002).

accompanied by an increase in winds from the south, a corresponding increase in mean temperatures, increased precipitation, and more storm activity. These indices accurately describe the weather patterns observed in the eastern part of the North Pacific, particularly during winter.

In this eastern part of the North Pacific, climatic division is determined by the main trajectory of MPHs. We can distinguish, from north to south:

- The region to the north of the mean trajectory of MPHs (and therefore to the north of the 'Hawaiian' AA) (i.e., the north-east Pacific which essentially harbours depressions associated with MPHs). This is roughly the domain of the so-called 'Aleutian' Low.

- The region located on the main trajectory of MPHs, and in the statistical area of the 'Hawaiian' AA, a region corresponding to the central and eastern Pacific. On the American coast, this region extends somewhat into California (both the state and the Baja California peninsula) and western Mexico.
- Further south, the situation is more complex due to seasonal changes. In winter (Figures 85 and 107), the Pacific off Central America becomes an extension of the North Atlantic space. In summer (Figures 86 and 108), the northward shift of the surface line of the ME, impelled by the Panamanian–Mexican monsoon, annexes this region to the South Pacific space.

13.2.2 The north-eastern sector of the North Pacific

MPHs from the Arctic arriving by way of the Bering Sea have to pass round the mountains of Alaska, so the Gulf of Alaska and the coasts of British Columbia are relatively protected from direct advections of cold air. Consequently, these regions are known for the relative mildness of their climate. However, this does not mean that direct invasions of cold air are ruled out: in January 1993, for example, *Environnement Canada* reported '*Arctic cold in the west*', bringing 14 successive days of snow to British Columbia, '*even in the mild city of Victoria*', with '*almost absolute temperature minima*' (in *Perspectives Climatiques*, 1993, vol. 15, No. 2).

This unexpectedly mild climate on the Pacific coast is apparent if we compare mean temperature data (cf. Normals 1961–1990, WMO), especially in the winter: at Point Barrow, at about 70°N, to the north of the Alaskan mountain ranges and on the edge of the glacial Arctic Ocean; at Fort Nelson, British Columbia, at about 60°N on the plain to the east of the Rockies; and at Yakutat, Alaska, on the Pacific coast below Mount Logan:

	December	January	February
Point Barrow	−24.0°C	−26.8°C	−27.9°C
Fort Nelson	−20.6°C	−22.4°C	−17.2°C
Yakutat	−2.2°C	−2.6°C	−1.9°C

Southern Alaska, near the Pacific, can enjoy, thanks to the local relief, average temperatures at least 20°C higher, a considerable benefit on the mean annual scale! It is important to remember this particular climatic situation, since, when Alaska is mentioned, most people (and especially tabloid readers) will imagine only bitterly cold conditions like those to the north and east of the mountains (within the Atlantic space). On the Pacific side of Alaska, it can be relatively mild. Ignorance of the real climatic conditions in the coastal areas of Alaska may lead some to bemoan the '*plight of the polar bear*', since everyone is assuming some extraordinary climatic upheaval ... which isn't happening!

Recent climatic evolution in this north-east Pacific sector is quite similar to what has been happening in the north-east Atlantic, although there is some amplification due to the presence of relief, which offers protection from cold air from the north,

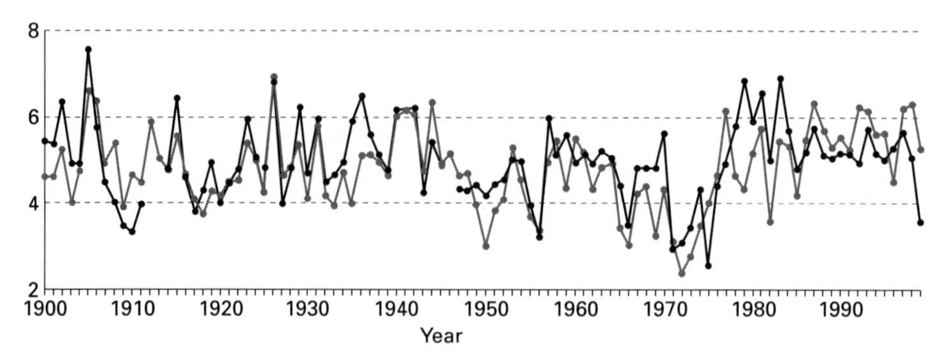

Figure 113. Mean annual temperature (°C) at Juneau, east of the Gulf of Alaska (grey line), and at Kodiak Island on the western side of the Gulf of Alaska (dark line), from 1900–1999. From Daly (2004).

and channels and lifts warm air from the south. The main features here are increased temperature and precipitation, and a decrease in pressure, with more storminess.

Increase in sea and air temperatures

Warming in Alaska was pointed out in 1990 by Trenberth, who considered that '*the largest increases in surface temperatures over the northern hemisphere in the decade prior to 1988 were in Alaska*'. This warming, dynamic in origin, has also been observed along the west coast of the USA and Canada (Morgan and Pocklington, 1995). Whitfield (2003) stated that '*temperatures have changed more in Alaska over the past 30 years than they have anywhere else on Earth*'. Figure 113 shows this recent warming for two weather stations located on either side of the Gulf of Alaska, and particularly the rapid rise which followed the trough of the 1970s. This curve also shows that this rise is not as exceptional as the literature on this subject would have us believe, since higher temperatures may be found, especially for the beginning of the 20th century.

The warming of the air is paralleled by that of sea surface waters, since the air flowing from the south also intensifies the warm Alaska Current, as does the stronger North Pacific Drift. In 1998, the sea surface temperature rose by 2°C above the average (Hunt *et al.*, 1999). The effects of this warming of both air and water spilled over from the Gulf of Alaska onto the Bering Sea, where '*the change of climate after 1976 was part of a larger North Pacific shift*' (Overland and Stabeno, 2004), with the same consequences (as in the Barents Sea) for the thickness of the ice sheets, warmed from above by the air, and from below by the water.

Increase in precipitation

Precipitation is being increased by more intense advection of warm, moist air from the south, in the depressionary corridor between the leading edge of MPHs and the Rockies, and by the lifting of air by the relief. In the States of Oregon and Washing-

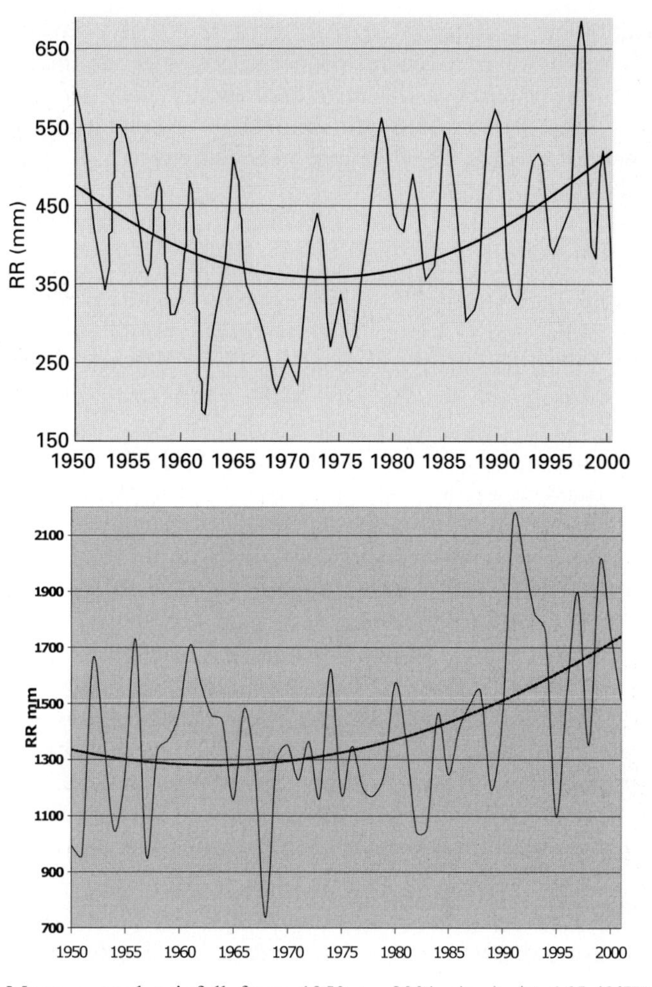

Figure 114. Mean annual rainfall from 1950 to 2001. (*top*) At 165.43°W/64.5°N (Nome, Alaska–Bering). (*bottom*) At 134.57°W/58.35°N (Juneau, Alaska).
From Favre (2004).

ton, considerable increases have been noted, with a trend of +20% for the period 1900–1994 (Karl *et al.*, 1996). A similar situation pertains in British Columbia, on the southern coast of Alaska, where the mountain barrier is responsible for higher rainfall values (Figure 114, bottom), and in the Bering Strait (Figure 114, top).

The greater availability of precipitable water is also observed at altitude, via an increase in snowfall events. For example, on Mount Logan, a study embracing the last three centuries shows that '*there is evidence for a recent acceleration in the rate of snow accumulation*', with, interestingly, '*an even more rapid increase ... over the period 1976–2000*' (Moore *et al.*, 2002). This confirms the increase observed since the 1970s (Figure 114). Additionally, Pfeffer *et al.* (2000) reported that certain glaciers were exhibiting '*a dramatic retreat*' as a result of exposure and local warming. This is not however a general phenomenon, since, as the same authors point out, '*neighboring glaciers in the same area do not show the same dramatic behavior*'. This is because increased rain and snow events tend to favour an increase in the mass of glaciers (as is the case in Scandinavia).

Decrease in surface pressure

Analysed statistically, the 'Aleutian' Low is now larger, and its pressure was lower, in the summer by 2 hPa, and in winter '*by a remarkable 9 hPa*', between 1961 and 1988 (Flohn *et al.*, 1990). This analysis is confirmed by Trenberth (1990), who finds '*a deeper and eastward-shifted Aleutian Low in the winter half-year*', and by Favre (2002) (see Figure 112, right). Figure 115(left) shows that, despite considerable interannual variations (associated with the passage of MPHs through the Bering Sea), there is a trend towards lower pressures in the Aleutian region. The boundary of this change of sign in the evolution of pressures, a result of the greater size of the AA, lies at about latitude 30°N (Figure 115, right).

During the period 1948–1998, '*a clear upward trend in cyclone frequency*' was observed (Graham and Diaz, 2001), and, as Figure 116 shows, '*a return toward higher values in latter years*', after a temporary diminution in the 1990s (cf. Figure 110). Statistically, there has been '*a pronounced downward trend of approximately 4.9 hPa over 50 years*' in the North Pacific (Graham and Diaz, 2001). This fall in pressure has been most pronounced to the north of latitude 30°N in the Gulf of Alaska, along the base of the coastal mountain ranges, in the Aleutian region, and in the Bering Sea (Figure 115).

Increase in storminess

Cyclonic activity having shown a remarkable increase, '*the frequency and intensity of extreme cyclones has increased markedly, with associated upward trends in extreme surface winds between 25°N and 40°N*' (Graham and Diaz, 2001). Deep lows are now about 50% more frequent, minimum central pressures of lows have gone down by about 4 or 5 hPa, and the associated extreme winds and vorticity have increased by 10–15%. Note the '*rotation in the preferred direction of extreme winds from north-west to south-west in the Gulf of Alaska*' (Graham and Diaz, 2001). This rotation confirms the augmentation in the intensity of southerly fluxes to the fore of the reinforced MPHs.

Symptomatic of this increase in wind speeds and the effect upon sea surface waters (Mantua, 2004), and also of increased storminess, '*there has been a progressive increase in wave heights during recent decades*' (Allan and Komar, 2000). This increase has been most marked in the coastal waters off the state of Washington. Sea conditions have therefore become more difficult in the North Pacific, '*with extreme wave heights increasing in the order of 1–2 m (20–30% of the long-term mean)*' (Graham and Diaz, 2001).

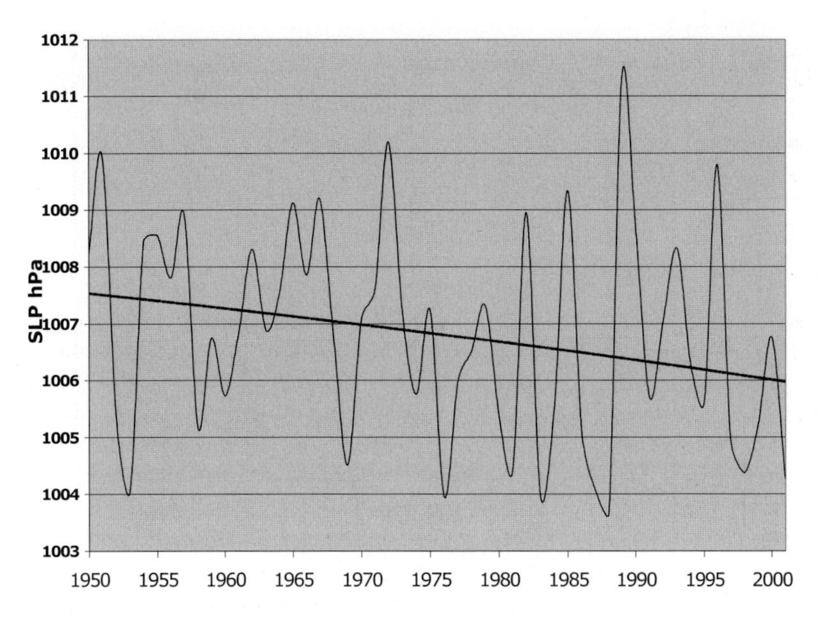

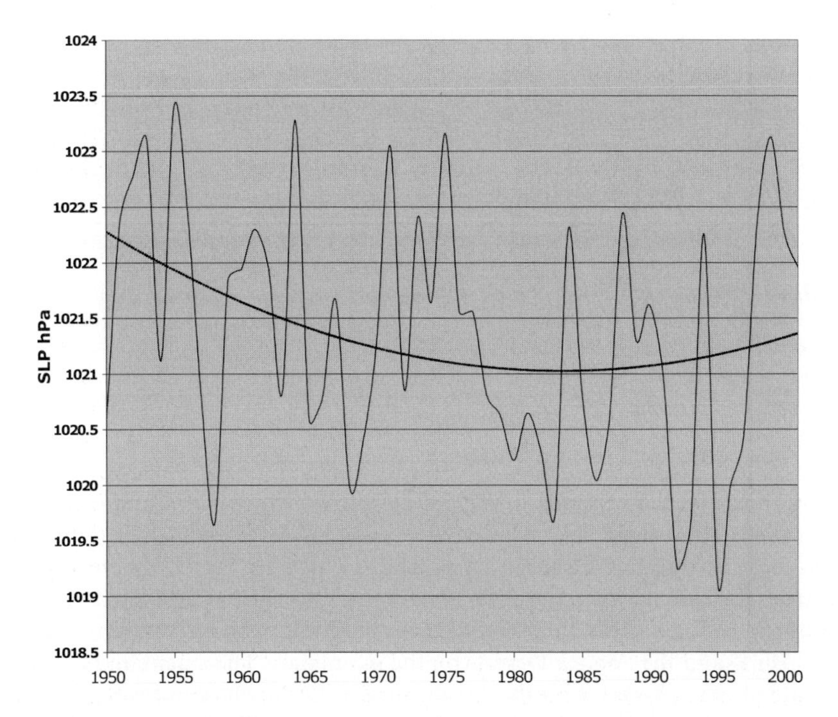

Figure 115. Mean annual sea level pressure from 1950–2001. (*top*) At 155°W/55°N, southwest of the Aleutian Peninsula, eastern side of the Gulf of Alaska. (*bottom*) At 140°W/30°N, southern part of the Gulf of Alaska.
From Favre (2004) (CDC/NCEP–NCAR data).

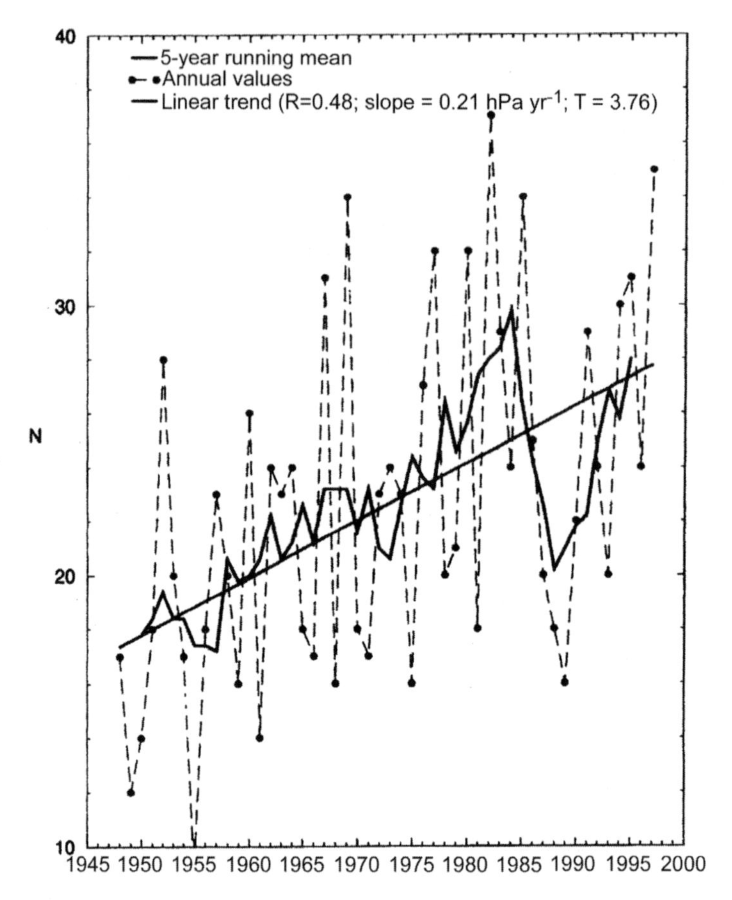

Figure 116. Frequency of North Pacific winter cyclones with minimal central pressure less than 975 hPa from 1948–1998: raw data, 5-year running average, and linear trend.
From Graham and Diaz (2001).

13.2.3 The central and eastern Pacific

This sector lies beneath the main MPH trajectories, which 'telescope' here to form the so-called 'Hawaiian' AA. The eastern edge of this AA is in California and Mexico, beyond the Baja California Peninsula, which air can easily cross, as far as the western flank of the Sierra Madre. Northern California is still visited in winter by low-pressure corridors, but they are filled in the south over the Vizcaino Desert (on the peninsula) and the Sonora Desert (on the mainland). The extent and the position in latitude of the AA vary with the seasons and with the phases of the NPO. A high or positive phase is characterised by higher pressures within the more extensive AA, shifted southwards over the central and eastern Pacific. The current climatic evolution of this sector involves increased pressures, lower temperatures at sea (but not on land), and both rising and falling rainfall figures.

Increase in surface pressure

The greater vigour of MPHs has led recently to an increase in pressure, already observed by Trenberth (1990), who noted that '*more and colder air is migrating southwards across the North Pacific*'.

An analysis of pressures reveals that the latitude where sign change occurs, between the decrease to the north (Figure 115) and the increase to the south within the AA, lies at about 30°N. Barbier (2004) showed (Figure 117, top) that, off the coast of southern California, around latitude 25°N, where low-pressure corridors frequently arrive, the rise has really only been in progress since the 1980s. Since the low-pressure corridors associated with MPHs cannot easily reach latitudes 14–15°N, the rise is more marked off Mexico and the Isthmus, at about 10–12°N (Figure 117, bottom), since the 1970s, and it occurs as far away as the Trade Discontinuity (TD). Here, the rise in pressure is a remarkably continuous phenomenon (Figure 117, bottom), and the rise is of the order of 1 hPa, a considerable figure given this tropical latitude.

Decrease in oceanic temperature/rise in continental temperature

Temperature falls under the influence of MPHs and stronger north winds which bring '*a rise to colder-than-normal conditions in the central and western North Pacific trend*' (Trenberth, 1990). The progress of this cooling in the western and central North Pacific was described by Gershunov *et al.* (1999), and by Graham and Diaz (2001), who commented on the persistence of low sea temperatures from the western Pacific to south of latitude 30°N.

Cold episodes accompany the most vigorous MPHs (e.g., California in 1999 was said to be '*without a summer*' by Schwing and Moore (2000)). Similarly, after a series of intense upwellings associated with more energetic north winds from late 1998 onwards, '*surface anomalies were as much as 3–4°C below normal during the 1999 upwelling season*'. As a consequence, the coastal upwelling in 1999 off California was '*the strongest in the 54-year record sea surface temperatures*' (Schwing and Moore, 2000).

However, this evolution is not the same over the land, where conversely the temperature is rising. This rise has been calculated for the long term at 2°C, according to Karl *et al.* (1996). We must bear in mind here the effect of higher pressures (greater anticyclonic stability within the AA), and even the local urban effect (Figure 48, Chapter 10), beneath a more assertive nascent trade inversion (TI), which discourages updrafts and the upward dissipation of pollution. Another factor to consider may be the influence of air returning from the south on the leading edges of MPHs, before they merge with the AA.

Increase/decrease in precipitation

Greater anticyclonic stability inhibits precipitation. Lower rainfall values for California achieved a coefficient of 20% for the period 1900–1994 (Karl *et al.*, 1996). However, the rise in pressure caused by more powerful MPHs suggests also

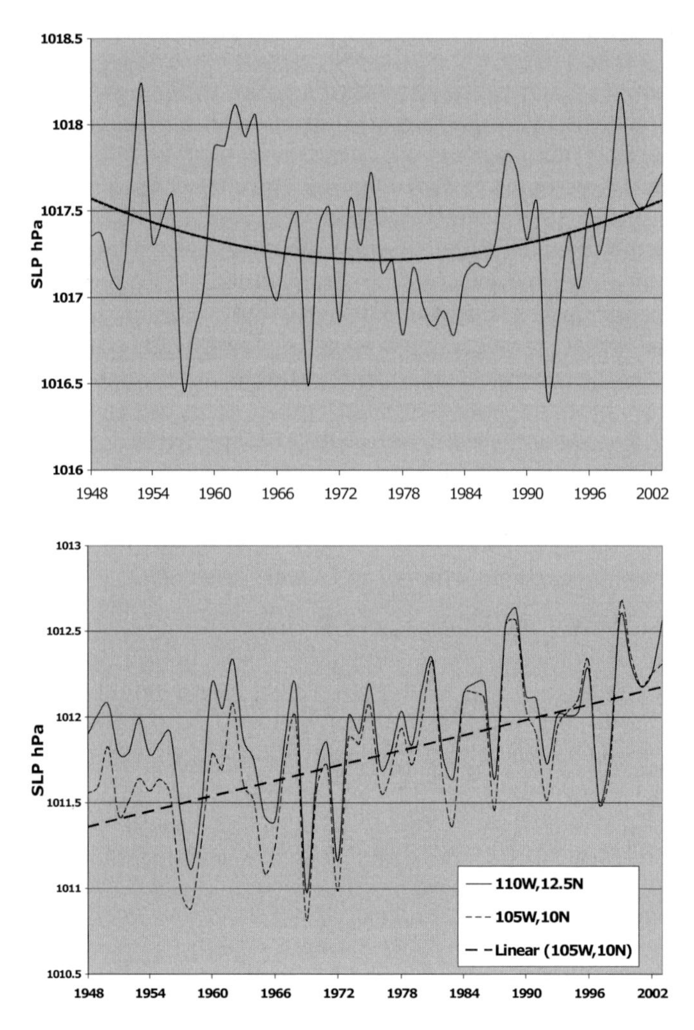

Figure 117. Mean sea level pressure (hPa). (*top*) Near 122.5°W/25°N, off the southern Baja California Peninsula (with polynomial trend). (*bottom*) Near 110°W/12.5°N, and 105°W/10°N (with linear trend), off the southern coast of Mexico, from 1948–2003.
From Barbier (2004) (CDC/NCEP-NCAR data).

that, episodically and mainly in the winter, vigorous MPHs are forcing their way further southwards within the agglutination, introducing active 'cold fronts'. Therefore, disturbances have recently migrated further to the south, and storms have been more frequent (Graham and Diaz, 2001). This has resulted in cold snaps, torrential rain and even snow, and storms with sometimes dramatic consequences, among the worst of which have been landslips in hilly areas. Now, although rainfall has recently shown a marked increase in northern California, as shown in Figure 118, it has been unchanged or has diminished in the south, in association with

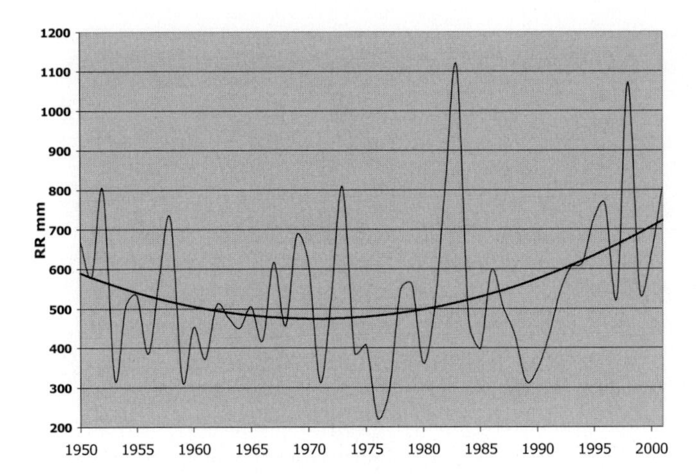

Figure 118. Annual rainfall (mm) evolution from 1950–2001 at San Francisco, 122.38°W/ 37.62°N.
From Favre (2004).

the migration of the AA. A comparison of Figure 114(bottom) and Figure 118 will show that, in these circumstances, when there is more rain in the south on the west coast, values are lower to the north (Favre, 2004). This was very much the case in 1973, 1983, and 1998, as a result of deep incursions by low-pressure corridors to the fore of MPHs (negative H index and positive L index, Figure 112). This kind of phenomenon has already been observed for the coastal area of Morocco (see Chapter 12).

13.2.4 The North Pacific off Central America

This domain is not in the strict sense a part of the North Pacific aerological unit, but it is geographically convenient to include it here. It experiences two well-contrasted seasons, during each of which it belongs to a different aerological space.

In northern winter (Figures 85 and 107)

The boundary with the Pacific unit is formed by the TD, with the preceding domain, that of the AA and the trade, extending as far as the south of the western Sierra Madre. The southern boundary is the ME, which is situated approximately over the geographical equator. In fact, during this season, the 'Central American North Pacific' forms part of the North Atlantic space, since it is then swept by the Atlantic maritime trade and experiences accelerated northerly winds (*northers*, cf. Figure 99, Chapter 12).

As is the case to the north of the Isthmus, in the Atlantic, the temperature has been falling since 1980 (Barbier, 2004). The rise in pressure is clear and regular, of the order of 1–1.5 hPa, since the 1970s (Figure 119). This domain may well be

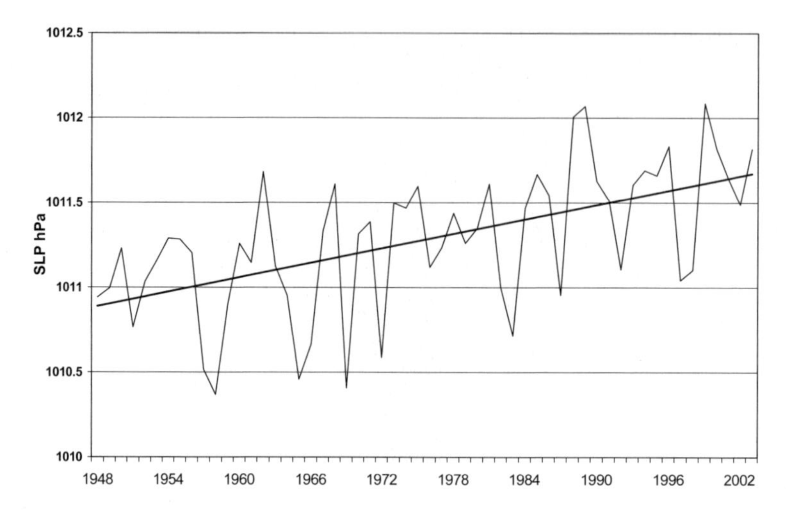

Figure 119. Mean annual sea level pressure (hPa) south of the American Isthmus, from 100°W to 80°W, and from the southern coast of the Isthmus to 2.5°N.
From Barbier (2004) (CDC/NCEP–NCAR data).

extended southwards during this season, as a result of greater impetus from the north, with a southward shift of the ME and a subsequent El Niño episode (cf. Section 13.3).

In northern summer (Figures 86 and 108)

This domain is part of the South Pacific aerological space. During this period it is invaded by the Panamanian–Mexican monsoon, fed by the southern maritime trade. The trade, in turn, is supplied by strong MPHs coming up against the Andean chain and forming the 'Easter Island' AA. Figure 120 shows the very clear deviation of the southerly flux crossing the equator, as a monsoon, traversing the Isthmus beneath the structure of the inclined meteorological equator (IME). The VME moves progressively further north, in step with the strengthening southern dynamic due to MPHs, and the reinforced trade pushes the confluence/convergence back northwards. Another MPH follows hard upon the previous one (Figure 120). The Panamanian–Mexican monsoon blows along the base of the western Sierra Madre range, up to the southern end of the Baja California Peninsula, where rain is a feature of the summer.

During the summer, there is an increase in the number of Pacific cyclones, or *cordonazos* (Figure 65, Chapter 11), connected with the greater frequency of MPHs in this region; the leading edges of these MPHs greatly deepen the lows along the surface line of the ME. The cyclones then develop in the low-pressure areas and move north-eastwards, towards the Baja California Peninsula, the Sea of Cortez, and the Sonora Desert, within the low-pressure corridor originating from the temporary encounter between the peripheral corridor and the IME (cf. Figure 108). So, the

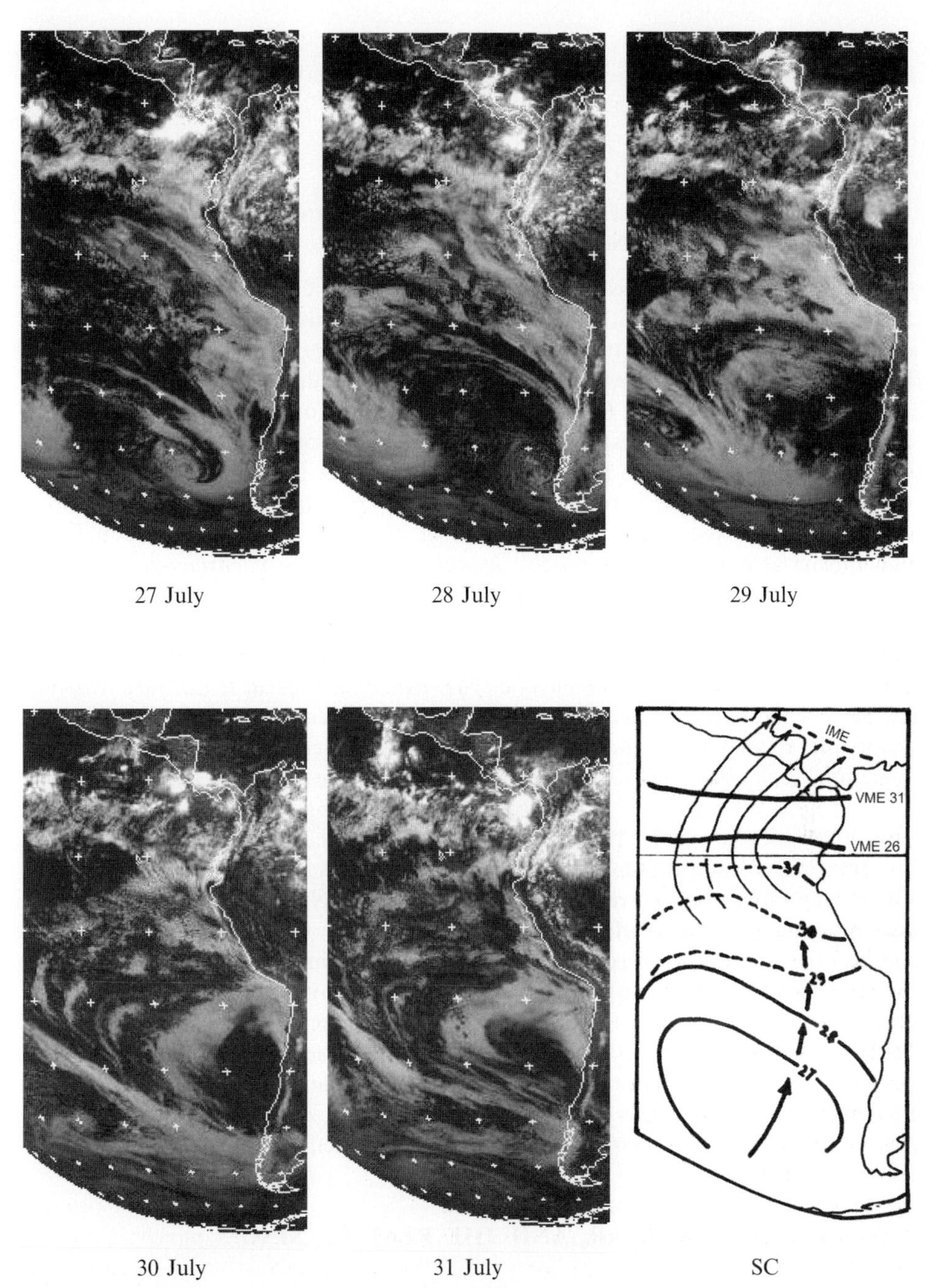

Figure 120. Supply of the Panamanian–Mexican monsoon by the southern maritime trade, itself supplied by MPHs encountering the Andean Cordillera. 27–31 July 1999 (visible, *GOES 08*/NOAA, 18 h).

Notes: SC = summary chart – path of MPH from 27–31 July, and VME from 26–31 July.

number of cyclones increases within 'the cordonazo triangle' (Barbier, 2004), because it is here that MPHs offer no protection to the coastal areas, but instead, push tropical cyclones towards the land (cf. Leroux, 1996a; Figure 121).

13.2.5 Conclusion: The greenhouse effect does not control climatic evolution on the eastern side of the North Pacific aerological unit

Climatic evolution on the eastern side of the North Pacific aerological unit is subject, as is the case in the North Atlantic, to **rigorous organisation, orchestrated by the dynamic of MPHs**. There is no need here to repeat the previously stated conclusions, as they apply to both of these neighbouring units of circulation (cf. Chapter 12).

All the while the role of the dynamic of MPHs is ignored, reasons will be sought elsewhere, as already suggested (e.g., in the variations of those 'indispensable' sea surface temperatures), even though they themselves are but consequences (i.e., co-variations) of the same aerological dynamic (or possibly another dynamic in a different unit). Again, in accordance with the *credo*, causes are sought in phenomena at altitude: '*A reasonable hypothesis is that the intensification of North Pacific winter cyclones has resulted from the increasing upper tropospheric zonal winds*' (Graham and Diaz, 2000). However, the reason why these high-altitude zonal winds might have changed is not given (unless it is connected with what is happening at the surface, since we read that '*one possibility is changes in tropical sea surface temperatures ...*' – and back to square one!). And of course, it comes as no surprise (it is so automatic) to see the greenhouse effect brought in: for many commentators, climatic variations such as warming, lower pressures and changes in storm activity can only be (and need we repeat it?) a '*consequence of anthropogenic climate change*'! – an answer of ignorance or laziness!

But since it results initially from cooling in the Arctic, which gives MPHs greater strength and frequency, **current climatic evolution in the North Pacific owes absolutely nothing to the greenhouse effect.**

Another reason is offered in an attempt to 'explain' climatic evolution in the North Pacific: one that might be expected (at least from a geographical point of view), since it involves El Niño, or the El Niño–Southern Oscillation (ENSO). We cannot conclude this section without mentioning this 'phenomenon', but let us put it its proper place in the scheme of things, as it has become so fashionable that some have seen it as, in the words of Lenoir (2001), the '*Ruler of the world*' ... though the title has now been won from it by the greenhouse effect!

13.3 EL NIÑO THE STAR, AND THE REAL EL NIÑO

To mention El Niño is to invoke a 'super-star' of the laboratories, the journals, and the media. It will not be practicable to treat this subject at length, since there is so much to be said about it. We shall limit ourselves here to a discussion of the essentials: studying a *statistical* phenomenon in the context of dynamic reality.

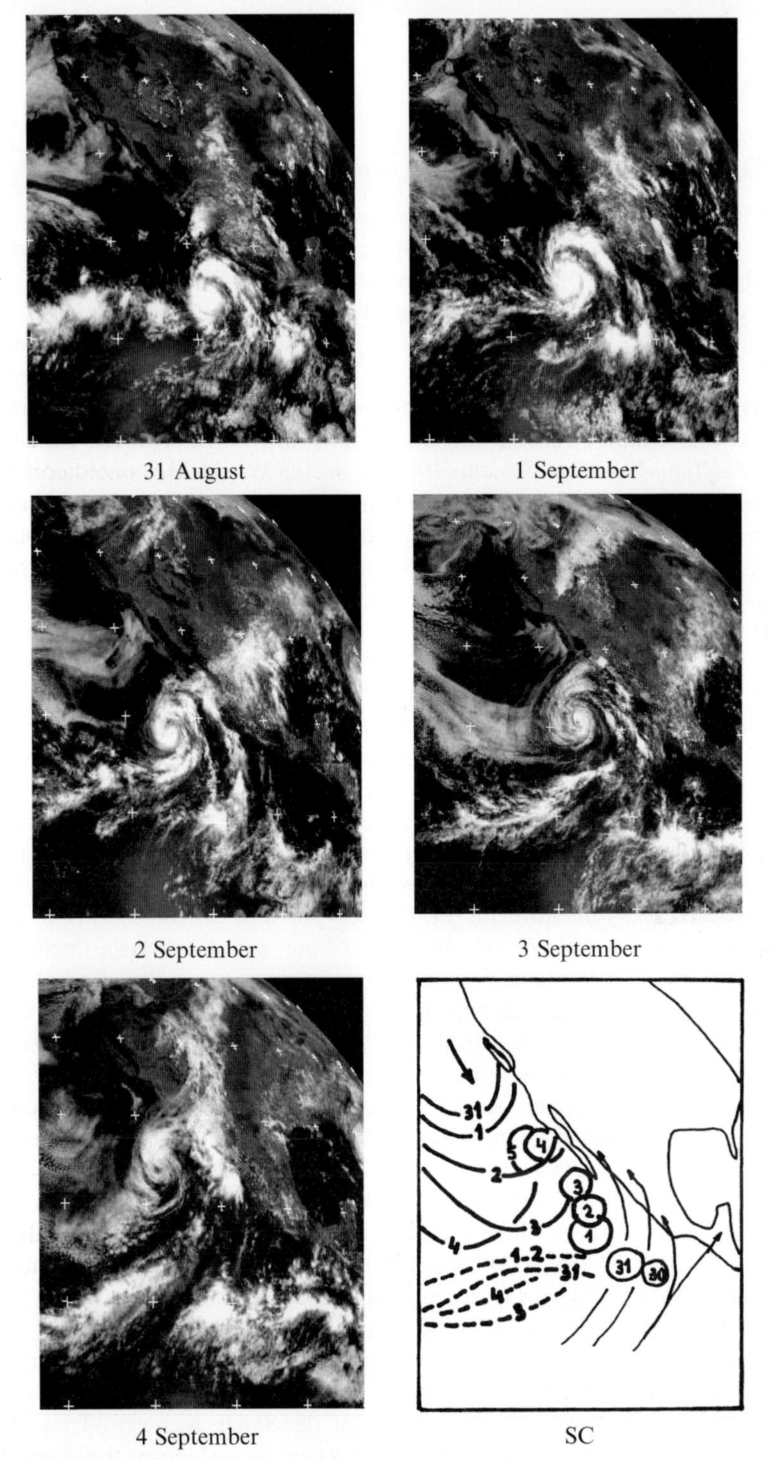

Figure 121. Dynamics of a *cordonazo* off the Mexican coast. 31 August–4 September 2004 (*GOES 10*, visible, 18h, NOAA).

A tropical depression, having appeared on 30 August inside the VME, off the southern Mexican coast and the Acapulco region, begins to move slowly northward. At the same time, an MPH arrives from the north along the Californian Coast, inside the AA, which limits the development of clouds on its leading edge. The encounter occurs on 3 September, and the *cordonazo* becomes a part of the peripheral corridor, and like all lows in the same situation, moves northward, towards California. On 5 September the low overwhelms and follows the MPH in its westward path.

Notes: SC = summary chart: continuous line – leading edge of the MPH; broken line – ME; circle – location of the tropical depression/*cordonazo*; fine arrow – Mexican monsoon.

13.3.1 El Niño, super-star: 'Ruler of the world'

Since the El Niño episode of 1972–1973 (contemporaneous with the 'climatic turning point' of that century and the beginning of the Sahel drought), and, more memorably, the episode of 1982–1983, called '*El Niño of the century*', followed by the 1986–1987 event, this phenomenon has been held to be responsible for climatic upheavals, not only in the Pacific, but also in the whole of the tropical region, and in temperate latitudes. El Niño, the 'super-star', has thus become the 'Ruler of the world'. Its reputation as such was established principally by an article written by Ropelewski and Halpert (1987), and by an edition, dealing with the period 1984–1986, of *The Global Climate System* (1987) from the World Meteorological Organisation (WMO–WCDP, CSM R84-86), which contained this as yet unpublished article. Pages 13–40 were dedicated to the ENSO and its 'influence' on the 'rest of the world'. With the *cachet* of the WMO upon it, the omnipotence of this phenomenon could only become more firmly established, and embellished.

Then along came El Niño of 1991–1992, and that of 1997–1998, on which occasion it was labelled, because of its particular amplitude, the '*climatic event of the century*'. Now El Niño was even more likely, it was thought, to be directly responsible for climatic catastrophes the world over. In its publication devoted to the ENSO event of 1997–1998, the WMO stated that this event had led it to '*develop a strategy within the framework of the International Decade for Natural Disaster Reduction (IDNDR) to prevent, mitigate, and rehabilitate the damage caused by the El Niño phenomenon*'. As a consequence, El Niño had now changed its status from mere '*climatic troublemaker*' to that of '*scourge of the world*' (Delécluse, 1998). The WMO attributed many phenomena (or 'impacts', as they called them) to El Niño, in South and Central America, North America, China, equatorial Asia/Pacific, and in southern and eastern Africa, reaffirming the existing conclusions of the 1987 publication (CSM R84-86). The adverse effects of El Niño (1997–1998) were even put into figures in a 'global assessment', citing a '*direct loss*' of '*US$ 34 billion; mortality: 24 thousand; morbidity: 533 thousand; etc ...*'!

Now, nothing can happen anywhere in the world without El Niño being held responsible for it: as the IPCC (2001) put it: '*El Niño and La Niña events have a worldwide impact on weather and climate*'. Publications never cease to assert its 'nuisance value', with El Niño creating record temperatures, heavy rain and flooding, droughts and fires (memorably in the forests of Indonesia), cyclones, and even '*local outbreaks of certain diseases such as malaria*'. El Niño was even said to be the determining factor of mean global record temperatures in 1997, and the question has even been asked: '*Is it true that El Niño years are good wine years in Europe*?' (cf. La Lettre de Medias, No. 12, 2003). Regional impacts via 'teleconnections' have included: the weather in Europe and the Mediterranean basin (Ricard, 2003); rainfall in southern Africa (Trzaska, 2003); rainfall for the Sahel, western Africa (Janicot, 2003); the extent of ice cover in the Arctic and Antarctic (Gloersen, 1995; Kwok and Comiso, 2002), etc ...

There has been a particular focus on the USA, where climatic parameters have been associated with El Niño (e.g., rainfall (Mason and Goddard, 2001), snow

(Smith and O'Brien, 2001), and even more extreme meteorological events such as waves of cold (Barsugli *et al.*, 1999)). An analysis of the destruction wrought by hurricanes found '*La Niña years exhibiting much more damage*' (Pielke and Landsea, 1999), and a study of weather impacts due to El Niño 1997–1998 revealed that '*the net economic effect was surprisingly positive*' (Changnon, 1999). Furthermore, research has even been done on the possibility of the 1877–1878 Niño being involved with the 1878 epidemic of yellow fever in the southern part of the USA (Diaz and McCabe, 1998)!

The newspapers, supposedly purveying 'factual' information, follow suit, with '*The enfant terrible of the Pacific!*', proclaiming that '*The climatic upheavals caused by that phenomenon of nature, El Niño, have left in their wake a formidable chain of catastrophes – droughts, floods, and storms – from one side of the planet to the other*' (*Le Monde*, March 7, 1998). With 'scoops' like this, it is hardly worth delving further into what horrors the sensationalist magazines offer. Some so-called scientific journals have also dwelt upon '*The whims of El Niño*', and how '*El Niño is causing the climatic machine of the Pacific Ocean to malfunction, and is destabilising the planet, and even its international trade*' (*La Recherche*, 307, 1998). Moreover, some 'scientists' have even announced that the ENSO is the '*keystone of global climatology*' (Pagney, 1994), which means that without the ENSO there is no hope of understanding the climate, since all planetary atmospheric phenomena are controlled by 'a minor coastal current'(!), itself only a local consequence of a dynamic which seems as yet unexplained!

Where is the sense in this 'rush of blood to the head' which has produced such a cascade of pronouncements? Exactly what *is* the real Niño, hidden behind its 'superstar' image? Let us first look at the 'traditional' Niño of the American equatorial Pacific.

13.3.2 The real Niño of the eastern Pacific

Features of an El Niño event

It was in the eastern Pacific that this marine phenomenon was first observed, as Aceituno (1992) recalls: '*The term El Niño referred originally to a relatively weak and warm southward oceanic current that develops almost annually along the coast of southern Ecuador and northern Peru around Christmas*'. The maritime trade drives the Humboldt (or Peru) Current northwards, at first along the Chilean coast, and then the coasts of Peru and Ecuador, causing cold waters to well up from deep down (the 'upwelling' phenomenon). Every year, a weak coastal current flows southwards towards northern Peru (Wyrtki, 1979), taking advantage of the weakened summer Peru Current and its associated upwelling. Warm water from the north, extending from the Equatorial Counter-Current (Figure 107), temporarily replaces the cold water delivered by the upwelling. Episodically, the El Niño current pushes farther southwards, an event not appreciated by fish and birds, and not least by the people who catch the fish. The same cannot be said for local farmers.

At the same time, rain, sometimes heavy and even torrential, falls in coastal areas of Ecuador and northern Peru, and on the Galapagos Islands. These regions, south of the Equator, are usually very short of rain, because of the presence above them of a TI (Figure 25), which is particularly unproductive and almost permanent. As an illustration, Guayaquil in Ecuador (at 2deg12S) receives almost no rain from June to December, when it is beneath the inversion. However, when the VME, with its plentiful rains, is present, figures for the local weather station are 217 mm in January, 189 mm in February, and 231 mm in March. Just south of Guayaquil, at Lobitos in northern Peru (at 4deg27S), it almost never rains, and the annual total is no more than 11 mm, because this weather station sits permanently beneath the TI, which allows absolutely no updrafts. There is therefore a very abrupt pluviometric transition here, since the VME does not move far from the geographical equator (Figure 107). It only needs an extremely small modification (i.e., a slight displacement of the VME) to cause major disruption of normal local conditions, and to bring the rain. This rain is particularly welcome, and, traditionally, an El Niño year has been justifiably called an *año de abundancia*, a 'year of plenty'. It is hard to imagine why God-fearing folk could think of El Niño as anything but beneficial, and as a 'gift from Heaven'. However, El Niño episodes are now associated with 'catastrophe' because of the growth of the commercial fishing industry (an international business) and the multiplication of the phenomenon, with consequent media 'hype'.

The southward shift of the vertical meteorological equator

Such a pluviogenic situation occurs when the ME is shifted well to the south. Figure 122 shows how, in the Gulf of Panama, the effect of the proximity of the ME is a more or less pronounced fall in pressure at the time of El Niño; the most important examples having been those of 1965–1966, 1982–1983 and 1997–1998 (with the other El Niño events showing, but in a less obvious way within the pressure field). Note too that, generally speaking, there has been a progressive rise in pressure since the 1970s. Remember that in this sector, and during this season, we are dealing with the Atlantic unit (Figure 85), and it is therefore the northern maritime trade, issuing from American MPHs, which causes this movement towards the south.

Figure 123 clearly shows the North Atlantic being invaded by an enormous MPH, which is immediately reinforced by another MPH, and then another, in uninterrupted succession across North America. Pulse lines on the southern edges of the MPHs reveal the flow of cold air to the south-west, towards the Caribbean, and on across the Isthmus into the Pacific. Cloud formations along the leading edge of the pulsation (representing the boundary between the hemispheres, and therefore by definition the ME) mark out the southward progress of the air from the north through successive pulses. The VME now delivers rain to the Galapagos Islands and the usually parched coastal area, with values (as in 1988) of the order of 10 times greater than normal (the 'normal' being, remember, close to zero). These rains can sometimes be severe, bringing temporary flooding, though they do replenish the water table.

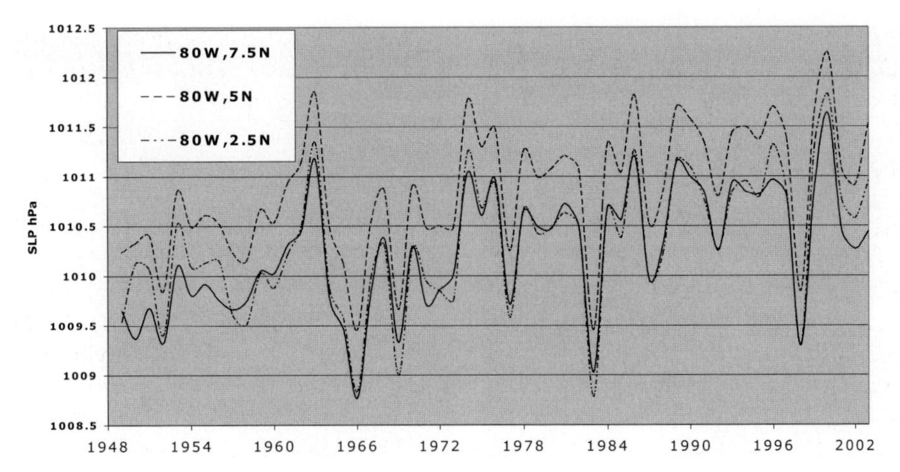

Figure 122. Mean monthly sea level pressure (hPa) in the Gulf of Panama, near the 80°W meridian (7.5°N, 5°N, 2.5°N), from 1948–2003, during the winter months December–January–February.
From Barbier (2004) (CDC/NCEP–NCAR data).

The southward shift of the ME is only possible because resistance from the southern meteorological hemisphere is weaker, a situation which normally occurs in southern summer. This was the case on 4–8 February 1998; an analysis of the situation shows that no powerful southern MPH had been present to reinforce the 'Easter Island' AA, and by extension the southern maritime trade over the east of the South Pacific. This weakening in the south leads to a diminution in the intensity of the upwelling and the Peru Current, allowing warm equatorial water to move southwards (the El Niño current).

Normally, rainfall coincides with the presence of warm water, and so it is not surprising that this rain is attributed to a rise in sea temperatures. However, it is well known that this is not a sufficient condition, since (need we repeat it?) pluviogenesis has its origins in the air and not in water (cf. Chapter 11). The determining factor here is the presence of the VME, a structure which encourages the vertical development of clouds. However, this 'warm water/rain' coincidence or co-variation is not always certain, as Deser and Wallace (1987) pointed out: 'not all the warm episodes were associated with an intensified El Niño current' (Aceituno, 1992).

El Niño events are provoked by a strengthening of northern MPHs

In the northern hemisphere, powerful MPHs lead to severe winters. Given the common, single origin of MPHs, wintry conditions propagate from the Arctic into both the Atlantic and Pacific units. So the winter of the El Niño event of 1997–1998 began with a blizzard over Colorado and the Great Plains, in late October 1997, as low temperatures were recorded across North America. It was in early January 1998 that the worst ever 'ice storm' hit Canada and the north-eastern USA (cf. Chapter 10).

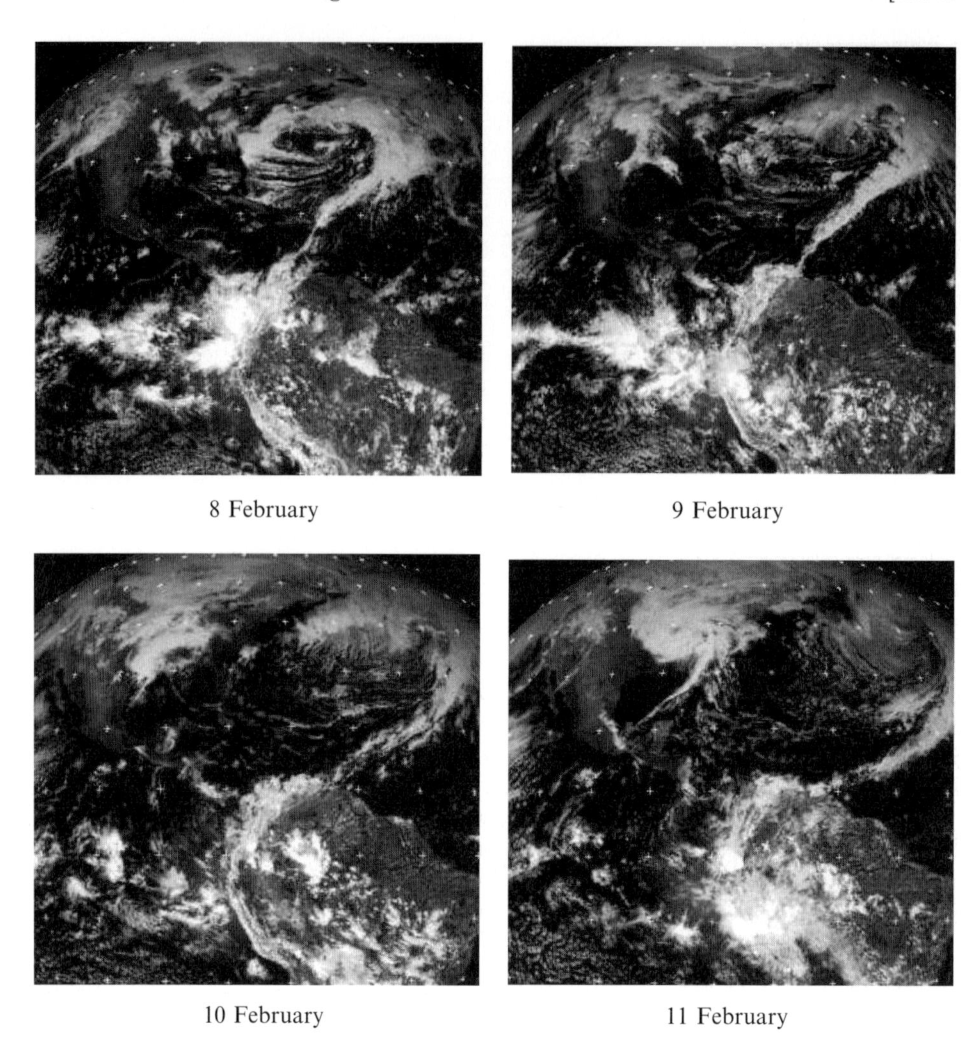

Figure 123. Dynamics of the meteorological El Niño in the eastern equatorial Pacific. 8–14 February 1998 (*GOES 08*, visible, 18h, NOAA).

On 8 February the leading edge of the first MPH has already reached northern Colombia. Polar air continues to pass through Central America (as winds called *nortes*, or locally *tehuantepecer*), a crossing started on the 6th by this MPH. During the following days the MPH moves eastward and polar air is spread out south-westward by successive pulsations (clearly visible on the images). At the same time, the ME is pushed southward and reaches, on the 11th and 12th, latitude 10°S, north of Lima (Peru). A second MPH, having appeared on the 8th over Canada, reaches the Gulf of Mexico on the 11th, and then polar air, after overtaking the Sierra Madre (on the 11th it is still stopped by it, (see Figure 116, Chapter 11)), strengthens the northern pulsations. One of them can be clearly made out on the 13th near the equator. The first MPH, having now arrived north of Brazil (since the 12th–13th), supplies the Amazonian monsoon, as clearly shown on the 14th by the two pulse lines over the basin.

Notes: SC = summary chart: continuous line – leading edge of the first MPH; broken line – leading edge of the second MPH; broken and bold line – ME; dotted line – southward pulsation of northern air.

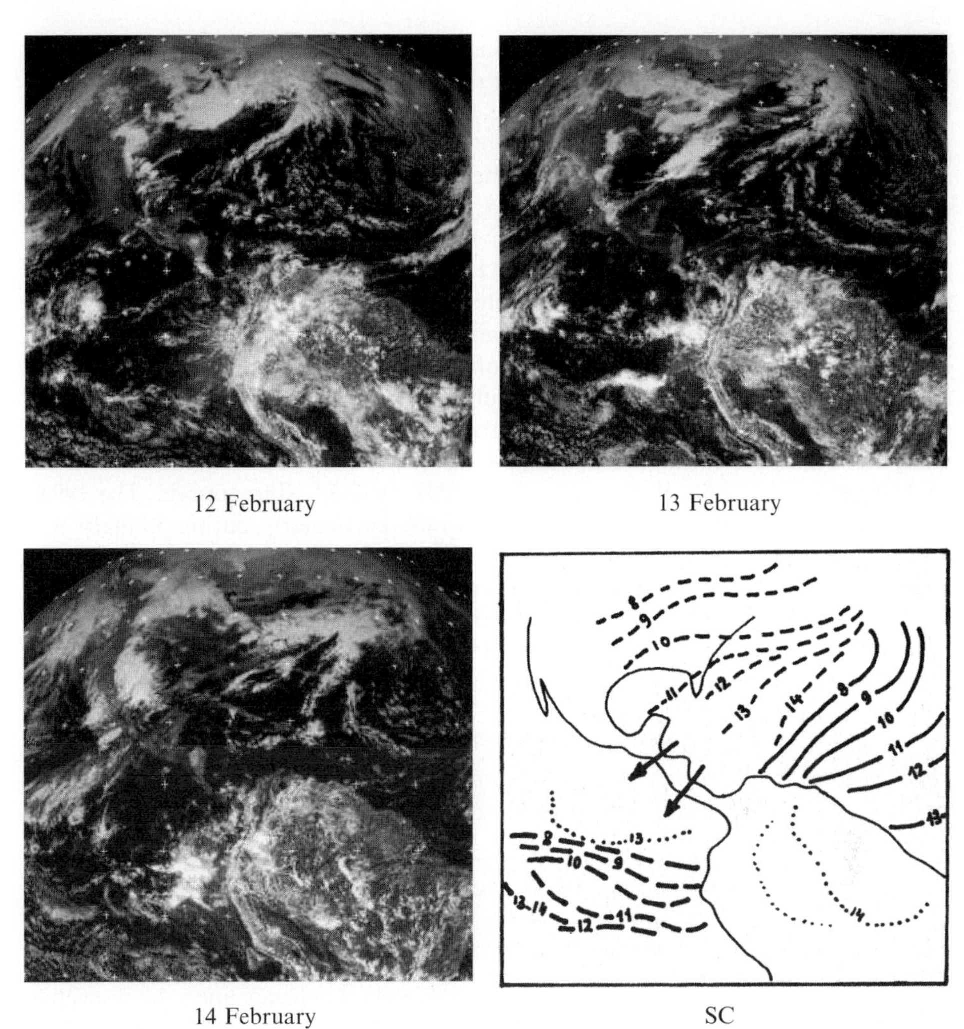

12 February 13 February

14 February SC

Then mean temperatures rose in these regions, as MPH trajectories came further south and they moved more directly towards the Gulf of Mexico, causing intense flows of returning air from the south. However, the south-eastern part of the USA was cold, with near-record rainfall, and Florida experienced the most destructive tornadoes in its history on 23 February. Adverse weather also struck the Pacific side of the country, with violent storms on the west coast, severe thunderstorms in January and February, and torrential rain and flooding in California (and on the Sierra Nevada, heavy snow), particularly during the first two weeks of February 1998.

Usually, at the time of El Niño events, and to varying degrees, strong winds and even intensely disturbed weather are experienced in more northerly latitudes in both the Atlantic and the Pacific. In the northern part of Baja California, in Mexico, for example, '*during El Niño events, above normal precipitation occurs largely in February and March*' (Minnich *et al.*, 2000). Figure 118 clearly shows this co-variation in the north of California, especially at the time of the strong El Niño events of 1973, 1983, and 1998. This variation is evidence of the vigour of MPHs when their meridional displacement is strongly shifted southwards (cf. Figure 111, bottom). A case in point is the extreme cold and severe weather which affected the southern USA and the Gulf of Mexico between December 1982 and April 1983. The Galapagos Islands suffered intense thunderstorms and extraordinarily violent, gusting north winds, with low temperatures and generally perturbed weather (Chenoweth, 1996). California, in turn, had torrential rain; this co-variation with the El Niño event was noted by Schonher and Nicholson (1989), who found that, of 11 events from 1950–1982, 9 were accompanied by heavier than usual rainfall in California. Similar findings were made for Texas, for the area around the Gulf of Mexico, and for Central America, by Halpert *et al.* (1993), in the case of the 1991–1992 El Niño episode. The 1993–1994 event coincided with very cold conditions, particularly during January and February, on the eastern side of North America, where it was the coldest January since 1920. Quebec recorded its lowest ever temperature, and the cold spell was felt as far south as the Gulf of Mexico and Cuba. At the same time, there was a rise in pressure in the 'Hawaiian' AA (WMO, 1995). Statistically, pressures were higher across North America, while the 'Icelandic' and 'Aleutian' Lows were deeper than usual (Niebauer, 1988; Gloersen, 1995).

It is therefore unquestionable that the El Niño episodes in the eastern Pacific, phenomena belonging to the North Atlantic aerological space, are **controlled by the strength of Arctic MPHs, and by reinforced northern trade winds** (WMO, 1995), **pushing the ME southwards.**

There is nothing very surprising here in meteorological terms . . .

It is worth remembering that position of the ME always depends upon the respective strengths of each of the meteorological hemispheres (cf. Chapter 8). During the same season (southern summer), the region where an El Niño event may be taking place is not the only one to experience the fluctuations in latitude of the fundamental pluviogenic structure of the ME, in its forms of the active VME and IME structures (cf. Figures 26 and 30, Chapter 8; Leroux, 1996a):

- To the east of the barrier formed by the Andes (Figure 85), in the Amazon Basin, the surface line of the meteorological equator (IME) is shifted to the south of the forest, and reaches the northern edge of the Gran Chaco depression. Here, the Amazonian monsoon is pushed southwards by MPHs and northern trades, and drawn in by the thermal low which has deepened over the continent south of the forest. At the same time, to the west of the Andes, over the near Pacific off Peru, the differential is maintained through the absence of a thermal low, and because of the unaccustomed vigour of the southern maritime trade which brings air northwards from southern MPHs held by the Andean range. This strong trade influence maintains the VME in its mean winter position near the equator (Figure 107).

- To the east of the Amazon Basin (Figure 85), in the Brazilian *Nordeste*, the ME (here, the active IME) may be pushed southwards by a reinforced Amazonian monsoon, bringing rain; or it may stagnate further to the north, its absence leading to severe drought conditions, endemic in this *Sertão* region (cf. Chapter 12). This situation has its analogies with the El Niño pattern (being subject to the same impetus), but in this case a 'descent' of the ME is not considered a catastrophe – quite the opposite, in fact!
- In Mozambique, Zimbabwe, and northern Madagascar (Figure 26), the southward migration of the ME (in the form of the active IME) is more or less early, and more or less stable during the summer months (DJFM), and its retreat northwards is to some degree belated. Because this structure brings abundant precipitation, the quality of the rainy season is closely linked to the dynamic of the ME. The cyclone season in the Indian Ocean east of Madagascar and in the Mozambique Channel also owes its particular characteristics to the presence of this structure, which favours updrafts (Leroux, 1983, 2001).
- In northern Australia (Figure 26), the strength of the Australian monsoon (also drawn southwards by the Australian thermal low) determines the degree of penetration of the flux, and of the IME, towards the heart of the continent, and dictates the quality of rainfall. The same is true of the cyclone season, with cyclones forming over the ocean to both the east and west of Australia within the VME structure, thence moving onto the continent, following the IME structure (Figures 107 and 125).

An occurrence of El Niño is therefore really of no great import when we study it in the context of the tropical weather dynamic in the same season, that of southern summer. We might also look at the variability in the position of the ME in northern summer (cf. Figure 26); this variability is also evidence of the activity of the Mexican monsoon, the Atlantic monsoon (involving sub-Saharan Africa), the Indian monsoon, and the Chinese monsoon. We ought therefore to be aiming to put the ENSO back in its rightful place in the context of the tropical (and general) dynamic, and recognising that that place is a very modest one!

El Niño events and the northern 'oscillations' (NAO/NAD and NPO/PDO)

The indices which have been worked out to increase understanding of the northern dynamic reveal the occurrence of an El Niño event: directly in the case of the Atlantic indices, since they involve the relevant unit; and indirectly for the Pacific indices, since they refer to the adjacent unit, although it is still subject to the same initial impetus.

An El Niño event corresponds therefore, meteorologically, to a positive NAO/NAD and NPO/PDO.

A purely statistical analysis (Gershunov *et al.*, 1999) reveals the coincidence of a 'canonical El Niño' with:

- An abnormally strong anticyclone over Canada (i.e., more vigorous MPHs, and a colder and rainier winter in the southern United States, with MPHs frequently reaching the Gulf of Mexico). These characteristics, together with a deeper 'Icelandic' Low, are features of a positive NAO.
- In the Pacific, intense disturbances, shifted further south, higher pressures within the 'Hawaiian' AA, which is also further to the south, and a deeper 'Aleutian' Low. The north-west and central Pacific are cold, though it is warmer in the east where air comes up from the south along the coasts of America. All these parameters are characteristic of a positive NPO/PDO.

Conversely, a 'canonical La Niña' corresponds to a 'low phase of the NPO', and to a negative (or low) phase of the NAO.

The NAO/NAD/NPO/PDO have been in their positive modes since the 1970s. The frequency of El Niño episodes has also been increasing since then:

- 1972–1973, 1975–1976, 1982–1983, 1984–1985 ('*the aborted Niño*' (WMO, 1987)).
- 1986–1987 (perhaps even 1986–1988: '*the semi-permanent Niño*').
- 1989–1990 ('*ENSO-like conditions*' (Janowiak, 1990)).
- 1991–1992 or 1991–1995 ('*the longest lasting Niño of the 20th century*').
- 1997–1998 ('*the greatest Niño of the 20th century*' (WMO, 1999)).
- 2002–2003, which event has just come to an end, being less vigorous than its predecessor (WMO, *Info-Niño*, August 2004).

Consequently, the increase in the frequency of El Niño events is undoubtedly linked:

- **to a southward shift in the position of the ME;**
- **thereby to cooling in the Arctic and an increase in the strength and frequency of northern MPHs; and**
- **should be seen in the context of a rapid mode of general circulation.**

It is, without a doubt, at variance with the presumed greenhouse effect.

13.3.3 El Niño and the Southern Oscillation: ENSO

Mean features of the ENSO

Another peculiarity of the tropical Pacific was noted by Walker and Bliss (1932), and designated the 'Southern Oscillation' (SO), corresponding to the Northern Oscillation (NAO and NPO). This oscillation, an attempt to explain and predict the rains of the Indian monsoon, though it has never itself been explained, occurs in a different season and belongs to a different unit of circulation. Since its beginnings, it has been much simplified, and it is characterised, mainly but not entirely, by a 'see-sawing' of surface pressures between the two sides of the tropical Pacific. Progressively, this oscillation has been whittled down to mean only this difference in pressures, and an index, the SOi, has been established for it. The most commonly used form of the SOi is that which deals with the differences in pressure between Tahiti (in the Society Islands) and Darwin, Australia:

- If pressure is relatively high in Tahiti, but lower in Darwin (positive SOi), the surface temperature of the central Pacific Ocean near the equator is low. This is known as a *cold* episode, or otherwise the anti-Niño, or La Niña. This corresponds in fact to a situation which might be considered normal.
- When the SOi is negative, pressure is relatively low in Tahiti, but higher in Darwin, and warmer waters move farther east: the surface temperature of the central and eastern Pacific Ocean near the equator rises. This is known as a *warm* episode, and is related to El Niño. Simultaneously, the sea level falls in the western Pacific, and rises on the eastern side, and warm water flows along the coasts of equatorial America, most notably southwards along the Peruvian coast (feeding into the El Niño current).

It should be recalled that, in Pacific equatorial latitudes, the mean situation involves the proximity of the VME, near to the geographical equator except in the west near Australia (Figures 107 and 125). Beneath the VME flow the warm waters of the Equatorial Counter-Current on their way eastwards, between the cooler North Equatorial Current and South Equatorial Current, moving westwards. There is nothing exceptional about this situation, and it is paralleled, with minor differences, in northern winter (DJF) in the tropical Atlantic and the tropical Indian Ocean (Figure 26). Any major modifications which may occur are therefore associated with a general migration of these aerological and oceanic components.

Figure 124, with curves for seasonal pressure based on averages for months from December to February, shows in a general way how the El Niño events mentioned above correspond more or less closely to a small difference in pressure between Tahiti (lower) and Darwin (higher), with a few anomalies mainly due to the persistence of the event across two calendar years. The closing of the two curves (negative SOi) raises the question of why pressure should fall in Tahiti and rise at the same time in Darwin. Another question is that of the meteorological significance of these differing behaviours, since these two weather stations are located, during El Niño episodes (i.e., seasonally) in two different units of circulation. **There is therefore no direct physical link between them**: Darwin is subject during this season to the northern winter dynamic, while Tahiti experiences the southern summer dynamic.

The first person to link the SO and El Niño was J. Bjerknes (1969), but at the time he restricted the term El Niño only to warm episodes on the coasts of Ecuador and Peru. He also suggested (but did not demonstrate) the existence of a cellular zonal circulation associated with the oscillation, with subsident air in the eastern Pacific and ascending air over Indonesia, with easterlies in the lower troposphere and westerlies in the upper troposphere. He called this circulation 'the Walker cell', of which more later.

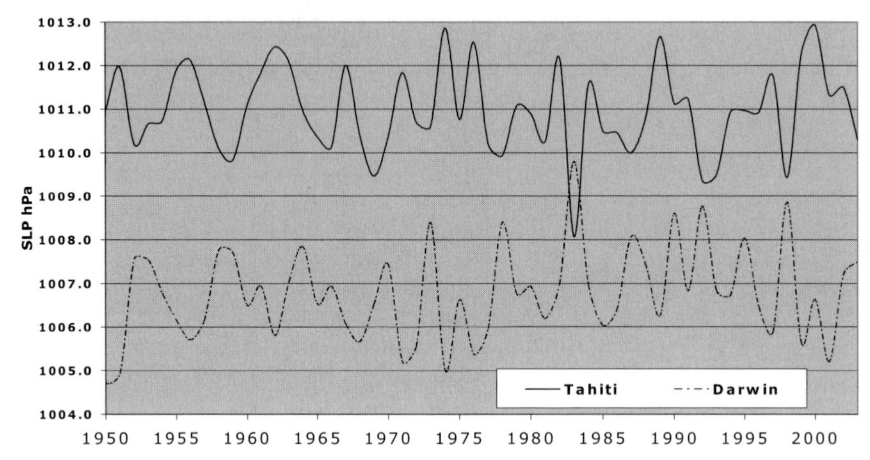

Figure 124. Mean seasonal sea level pressure (hPa), for December–January–February, at Tahiti, Society Islands, 17°33′S/149°37′W, and at Darwin, Northern Australia, 12°25′S/130°53′E.
Data from the Bureau of Meteorology, Australia, and WMO.

El Niño and the SO were linked together using the acronym ENSO in the 1980s. This new appellation aimed to associate the 'traditional' El Niño of the South American coasts (Niños 1 and 2) with phenomena observed in the tropical central Pacific (Niños 3 and 4, in the equatorial band between 5°N and 5°S, and between 90°W and 160°E). The result was '*considerable confusion, and past attempts to define El Niño have not led to general acceptance*' (Trenberth, 1997). We shall not attempt here to unravel these tangled definitions, which involve the fact that El Niño and the SO belong, briefly, totally or partially to different units of circulation. Neither do we intend here to go into detail about the carving up of this phenomenon into rectangles labelled Niños 3 and 4: it should be studied in its entirety, taking account of all its components and the way in which it works, and not through artificial windows. These distinctions do not however involve the Equatorial Counter-Current and the El Niño current, since the latter is an extension of the former. It will be more useful to examine the relevance of the ENSO in terms of the dynamic of the North and South Pacific, as the equatorial space in question (10° wide, straddling the equator) is not an isolated entity.

Figure 125 shows the aerological components of the ENSO. From the vast domain of the Pacific, we have already dealt with the Central American sector, where (the 'traditional') El Niño is subject to the dynamic of the Atlantic unit (Figure 123) and to southern 'resistance' in the vicinity of the coast of South America.

The increasing northern dynamic

The ENSO phenomenon implies the intervention in the Pacific space of powerful dynamical factors, capable of causing the migration – partially or as a whole – of

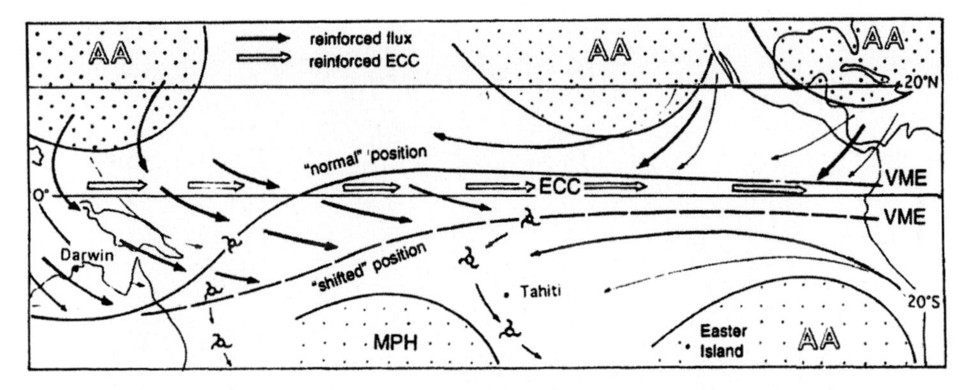

Figure 125. Aerological components of the ENSO in northern winter in the tropical Pacific Ocean.

the ME and its associated oceanic and aerological phenomena. The dynamic of the northern hemisphere involves a threefold impetus from the north:

- from Atlantic MPHs, as already mentioned (Niños 1 and 2);
- at the centre of the Pacific, from MPHs entering the 'Hawaiian' AA (more directly involved with Niño 3); and
- in the west, from Asian MPHs supplying the 'Philippines' AA, the north-easterly trade and thereafter the Australian monsoon (and also controlling Niño 4).

These three driving forces (impulsions) stem from the same initial cause, but they do not necessarily appear at exactly the same time, since cold air cannot be ejected from the polar regions in every direction at once.

The northern dynamic, with its three impulsions, has more or less well co-ordinated zonal and meridional effects (Figure 125).

More powerful Asian MPHs raise pressures over the western Pacific, the place of origin of the Equatorial Counter-Current, and, together with the reinforced Australian monsoon with its strong westerly component, drive back the surface water towards the east. Figure 124 makes the point that, at Darwin, an El Niño episode coincides exactly with a marked rise in pressure in the Australian monsoon flux.

Figure 126(bottom) shows that this rise in pressure is felt to the north of Darwin, at the latitude of the 'Philippines' AA, where again we see definite peaks in pressure for 1973, 1983, 1987, 1992, and 1998 (i.e., El Niño years). These curves also reveal the continued rise in pressure, which is also observed even further north in eastern China (Figure 126, top).

Tracking down the source of the Australian monsoon, and of the 'Philippines' AA which supplies it, we find ourselves in the heart of China, and thence in Inner Mongolia, in the Gobi Desert (Figure 127), destination of MPHs which have crossed central Asia. Pressure has been rising markedly here since the period 1960–1970, by

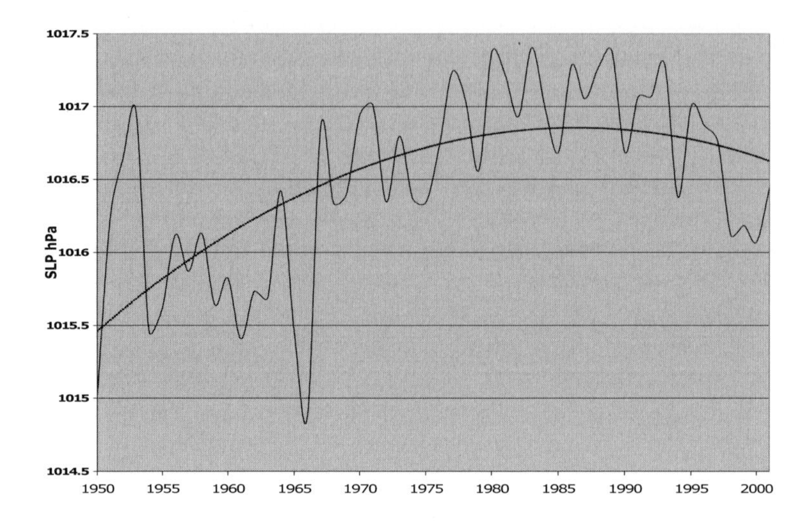

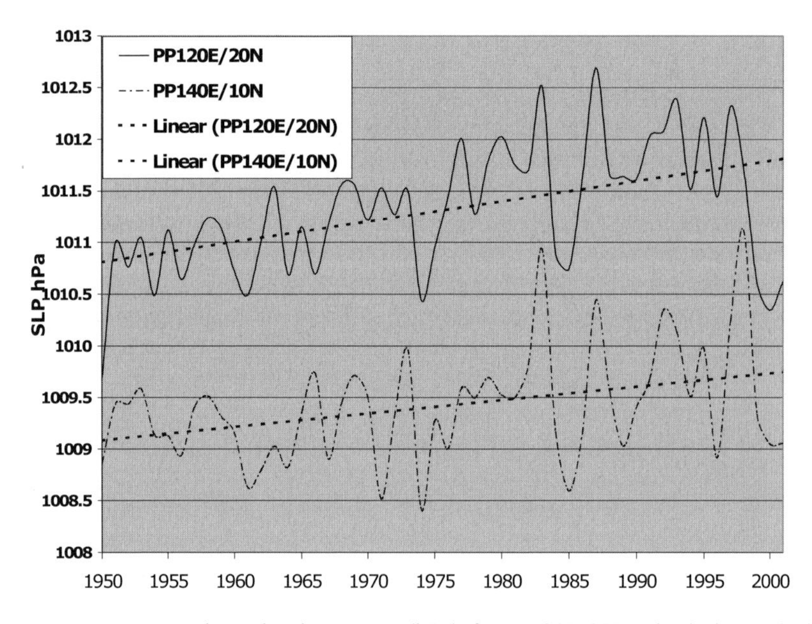

Figure 126. Mean annual sea level pressure (hPa) from 1950–2001. (*top*) At 120°E/32.5°N near Nanking (China). (*bottom*) At 120°E/20°N in the Luzon Strait, South of Taiwan, and at 140°E/10°N between the Philippines and the Caroline Islands.
From CDC/NCEP–NCAR data.

average annual values of more than 7 hPa, a considerable figure. Pressure has been remarkably high since the 1980s. At the same time, temperatures have been decreasing regularly (Figure 127, bottom). These observations confirm the finding of Favre and Gershunov (2004): a pronounced rise in anticyclonic activity in the western

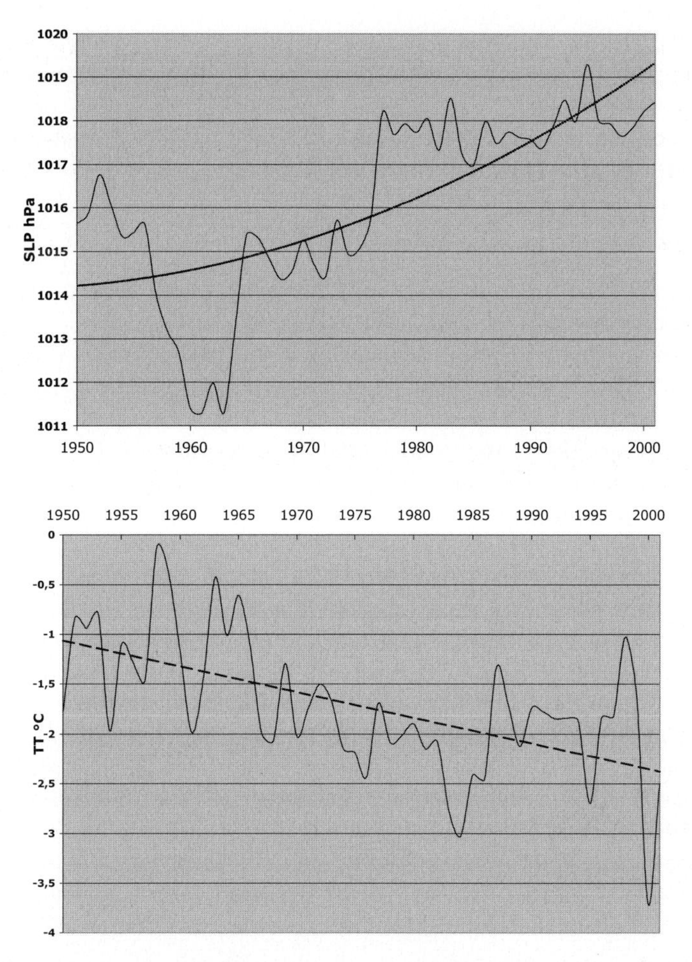

Figure 127. Surface pressure and temperature at 100°E/40°N in the Gobi Desert (China), from 1950–2001. (*top*) Mean annual sea level pressure (hPa). (*bottom*) Mean annual temperature (°C).
From CDC/NCEP–NCAR data.

Pacific (i.e., an increase in the strength of MPHs and higher pressures for the 'Philippines' AA).

The strong northern dynamic tends, especially in the west, to reposition northern trades further south, together with the North Equatorial Current. The VME is in turn shifted southwards. With it goes the Equatorial Counter-Current, flowing eastwards beneath lows. The rise in pressure over the western side of the ocean (Figure 126) accelerates the eastbound flow of the Equatorial Counter-Current, and the force of the Australian monsoon blowing east. This 'slippage' of the MPHs, the AA, the trades, the Australian monsoon, and the VME, brings the warmer waters of the Equatorial Counter-Current farther south. The same thing occurs, to the east of the 'Philippines' AA (rising pressure), in the case of falling atmospheric pressure associated with the surface VME, and the consequent rise in sea level (cf. Chapter 14).

So the recent increase in the frequency of El Niño episodes is easily explained: it is associated with greater dynamism in the north.

The contribution of the southern dynamic

Figure 124 shows the fall in pressure in Tahiti when an El Niño event occurs. The fact of this fall may mean that the VME is close by, or that there is a simultaneous weakening in southern MPHs, encouraging the VME to move southwards, the repositioning of the VME therefore being favoured by lessened resistance from the southern hemisphere.

Figure 128(top) shows the recent variation in pressure for Easter Island during the period (December to February) of the occurrence of El Niño events. With a trend towards higher pressures since the 1970s, the curve shows a pronounced fall at the time of certain El Niño events, very clear examples being those of 1972–1973 and 1982–1983, while the El Niño events of 1997–1998 and 2002–2003, occurring during the high-pressure phase, are much less obvious. Because of the latitude of Easter Island this curve is more representative of the evolution of the southern dynamic than of that of Tahiti (Figure 124), where all El Niño events coincide with a fall in pressure, but where the recent rise (as at Easter Island) in southern pressures (from the increase in the strength of southern MPHs, cf. Chapter 10, Figure 53) is not in evidence.

Note that the trade issuing from the 'Easter Island' AA (to the east of the island) comes to Tahiti (Figure 125), as does a cooler and faster trade originating in MPHs which do not reach the coasts of South America. These latter MPHs or AAs are known by the local weather service (Météo-France) as the 'Kermadec anticyclone', located on the west of the island. Moreover, pressure at Tahiti is affected during El Niño events, because of its latitude, by the proximity of the low-pressure corridor within which the ME lies. Consequently, the data from Tahiti (which mix causes and consequences) do not really constitute an ideal reference for the study of the ENSO phenomenon.

Figure 128(bottom) also shows the southern rise in pressure (on the mean annual scale) since the 1970s, off northern Chile.

Cyclogenesis in the south-east Pacific

The displacement of the ME in the South Pacific modifies – temporarily – the conditions for cyclogenesis. The south-east Pacific shares with the South Atlantic the particular characteristic of being spared from tropical cyclones, a result of both the anticyclonic situation ('Easter Island' AA and/or 'St. Helena' AA) and the absence of the vertical structure of the VME, which remains north of the equator. In fact, most tropical depressions initially form within this structure, and they may thereafter become tropical cyclones – always assuming that this structure is far enough from the geographical equator for vorticity to develop (as a function of

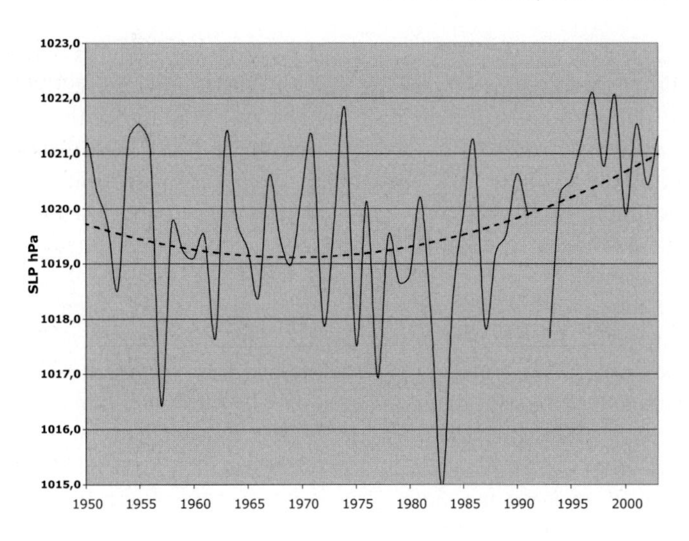

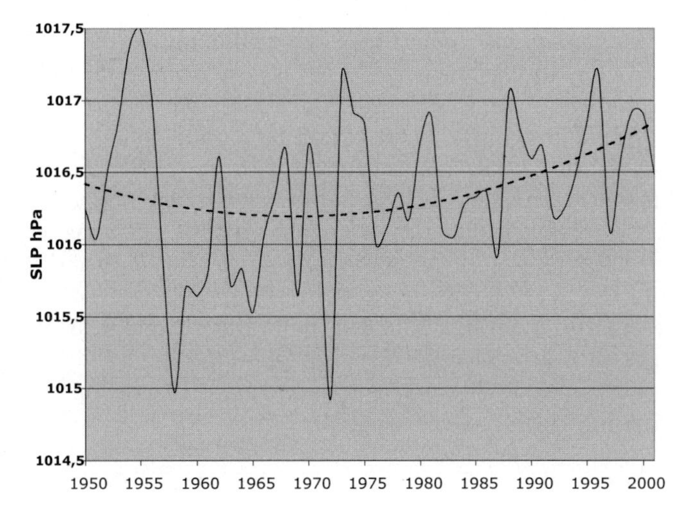

Figure 128. (*top*) Mean summer sea level pressure (hPa) at Easter Island Mataveri, Chile (27°10′S/109°26′W), during the summer months of December–January–February, from 1950–2003. (*bottom*) Mean annual sea level pressure (hPa) at 80°W/20°S, off Iquique (Northern Chile), from 1950–2001.
(*top*) From UCAR/WMO data. (*bottom*) From CDC/NCEP–NCAR data.

latitude). Cyclones may also form to the fore of MPHs, in the low-pressure corridor which they engender, and which draw in energy from their surroundings. It is well known (cf. Chapter 11), although it bears repetition, that the presence of warm tropical sea water is not one of the necessary conditions for cyclogenesis; it *is* essential that the tropical fluxes are energy-rich. The VME normally stays close to the equator in the eastern Pacific (Figure 125, 'normal position'). As the VME moves southwards, the onset of cyclogenesis is made possible within it, as witnessed by the (regular) summer cyclonic activity to the east of Australia (Figure 125).

During El Niño periods, the displacement of the ME (under the impetus of the Australian monsoon) brings its associated, abundant rains eastwards (Figure 125, 'shifted position'), and there is a consequent decrease in precipitation in regions which are normally well watered. This displacement certainly does not bring 'drought', and neither is it responsible for spontaneous forest fires, as some have been quick to claim. Have they deliberately forgotten that the Indonesian authorities' policy of resettlement, to relieve the pressure on some overpopulated islands, involves the burning of forest areas? The movement of the VME also causes the eastward displacement of the cyclonic region of eastern Australia. Simultaneously, cyclogenesis has appeared in areas usually free from it (e.g., in Polynesia). Between February and April 1983, six cyclones crossed this region (e.g., Cyclone Veena, at Bora-Bora). The El Niño of 1997–1998 is also associated with '*a record number of tropical cyclones for the austral summer*' east of the International Date Line (WMO, 2004).

These cyclones, born within the VME structure and moving westwards, and also southwards (obeying the geostrophic force, at a maximum on their southern flanks), were then drawn in by the leading edges of MPHs driving into the tropical zone and deflected by them towards the south-east (Figure 125). One has to look right back to the beginning of the 20th century, to the period 1903–1906, to find similarly powerful phenomena, which, then, were not connected with any presumed 'warming'! The formation of these unaccustomed cyclones has, temporarily, 'upped' their numbers worldwide ... If the ENSO phenomenon coincides with a rise in sea temperatures, it is easy to assume that this rise is '*responsible for an increase in cyclones*'! Remember, though, that this rise in sea temperatures is a result of the replacement of the cooler waters of the South Equatorial Current by warmer waters as the Equatorial Counter-Current slips southwards, this movement being itself (like that of the VME, which encourages cyclogenesis) a consequence of cooling in the northern hemisphere. A point worth remembering!

13.3.4 The presumed causes of El Niño

The El Niño phenomenon is attributed to a multitude of different causes, which suggests that direct observation is not always to the fore. If we are to believe, for example, McPhaden (1999), writing about the '*child prodigy of 1997–1998*', it would seem that several factors could have contributed to the vigour of this Niño: '*one is chaos ... / ... a related issue is that of weather "noise" ... / ... one notable source of weather in the tropics is the Madden–Julian Oscillation ... / ... another possibility is interacting with the Pacific Decadal Oscillation ... / ... Global warming trends are yet another possible influence ...*'. In addition, of course, the possibility is not excluded that, as well as seasonal variability, '*unconsidered processes may also be involved*' (McPhaden, 1999). This is undoubtedly the best view of the origin of the process, since after such an investigation, it seems even more magical in origin!

However, there is **no reference to any actually observed facts**. Given that an El Niño episode, whether it happens in the east (as El Niño) or across the whole of the tropical Pacific (ENSO), is associated with the southward migration of the VME, the reasons for such a migration should be sought with reference to factors which can affect the frequency/strength of northern (and southern) MPHs. The rest is mere detail.

Long-term trend: rapid general circulation

The greater frequency of El Niño events, discussed above, since the event of 1972–1973, involving positive indices of the NAO/NAD and NPO/PDO 'oscillations' and the general upward trend in pressure curves, has been subject to a slow evolution since the 1970s. The rise in pressure is in the north, and easily explains the southward shift of the ME, the principal agent of El Niño. There is, however, also a rise in pressure in the south, albeit less pronounced. We are therefore not dealing only with a unilateral 'dilatation' of the northern meteorological hemisphere, encroaching step by step upon its southern counterpart: the latter is by no means passive.

The increase in pressure in the north, which is associated with cooling in the Arctic, and the increase in pressure in the south, associated with cooling in the Antarctic, place the current evolution in the Pacific space as in the other units of circulation, of course **in a context of rapid circulation** (Figure 33). In such a context, circulation is accelerated in both the north and the south, but the northern hemisphere still has the upper hand, and trespasses onto the southern (cf. Chapter 9). So, **the ME progressively edges southwards** (Figure 34).

Putting these phenomena in the context of general circulation allows us to relate phenomena which are remote from each other, not always in the first degree (different units, different seasons), an immediate example being the ENSO and polar ice, though this time in the right direction, that dictated by general circulation, here in the lower layers. We can also relate (this time in the second degree) the ENSO to rainfall in the Sahel, or even in India, since an ENSO episode is more likely to happen when the VME is already located in a more southerly position. Such a situation discourages rain in the Sahel, and in India, because the respective monsoons and associated pluviogenic structures do not move very far northwards. On the other hand the southward shift benefits the Brazilian *Nordeste*, and also leads to a stagnation in tropical cyclonic activity in the North Atlantic, since the VME is now near the Equator, and vorticity is discouraged. In the opposite situation, the episodes of La Niña (when the VME has moved northwards), see more intense cyclonic activity, as occurred in summer 2004.

Rapid evolutions: volcanic eruptions

Within this long-term trend there appear sudden 'jolts', coinciding with El Niño events. One of the possible causes of these short-lived changes, which must be able to influence the strength of MPHs, directly and rapidly, and often in a single hemisphere, is volcanism (cf. Chapter 6).

Handler and Ansager (1990) considered that, among the phenomena linked with El Niño episodes, '*volcanic aerosols from low latitudes are the immediate and only cause of an ENSO event*'. Mass and Portman (1989) limited this relationship to major volcanic eruptions only. For their part, Robock and Free (1995) thought that '*there is no obvious link between volcanic eruptions and El Niño events*'. The relationship is still a matter of controversy. Adams *et al.* (2003) re-opened the debate when they demonstrated '*a significant, multi-year, El Niño-like response to explosive tropical volcanic forcing over the past several centuries*'. However, their statistical analysis did not specify just how such a relationship might exist; it is evident that no relationship between the effects of volcanic aerosols and climatic consequences can be established directly without the intermediary of general circulation.

Figure 129 illustrates in diagrammatic form the progression of the aerological dynamic associated with the eruption of Mount Pinatubo (14–15 June 1991). By July 1991, the aerosols projected into the stratosphere lay above the tropical zone, and remained within the tropical easterly circulatory ring until the end of 1991 (Halpert *et al.*, 1993; Sato *et al.*, 1993; WMO, 1995). They spread mainly northwards, reaching their maximum concentration in March–April 1992 over the North Pole, with few aerosols over the South Pole, these being from Mount Hudson in the Andes of southern Chile. The aerosols attenuated the effectiveness of solar radiation, (which was already weak because of the low angle of incidence), with a consequent increase in the polar thermal deficit. Reinforced MPHs intensified the circulation of cold air towards the tropical zone, and, in turn, the AAs were reinforced (and shifted southwards in the process). Tropical circulation in the north, now stronger, pushed the ME south, setting up the conditions for the development of an El Niño episode.

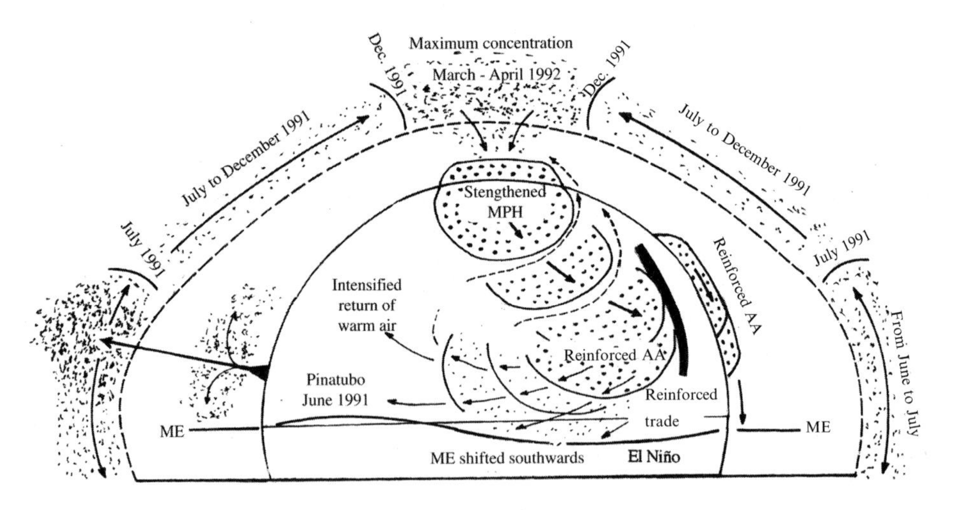

Figure 129. The progression of the climatic impacts of a volcanic eruption. From the Mount Pinatubo (Philippines) eruption, June 1991, to the El Niño event of March–April 1992. From Leroux (2001).

Note that the El Niño event which developed in the Pacific from December 1991 to April 1992 reached its maximum intensity during March–April 1992, at the same time that the concentration of volcanic aerosols from Pinatubo was at its greatest above the North Pole. Similar co-variations may be found, for example, in the case of the eruptions of Awu (1966), Hekla (1971), El Chichon (1982), and Nevado del Ruiz (1985). A close examination of these events confirms the importance of the dynamic of MPHs. However, such a close co-variation with El Niño episodes is not always observed (and/or needs to be better documented).

Other causes have been suggested, among them the effects of solar activity, for example by Landscheidt (2000, 2001, 2003), who considered that the NAO, the NPO, and the ENSO were directly influenced by variations in the intensity of the Sun's radiation. There are of course others who would involve the greenhouse effect. We shall come back to this.

13.4 CONCLUSION: THE ENSO, KEY TO THE RECENT EVOLUTION OF 'CLIMATOLOGY'...

El Niño is, therefore, not an 'extra-ordinary' phenomenon (i.e., removed from everyday meteorology). The way it has been dealt with, though, says much about the recent development of 'climatology', especially in its 'popular' form!

13.4.1 El Niño: an ordinary phenomenon of the tropical Pacific

The El Niño phenomenon has, firstly, an oceanic component and a meteorological component. The oceanic component is, locally, the warm El Niño current, which temporarily replaces the cold waters of the upwelling. More or less simultaneously, rain falls upon normally rainless regions (beneath the TI), though they are not dry regions, because they lie within the lower stratum of the maritime trade. The meteorological component at the origin of these rains is the VME, which is at this time shifted southwards by the northern dynamic.

The extension in size of the tropical Pacific, in the immediate vicinity of the equator, and the association with the SO (hence the acronym 'ENSO'), has not supplied an explanation of the initially observed phenomenon; instead, it has served rather to increase confusion. The 'oscillation' is still unexplained, and the origins of El Niño remain, strangely enough, shrouded in mystery.

Meteorologically, there is no 'mystery' surrounding these phenomena. The ME, terminus of pulses from both north and south, is constantly on the move, and its southward migration, like that of its northern counterpart, is observed on daily, seasonal, and even interannual scales, in various places in the tropical zone. So we are looking at something quite ordinary here; at least, ordinary in the eyes of the meteorologists!

Some oceanography: there is also an Equatorial Counter-Current in the Atlantic, which becomes the Guinea Current and brings warm waters into the Bight of Biafra (Figures 85 and 86). In the Indian Ocean, too, there is an Equatorial

Counter-Current in northern winter, flowing from the Somali and Kenyan coasts towards Indonesia. Nothing extraordinary there, either! Did we not hear in 1984 of '*an Atlantic El Niño*' (WMO, 1987)?

Whatever this phenomenon is called, be it El Niño or the ENSO, there is, as we have seen, nothing 'phenomenal' or peculiar about it. Or perhaps everything that results from variations in the activity of the ME is peculiar!

To recapitulate on the essential points:

- an **El Niño event is a local phenomenon, not a global one**;
- it is responsible only for local climatic consequences, and not for global ones;
- it does not 'rule the world', nor even itself;
- in fact, it is itself a consequence, and one of many it is not a cause; and
- it stands at the very end of a chain of processes, and not at the beginning.

So it is neither 'unique' nor 'sensational' as a phenomenon. It is merely the **local consequence (as El Niño) or the regional consequence (as the ENSO) of the dynamic of meridional exchanges originating near the poles and reaching deep within the tropical zone**. Hard luck for the tabloids! And a shame, too, for the IPCC, whose literature on the subject thus qualifies as 'sensationalist' reporting, if they say, for example, that the ENSO '*consistently affects regional variations of precipitation and temperature over much of the tropics, sub-tropics and some mid-latitude areas*' (IPCC, 2001)!

13.4.2 'Chicken or egg' meteorology: the 'Walker cell'...

The facts (either closely or remotely) associated with El Niño are not presented within the context of a more general organisation, or coherent schema of general circulation (cf. Chapters 7 and 8). The watchword is 'teleconnections' (the *tele-* is the operative part, in fact), and with it, facts may be presented in any order or direction, this way round or that way round, without the need being felt to suggest scientifically based relationships in which cause and effect are unambiguously described. This 'chicken or egg' approach means that links are sometimes chosen to fit in with what is to be demonstrated, and any respect for the reality of phenomena goes out of the window! Such un-meteorological thinking is behind the pattern of cellular zonal circulation known as the 'Walker cell', a schema created in order to 'explain' the ENSO.

Walker did not invent the cell which bears his name: what he did was to suggest the idea of the SO, upon which the concept of the zonal cell was based. The first person to propose the existence of such a cell was J. Bjerknes (1969), without however suggesting the general pattern involved. According to Bjerknes, there was subsidence over the eastern Pacific, with ascending air in the west (Indonesia), and easterly winds (trades) in the lower layers, with westerly winds in the upper troposphere. This theory, please note, required (and still requires?) that the absence of rain in the Peruvian coastal desert is the result of air moving down from upper layers (even here, close to the equator!). This definition represents 'normal' zonal circulation, now known as La Niña. Such a circulatory cell '*tends to be less pronounced, or works in the opposite direction, during El Niño events*' (Beltrando, 1995). The concept was progressively extended to the whole of the tropical belt, and the number of cells multiplied, with as many as seven around the equator (Krishnamurti *et al.*, 1973). Successive simplifications have led to the following pattern: '*Ascending air is found over warmer regions and subsident air over cooler regions*' (Fontaine, 1991).

A zonal cell highly improbable

From the outset, *this zonal 'Walker cell' circulation*, said to be situated near the equator, is highly improbable (Leroux, 1994a)!

- Deep within the tropical zone, at altitude, circulation flows not from the west, but from the east, and the Tropical Easterly Jet (TEJ) flows high above, directly along the VME. A westerly can occur here only if a talweg moving in from middle latitudes projects well into the tropical zone. This would also happen if the extra-tropical westerly jet moved in to replace the easterly jet at the equator, which would involve a seasonal shift of considerable amplitude (cf. Figure 30, Chapter 8), and this displacement would also, of course, involve the VME, which would be displaced even further!
- Updrafts and downdrafts are said to be occurring over the ocean, as a result of sea temperatures. **This is nothing but a physical impossibility**! It should be remembered that, across an ocean surface, temperature differences are minimal, especially in the vicinity of the equator, and the terms 'warm' and 'cold' have very limited meaning in this area. In any case, there is no way in which they could engender vertical movements, either upward or downward, and certainly not throughout the troposphere (at speeds which have never been specified!). If this were the case, anyone standing upon the *inlandsis* would be flattened, while updrafts over the Sahara would move at dizzying speeds in the summer; indeed, they could be so strong that everything would fly up to the tropopause, since ground temperatures can be so high (especially when compared with sea temperatures)! However, these updrafts, a result of which is the Saharan thermal low, do not rise beyond 2,000 m, and above, the tropical structure remains unchanged, and indeed, the normal easterly circulation may well be strengthened. Need we remind ourselves that when the *harmattan* blows from the land onto the Atlantic Ocean, it does not 'plunge' downwards over the Mauritanian upwelling, but continues on its way across the ocean above the lower stratum of the maritime trade, with its load of dust (and sometimes locusts), carried far beyond the Cape Verde Islands, towards the Caribbean (Leroux, 1983, 2001)?
- In order for subsident air to reach ground level, trade circulation would have to be homogeneous across the whole of the troposphere; if this were true, there would be no TI to separate vertically two strata of different origins and natures. The 'Walker' schema does not take the stratification of the trade into account, and therefore overlooks specificities of tropical circulation. The most serious omission is that of the TI (at approximately 800–1,000 m) which, being unpro-

ductive, is the root cause of the coastal desert of Peru (or of California, Mauritania, or Namibia).

- The existence of this cell, which is just an unproven hypothesis, has never been verified by observation. Yet, to cap it all, we are told that it can '*work in the opposite direction*' (Beltrando, 1995)! This would mean that, during an El Niño event, the trade would be blowing eastwards(!), reversing its normal behaviour (but subject to what physical logic?); and at the same time the TEJ would finally be moving ... in the right direction! Eventually, everyone stops thinking rationally about this (especially those using this theory, who never back up their propositions with any concrete examples)! ... Though it would be interesting to see a trade wind so 'disoriented' that it would retrace its steps, and it would be even more interesting to discover what factor might be capable of setting the laws of circulation into reverse!

The fact is that the concept of the imaginary 'Walker cell', originally a simple hypothesis, has become, through the abandonment of meteorological logic, a simplistic amalgam of:

- undemonstrated 'links', for example between cold sea water and the absence of rain, and/or warm water and the presence of rain (again we meet the famous 'T/R equation');
- erroneous links between subsident and/or ascending air above the ocean, a *milieu* where thermal conditions are inappropriate to vertical movements of this nature; and
- the effects of events widely scattered in both space and time but brought together, statistically, in the vicinity of the Equator.

There are, therefore, no 'Walker cells'. The causes of the real meteorological events which this imagined circulation is supposed to explain are sufficiently evident to be recognised for what they are: always assuming that there is actual observation of reality (though the procedure seems to be falling into disuse).

The reality of meteorological facts

Actual observation of reality shows that, during El Niño events (cf. Figure 125):

- The abundant rain in Ecuador and Peru is the result of the southward shift of the pluviogenic structure of the VME.
- In the lower layers, the 'appearance' of a westerly flux across the western equatorial Pacific does not mean that the trade has 'changed direction' but that the Australian monsoon has gained strength, and is pushing the VME further south and east.
- The lack of rain in Indonesia and northern Australia, and to the west of the International Date Line (WMO, 2004), is a result of a rise in pressure (leading to a fall in sea level and to the waters of the Equatorial Counter-Current being

impelled eastwards). This rise in pressure is associated with the strength of Asian winter MPHs.

- Increased rainfall to the east of the International Date Line is a result of the southwards movement of the VME with its abundant rains; falling pressure here is associated with a rise in sea level.
- The appearance of tropical cyclones in the South Pacific is associated with the general shift of the VME, presenting (temporarily) favourable conditions for cyclogenesis.
- Finally, an El Niño episode is nothing more than a southward extension of the North Pacific aerological unit, be it brief or prolonged, localised or more general.

However, this has not prevented 'Walker cells' from popping up everywhere! For example, in Africa:

- In equatorial East Africa: '*the zonal cellular* ('*Walker*'-*type*) *circulation, during rainy spells* (*vs. dry*), *is characterised by the reduction* (*vs. intensification*) *of the subsident branch over the western Indian Ocean and eastern Africa, and of the ascending branch over the eastern part of the ocean*' (Beltrando, 1990).
- In West Africa, drought may be associated with '*a strengthening and/or eastward displacement of the subsident branch of the Atlantic* '*Walker*'-*type cell*' (Janicot and Fontaine, 1993).
- In tropical East Africa: '*In October–December ... rainfall in southern East Africa* (*Tanzania, Malawi, Zambia, Madagascar*) *is excessive* (*vs. deficient*) *during those times when the zonal Walker circulation in the Indian Ocean is weaker* (*vs. stronger*) *than normal*' (Richard, 1993).

These statements are accompanied by diagrams, with arrows pointing in all directions, illustrating the simplistic principle that 'rain = ascending air', and 'lack of rain = descending air'! There has, of course, been no radiosonde investigation to confirm the existence of Janicot's hypothetical descending branch (1992, 1993), over the West African upwelling, although there have been investigations which might be referred to: from Nouadhibou, Sal (Cape Verde Islands), and Dakar. These sonde results will undoubtedly not 'paint the same picture', but will rather reveal the presence of a vigorous TI, invalidating the famous old schema! Also, no references are made to aerological weather data for East Africa, even though there are plenty of them. Neither do we find reference, in these schemas, to the two structures of the ME (the IME and the VME), or to the low-level jet over eastern Africa. Where is the Somali monsoon (*kusi*) mentioned? Or the Madagascar monsoon, or the TI (see above)? Or, in fact, anything which might count against such unfounded flights of fancy? There seems to be no attempt to take account of the real vertical aerological structure above Africa, as it has been described on the 250 meteorological charts in the Atlas section of *The Meteorology and Climate of Tropical Africa* (Leroux, 1983, 2001).

Given all this 'freedom of thought', one might even go on to imagine, draw up, and publish a schema which 'explains' aridity in South Africa: '*The rain of the summer Indian monsoon is supplied by a flux in the lower layers, from the direction of the Arabian Sea. The return flux in the upper layers heads for South Africa, where it causes strong downdrafts*' (Sadourny, 1994). This schema, then, has strong updrafts (and clouds) over northern India; an arrow points directly (at altitude) from the Tropic of Cancer towards the Tropic of Capricorn, crossing the TEJ and/or the zone of turbulence above the VME with no ill effects, and plunges unerringly towards its target in southern Africa, during its dry season (fortunately)! As if this were not enough, Sadourny adds that, '*in the winter, the situation is reversed . . .*'! The fact that the author, who is presenting someone else's half-baked theory, is a former director of the *Laboratoire de Météorologie Dynamique* (LMD), seems to give him *carte blanche*! The LMD, a great proponent of the greenhouse effect scenario, seems oblivious to MPHs: some of its staff even vehemently deny their existence ... but conversely clearly see the direct exchanges in upper levels between India and south of Africa!

13.4.3 A decisive moment in the 'hijacking' of meteorology by models...

The treatment of El Niño represents one stage in the 'hijacking' of the discipline of climatology by climatic modelling, a situation denounced by Dady (1995). This approach to the subject shows **the triumph of the statistical methodology imposed by models**.

A defining moment in this process, to cut a long story short, was an article by Ropelewski and Halpert (1987), published in full by the WMO (1987). This article, like many others before and since, '*involves the direct digitising of the same equations, and ignores the physics of major atmospheric systems*' (Dady, 1995). An analysis of this sort can easily lead to the conclusion that '*major ENSO episodes ... lead to massive displacements of the rainfall regions of the tropics, bringing drought to vast areas and torrential rains to otherwise arid regions. The related atmospheric circulation anomalies extend deep into the extra-tropics ...*' (WMO, 1987), meaning that the ENSO is now 'a global event'.

Reference is made to tropical phenomena, but no dynamical explanation of them is included (nor, *a fortiori*, analysed: pluviogenic factors have lost their importance!). Anomalies of circulation are invoked for extra-tropical areas, but no details are specified, and neither are the mechanisms explained through which a minor equatorial event might influence a polar or temperate phenomenon with very different causes! The analysed data have their origin in meteorology, but their interpretation has now nothing to do with meteorology!

The WMO gave its official blessing in 1987 to this vision of phenomena, in the context principally of an '*ENSO and worldwide regional precipitation response*', thereby establishing an illusory 'causal link', completely undemonstrated in reality, and the analysis being of course a purely statistical one.

So, in the 1980s, **links were proposed which drew 'climatology' closer to magic, and rapidly distanced it from meteorology**!

This computer-based, statistical approach, with its ever more remote teleconnections, will prevail; the point of view of those meteorologists who rely on the direct observation of phenomena will be supplanted by theoretical and hypothetical modelling. Those interested in real atmospheric phenomena seem to be a dying breed (or they are well hidden). Under the heading of 'climatology', the data files are fed with the climatic parameters, and the ENSO becomes responsible for everything that happens ... except, of course, for whatever isn't in the computers' memory banks! Why not introduce into the system data about road and rail accidents, the incidence of rabies or toothache, the divorce rate and domestic rows? The ENSO would surely be held responsible for most of these parameters! It has become normal to cast the net wide, and to attribute to some selected phenomenon the responsibility for all modern and recent phenomena, the world over! This situation of *tout permis* brooks no argument, as soon as it is stated that 'this has been through the model'. A statistical link which is a simple coincidence or co-variation becomes *ipso facto* a physical link of cause and effect, even though neither the cause nor the effect are specified! To cap it all, there are those who consider such specification to be 'no use', since it is all built into the models anyway!

Modellers challenged on this point claim, in fact, that physical links are 'understood', and included in the programming. There is such faith in the models that the question of what physical links are involved, and their relevance to the subject, is never asked; does the image within the model match reality? (cf. Chapter 7). For example (and there are countless such examples), Gloersen (1995) examined the '*modulation of hemispheric sea ice cover by ENSO events*'. He repeated the credo: '*ENSO events ... are the largest source of interannual variability of temperature and precipitation on a global scale*', and proposed that the extent of sea ice, particularly in the Arctic, represented a '*response*' to variations in the ENSO. No reservations were expressed, and the ENSO is clearly stated to be responsible for variations in sea ice cover! Even though, as Lenoir (2001) pointed out, '*the two maxima of the anomalies in sea ice extension occurred respectively 4 and 5 months before those of the ENSO index*'. This pre-dating dispels any argument about the order of events, and confirms the important role of polar latitudes and MPHs in influencing the ENSO phenomenon. However, a statistically based analysis is not concerned with such details, meaning that 'climatology' no longer has to follow the path of the logic of phenomena!

In this climate of opinion, the way is open for the greenhouse effect. The changeover from one to the other is a natural progression ...

13.4.4 From ENSO to greenhouse effect... a 'natural progression'!

The (meteorologically) illogical way in which the El Niño phenomenon is approached goes some way towards answering the question asked in the conclusion to Chapter 4. That question was: what happened in the 1980s, and particularly in 1985, to tip the balance of explanations so firmly towards the greenhouse effect?

The ease with which the ENSO became 'ruler of the world' can only be understood with reference to the ground lost by meteorology to computer-based methods. When referring to real things is no longer thought necessary, 'anything goes'! This approach even receives official backing, from within the WMO, and from the IPCC too! There is just one further small step to be made ... and it will not take long.

Now, it appears that the ENSO and the greenhouse effect are in competition, especially if we are to believe the popular science in the media. Responsibility for the world's weather is now divided between:

- The greenhouse effect, said to be influencing the ENSO: '*If the average temperature is rising in the tropics, it is difficult to imagine that the nature of the El Niño phenomenon will not be affected*' (Le Treut, in *Le Monde*, 18 November 2000).
- The ENSO, also supposed to be responsible for increased warming: '*the rise in global temperature since 1976 could be linked to the increasing frequency of El Niño events. These could be raising the temperature of the Pacific Ocean, with consequences for the world's climate*' (Juillet-Leclerc, in *Le Monde*, 1 December 2000).

A difficult choice to make!

However, the greenhouse effect is a tough contender, and the IPCC was quick to come down on its side:

- Firstly by considering that there was probably no major modification of the ENSO going on: '*current projections show little change or a small increase in amplitude for El Niño events over the next 100 years*' (IPCC, 2001, p. 16).
- Further, by estimating, in spite of everything, that '*even with little or no change in El Niño amplitude, global warming is likely to lead to greater extremes of drying and heavy rainfall and increase the risk of drought and floods that occur with El Niño events in many different regions*' (IPCC, 2001, p. 16)!

The game is over: the ENSO is dethroned, and the new 'ruler of the world' is the greenhouse effect!

- So the ENSO has been neutralised: it will not change, at least before 2100!
- Even if global warming is going to get worse, the ENSO will not (though it is not clear why, given the IPCC's article of faith on this point).
- Extreme phenomena attributed to the ENSO are going to get worse.
- The prime mover will be the greenhouse effect, and not the ENSO!

With this (scientific!) sleight-of-hand, the IPCC does away with the ENSO, but transfers the extreme phenomena associated with it (droughts, torrential rains, floods, etc.) to the greenhouse effect: and on a global scale! The same old clichés can be recycled – all that is needed is to change the name of the culprit!

This *volte-face* is interesting for more than one reason. Consider the following points:

- The IPCC, which cannot predict the ENSO, still does not know the causes of its variations.
- The IPCC therefore does not know that **an ENSO event runs counter to the greenhouse effect scenario, which would encourage instead the so-called La Niña phenomenon**.
- The IPCC still attributes effects to the ENSO which it cannot cause. This means that the phenomenon remains just as mysterious as before.
- The real causes of the variations in extreme meteorological phenomena, such as drought and heavy rain, and the rest, are still not clear.

The promotion of the greenhouse effect to climatic 'stardom' can be taken no more seriously, therefore, than the similar fuss in the past over the ENSO: indeed, here we have two absolutely contradictory phenomena, both thought to be responsible for the same climatic effects!

The 'climatological' community seems to have expressed little objection, and the media even less, to the shifting of the 'blame' from the ENSO to the greenhouse effect. Sometimes the two 'climatic culprits' conspire, each implicated in its turn in dramatic events.

This situation highlights the surprising versatility of certain people in the so-called 'scientific' world. It is often these same 'scientists' who swing from one explanation to the other, confirming **the poverty of the meteorological content of the 'greenhouse effect dossier'**!

The lessons of the observation of real facts in the aerological units: Conclusion

You're spinning us a yarn, said the Fulani shepherd disdainfully.
– Yes, replied the crocodile hunter, but if everyone spins it long enough, it will start to sound like the truth!

J. and J. Tharaud. *La Randonnée de Samba Diouf*, Fayard, Paris, 1927.

The North Atlantic and North Pacific units of circulation do not constitute the whole of world circulation, but it is not the intention of this book to cover the subject exhaustively.

In the northern hemisphere, the 'central Europe/western Asia/northern Africa' unit (Figure 26) remains to be analysed. We have already mentioned certain sectors within this space (e.g., the Mediterranean and northern Africa). Also, as stated, MPHs on the *Scandinavian* and *Russian* trajectories bring cold air and high pressure to the Balkans and the eastern basin of the Mediterranean (Chapter 11). The pressure curve for Constantza (Romania) is therefore quite remarkable, revealing, for the 1970s, a considerable rise in average annual pressure of 4 hPa (Aubert, 1994). Schönwiese and Rapp (1997) showed that, during the century between 1891 and 1990, temperatures fell by 1°C in Scandinavia and beyond, into central Europe, on the MPH trajectory; while in the Ukraine and southern Russia, they rose by about 2°C, on the trajectory of the cyclonic air returning from the south. This evolution was confirmed by Litynski (2000) for 1931–1960 and 1961–1990.

Recent studies highlighted at www.co2science.org show that the climate of northern Europe is quite unlike that predicted by models. For example, Raspopov *et al.* (2004) pointed out that the temperature of the Kola Peninsula exhibited two equal maxima around 1950 and 1990, but in the last decade, the temperature has fallen. Koslov and Berlina (2002) showed that, in this peninsula, between 1930 and 1998, the snow-free period decreased in length by 20 days and the ice-free period for lakes decreased by 15 days. In the same period, the growing season for vegetation became shorter, with the onset of spring happening later, and autumn coming

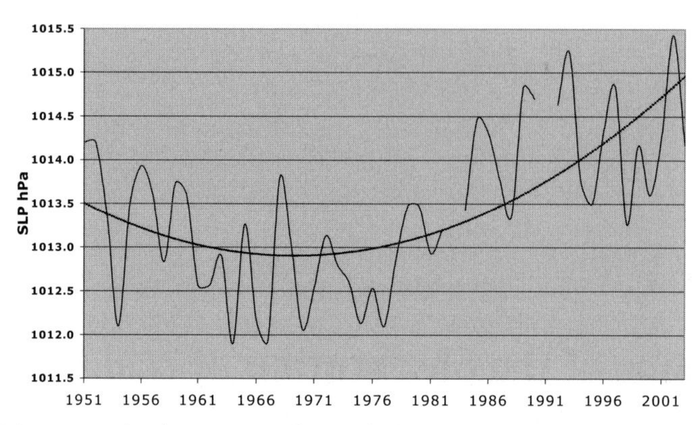

Figure 130. Mean sea level pressure (hPa) during winter (JJA) at Agalega, Mauritius (10°26′S/56°45′E), from 1951–2003.
From UCAR and WMO data.

earlier. Eschenbach's analysis (2002) for five stations in the north of the Scandinavian peninsula during the 20th century is thus confirmed; the curves in fact show that the warmest period occurred during the 1930s, and, although the temperature has recently shown an increase during the period 1981–2001, it does not rival the mid-century thermal optimum.

The evolution of this unit has been identical to that of the other aerological units of the northern hemisphere.

Data are less plentiful for the southern hemisphere. We have noted that the extremities of continents where MPHs arrive directly from Antarctica are cooling (Figure 53, Chapter 10), and that there is a general rise in pressure off the Atlantic coast of southern Africa, in the 'St. Helena' AA (Figure 83), and off South America in the Pacific, in the 'Easter Island' AA (Figure 128). Figure 130 shows the same evolution since the 1970s for the Indian Ocean (Mauritius (i.e., in the 'Mascarene' AA)). These examples, albeit limited, nevertheless confirm (as already underlined) that the southern hemisphere, like its northern counterpart, is moving into a mode of rapid circulation (Figure 33).

This rapid circulation mode is characterised (cf. Chapter 9) mainly by an intensification in meridional exchanges from the poles to the tropical zone, and from the tropics back towards high latitudes. Within each unit of circulation, climatic contrasts (especially thermal) are widening between their western and eastern flanks. This means that any average value established for a given unit will become less valid as meridional exchanges accelerate.

NO 'ATLANTIC CLIMATE', NO 'PACIFIC CLIMATE', NOT EVEN A 'GLOBAL CLIMATE'

What can be said about climatic evolution in the North Atlantic unit? Scandinavians may well talk about 'warming', but North Americans will contradict them, saying it's getting colder. A 'mean Atlantic temperature' has no climatic significance. A 'mean rainfall figure' would also convey little, and neither would 'mean pressure'. So **there is not really an 'Atlantic climate'**. We might, though, acknowledge the fact that the weather is becoming more irregular and more violent.

Similarly, there is no 'Pacific climate', and neither is there a 'South Atlantic climate', nor an 'Indian Ocean climate', and so on ... An 'average climate' for the six aerological units (i.e., a 'global climate'), is therefore meaningless. **There is no 'global climate', and so there can be no global climatic (or thermal, or pluviometric) evolution.**

The IPCC's official mean thermal curve (Figure 46) has no meaning, in climatic terms. All it really represents are arithmetical mean values – totally artificial – for disparate observational data, which are dissimilar and contradictory.

So where does this positive value for the temperature anomaly, especially since the 1980s, come from? Is it a result of calculations of mean temperatures? Or of the way in which the thermal anomaly is determined? Or of the fact that cold air causes a return flow of warm air, which may diminish its own impact on thermometers? Is it because the spaces covered by cold air are not as large as those covered by the returning warm air? Is it the intensity of the transfers of tropical warm air towards regions which are normally colder? These are all questions that need answering. But whatever is going on, **this curve does not represent global thermal evolution.**

One cannot give credence, in the first degree, to the IPCC/WMO official mean thermal curve (Figures 46 (Chapter 10) and 131), just as one cannot ascribe an absolute and immediate value to any observational datum, or reach automatic conclusions about some supposed 'warming'. Such a curve cannot, in fact, represent a mean thermal evolution which does not exist! Perhaps though, *à la rigueur*, it could indicate that some evolution is occurring. But which?

THERE IS NO PROOF OF THE SUPPOSED 'GLOBAL WARMING'

The time has come to offer a final answer to the question asked in Chapter 10. There, it was pointed out that *only in the most recent decades (the presumed Recent Climatic Optimum (RCO))* can one try to locate the IPCC scenario.

Also, this supposed warming during the RCO:

- is 'illustrated' by *only one* 'official' mean temperature curve;
- might simply be due to an urban effect; and
- is clearly rebutted by satellite and sonde observations.

It was also stated that the climatic significance of the mean global curve had not been demonstrated, and that only one course remained if the validity of the IPCC scenario

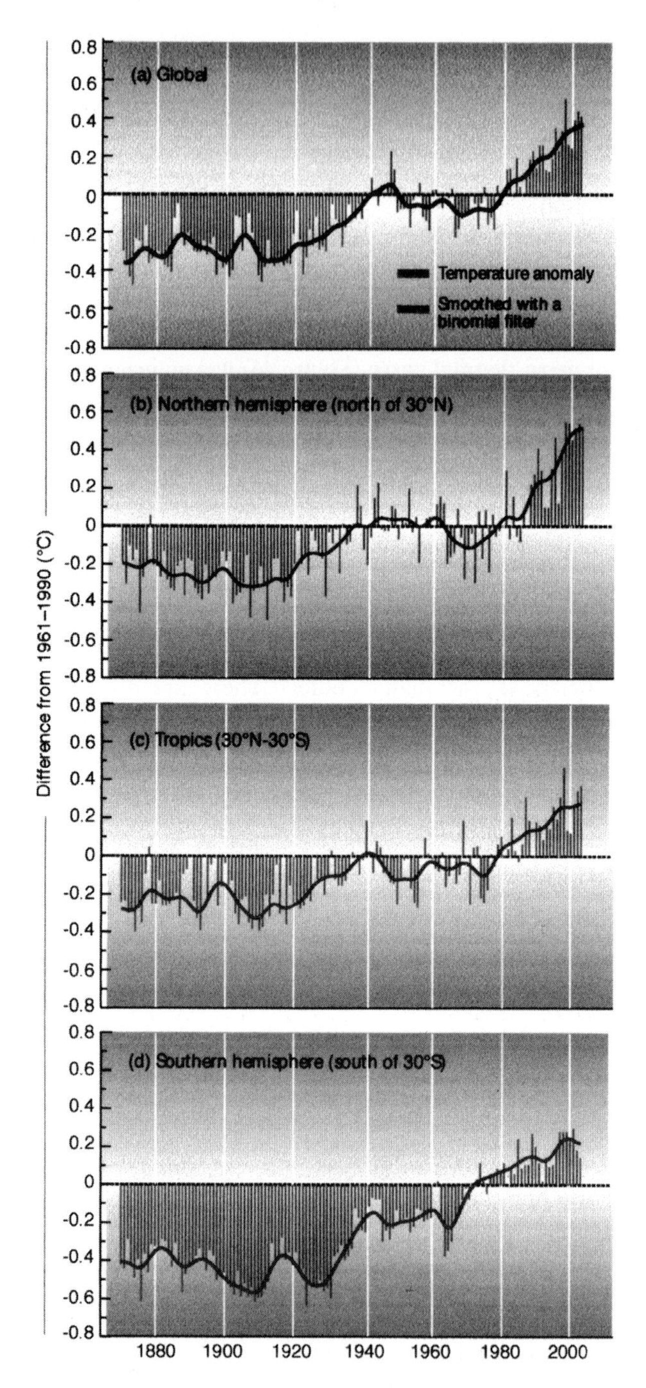

Figure 131. Combined annual land and sea surface temperature anomalies (°C) from 1861–2003; departures from the average in the 1961–1990 base period.

Sources: Hadley Centre (Met. Office) and Climatic Research Unit (University of East Anglia), in WMO/WCP (2004).

were to be determined: an analysis of the intervention of the aerological dynamic. This has now been done.

The dynamical factor is responsible for different and contrasting regional weather behaviours within each unit, and it invalidates the notion of a 'mean climate'.

This factor points to one fundamental conclusion: the IPCC is relying exclusively upon the official thermal curve (Figures 46 and 131) to back its claim of global warming. **Since this curve is not a reference, and signifies nothing in climate terms, there is absolutely no proof that our planet is getting warmer.**

AN INTENSITY INDEX FOR MERIDIONAL EXCHANGES

According to the modes of general circulation and the indices within the aerological units:

- With slow general circulation (i.e., when the NAO/NAD and NPO/PDO indices are negative), meridional exchanges are weaker, as are thermal contrasts: temperature anomalies are also slight, and the mean temperature curve remains relatively close to normal.
- With rapid general circulation (i.e., when the NAO/NAD and NPO/PDO indices are positive), meridional exchanges are intense, and thermal contrasts wide: temperature anomalies are larger, and depart from the normal.

Unless some thermal criterion can be established, **the 'famous' IPCC curve (Figures 46 and 131), without climatic significance, might be used as an index of the intensity of general circulation**.

One effect of the acceleration in exchanges which has been observed since the 1970s, with lower temperatures at the poles, has been to transfer more perceptible and latent heat towards higher latitudes, thereby conferring upon them a mean thermal benefit. **So the resultant 'warming' is only an artefact, a mathematical result, of the aerological dynamic**. This 'heat' is really associated dynamically with accelerated advection, which itself stems from an initial cooling.

Figure 131(b) and (d) shows, via the overall similarity of the curves for the northern and southern hemispheres, that general circulation has recently (since the 1970s) entered a rapid scenario, though this is more clearly seen since the 1980s. This scenario means that the northern hemisphere will gradually 'take the initiative', and by its enlargement cause a southward shift of the ME. The differences in the curves require a more detailed analysis.

Let us examine the stages in this dynamical process in the northern hemisphere, north of latitude 30°N (Figure 131(b)).

- We begin by looking at temperature evolution in the Arctic (Figures 54 and 55, Chapter 10), since circulation originates there: the Arctic experienced its highest temperatures in the 1930s and 1940s. Of course, mean temperatures in the Arctic should be treated with caution during this period also, but they are less open to question during this period of slow circulation (i.e., more 'representative') than they might be during a rapid circulation period (cf. Chapters 8 and 9).

- The resultant cold was transmitted by MPHs towards lower latitudes. Figure 131(b) reveals this (as yet modest) transfer; mean temperatures then suggested the 'return of the Little Ice Age'.
- During the 1970s and 1980s, more intense cold and the increased power of MPHs brought about a stronger return flow of tropical warmth to the north. The thermal cold/warm balance tipped more and more towards warmth, and the resulting 'mean' temperature rose.
- Ever more intense advections of warm air now reached high latitudes, and they 'masked' the perception of the phenomenon, since they affected even the Arctic temperature, established on the scale of means (cf. Figure 55). However, the Arctic 'was not warming up' everywhere (cf. Figures 93, 94, and 95, Chapter 12), and the process of cooling continued, and meridional transfers intensified, as the evolution of the weather showed. Let us stress once more that, the more the Arctic 'warms up', the more intense the advections of warm air from the south (responsible for this warming) will be, and the more powerful MPHs will be, meaning a more severe initial 'coldness'.

Since the 1980s, this illusory warming has been greater in the northern hemisphere (+0.71°C in 2003) than in the southern (+0.15°C in 2003). This must therefore be symptomatic of greater activity in the northern meteorological hemisphere, which encroaches upon its southern counterpart. The 'record' held by the northern hemisphere in 1998, an El Niño year, is thus logical, since it reflects the increased strength of northern MPHs, a greater enlargement of the northern meteorological hemisphere, and more intense warm air flows returning north. Of course, and it is worth repeating, El Niño is not responsible for the intensification of meridional exchanges!

CLIMATIC EVOLUTION, APPARENT AND REAL...

To confirm these processes, we require an objective analysis of polar temperature data, which would offer an insight into the real evolution of temperatures, without recourse to the 'mush' of mean values which mask this evolution. It would be particularly necessary for this analysis to differentiate between regional temperatures as a function of the dynamic (i.e., as a function of the origin of the fluxes).

Meanwhile, though, it is the IPCC (2001) which – involuntarily – provides confirmation of these processes. It confirms in particular one 'delicate' point of this argument: the continuation of cooling in the Arctic at the time of the climatic turning point of the 1970s, while temperatures were apparently rising (Figure 131).

Looking back now to Figure 20(a) (Chapter 6), let us first of all discard the curve showing the imaginary warming, and adhere to the effects of natural causes, both solar and volcanic:

- According to these causes (Figure 20(a)), temperature culminates between the 1920s and the 1960s, and thereafter falls.
- These natural (essentially solar) factors encourage rapid warming in the early 20th century, between 1920 and 1940 (Figure 131).
- These natural factors also explain the CCO during the period 1940–1960 (Figure 131).
- The CCO came earlier and was shorter in the Arctic (Figures 54 and 55, Chapter 10), since these factors show themselves early and with greater acuity at high latitudes.
- Later temperature measurements in the Antarctic do not lead to the same conclusions, but the long-term data in Figure 53 (Chapter 10), especially from Cape Town, where the CCO was quite evident, partly compensate for this gap.
- The subsequent fall in temperature between 1960 and 1980 (Figure 131) also reflects the influence of the unfavourable factors of that period (Figure 20(a), Chapter 6).
- From the 1980s onwards, average temperatures suggest a mean 'warming', especially in the northern hemisphere, in contradiction to the natural factors, and the curves diverge, cooling (Figure 95, Chapter 12) or warming (Figure 131). However, just in time to rescue the IPCC's illogical approach (Figure 20(b), Chapter 6), there appears a *deus ex machina* which had not yet previously shown itself (and why not?): the greenhouse effect! The reason: the absolute necessity to 'stick with' the rising thermal curve (Figure 131)!

However, as the aerological dynamic proves, **the recent 'global warming' is mere fiction**, and the efforts of the IPCC to promote this fiction, come what may, are as pathetic as they are pointless.

For the claimed 'global warming' is the result of the arithmetical mean (apparent evolution) of two different thermal behaviours (real evolutions):

- cooling, which is subject to natural factors, acting along the trajectories of the MPHs which propagate them; and
- warming, which is the 'paradoxical' result of transfers of warm air caused by the increased power of MPHs, itself linked to the initial cooling.

If we allow the conclusions of the IPCC about the essential role of natural factors (Figure 20(a), Chapter 6):

- The cooling associated with the solar and volcanic factors will continue, affecting high latitudes in particular; the strength of MPHs will increase; the imaginary temperature rise will continue; the notional 'apparent global warming' (which the 'official' curve supports) will be more and more important, since this claimed warming is only apparent.

- At the same time, waves of cold along the paths of MPHs will become more and more severe, and the weather will become ever more violent and irregular.
- We will be less and less prepared, especially in regions where cooling is already in effect, for the reality of the climate of tomorrow!

Perhaps we will finally realise that **we must not confuse the apparent and the real evolution of the climate, and that the supposed 'global warming' is only an artifice, a statistical fiction**!

When, though, will we be ready to realise that climate described by the models is far from the real climate? Do we have still to wait until natural factors force us to understand?

14

The observational facts: Sea level and circulation

The most unnerving thing about the future is that it is unknown: so any prediction may be welcome. Most predictions are quickly forgotten, since they have served their purpose, which is not to be accurate, but to distract us for a while ...

Didier Norton, *Pour La Science*, No. 170, 1991.

One of the threats constantly aired by the International Panel on Climate Change (IPCC) and the catastrophists is that of a dramatic rise in sea levels. This is directly linked to the supposed 'warming', via the key formulae: 'the temperature rises/the water increases in volume, and/or the ice melts/sea levels rise'. The Flood awaits us again.

There is really no justification for suggesting this, since, as has already been pointed out, the Earth is not warming up generally. There is therefore no reason for sea levels to be rising everywhere, if the cause of this is supposed to be the expansion of the water and the melting of the ice. However, this subject has become so much a part of the greenhouse effect debate – for scientists and also the media – that it needs to be discussed here, if only briefly: it is a falsehood, dishonestly used.

Progressively, and especially as meteorology has drawn away from subjects proper to it, oceanography has become more prominent in the 'explanation' of phenomena. For example (see above), sea surface temperatures have become a veritable 'universal panacea', brought in to explain both drought and rain, in the tropics as well as in the temperate zone. Oceanic circulation has become a possible cause of (incredibly rapid) climate change, with hypothetical slowing, or even interruption, of thermohaline circulation or the Gulf Stream likely to trigger waves of cold, or even glaciation!

It therefore seems useful to look at one of the main claims made by the IPCC, one which gives small island states 'threatened with disappearance' a deal of political clout at IPCC meetings, and which provides the media with endless tear-jerking stories about floods, deluges, and giant waves: spectacular stuff – for which humanity is blamed!

14.1 SEA LEVEL RISE?

The idea of a 'sea level' is essentially a relative one, since, far from being uniform, there are differences in 'level' of several tens of metres. This is the result of two contrary movements: global variations in sea levels (eustatism) and variations in the level of the Earth's surface (isostasy). Continental subsidence has the same effect as a eustatic rise in sea level, and '*in coastal regions, subsidence is more frequent than uplift*' (Pirazzoli, 1996).

Whatever the rash pronouncements may say, **there are no real certainties about the evolution of sea levels**. Global estimates based upon data from tide gauges show:

- a rise of between 1 and 1.5 mm per year (13% of results);
- a rise of between 1.5 and 2.4 mm per year (17%);
- a rise of more than 2.4 mm per year (21%);
- a rise of between 0.1 and 1 mm per year (20.5%);
- a relative stability (1%); and
- a relative fall in sea levels (27.5%) (Pirazzoli, 1996).

Given these readings, with 48.5% showing no rise or a rise of less than 1 mm/year, the average figure of the order of 1 mm/year has very little significance.

It would seem from tide gauge measurements that sea levels have risen, very gradually, by about 15 cm over the last hundred years, though even this is not certain. The consensus is that no study claims that levels have fallen, while there is no proof of any global acceleration (Pirazzoli, 1996). It may even be, as Cabanes *et al.* (2001) state, that '*the 20th century sea level rise estimated from tide gauge records may have been overestimated*'. The presumed rise might therefore be less than 15 cm during the last century.

According to the models and the IPCC, '*global average temperature and sea level are projected to rise under all IPCC SRES scenarios*' (IPCC, 2001). Previous estimates from the IPCC were much more dramatic, dealing in metres, but they were soon revised downwards, and now '*global mean sea level is projected to rise by 0.09–0.88 m between 1990 and 2100*' (IPCC, 2001). Since these latter values (9 cm to 88 cm), with a probable median value of 47 cm, run the risk of appearing somewhat inconsiderable over a century, the IPCC does not hesitate to envisage phenomena on the scale of the next millennium! This rise in sea levels is, according to the IPCC (2001), '*primarily due to thermal expansion and loss of mass from glaciers and ice caps*'.

In spite of the extremely modest rise revealed by the observations, and the lack of certainty and the absence of any link with non-existent 'global warming', the supposed 'threat' is (says the IPCC) a permanent one for '*coastal areas and small islands*'. What is the truth of all this?

14.1.1 Thermal expansion of seawater?

The first reason put forward to explain a future rise in sea levels is a rise in the temperature of the water, itself a result of the warming of the air (according to the IPCC hypothesis). This boils down to the following process: '*the air warms up/the water, warmed by the air, expands/sea levels rise*'. A rise in air temperature of 1°C is thought to bring about an elevation of 20 cm in a 200 m vertical section of water (according to Oerlemans, in Paskoff, 2001). Can it be this simple and immediate?

Hypotheses and observations

If this relationship is real, it should apply on all timescales: firstly, seasonally, the '*seasonal steric signal being the consequence of warming and cooling of ocean waters as summer and winter wane respectively*' (Ménard, 2003). For example, at Brest, which boasts the longest standing tide gauge series in France and is an international reference, average air temperature is 16.0°C in July–August, and 6.2°C in February, a mean amplitude of 9.8°C. If we apply the relationship described above, the variation in water level should be 200 cm for the annual amplitude of about 10°C. However, during the year (Figure 132), the mean level of the ocean is higher in October–December (by a maximum of +8 cm), and lower in March–August (mean minimum: 4 cm). The observed difference in level for the year is therefore of the order of 12 cm (not 2 m!).

Figure 132 also illustrates the fact that variations in water levels and temperatures are not synchronous at Brest: the highest (positive) values do not occur in summer, but in autumn–winter. Moreover, if we look only at extreme values, and if we evoke in particular the considerable thermal inertia of the air compared with the water, if there were a close link between air temperature and water level, we

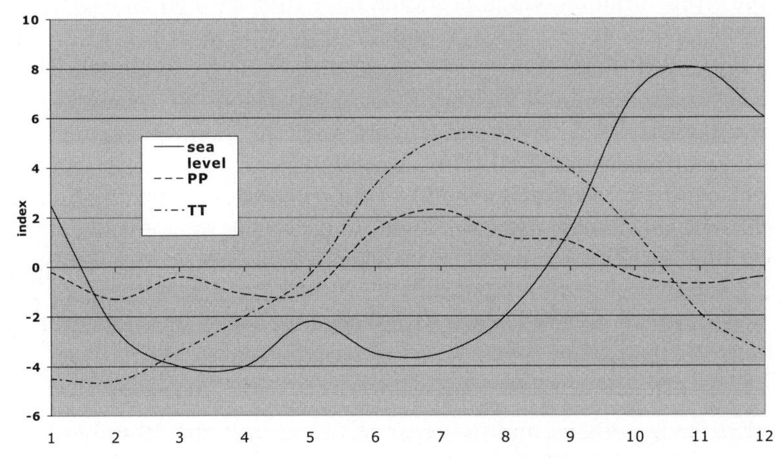

Figure 132. Comparison at Brest, France (48°27′N, 04°25′W) of mean monthly: sea level (index 1 = 1 cm; 0 = mean annual level); mean temperature (index 1 = 1°C; 0 = 10.8°C = annual average); and sea level pressure (index 1 = 1 hPa; 0 = 16.5 hPa = annual average).

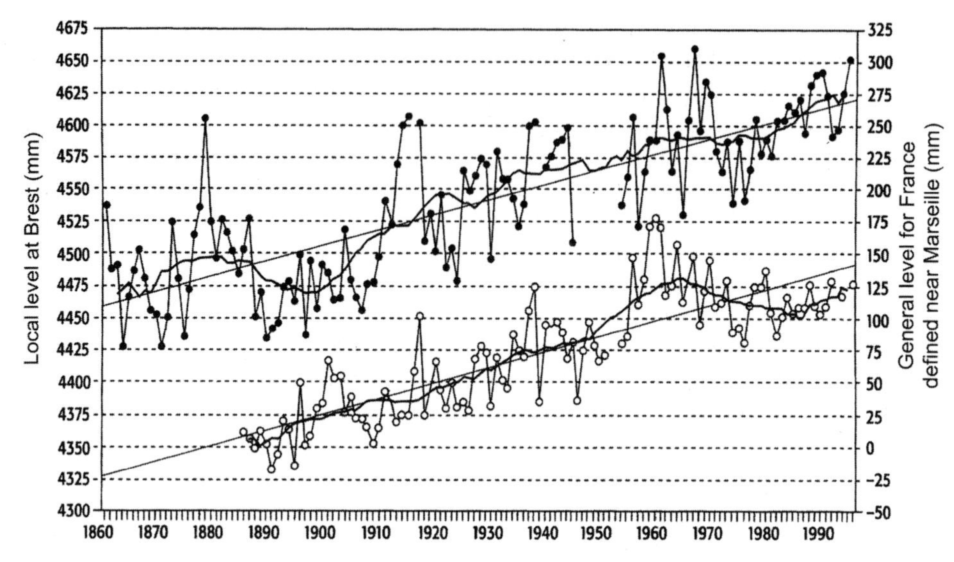

Figure 133. Mean sea level at Brest (*top*) and at Marseille (*bottom*), France, from 1860–2000 Left axis – local level at Brest (mm); right axis – general level for France, or *NGF* (mm), defined near Marseille.
From Pirazzoli (2001).

might suppose that, on average, a rise of 1°C would correspond to a rise of about 1.2 cm in water level.

Still in Brest, let us pursue these schematic suppositions, this time on the annual scale.

- With the given variation, sea level should have risen by 0.96 cm over the period 1860–2000, assuming the IPCC's rise in temperature of 0.8°C (Figure 46, Chapter 10). Less than 1 cm in 140 years, or 0.68 cm in 100 years! This is an absolutely insignificant and unmeasurable value! The observed rise at Brest for the same period, from 1860–2000 (Figure 133) is of the order of 15 cm (i.e., 1.24 mm (±0.03 mm) per year) (Paskoff, 2001).
- With our figure of 1.2 cm per 1°C and a supposed mean temperature rise of 3°C by 2100, the level would rise by 3.6 cm, and even if we allow a temperature increase of 6°C, the rise in sea level might reach 7.2 cm in a century. But, applying the hypothetical relationship '1°C/20 cm', this rise would be between 0.6 m and 1.2 m, an estimate in line with that of the IPCC, but not agreeing with the observations made at Brest.

So we see that the hypotheses and the observations do not match, and even a quite modest rise in sea level does not seem closely linked to temperature. Of course, a prediction based on seasonal variation alone might be misleading, but then, why would a supposed relationship between air and water not work on this scale, yet be functional on others? It is quite obvious that such simplistic relationships cannot be satisfactory. Remember also that, if there is to be a global expansion of ocean waters, there must be a global warming of the air, a situation which has not been demonstrated! Moreover, when studying tide gauge values, we must take account of the effect of continental subsidence; Lambeck and Johnston (1995) find for the Brest area a rate of subsidence of 0.9 mm per year (in Paskoff, 2001). The 'real' sea level rise at Brest, then, between 1860 and 2000, should be 1.24 mm – 0.9 mm = 0.34 mm per year, or 4.76 cm in 140 years (not 15 cm)! At Marseille, subsidence is estimated at 0.8 mm, and the rise in sea level is about 1.20 mm (±0.07 mm) per year (Figure 133). The real rise then should be 1.2 mm – 0.8 mm = 0.4 mm per year, and over a century, 4 cm (not 12 cm)! These are derisory values (and other phenomena are at work, too), and they are quite incapable of bringing about some 'great upheaval'. **They certainly do not permit us to envisage some catastrophic scenario**!

Thermal exchanges air/water

Thermal exchanges at the air/water interface are complex, and any thermal expansion will take place near the surface. Normally, the link between sea surface temperatures and the overlying air (SST/air) is brought in to 'explain', for example, drought in the Sahel or in North America, but the inverse relationship between air and water is less frequently invoked. Warm air moving in above cold water is stabilised, and a state of equilibrium is soon achieved, evidence of which is seen in the occurrence of radiation fog above upwellings. The surface water, warmed by the Sun, is lighter and does not readily transmit its heat to lower layers. Transmission is however easier in the case of surface cooling, which causes surface water to move downwards, and to be replaced. Moum and Caldwell (1994) confirm the role of cold air, and also mention surface mixing: '*Vertical mixing* (*of heat*) *is dominated by three influences: nocturnal cooling, squalls, and westerly wind bursts*'. As Stephenson (2001) puts it: '*The warming of the ocean is not a simple business ... the main thing to bear in mind is that the ocean is not warmed by the air above it*'. About 3% of solar radiation reaches a depth of some 100 m, but infrared radiation (cf. the greenhouse effect) penetrates to only a few millimetres – and it is precisely within these few millimetres that evaporation occurs, transferring latent heat to the air. It is therefore impossible to see how a supposed warming of the air (increasing evaporation) might warm up the ocean! The same questions have arisen in exchanges of views in *Physics Today*, where discussions revolve around the complexity of thermal exchanges between air and seawater, revealing the essential points of the debate on this subject, citing direct radiation and counter-radiation, and exchanges at the interface. Whitten (2001) stresses that '*warming the oceans is never explained* (*no phase change is involved*)'; and Herman *et al.* (2002) conclude that '*there is no way that the ocean surface can warm without a resulting warming of the overlying atmosphere*'.

What we do know is that some areas are warming, while others cool; but nothing is said, especially by the IPCC, about the latter, where, theoretically, sea levels should be falling ... But isn't the idea supposed to be that there is *global* warming? ... and so, *ipso facto*, levels are rising 'globally'! As in other fields, uncertainties seem rapidly to have evaporated. In 1987, Wigley and Raper stated that '*our knowledge of how deep-ocean temperatures have varied in recent decades and of the physical processes that control any such variations is still rudimentary*', and they estimated that the contribution of thermal expansion would be only 2.5 cm, of the supposed 10–15 cm for the period 1880–1980. The IPCC now considers thermal expansion to be the major factor, predicting '*0.11–0.43 m, accelerating through the 21st century*' (IPCC, 2001) (i.e., half of the predicted rise).

The Intergovernmental Oceanographic Commission (Global Sea Level Observing System, GLOSS, in Woodworth *et al.*, 2003), maintains a network of 290 stations with back-up from satellite altimetry, since 1992 mainly from *TOPEX/Poseidon*, and from 2001, *Jason 1*. So we may expect more information in the future, hopefully based on objective analyses. **Whatever the situation may be, we are not in imminent danger of drowning.**

14.1.2 The meteorological factor

There are other, meteorological, factors which determine sea levels, principally atmospheric pressure and wind. The surface of the ocean reacts to variations in air pressure like a kind of reversed barometer. A rise/fall in pressure of about 1 hPa causes the surface to fall/rise by 1 cm. The effect of wind, dynamically combined with that of pressure within a disturbance, depends on its direction (which may cause water to 'pile up' off the coast) and on its speed and the duration of the phenomenon.

The wind

Figure 132 shows that, at Brest, highest mean pressures occur in summer (18.8 hPa in July), keeping the sea level at its lowest, while in winter, the level rises as pressure falls (15.7 hPa in November). However, these modest pressure differences are not the main factor, since average minimum pressure occurs in April (15.4 hPa), without any corresponding rise in sea level (Figure 132). The determining factor in high sea levels (*surcotes*) at Brest is the wind coming in from the south or south-west. Because of the geographical situation of Brest, these winds are '*the most frequent prevailing winds throughout the year*' (Bouligand, 2000).

These southerly winds have their origin in the low-pressure corridor to the fore of Mobile Polar Highs (MPHs), and so, as a disturbance passes, pressure falls and strong winds blow.

For this reason:

- The highest sea levels occur at Brest in the winter, the stormiest season (Figure 132).
- The level continues to rise at Brest (Figure 133), where atmospheric pressure is falling (Figure 134) because of deeper low-pressure corridors to the fore of more vigorous MPHs, where south-westerly cyclonic winds (*piling-up winds*) are

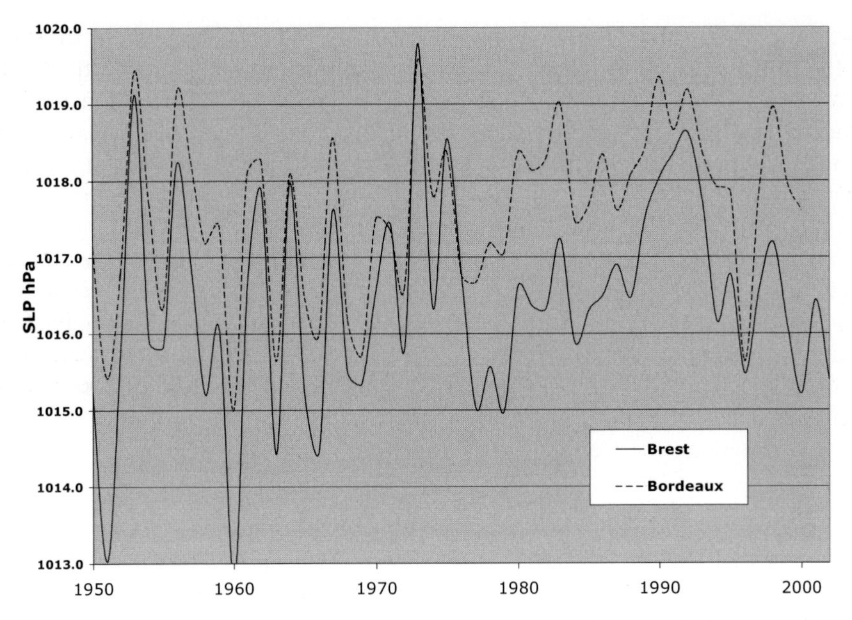

Figure 134. Mean sea level pressures (hPa) at Brest (48°27′N, 4°25′W) and at Bordeaux (44°50′N, 0°42′W) from 1950–2002.
From Météo-France data.

stronger, and where storms are more intense and wave heights are increasing (cf. Chapter 12), together with an increase in rainfall (Figure 66, Chapter 11).

- Meanwhile, the level of the Mediterranean has fallen slightly at Marseille since the 1970s (Figure 133), where pressure is rising (Figure 69, Chapter 11), where north winds are becoming stronger, and rainfall is decreasing (Figure 66, Chapter 11). This does not exclude occasional notable *surcotes* when southerly winds are particularly strong.

So the simplistic relationship 'warm air/rising sea level' is far from applicable. The atmospheric factor is fundamental where sea level variations are concerned, not only for mean levels but also for actual levels during the incidence of *surcotes*. If the level occasionally rises at Brest, it is because of the recent dynamic of the weather in this part of the North Atlantic aerological unit, where the weather is indeed becoming more irregular and disturbed, and strong winds and storms are more frequent (cf. Chapter 12). From the observations made at Brest (Figure 133), let us subtract the effects of subsidence (see above), falling pressure (Figure 134), and the aerological dynamic of a region where the frequency of incoming winds is in fact increasing ... what then remains of this presumed 'warming'? Its contribution seems to have faded away almost to nothing! It might even be (see Cabanes *et al.*, 2001) that 'almost nothing' in this case is an over-estimate!

When Venice is flooded, as often happens, it is because the southerly *sirocco* wind, intensified by vigorous MPHs reaching the Mediterranean, drives the waters of the Adriatic northwards (Figure 70(a), Chapter 11). When pebbles from the beach litter the Promenade des Anglais at Nice, it is again due to the intensification of the cyclonic wind from the south, along the leading edge of MPHs (Figure 70(a)), channelled into the Gulf of Genoa by the relief of the Italian peninsula, causing what the local press calls *raz-de-marée* tides. And when the Vieux-Port of Marseille floods (now more often), the southerly wind is at work, building up the *surcote*, which prevents rainwater from draining away (Figure 70(c)). The low-lying areas of the Camargue, already sinking by 2 mm per year beneath the weight of the alluvium brought down by the Rhône, are also threatened by storm tides piled up by southerly winds (Sabatier and Provansal, 2002). When we see dramatic images in the media of flooding from Bangladesh, it has been caused by winds associated with cyclones and the immediate tidal and hurricane surge effects pushing northwards, amplified by the funnel-shaped Bay of Bengal. The sea level rises, hampering the outflow of rainwater flooding down along the Ganges, the Brahmaputra, and the Meghna.

Atmospheric pressure

Atmospheric pressure lowers the level of the sea beneath MPHs and Anticyclonic Agglutinations (AAs), and allows it to rise beneath low-pressure corridors and closed lows. Take as an example the situation at Tuvalu, an island about 5° south of the equator, which, according to Daly (2004), '*egged on by Greenpeace, is the most strident of the small Pacific Island countries claiming they are being swamped by rising seas*'. Tuvalu's government is seeking compensation under the Climate Change Convention of 1992. The curve for the evolution of the level of the Pacific Ocean at Tuvalu (Figure 135) clearly refutes these uninformed claims. It shows clearly that sea levels here are not rising: as the Australian National Tidal Facility stated in March 2002, '*the historical record shows no visual evidence of any acceleration in sea level trends*'.

In fact, this curve actually reveals noticeable falls in sea level in 1983, 1987, 1992, and 1998 (i.e., during El Niño events, with values of more than −50 cm at the time of the three most marked episodes). These falls in sea level correspond closely to episodes of high pressure (moving in from the north), characteristic of the equatorial western Pacific at times when the Australian monsoon is reinforced and further invigorated by the northern dynamic (cf. Figures 124, 125, and 126, Chapter 13).

It is quite amazing that such curves (Figure 135), from competent authorities (NTF), confirmed for neighbouring islands (Nauru, Solomons), and freely available on sites such as Daly's, still do not figure in the data from the IPCC. Neither are they found in apparently serious publications on sea level, some concentrating on the Pacific, and notably those analysing data from *TOPEX/Poseidon*. Where are they in the media reports about the 'doomed Pacific islanders'? But a great deal of publicity was stirred up (especially by the BBC) when the inhabitants of Tuvalu 'decided' to evacuate their islands – in October 2001, just before (and it was certainly a coincidence!) a UNO climate meeting in Marrakech.

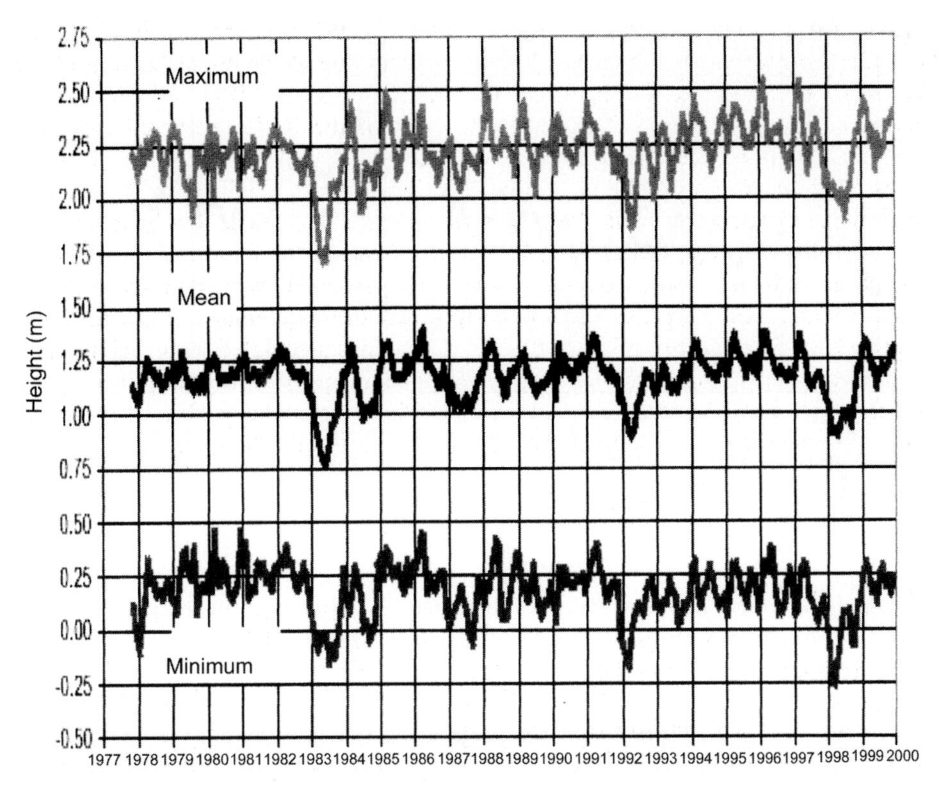

Figure 135. Monthly sea levels at Funafuti, Tuvalu (5°S, 175°E), from 1978–2000, from the National Tidal Facility (NTF), Australia.
From Daly (2004).

14.1.3 '2.5 mm per year'?

Now, the curve in Figure 136 has been published everywhere, and especially by the IPCC during the period 1993–1998 (TAR/IPCC, 2001, sea level, Figure 11-8, p. 663). It is supposed to represent the 'mean' sea level across the planet (between 66°N and S), according to observations by the *TOPEX/Poseidon* satellite since 1993.

The observational facts

This curve shows a (quite pronounced) 'mean' or global rise in sea level, but only for 1997–1998 (i.e., during a particularly vigorous El Niño episode). So the localised rise (in the equatorial Pacific) is 'spread out' across all the oceans, its 'averaged' value representing in Figure 136 just a few millimetres. The conclusion from this curve is '*a very strong correlation ... confirming the important role of thermal expansion of ocean surface water*' (Paskoff, 2001), or, '*the thermal expansion due to the warming of the oceans explains almost all of this augmentation ...*' (Cabanes *et al.*, 2002).

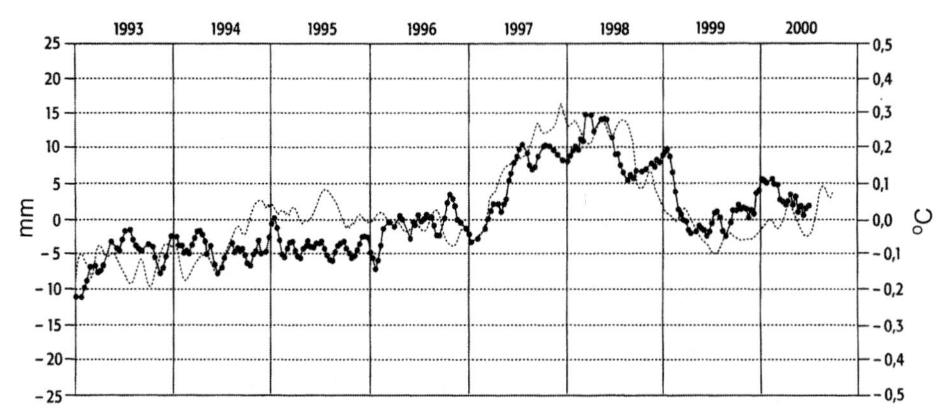

Figure 136. Comparison between global sea level (mm) (points and thick line), from *TOPEX/Poseidon* measurements, and mean sea surface temperatures (°C) (thin dotted line), from January 1993 to June 2000.
After Paskoff (2001).

There is of course no 'correlation', but merely a co-variation of temperature and sea level: there **is no progressive rise** of '2.5 mm per year' (Cabanes *et al.*, 2002), but only **a temporary rise, very limited in space and time**, starting in 1997, and then returning to something like its former value. Curves by Cabanes *et al.* (2002) for each ocean also show clearly that **the conjectured rise is essentially in the tropical Pacific**, there being no such trend in the North and South Atlantic, the southern Indian Ocean or the North Pacific. This lack of a rising trend is confirmed for the South Pacific by Hannah (2004). **These curves show the absurdity of an average, and the pointlessness of spreading values across the generality of the oceans** but it is still shamelessly done!

This evolution is in fact linked to the aerological dynamic, mainly in the equatorial Pacific (Chapter 13). An El Niño episode is characterised by the intervention of the meteorological equator (VME – vertical meteorological equator), displaced southwards by heightened activity in northern MPHs (i.e., by higher pressure in the northern AA, lowering the level of the ocean in the northern and western Pacific (Figure 135)). Such a rise in pressure may also occur at the same time within the southern trade and the corresponding AA (Figure 128, Chapter 13). However, when the VME migrates, updrafts and resulting lower pressures cause a considerable rise in sea level below that structure. This rise is also favoured by highs to the north and south, and by the confluence of trades which cause warm air and warm water to converge around the axis of the lows. As sea levels beneath strengthened AAs fall markedly, the VME, where updrafts are intensified and low pressure accentuated, sees a corresponding, considerable rise in sea level, which may attain '*more than 20 cm*' (Menard, 2003) and certainly even more, especially in the equatorial eastern Pacific. Meanwhile, there is a fall in the western Pacific of more than 50 cm (as shown in Figure 135). So we observe a simple co-variation of maximum sea temperatures and maximum sea levels. The cause of the rise is an aerological rather than a thermal one, linked to the intensification of circulation, the presence of the VME and its lows, and the confluence/convergence of reinforced tropical fluxes in the lower layers.

A new 'coup'!

However, there is no mention in the literature of the ME, which seems, like the cause of the rise, unknown. Neither is any account taken of a variation in atmospheric pressure. The simple aim seems to be to associate a 'mean' altimetric curve and a 'mean' thermal curve (Figure 136), the (repeated) postulate of the IPCC being that '*thermal expansion of water is by far the main cause of the rise in sea levels*' (Cabanes *et al.*, 2002). In spite of the fact that '*the main, and previously unpublished, conclusion of this study is the great geographical variability of this evolution*' (Cabanes *et al.*, 2002) and, although only a local and normal rise beneath the VME is involved, there is no hesitation in passing everything though the 'blender' to arrive at a 'mean' or 'planetary' level which is quite meaningless! With only six years' worth of satellite measurements, the IPCC (2001) hastens to endorse the idea that '*analysis of TOPEX/Poseidon data suggests a rate of sea level rise during the 1990s greater than the mean rate of rise for much of the 20th century*'! Three years on, in spite of great spatial disparity, ignorance of the real (dynamical) cause of the rise, the paucity of satellite observations, and the absence of reference (e.g., as comparison to that of El Niño 1982–1983), Cabanes *et al.* (2002) are quick to claim that '*It is proved beyond doubt: over the last nine years at least, mean sea levels have risen much more rapidly than they have during the last five or six thousand years*'!

So it's true, 'beyond doubt': now the 'hockey stick' is shattered, this is THE proof, so long sought! Warming has accelerated rising sea levels! But doesn't Figure 136 show **absolutely no rise in 'global' sea levels**? What it does underline is a lack of understanding of the causes of the temporary rise (i.e., the real dynamical processes leading to an El Niño event) ... Cazenave now agrees that '... *in 1997, the large anomaly seen by TOPEX was not due to thermal expansion*' (pers. commun., 2004), but she suggests that '*it could be due to a deficit of water on the continents (caused by a deficit of precipitation)*'! This again confirms perfectly how little is known about elementary meteorological mechanisms!

Where is the science in this? *TOPEX/Poseidon* could be used for the scientific analysis of a particular phenomenon, but its data are misused for a parody of science, for the purposes of propaganda, particularly, as it seems, to justify the launching of satellites! '*2.5 mm per year*' even appeared as a title in the so-called scientific journal *La Recherche* (2002, p. 64), above the assertion of '*a definite rising trend, associated with global warming*'. Not only is this doubly incorrect, but it is rather pathetic, which is apparently the most important thing! The journal *La Météorologie* had the same 'scoop' in 2003: '*a rise in sea levels on a planetary scale has been going on for several decades ... the most logical explanation for this phenomenon is that of planetary climatic warming*' (Cazenave, 2003). This time, the curve was for the period 1993–2003 (i.e., it also showed the rise associated with the El Niño event of 2002–2003, which prevented the return to 'normality' which began after

El Niño 1997–1998 (Figure 136). Three journals (at least), three editorial committees, but three times the same falsehood: what use will the new satellite *Jason 1* be, if this is the state of scientific 'objectivity'?

Pacific islanders, and those who live elsewhere, can benefit from careful observations (Figure 135) and serious studies like those carried out, for example, by Mörner *et al.* (2004) for the Maldives. An interesting point made by this study was that '*at about 1,000–800 BP, the people of the Maldives survived a sea level higher by about 50–60 cm*'. Another of its observations was that, contrary to the IPCC's suggestions, '*a general fall in sea level occurred some 30 years ago*'. At the origin of this fall, of the order of 20–30 cm, may be increased evaporation caused by stronger trade winds from the north-east, with associated higher atmospheric pressure in the north. Consequently, and the conclusion is categorical, '*there seems no longer to be any reason to condemn the Maldives to flooding in the near future*' (Mörner *et al.*, 2004).

14.1.4 Melting ice?

With the only 'proof' of the supposed recent rise in sea level being the famous curve of Figure 136, the one thing left to worry about, according to the predictions of the IPCC, is the melting of the ice of continental and mountain glaciers. Let us remind ourselves, to keep a sense of the scale of the fearsome phenomena to come, that we are talking about only one half of the probable hypothetical rise of 47 cm (i.e., a presumed 'global' rise of 23.5 cm by 2100)!

The Antarctic

Let us first deal with the greatest 'threat', that of the Antarctic, representing 90% of the cryosphere. Its hypothetical melting could mean a sea level rise of some 70 m.

The eastern part of the Antarctic covers 10 million km^2, and carries 85% of the volume of the Antarctic ice which may be as much as 4.8 km deep. The temperature there varies (depending on latitude and altitude) between -70°C and -50°C (except on the more low-lying margins, cf. Figures 51 and 52, Chapter 10). A presumed 'warming' could not affect the great mass of the ice; however, the question does not arise, since (see Chapter 10) cooling is observed here.

The ice of the Antarctic is remarkably stable: '*the main mass of the Antarctic ice cap has remained unmelted ever since it was formed, 60 million years ago*' (Postel-Vinay, 2002), or even earlier; it has remained almost unchanged '*since the end of the last glacial period*' (Lorius, 1983) (i.e., for at least 20,000 years (cf. Chapter 9)). Satellite observations show that, during the period 1979–1999, when the greatest temperature rise is said to have occurred, the ice sheets around the Antarctic generally became thicker (Parkinson, 2002).

In the western Antarctic, the West Antarctic Ice Sheet (WAIS) stretches away from the foot of the mountains to form tongues of floating ice around 400 m thick. There are two principal ice shelves, the Filchner-Ronne Shelf of the Weddell Sea, facing the Atlantic, and the Ross Shelf, facing the Pacific. In common with the ice sheets, these ice platforms float, and they could melt in their entirety without any effect on sea levels. This western part of the continent experiences a meteorological situation all its own (Figure 38, Chapter 9), as it extends towards the southern extremity of South America. This means that it receives warm, advected cyclonic air from the leading edges of MPHs, promoting warming of the Peninsula (Chapter 10). This warming is affected by cooling in the eastern Antarctic and by the strengths of consecutive MPHs. The acceleration of meridional exchanges cannot have only this localised warming, but it also causes increased precipitation, and in particular large amounts of snow. This must be taken into account when discussing the 'calving' of glaciers, a sight much talked about, as glaciers meet the sea. However, this should not be naively construed as evidence of current events, since '*it takes about 1,200 years for ice from the centre of the West Antarctic ice cap to make its way to the calving front*' (Postel-Vinay, 2002).

From the Antarctic Peninsula come exciting media images of crumbling ice, said to be '*the most visible sign of the warming of our planet*' (*Le Monde*, 2002). It is presented as a considerable threat: '*The West Antarctic ice sheet disintegrates . . . the worst consequence of warming and the least predictable*' (*The New York Times*, 2002). A similar fuss was made about the dislocation of the Larsen Ice Shelf in March 2002, when a slab of the order of 3,275 km^2, and 200 m thick, weighing 720 billion tonnes, detached itself from the north-eastern tip of the Antarctic Peninsula (at about 65°S). It is worth mentioning that this so-called 'exceptional' phenomenon, labelled a 'spectacular break-up', had already occurred before in 1995! Moreover, the largest known 'ice island', measuring 31,000 km^2, 335 km by 97 km, broke away from the Filchner Shelf in . . . 1956! So 'B-15', an ice island detached from the Ross Ice Shelf in March 2000, 295 km by 37 km, was not, as Long *et al.* (2002) commented, '*the largest ever observed*'.

Figure 137 shows that the dislocation of ice around the Antarctic is commonplace: accumulated ice, constantly renewed, has to go somewhere, someday. Satellite imaging was instrumental in the study in October 1967 of '*the disappearance of a glacier, around 69°30′S, 1°W . . . a tongue of ice called Trolltunga*'. On 11 December 1973, *Landsat* provided an image of this iceberg, measuring 103 km by 60 km, at about 77°S, 48°W, with an area of 4,650 km^2 and a volume of 11,625 km^3 (i.e., 9 times as much water as is consumed in the USA in a year). In February 1975 the ice island still measured 92 km by 50 km, and in May 1978, 71 km by 31 km. Trolltunga was tracked until June 1978, beyond latitude 48°S, and even now its dimensions were 56 km by 23 km. In another example, from 1999, 26 giant icebergs were tracked by satellite by the US National Ice Center as they wandered in Antarctic waters. Among them were icebergs A-38A (86.4 km long), A-38B (75.6 km), D-15 (93.6 km), B-9B (81 km), and B-10, tracked from January 1992, which separated into two in June 1995: fragment B-10A (66 km long) was drifting in the winter of 1999 off Cape Horn, where it presented a danger to shipping.

People have long been impressed by the possibilities of the fabulous amounts of fresh water locked up in icebergs. Indeed, in the 18th century, Captain Cook replenished his water supplies in the South Pacific, beyond the Antarctic Circle, from icebergs. John Isaacs, of the Scripps Institution (San Diego), was the first person

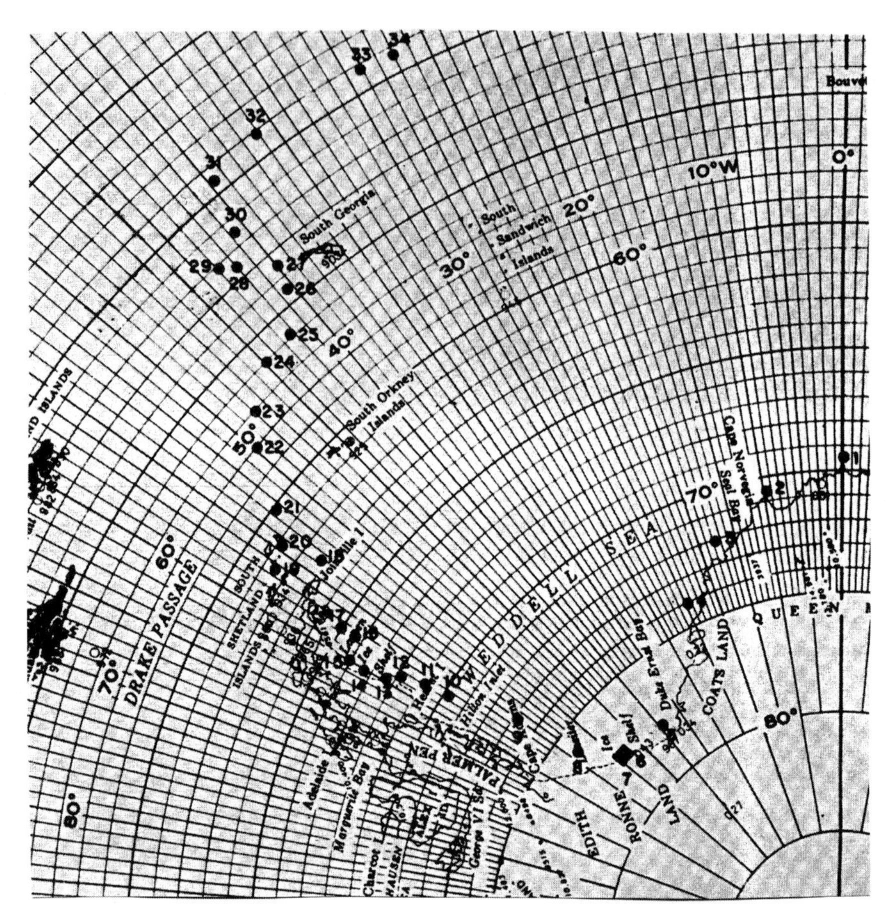

Figure 137. Track of the giant iceberg Trolltunga, from 8 March 1967–29 May 1978. Some locations of Trolltunga: (1) 8 March 1967; (2) 11 October 1967; (4) 10 January 1968; (6) 15 January 1970; (7) 26 January 1975; (10) 30 June 1975; (14) 10 March 1976; (16) 31 January 1977; (27) 2 March 1978; (34) 29 May 1978.
From E.P. McClain (1978).

to suggest, in the 1950s, that Antarctic icebergs be towed to California, to alleviate drought. This idea was taken up again in 1973 by Hult and Ostrander, then abandoned, and later redirected, with Australia as the destination. In 1977 Al Faisal proposed a similar scheme for Saudi Arabia. However, problems with towing, the insulation of the ice, and storage of the water, etc., caused these projects to be dropped, though in some minds they have only been postponed.

So the dispersal of ice from the shelves is no unusual phenomenon, and the Peninsula is not the whole of the Antarctic. The consensus is that the Antarctic presents no immediate threat, and certainly not in its eastern part, where '*it is still not possible to determine even the sign of ice sheet mass balance*' (Rignot and Thomas, 2002). In the WAIS too, as the IPCC itself (2001) says, '*it is widely agreed that major loss of grounded ice and accelerated sea level rise, is very unlikely during the 21st century*'. Those who can envisage the South Polar ice melting, in spite of everything, still consider that, if the danger is not imminent, it could be in the next century, or even the next millennium!

Greenland

Greenland's ice extends down to about 61°N, approximately the latitude of Oslo (Norway), Helsinki (Finland), and Saint Petersburg (Russia). The reason why it comes so far south is found in the disposition of Greenland's relief, which surrounds a central 'bowl' in which ice has accumulated to a depth of over 3,000 m. It escapes to the ocean through mountain gaps. Abundant precipitation means that glaciers in Greenland are better supplied than their Antarctic counterparts, and therefore they move more rapidly. So the ice is doubly protected by the relief, which raises its level, most of the ice surface being at an altitude of more than 2,000 m where the air remains cold, even in summer; and around the perimeter, the ocean is not present to cause calving.

This *inlandsis* is a very ancient feature, dating from the Tertiary period, and the ice has managed to survive episodes of warmth surpassing today's (Chapter 9). Also, even though the *inlandsis* represents just 10% of the cryosphere, and in spite of the presumption that '*the Greenland ice sheet is the most vulnerable to climate warming*' (IPCC, 2001), this ice represents no danger right now. This does not prevent the IPCC from telling us that '*ice sheet models project that a local warming of more than 3°C, if sustained for millennia, would lead to virtually a complete melting of the Greenland ice sheet with a resulting sea level rise of about 7 m*' (IPCC, 2001). '3°C'? 'Millennia'? This sounds more like a truism, or a shot in the dark, than a prediction: science fiction?

The Program for Arctic Regional Climate Assessment (PARCA) notes that '*higher elevation parts of the ice sheet are in overall balance, with local areas of quite rapid thickening or thinning*', while '*many coastal regions thinned considerably during the 1990s*' (Thomas, 2001). Since the sea does not work directly upon this ice cap, melting is the result of higher air temperature, though the question of altitude remains. Figures 94 and 95 (Chapter 12) show declining temperatures in the lower layers around Greenland and along its coasts, as Chylek *et al.* (2004), confirm. This cooling in the lowest layers, which accompanies the passage of more vigorous MPHs, intensifies the return of cyclonic warm air from the south, above the lenticular cold mass. This advected warm air might well be responsible for limited melting around the coastal relief. This warming, though, seems not to have much depth, since, at 3,000 m, '*at the summit of the Greenland ice sheet, the summer average temperature has decreased at the rate of 2.2°C per decade since the beginning of the measurements in 1987*' (Chylek *et al.*, 2004). Advected cyclonic air (Figures 85 and 86, Chapter 12) increases precipitation and the accumulation of snow on the roof of Greenland, which gains thereby in ice mass, though with considerable interannual variability (Thomas, 2001).

In summary, the ice of Greenland, which survived the Holocene Climatic Optimum (HCO) and the warm periods preceding our own, is not about to disappear. It certainly presents no threat to us. All that remains now in the spectrum of catastrophes are the mountain glaciers.

Mountain glaciers

Mountain glaciers hold about one thousandth of the total world volume of ice. Even if they melted completely (which is highly unlikely), sea levels would rise by a negligible amount. If all the glaciers in the Alps melted, the sea would rise by about 1 mm! But this won't happen. All mountain glaciers are not melting.

It is a common belief that '*most glaciers around the world are melting*'. Courtney (2004), referring to Kiefer *et al.* (2000), plays down this suggestion, reminding us of the fact that of the world's 160,000 glaciers, '*only 67,000 (42%) have been inventoried to any degree*'. The usual criterion for evaluating the condition of glaciers is the length of the glacial tongue (advancing or retreating), an indicator of climatic evolution. Bezinge (1999) reminds us that, in Roman times, and even earlier, glaciers in the Alps were more modest affairs than they are today '*... and humankind had little to do with that*'. Porter (1989) has described the marked retreat of glaciers during the Medieval Optimum. Vivian (2002) stated that, in the ancient Alps, glaciers '*exhibited much greater fluctuations than are seen nowadays*'. Vivian also stressed that the second-half of the 20th century was characterised '*by stabilisation and advance, rather than retreat, of glaciers*'.

Any report of a glacier in retreat is immediately seized upon. For example, in Alaska, which enters the limelight if there are no reported 'threats' from elsewhere: '*as regards continental ice, the main contribution to rising sea levels seems to be that of the mountain glaciers at the edge of southern Alaska*'! (Postel-Vinay, 2002), to which Rignot (2004) has recently added those of Patagonia. '*A dramatic retreat*' of Alaskan glaciers exposed to the south is reported (Pfeffer *et al.*, 2000). However, it is not stated that this does not represent a general trend, mainly because of a marked increase in snow accumulation (Moore *et al.*, 2002), itself a result of much higher precipitation levels (Chapter 13). Generalisation, though, is the order of the day: for example, Hansen wrote in 2003: '*all the glaciers in America's Glacier National Park are retreating inexorably to their final demise*'. Rushing to conclusions like this, people too often (conveniently?) forget to mention those glaciers which are not in retreat, or are stable, or even advancing. An article entitled *Glaciers are growing around the world, including the United States* (www.iceagenow.com, 2003) is therefore very useful as a counterweight to the not very real scenario usually presented. **Glaciers are not generally in retreat – far from it**.

The ice mass balance

Although the condition of glaciers is estimated from the length of the glacial tongue, it seems that their variations result only from the dynamic of the ice, and should not be seen as meaningful climatic markers, while, as Wagnon and Vincent (2004) put it, '*annual fluctuations in glacier volumes (involving the mass budget) are direct*

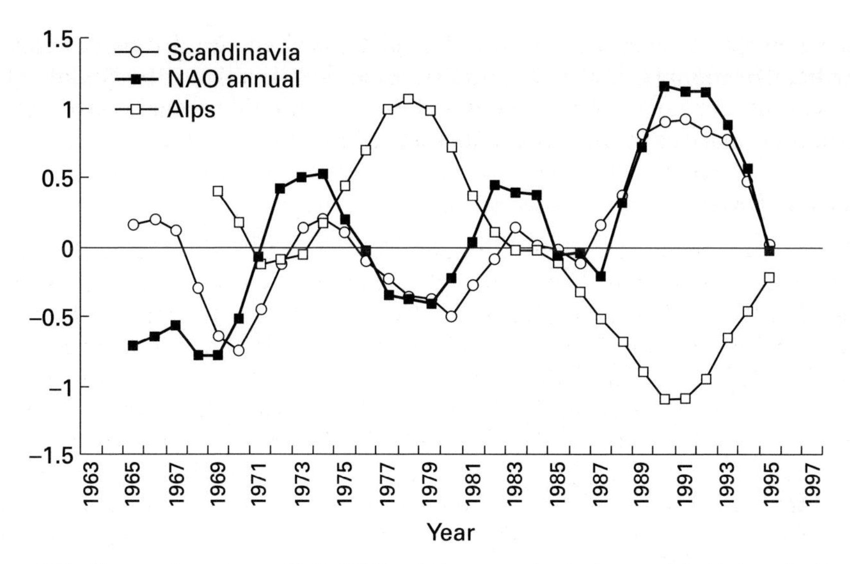

Figure 138. Comparison, from 1967–1997, of the mean ice balance of 9 Alpine glaciers and 7 Scandinavian glaciers, and of the annual index of the North Atlantic Oscillation (NAO). From Six *et al.* (2001).

consequences of climate variations'. The mass budget (the balance between precipitation received (accumulation) and melting due to energy flux at the surface), is a better indicator of the state of glaciers. But such investigations have been carried out for very few glaciers. Braithwaite and Zhang (2000) remarked that '*only about 200 existed with mass balance data for an entire single year ... 115 had a mass balance record of five years ... with both winter and summer mass balances the number dropped to 79*', and finally '*only 42 glaciers had been inventoried for ten years*'. Courtney (2004) was right to state that '*42 out of 160,000 is very far from adequate sampling to permit any meaningful statistical analysis*'. This total was increased to 280 glaciers (Dyurgerov, 2002), but there are wide disparities in the quality of the data, and most of the data periods are short.

To sum up, '*there is no obvious common or global trend in conditions of increasing glacier melt in recent years, and the data mainly reflect variations on intra- and inter-regional scales*' (Braithwaite, 2002). Therefore, we cannot begin to claim that 'it's getting hotter, so the glaciers are melting'! Such things are not usually simple binary relationships or co-variations, and it is quite inappropriate to lump together all the world's glaciers: each one is a unique entity, and its behaviour reflects not some supposed global evolution, but specific local geographical and aerological conditions. As Figure 138 shows, the lives of glaciers in the Alps and in Scandinavia are closely associated with conditions within the North Atlantic aerological space (and how could it be otherwise?). Mass budget variations suggest that, for the period 1967–1997, when Alpine glaciers were less well supplied, those in Scandinavia were '*putting on weight*', and vice versa (Reynaud, 2003). At times when AAs affect the area around the Alps, inhibiting precipitation/accumulation and favouring insolation/melting, precipitable potential is diverted towards Scandinavia by cyclonic circulation from the south (Vigouroux, 2004). Then, the Scandinavian glaciers receive enhanced precipitation. The maintenance of anticyclonic stability over Europe, and especially over its northern edge, is linked with the behaviour of the North Atlantic Oscillation/North Atlantic Dynamic (NAO/NAD) index, and more closely with its positive mode, which has determined the character of recent decades, since the 1970s (Chapter 12). A positive index which is, we still highlight it, associated with a strengthening of MPHs.

Similarly, before bringing supposed climatic warming into the equation as a factor in, for example, '*the retreat of glaciers in the tropical Andes*' (Francou and Wagnon, 2004), or elsewhere, we ought to take a close look at the dynamic, both regional and specific, of precipitation for each glacier. As far as the central Andes are concerned, the link with El Niño, and the '*accelerated retreat of the glaciers since the 1970s*' (Francou and Wagnon, 2004), prove, if proof is needed, that the evolution of the ice is but one element of a general co-variation (cf. Chapter 13); and it must be remembered that no one element can be considered in isolation. This general co-variation involves principally the migration of the VME and its associated updrafts, important factors in the distribution of precipitation in the tropical zone. Now, since the 1970s, in the Andes as well as in the Pacific or in Africa, the VME is shifting slowly southwards, with various effects depending upon local conditions. The evolution of glaciers in Africa (on Mounts Ruwenzori, Kenya, and Kilimanjaro) also indicates that, at these equatorial latitudes, the main factor is not temperature, but the supply of snow, which essentially occurs at these altitudes at the time of the presence of the VME. For Mount Kilimanjaro, Kaser *et al.* (2004) considered that '*climatological processes other than air temperature control the ice recession in a direct manner*'. Singer (TWTW, 2004) also points out that '*glaciers are not thermometers ... and neither Kilimanjaro – nor other glaciers – are the alleged icons of man-caused global warming*'.

So the fact that glaciers exhibit variations is neither a new nor an exceptional phenomenon, but it is evidence of climatic variability. Mountain glaciers undergo changes, as do other physical entities, since they are not immune from the vagaries of the climate; but this climate is not that envisaged by the IPCC.

14.1.5 Conclusion: Sea levels are not rising

Other parameters (tectonic, sedimentological, and hydrological) may be involved. Theoretical mean sea level (the 'zero point') depends on the variable contributions of rainfall, variations in the extent of watered areas, loss by evaporation, retention of water as permafrost, water tables, lakes, dams, irrigation ... However, to return to what was previously said about the major factors in this area: **tide gauge measurements show no significant rise in sea levels during the last century**, unless we consider the presumed '15 cm' as a worrying rise.

Satellite observations indicate no recent overall rise, but confirm localised changes in the tropical zone, mainly in the Pacific, and these are a result of the aerological factor.

The IPCC claims that, as the temperature of the air increases, that of the sea will follow suit, and sea levels will rise. The IPCC also says that thermal expansion is the main cause of the rise in sea levels. However, the temperature of the air is not increasing globally, and the temperature of the sea has therefore no reason to rise. What is more, at high latitudes, temperatures are not rising everywhere to cause ice to melt. **So, no global sea level rise will occur in the foreseeable future**. Consequently, in the words of A.-N. Mörner (2004), '*there is no fear of any massive future flooding as claimed in most global warming scenarios*'. Yet another of the IPCC's 'threats' melts away!

14.2 THERMOHALINE CIRCULATION

As already stated, oceanographers have often tried to help out the struggling meteorologists by putting forward dynamical explanations, obviously based upon their own, ocean-centred, discipline, but with strong climatic implications. A case in point is that of sea surface temperatures (themselves consequences), considered (thanks to computer-based studies) to be responsible for a multitude of phenomena, from droughts to floods (Chapter 11). So the existence of tropical coastal deserts is linked with low sea temperatures caused by rising cold waters (upwellings), while aerological stratification and, in particular, the trade inversion (TI), are forgotten, even though the unproductive character of the TI is really responsible for the lack of precipitation. Zonal 'Walker cells' are conjured up and associated with differences in sea surface temperatures, flouting the laws governing general circulation (Chapter 13). Also, the birth of tropical cyclones is linked with the temperature of the water, and El Niño events are nearly always said to be oceanic phenomena (Chapter 13), while the determining role of aerological structural conditions is swept aside.

Similar things are said about the influence of the Gulf Stream in the North Atlantic and the mild climate of western Europe; and also about the thermohaline circulation, supposed to be causing considerable climatic upsets. Remember that thermohaline circulation, associated with temperature and salinity, is generated by contrasts in the buoyancy of surface waters resulting from exchanges of heat and freshwater with the atmosphere, sea ice, and the run-off from the land. The subsequent powerful convection due to high densities at the ocean surface occurs mainly at high latitudes, though it can also be found at low latitudes if the effect of salinity outweighs that of temperature.

14.2.1 Heat causes cold!

A recent weather fad has led to a media flurry of articles and alarmist statements warning of the '*switching off of the Gulf Stream ... and sudden cooling or even glaciation, in Europe*'. This new fashion is the cherry on the cake of catastrophism. Here we have 'global warming' bringing about glaciation! This 'warm frost' or 'hot ice' scenario is the last thing we might have expected! Sensational – sadly yes –

though not for science! This 'scenario' or 'fable', featured in the disaster movie *The Day after Tomorrow*, merits close analysis.

The scenario

Bard (2004) posed the question of whether the climate could topple: '*could the warming of the planet lead paradoxically to rapid cooling in the northern hemisphere in the region of the Atlantic Ocean*?'; and later, '*observations ... give cause for concern*'! This is so astonishing, to say the least, that we must ask what has led up to it. Apparently, various threads, some ancient, have come together here.

The climatic dissymmetry of the two sides of the Atlantic has long been remarked upon, with a mean temperature difference of 15°C between the mild winters of western Europe and those in Canada. As a telling image of this difference, we can contemplate the blizzards which strike the eastern flank of North America, for example in New York, which lies at the same latitude as Porto (Portugal), Barcelona (Spain) and Naples (Italy)! The first explanation of this was an oceanographic one: a current 'flowing out' (as was first thought) of the Gulf of Mexico, the Gulf Stream (GS, Figures 85 and 86, Chapter 12). The first recognition of this current was in 1769, when Timothy Folger traced out the course of the Gulf Stream at the behest of Benjamin Franklin. The oceanographer Maury completed in 1847 a 'Chart of Winds and Currents', and in his book *The Physical Geography of the Sea and its Meteorology* (1855), he gave an early indication of the role of the Gulf Stream: '*it is the influence of this stream upon climate ... that clothes the shores of Albion in evergreen robes; while in the same latitude, on this side, the coasts of Labrador are fast bound in fetters of ice*' (in Seager *et al.*, 2002).

The recognition of the NAO (Chapter 12) was confirmation of the Atlantic dissymmetry, but, without an explanation for the NAO, no new insights were gained. Nowadays, it is generally agreed that the Gulf Stream brings subtropical warmth to more northerly areas, '*the heat dissipating into the atmosphere and warming the winds blowing across Europe*' (Rahmstorf, 1997). This approach to the phenomenon appears in Figure 139(top), and '*the transfer of this heat to Arctic air masses over the North Atlantic accounts for the anomalously warm climate enjoyed by Europe*' (Broecker, 1995).

Figure 139 illustrates various 'received ideas', principally:

- From a meteorological point of view, the existence of a supposed 'standing wave', partly the result, according to theory, of pressure differences between the ocean and the land, and of the distribution of relief on the land, in this case mainly the Rockies. Seager (2003) stated: '*it is clear that this sinuosity of atmospheric circulation is also involved*', bringing cold air eastwards across North America and warm air towards western Europe – always bearing in mind that, meteorologically, Figure 139 hardly reflects reality (Figure 60, Chapter 11, Figure 85 and 86, Chapter 12).
- From an oceanographical point of view, this figure is taken from the '*Global Conveyor Belt*' schema of Broecker. Focussing on the North Atlantic: '*warm upper waters flow northward, reaching the vicinity of Greenland, where the Arctic*

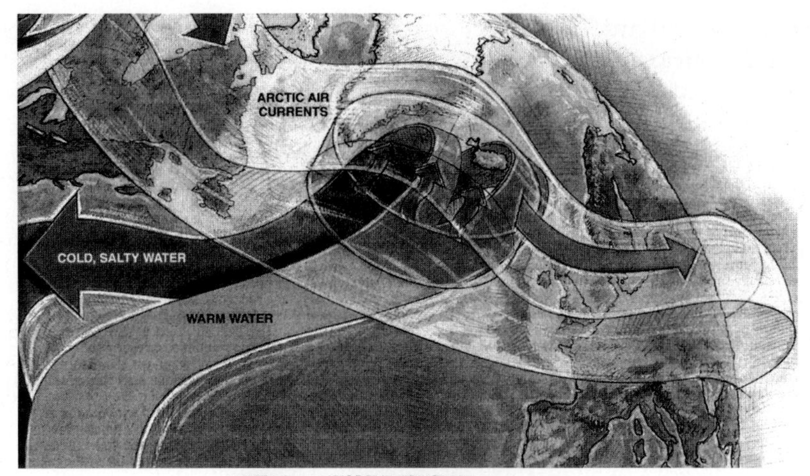

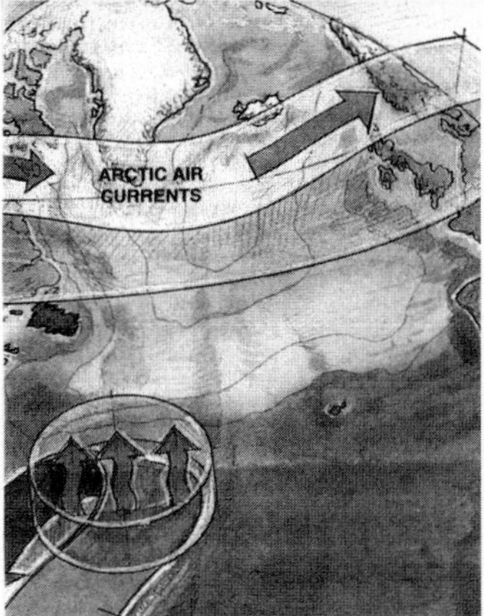

Figure 139. (*top*) Heat exchange, from warm water (Gulf Stream/North Atlantic Drift) to Arctic air currents, over the Northern Atlantic. (*bottom*) During glacial times, alternating conveyor operating at the latitude of southern Europe: no transfer of heat to Arctic air currents.
From Broecker (1995).

> *air cools them, allowing them to sink*' (Broecker, 1995). The sinking of cold waters into the depths makes the North Atlantic '*a crucial region of world oceanic circulation*' (Bard, 2004). Note also that the oceanic circulation is highly schematised, representing the North Atlantic especially as a cul-de-sac, having no apparent interaction with the Glacial Arctic Ocean.

The schema as described, transferring warm water and heat towards Europe, and causing water to sink to great depths, is nevertheless '*vulnerable to disruptions by injections of excess freshwater into the North Atlantic*' (Broecker, 1995). The freshwater is contributed by rain, run-off from the land, and melting ice. So '*by considerably modifying the water cycle ... warming of the planet could destabilise this conveyor belt*' (Bard, 2004), since lesser salinity alters the density of the water at the surface. Increased CO_2 (it's important, of course, to link the scenario to the greenhouse effect!) '*would lead to general cooling across the North Atlantic, accompanied by a marked diminution in the deep circulation of the Atlantic*' (Bard, 2004). Broecker (1995) considered that, as the conveyor slowed or halted, '*winter temperatures in the North Atlantic and its surrounding lands would abruptly fall by five or more degrees*'. Fellous (2003) stated that this 'climatic surprise' would cause mean temperatures in Europe to fall by 6°C from present levels, and the '*mild maritime climate of western Europe would be replaced by the rigorous climate of the regions at the same latitude across the Atlantic*'. Nobody tells us, though, about the aerological mechanism which will produce this upheaval! Such a scenario, completely divorced from any climatic reality, is like some child's game involving just swapping pieces at random: let's put Canada's climate in Europe! And as if that were not enough, Broecker (1995) fuelled the catastrophe with: '*the shift would occur in 10 years or less*'! A chill runs through us! According to Rahmstorf's model (Figure 139, bottom) illustrating a glacial period, cold waters would sink to the north of Bermuda (instead of near Greenland), and the warm current from the south would bring less warmth to Europe. The undulation of Arctic air would be shifted further to the north, above the ice ...

These suppositions make reference to past climates (Broecker, 1995; Duplessy, 1997; Delécluse, 2002; Clarke *et al.*, 2003), and especially to post-glacial cold episodes, along the lines of: '*sedimentologists and glaciologists think that this cooling (The Younger Dryas) was due to a sudden inrush of freshwater in the North Atlantic*' (Bard, 2004). This hypothesis is based on a large number of models which suggest '*that the circulation of heat and salt through the world's oceans can change suddenly, with drastic effects on the global climate*' (Broecker, 1995). We have already looked at these events (Chapters 9 and 13), stressing that, alongside and/or instead of these 'explanations' involving salt and freshwater, there is no 'aerial' explanation: which is surprising, since we are discussing climatic phenomena!

Many other aspects might be brought in, but let us confine ourselves to the points already mentioned. To sum it up: warming could diminish the thermohaline circulation, slow or stop the transatlantic current (Gulf Stream–North Atlantic Drift, GS–NA), and bring about a marked diminution in the warming effect in the North Atlantic, which would receive no more warmth from the GS–NA (Figure 139, bottom). This means that, if we allow the supposed mechanism, the effect of the modification of the thermohaline circulation would not be able to compensate for global warming; and cooling would be the result. Since this approach seems not a little 'complex' – let us examine the validity of this scenario point by point.

The density of surface water

'*The additional freshwater decreases the density ... a crucial point ... and this might topple the climatic system*' (Bard, 2004). What is the value of such an affirmation? In particular, what is the actual difference in density between 'freshened' seawater as opposed to salt water, bearing in mind that, as Stevenson (2001) put it, the fundamental physical factor affecting the density of seawater is temperature? If the density of freshwater is, by definition, 1, the density of seawater is 1.03/1.04 with average salinity of 35 g/l. In other words, a difference in density of a few hundredths. Also, (pure) freshwater freezes at 0°C (by definition), while seawater freezes at −1.8/−2°C. Would this difference in density be subtle enough at the temperatures found at these latitudes? Can the difference prevent water from cooling – especially at such a latitude – and from sinking? And – the fundamental question – is this minimal difference in density capable of bringing about the supposed climatic changes? If, as the hypothesis suggests, the temperature drops (and severely!), would the 'cooling' not cause the seawater to increase in density, finally making the surface waters sink (and provoking therefore the end of the hypothetical cooling)? Moreover – and this is why we are not looking at the 'Conveyor Belt' concept – the planetary deep water 'circuit' takes '*almost 2,000 years*' (Duplessy, 1997), a rhythm out of all proportion with the proposed scenario, which would require just a few years!

So, this hypothesis involving a cessation of deep water formation has its inaccuracies and paradoxes, and indeed goes against the reality of phenomena: for example, Wu *et al.* (2004) do not accept '*a decreasing trend of the North Atlantic thermohaline circulation*' but instead state that '*the thermohaline circulation unexpectedly shows an upward trend*'. Most importantly, there is a considerable disproportion of scale between phenomena, in the main concerning density, duration, and the amplitude of consequences. This hypothesis gives the impression that, as usual, the virtual scenario is founded upon a few equations and cells within models, with reality and the gamut of meteorological phenomena ignored.

Applying a modicum of common sense to this, could additional freshwater, mostly from the shrinking ice sheets, endanger the planet? There is a good way to test the validity of this scenario (it is indeed difficult to call it a 'hypothesis' in the scientific sense of the word): by considering the phenomena on a seasonal scale. Half of the Arctic ice sheet melts in summer ... the considerable volume of additional freshwater will lower salinity ... and if we follow the scenario, there will be immediate cooling, and consequently it should be cold in summer. So, **every year, in summer, a new and severe glaciation should be unleashed**!

14.2.2 Circulation in the Arctic

Now, what of the circulation in the North Atlantic and the Glacial Arctic Ocean? Figure 140 reminds us that the north-east Atlantic is not a 'cul-de-sac', as Figure 139 might suggest, but is, on the contrary, a through route for both cold and warm waters.

The Arctic Ocean is divided into two main basins by the Lomonosov Ridge: the Canadian basin and the Eurasian basin. Its ice covers 14 million km^2 in winter, and about half that in summer. The depth of the ice sheet exceeds 3 m in the western part. Some of this ice (about 2,000–3,000 km^3), moves away from the Arctic every year through the Fram Strait between Spitsbergen and Greenland with the cold East Greenland Current. Warm water moves northwards in the opposite direction with

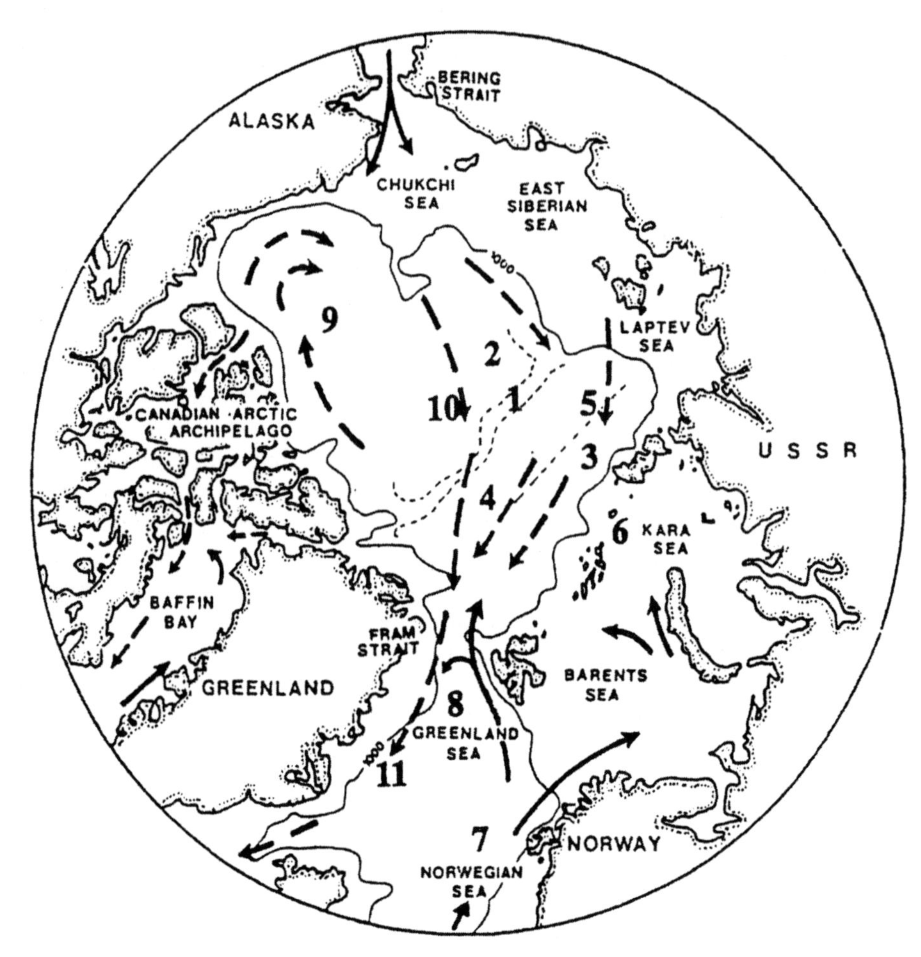

Figure 140. Main topographical structures and surface circulation of the Arctic Ocean and northern seas. 1 – Lomonosov Ridge; 2 – Canadian Basin; 3 – Eurasian Basin; 4 – Amundsen Basin; 5 – Gakkel Ridge; 6 – Sta Anna Rift; 7 – Norwegian Current; 8 – West Spitsbergen Current; 9 – Beaufort Gyre; 10 – Transpolar Drift; 11 – East Greenland Current.
From Houssais and Gascard (2001).

the West Spitsbergen Current, and north-eastwards with the Norwegian Current, mainly towards the Barents and Kara Seas. These warm salt waters from the Atlantic move into the Eurasian basin. At present, these warm waters are finding their way beyond the Eurasian basin, into the Canadian basin (Houssais and Gascard, 2001), a sign of the intensification of the Norwegian Current (Chapter 12).

Can this movement cease? If the water at the surface does not move downwards, then it will stay at the surface (obviously) – and so what becomes of it? Does this accumulation of warm water really lead to cooling? Is the level of the Arctic Ocean rising? There is in fact no reason why the advection of warm, salt water should cease, since the motor of the southerly current is not sinking cold water: in order for the inflow of warm water to stop, the North Atlantic Drift would have to be halted, as would the Gulf Stream, and 140 million m^3/s would cease to flow past Cape Hatteras! The shrinking area (Parkinson *et al.*, 1999) and thinning of the ice (Rothrock *et al.*, 1999) were confirmed by Rigor *et al.* (2000), who noted '*a trend toward a lengthening of the melt season in the eastern Arctic, but in the western Arctic a slight shortening of the melt season*'. These reductions of extent and thinning are not a sign of any slower movements, but rather of the accelerated flow of warm water into the Arctic basin, with, moreover, salinity lowering freezing levels. As a consequence (unless the storage capacity of the Arctic Ocean changes!), this addition causes an intensification of exchanges within the Arctic, and thereby an intensification in the return of cold water to the Atlantic via the East Greenland Current. This cold water will, sooner or later, play its part in the formation of deep water.

It is useful not to lose sight of the fact that sea surface circulation is itself a consequence, as a particular element in a general co-variation, and that it is largely driven by the circulation of the air. It is usually said that the North Equatorial Current supplies the Gulf Stream, with a lesser contribution from the South Equatorial Current, both currents being driven westwards by the trade circulation. However, **the main driving force is that of MPHs**, which, moving off the American continent, exert strong pressure upon surface waters and push them eastwards. The succession of MPHs and the associated low-pressure corridors cause the 'Gulf Stream system' (GS–NA), thought to be a schematically homogeneous one, to consist in reality of multiple 'arms', with varying trajectories across the Atlantic. So, to stop the current, we have to halt the trade wind (if we follow the usual concept), and this means doing away with MPHs, the real driving force behind ocean surface circulation.

The circulation of air and water

In any study of the weather, the obvious place to start is with the aerological aspect. However, it should not be turned on its head! For example, when we read that, in the period 1958–1997 '*an increase in the frequency of cyclones, and lower pressure at sea level, are connected with retreating sea ice in the Arctic*' (Houssais and Gascard, 2001), we see an effect (the retreating ice) becoming a cause (of increased storm intensity, cf. Chapter 12)!

Figure 139 shows that the scenario is based on that concept of the dynamical school according to which relief brings about the formation of a standing wave at altitude, which controls circulation at the surface. Here is the same idea from Seager *et al.* (2002): '*The Rocky Mountains play a major role. Analogous to an island in a stream, the Rockies set up a persistent wave in the winds*'. The authors have only half-seen what the 'major role' of the Rockies really is: '*south-westerlies bring warm maritime air into Europe and north-westerlies bring frigid continental air into north-eastern North America*' (Seager *et al.*, 2002). Their model does not in fact match the synoptic reality of circulation in the lower layers (cf. Figures 85 and 86, Chapter 12). We have already discussed this idea (cf. Figure 58, Chapter 11), and concluded that meridional exchanges of air and energy do not take place as in

Figure 139: not in America or Europe, nor in the North Atlantic. In fact, the vehicle of heat, even though ocean currents do play their part, is mainly the air, and it moves it rapidly via cyclonic circulation within the low-pressure corridors which form along the leading edges of MPHs. Ocean and air currents are therefore two components, exhibiting different speeds and intensities, of the same general movement. So zonally extended 'Arctic air' does not 'wait' to be warmed by its encounter with a warm sea current (Figure 139, top), and then move off towards Europe; it is subtropical air which transports, meridionally and much more rapidly, perceptible and latent heat northwards (cf. Figures 85 and 86, Chapter 12). As for the air of MPHs, its long journey takes it, feeding the trade winds as it goes, down to the tropics to draw energy before returning to the temperate zone to the fore of MPHs.

So, although its contribution is not a negligible one, the Gulf Stream (GS–NA) does not 'warm' Europe. Seager *et al.* (2003) recognised that '*atmospheric circulation is more important to understanding climate variability than is the ocean circulation*'. Therefore, hypothetically indeed, if the Gulf Stream were to slow down (but not halt), this would be of only secondary (or no) importance, and certainly would not lead to 'cooling'! If one wants to go looking for a cause of some supposed cooling (assuming a warm scenario), it should not be in some difference in density of surface waters, nor in the interruption of a current, but in a modification of general circulation caused by other factors, not confined to a particular sector (Chapters 8 and 9).

The way in which both air and ocean circulations work and move together means that we cannot consider one isolated element without taking into account the other elements of the *ensemble*. **The interdependence of phenomena has been constantly referred to**. It is an absurdity to envisage the interruption of deep currents or the Gulf Stream, since all other currents would have to disappear, together with the general circulation of the atmosphere! It is just as absurd to imagine, courtesy of a model, that '*the average worldwide warming could be about 3°C ... with, in the North Atlantic, a cooling trend of more than 4°C ...*' (Ganopolski and Rahmstorf, in Bard, 2004)! Such a 'simulation', based upon a weakened circulation at depth, showing a (temperature chart) virtual world coloured red, with the Atlantic standing out in dark blue, could only be the work of a model! Is it possible, though, to conceive of the existence of a cold region completely independent of circulation: where would fluxes come from, and where would the cold come from, and why would cold air stay put? How could a particular region have its own climatic evolution different from that of the warmer rest of the planet? The fundamental question here is: does a model have the ability to 'conceive of' anything?

For the circulation of air or water in the North Atlantic to be modified, everything else would have to be modified, and especially the mode (see Chapter 8) of general circulation.

- Any warming, which would attenuate the initial impulsion by MPHs, would slow the circulation of the air and ocean (Figure 35, Chapter 9). It is not easy to imagine how this (general) slowing could then cause 'cooling', since this hypothesis would require an attenuated polar thermal deficit. Where would the cold air come from to cause the dramatic drop in temperature?

- On the other hand, polar cooling, strengthening the initial impulsion by MPHs, (rapid circulation), increases the vigour of MPHs and thereby the intensity of the advection of warm air towards the pole, also strengthening ocean currents (Figure 33, Chapter 9). This is what is happening at present in the North Atlantic (Chapter 12), where the transporting of warm air and water is now more intense in the Norwegian Sea and beyond, both above and below the ice. Figure 139(bottom) therefore bears little resemblance to what is actually happening! Such a configuration, which would interrupt the supply of warm air and precipitable potential, is not applicable to the dynamic of past climates, since it could never have allowed the formation of the *inlandsis* (cf. Chapter 9), and it applies neither to a winter situation, nor to the evolution of the North Atlantic since the 1970s, with its intensification of meridional exchanges in air and ocean.

No, the climate cannot be 'toppling'!

The idea of thermohaline circulation causing localised cooling **is, meteorologically, a totally unlikely scenario.**

This scenario is based upon several hypotheses.

- Upon palaeoclimatology, particularly the interpretation of cold episodes which interrupted deglaciation. An interpretation favouring the appearance of large volumes of additional freshwater from melting icebergs and drained lakes is wrong (Chapter 9), since it does not explain past happenings and therefore we cannot see how it can explain the present, with its seasonal variation; or, *a fortiori*, the future.
- Upon the difference in density between brackish and salt water: a tiny difference supposed to stop water from sinking. The main factor determining density is temperature. Moreover, nothing is said about what becomes of those surface waters which do not descend into the depths (especially within a schema involving cooling).
- Upon an over-simplification of ocean circulation in the North Atlantic, with no indication as to whether the circulation will cease in the Arctic, or whether the inflow of cold water via the East Greenland–Labrador Current will also cease.
- Upon the transporting of heat by the Gulf Stream system, supposed to ensure the mildness of Europe's weather, even though climatic characteristics are determined by the aerological factor.
- Upon the presence of a standing wave at altitude, caused by the Rockies, responsible for the distribution of circulation within the lower layers (as taught by the dynamical school), a configuration not reflected in real circulation.
- Upon the supposed interruption of the Gulf Stream (but not other currents), its motion connected with that of the general circulation of the atmosphere, which certainly does not stop.

- Upon the hypothesis that the interruption of the Gulf Stream would bring about cooling, which could not happen in that event since the Gulf Stream is not responsible for the climate. Remember that, given a general warming hypothesis, the origin of the cold is unclear: theoretically, it should be warmest at the pole! Where would the cold come from?

Therefore, this scenario, based on thermohaline circulation, is not scientifically sound, and deals only in inconsistencies.

The existence of such an improbable hypothesis is evidence of a sad lack of knowledge about the actual dynamic of past and present climates in the region concerned. Before saying that the climate will 'topple' or that it is 'malfunctioning', one must first know about the 'normal' conditions governing its behaviour: this seems not to be the case here. Here is a scenario owing much to computer processing, to virtual rather than real meteorology, and it represents, finally, only a movie scenario!

14.3 CONCLUSION: THE SEA, A MICROCOSM OF THE WHOLE GREENHOUSE DEBATE

Sea levels are not rising, or are rising by a very small amount, and there is no recent acceleration. Those dwelling near the water's edge need have no fear.

The Antarctic is not melting, even if more of its vulnerable western part occasionally loses enormous icebergs – as it has since the mists of time. Like the southern *inlandsis*, Greenland has stood the test of time, experiencing great variations, because of its wide diversity of climate, and it represents no threat. Mountain glaciers continue, as they have always done, to retreat and advance, reflecting, with varying degrees of inertia, the particular climatic evolution of the areas where they are found.

The 'thermohaline circulation' scenario is unfounded, and, as common sense tells us, warmth does not give rise to cold! And the climate is not about to 'topple'! Instead, climate is evolving, and this evolution, common to all phenomena, is a matter of the intensity rather than of the nature of those phenomena.

Consequently, all is well; we need not have included a chapter on the ocean. For any possible variations in sea levels and thermohaline circulation are, according to the IPCC, linked to global warming – which isn't happening!

Why the analysis? **Because this subject is a very good example of the way in which climatology is interpreted** by the IPCC, some scientists, and the media, in the context of the greenhouse effect postulate.

Let us now examine some aspects of this question, and, since nobody can tell us whether this idea arose by chance, or whether it is the fruit of circumstance, or has been foisted upon us deliberately ... the pronoun 'they' in what follows is a non-specific one, and does not refer to anybody in particular.

- **They started with nothing** (shades of the greenhouse effect), since there is no proof of what they were proposing: as Cazenave *et al.* (2004) affirmed when discussing sea levels: '*we may hope to find, for the first time, one of the effects of climate change*'. Apart from the frequency and violence of storms, which are determined by the aerological factor (nothing to do with the greenhouse effect), the sea presents as much of a 'serious threat' to us in the future as it did during the last century.
- **They found a slogan, and repeated it over and over**: '*it's getting warmer, and sea levels are rising*'. Earlier, it was: '*it's getting warmer, the ice is melting, and sea levels are rising*', and then, armed with Archimedes' Principle, they arrived at the amicable co-existence of, on the one hand, the expansion of water, and on the other, the melting of the ice, and the predictions were far less alarming.
- **Everything revolved around the slogan**. Could it be that predictions of a big rise in temperature at high latitudes were invented just to melt that ice, and provoke the sea rising? However, the rise is not observed, except locally as a result of meridional exchanges.

  The coral reefs are turning white – the greenhouse effect is to blame! Of course, we cannot discount the idea of a heatwave as the cause, but other possibilities suggest themselves: an overdose of ultraviolet, a lack of nutritive minerals, pathogenic diseases as the 'malaria of oceans', or 'increasing pollution' (IPCC, 1992). The most likely explanation, though, is a thermal anomaly in Polynesia of +1° to 2°C per 100 m depth: '*this anomaly is a prelude to the El Niño phenomenon, as was the case ... in 1983–1984 and 1987*' (Rougerie *et al.*, 1992). If this is the case, and given that El Niño episodes have their origin in the southward shift of the VME, the withering of the coral is certainly not a result of the greenhouse effect, far from it. And recent analyses suggest that warming water is rather favourable to coral growth, as most tropicalists already thought! What was called '*the white death of the coral*' brought back memories of the '*Death in the Forest*' headlines, of the obligatory acid rain (i.e., pollution from factories and vehicles) killing the trees.
- **They reinforced the slogan with terrifying images**. This is where the media come in, and the images may not have much to do with their subject: for example, freshwater floods are used to illustrate 'rising' sea levels, and a photo of an iceberg will inevitably conjure up the *Titanic*. They play on fears, and, if no image is available, they invent one.

  Take this example from a section of *The Guardian* on 11 September 2004: beneath the heading '*The drowned world*' there appeared an image of Holland in 2020, and the text read: '*rising seas have flooded the fields of low-lying countries*' – only the tops of the windmills can be seen above the water! The magazine also tells us that '*icecaps will be melting, sea levels will be rising ... With global warming, coral reefs, and the life that thrives on them are dying ...*'; some of the future '*horrors we could face by 2020*'! Yes, 2020! Be very afraid, poor people! It's getting nearer! How many of the readers had the presence of mind, seeing this comparison between 2004 and 2020, to look *back* 16 years to 1988, and ask themselves what 'transformations' had come about during the same period of time, between 1988 and 2004? It might be of interest to recall that it was *The Guardian* which carried, in 2003, the statement by John Houghton comparing '*the threat of anthropogenic climate changes to weapons of mass destruction*' (quoted in Kondratyev, 2004). For these imaginings we can thank '*The Guardian*'s science editor', Tim Radford, so we can assume that it was supposed to be science! How urgent it has now become to disentangle real science from its media equivalent.

- **They 'wheel in' committed, obliging, or unsuspecting (but ingenuous) scientists**. The slogan now becomes (as in marketing) 'scientifically proven'!

  The way in which the *TOPEX/Poseidon* data have been used is illustrative of this approach. Scientifically, it is inappropriate to average out and 'globalise' disparate and contradictory local values. Yet the same procedure is used to 'prove' that sea levels are rising, in the same way that the 'official' curve is used to show rising temperatures (Figure 46, Chapter 10). The result will of course be the same: these curves do not apply to what they are being asked to prove. There was much dissent when satellite observations showed that the 'global' temperature was either not rising, or at most rising much less than the official curve suggested. However, the same reticence was not apparent in the case of the first altimetric satellite measurements, accepted everywhere. It seemed that the *TOPEX/Poseidon* satellite had been launched solely to prove that sea levels were rising! After the reception accorded to the 'hockey stick', a new 'proof' was needed right away! Was *Jason* then just another instrument of misinformation, or will its data be used for accurate analysis of the real physical phenomena causing sea levels to rise and fall?
- **They even invent an unforeseen catastrophe scenario**, completely divorced from reality! There is no limit to their imagination, since the real world no longer sets the margins of probability. Hollywood showed us the true nature of this scenario: a disaster movie is merely its ultimate triumph! What is regrettable is the confusion of the genres. Science? Science fiction? What proportion of readers, 'scientific' journals, or popular newspapers can tell the difference?

It is a good thing that this last chapter is about the sea. The way in which it is dealt with is symptomatic of many aspects of the debate today. It is indeed becoming increasingly difficult to distinguish between science and its media equivalent, given the frenzied scrambling for the next scoop. The models' simulations often have more in common with video games, as climatology recedes, and theory is disconnected from reality. Ignoring meteorological phenomena leads to an over-simplified, or naive and immature vision of climatology.

How simple: a few extra ppm of CO_2 and we burn, a little less salt and we freeze!

The analysis of what is going on in the sea provides an excellent bridge to our general conclusion.

15

General conclusion

Global warming is not anywhere near the most important problem facing the world.

B. Lomborg. *The Skeptical Environmentalist*, Cambridge University Press, 2001.

The dominance of such a false priority as 'greenhouse warming' not only compromises science, but may also hamper the socio-economic progress of the developing and industrially developed countries.

K.Y. Kondratyev. World Climate Change Conference, Moscow, 2004.

The question asked in the introduction to this book was: 'is the greenhouse effect/ global warming scenario a myth, or a reality?'

The answer is clear: THE 'GLOBAL WARMING' SCENARIO IS A MYTH.

15.1 THE ELEMENTS OF THE 'GREENHOUSE EFFECT/GLOBAL WARMING' SCENARIO

Let us review the elements of the case. The whole edifice, or fragile house of cards, rests upon one hypothesis from long ago: that increasing levels of emissive gases 'must' or 'will' cause global temperatures to rise.

This idea hung around for more than a century, without much acceptance, as a mere theory, the much more obvious role of water vapour brushing aside any thoughts of a significant contribution from other 'greenhouse' gases. There were substantial advances in the 1950s, and serious assaults upon the environment were answered by the growth of the ecologist philosophy and movement. As the ideas of the vulnerability of nature and its despoiling by Man's activities caught on, the debate passed from the scientific into the political arena. Conferences were organised, but the reality of the scenario was still not recognised, since in the 1970s the nuclear winter and the return of a Little Ice Age (LIA) were to the fore. The watershed occurred in the 1980s, especially as a result of the 1985 Villach

Conference, where the major traits of the scenario were proclaimed. The ensuing Greenhouse Panic led to the creation of the Intergovernmental Panel on Climate Change (IPCC) in 1988. The greenhouse effect, in the beginning a purely scientific notion, thus became an 'official' phenomenon discussed at an international political level, and **science receded more and more into the background as political and economic preoccupations took over**.

The IPCC brought out three successive reports, in 1990, 1995, and 2001. All three had much in common, as they revolved around the key idea that global temperature was on the increase. Humankind became responsible for the increase, now known as 'climate change', and it would cause sea levels to rise, and climatic phenomena to become more severe. The IPCC's predictions were based on climate models. The gamut of emissive gases slowly increased, as did the predicted temperature values. Humankind was put in the dock and blamed for climate change by the 1995 report. The number of phenomena said to be caused by the greenhouse effect increased, and in the end, all meteorological phenomena were included, especially the most violent. **So the greenhouse effect became the climate itself, in its entirety**!

15.2 THE RUSH TO 'PROOF', AND THE '*COUPS*'

The hypothesis upon which the greenhouse effect is based, particularly as regards greenhouse gases, has never been demonstrated in fact: there is therefore **no tangible and indisputable proof** that the IPCC scenario is really happening. THERE HAS BEEN NO PROGRESS SINCE 1985!

So there is a perpetual race for proof, and the IPCC seizes upon the slightest glimmering, relying strongly upon (dedicated) scientists and the media to disseminate its 'unquestionable proofs'.

- The first 'proof' (pre-IPCC), halfway through the 20th century, was the relationship between rising temperature (Figure 1, Chapter 2) and changing levels of CO_2 (Figure 2, Chapter 2). However, during the 1960s–1970s, temperatures began to fall, and the supposed link disappeared. So the greenhouse effect scenario was put on hold.
- The official curve of mean global temperature (Figure 46, Chapter 10) began to climb again (for a still undefined reason) during the 1980s. It now became the unassailable icon of the so-called global warming.
- In 1995, the IPCC produced its first coup, with the declaration that there exists a '*discernible human influence on climate*'. This (essentially media-led) *coup*, which caused much controversy, led to no proof of either the supposed human influence or of the reality of the anthropic greenhouse effect.
- Palaeoclimatic curves, and in particular that for the last 420,000 years (Figure 31, Chapter 9), showed an overall co-variation (mistakenly called a correlation) between levels of emissive gases (CO_2, CH_4) and the evolution of temperature.
- The so-called 'hockey stick' curve, the second coup of the IPCC, was used in

2001 to claim that temperatures are higher now than they have been for at least the last 1,000 years.
- *TOPEX/Poseidon* satellite measurements over the sea gave the IPCC in 2001, and in later studies, the opportunity to present its third coup, since there was now 'proof' that sea levels were now rising at an accelerated rate. We may expect this to be a major prop of the IPCC's fourth report, due in 2007.

Surely, in the field of science, the practice of riding on these tenuous coups, in the manner of the media, can only incur suspicion of the greenhouse effect scenario, and of the unscrupulous *modus operandi* of the IPCC trying to impress its views through the *Summary for Policymakers*, more political than scientific.

15.3 THE SO-CALLED 'PROOFS' ARE WORTHLESS

Analysis of the palaeoclimatic argument (Figure 31, Chapter 9) reveals **that the greenhouse effect scenario is not established for the long term**. The evolution of temperature is not determined by the evolution of emissive gases. It is the temperature which changes first, and the levels of greenhouse gases later. Moreover, it is by no means certain scientifically that past (estimated) levels and current (measured) levels are comparable, and that past greenhouse gas levels are representative of the composition of the atmosphere at the time. **So it cannot be claimed that greenhouse gas levels have never been as high as they are today**.

The 'hockey stick' curve (Figure 43, Chapter 10) has been heavily criticised, as it has been shown to be **a deliberate falsification of the climatic history** of the last millennium. Also, this is a well-documented period, and the (proven) existence of the Medieval Warm Period (MWP) shows that temperatures then were higher than they are nowadays. The authors of this famous curve were obliged to recognise in writing that their analysis contained errors. This 'affair of the shoddy stick' was particularly revealing of the state of thinking at the IPCC, and sadly, among some scientists ready to come up with 'proof at any price'. This curve also raised the question, as in palaeoclimatology itself, of the representativeness of past estimates, and the problems inherent in their juxtaposition with recent observations.

The official curve of mean global temperature (Figure 46, Chapter 10) is not representative of global climatic evolution, since such an evolution, on a dual scale (mean and global) is not possible. Some regions may become warmer, others may cool, and **a calculated mean of regional values going in different directions can only be an artefact**.

Also, the greenhouse effect scenario cannot be invoked to explain the evolution of the mean global thermal curve of temperature for recent times (a curve only considered for itself):

- it does not show the rise of 1918–1940, the CO_2 increase being comparatively too little compared with the mean temperature rise;

- neither does it show, *a fortiori*, the fall in the thermal curve of 1940–1970, when CO_2 levels were continuing to rise; and
- there could not be just one single period out of all the thousands of years studied (i.e., the 30 year long Recent Climatic Optimum (RCO)), when a hypothetical 'relationship' between temperature and CO_2 levels may be imagined.

This last supposed relationship, which must be considered, for want of better information, to be just a co-variation, is already disproved by Figure 50 (Chapter 10), which shows that the RCO is not linked to the greenhouse effect. An analysis of the dynamic of the aerological units suggests too that, within each unit, climatic behaviours are very diverse, particularly since the 1970s, and an overall thermal picture can only be an artefact. There can be no 'mean' expressed for such differing regions, and so there can be no 'mean Atlantic' temperature, or 'mean Pacific' temperature. Consequently, there can be no 'global' temperature. So **the official curve of mean global temperature is climatically meaningless**. The hemispheric curves might be useful, though, as an indicator of meridional exchanges in each meteorological hemisphere.

Measurements from the *TOPEX/Poseidon* and *Jason 1* satellites might well show an alleged recent acceleration in rising sea levels, which is then used to establish an (absurd) mean value for the whole of the Earth's oceans. In fact, this is a very localised effect in the equatorial Pacific, incorrectly averaged to the scale of the whole world. This 'mean' has no global significance, and **it cannot be claimed that sea levels have been rising at an accelerated rate during the last decade**. This localised rise has been caused by identifiable dynamical factors (the intervention of the vertical meteorological equator (VME)) which have nothing to do with, and run counter to, a supposed 'greenhouse effect'.

CONSEQUENTLY, THERE IS NO PROOF OF THE SUPPOSED 'GLOBAL WARMING'.

It cannot be claimed that the planet is warming up, and so any statement based on this false postulate is, climatologically, absolutely valueless. This is particularly true of most of the fictions/imaginings, gratuitous and intended only to frighten, which have come from Group II of the IPCC, who base them upon **an undemonstrated hypothesis**.

15.4 THE GREENHOUSE EFFECT SCENARIO AND THE WEATHER

The IPCC has progressively monopolised all climatic phenomena, especially those likely to involve catastrophe (from their point of view, the most 'interesting'): rain, floods, waves of cold, and snow in low-lying areas as well as the lack of snow in the mountains; the changing weather, extreme events, tropical and temperate disturbances, El Niño events ... in the end, everything. All is 'gathered in', and climatology thus becomes, as the IPCC reports succeed each other, ever simpler: **the greenhouse effect (i.e., humanity) is responsible for it all**!

Not content with appropriating these phenomena (whatever physical laws are defied), the IPCC tells us that they will get worse. In order to predict the future evolution of these phenomena, the models use schematic relationships based on simplistic 'equations' (of the type 'T/R') for rain or its absence. This approach does not reflect the real conditions for pluviogenesis, and cannot ensure correct prediction. What is more, since temperatures are not increasing globally, a 'global' prediction is meaningless.

Also, the FASTEX (non-)experiment in the North Atlantic, and the analysis of the conditions bringing about extreme rainfall in southern France or across the whole Mediterranean basin, are examples of deep insufficiencies in knowledge of the dynamic of the weather in middle latitudes. The same is true in the case of summer or winter droughts, whether in Europe or America, whether temperate or tropical. Shortcomings also permeate the meteorology of the tropics, particularly when we consider the supposed relationship 'sea surface temperature/convection' said to be instrumental in cyclogenesis, or the vertical structure of the meteorological equator (ME). This faulty understanding of meteorological phenomena cannot lead to reliable predictions about the evolution of the intensity of disturbances.

The examples analysed highlight the fact that the greenhouse effect does not control the dynamic of the weather, which, in a warm scenario, ought to be getting milder. However, observation does not bear this out. Since pluviogenesis is absolutely bound up with disturbed conditions, it is inevitable that prediction of this parameter is not possible (at least, not at present). The great disparity in the pluviometric evolution of the 20th century, with its supposed 'warming' trend (cf. Figure 46, Chapter 10), confirms this. A presumed 'global warming' has not (as predicted by the IPCC scenario) led to a general increase in rainfall values; it has not prevented severe droughts; nor has it tempered the violence of the weather.

Therefore, the greenhouse effect controls neither the dynamic of the weather, nor the evolution of the intensity of disturbances, nor the evolution of rainfall or drought, either now or a fortiori in the future.

15.5 MODELS AND THE CLIMATE OF THE FUTURE

Models, the initial principle of which was laid down exactly a century ago, became in the 1980s indispensable props of climatology. It is not actually necessary to use them to support a greenhouse effect scenario: the scenario was invented long before models came on the scene; a simple calculation suffices. But in giving validity to that scenario, they provided it with an illusory 'scientific' endorsement. Just like schools of thought before them, models cast their spell over climatology, and later, over the IPCC. Numerical modelling gave rise to new ways of analysing meteorological and climatological phenomena, and, like any new 'religion', it imposed a new mindset: 'reductionism', which, as Dady put it in 1995, led to a kind of '*scientific totalitarianism*' (Dady, 1995). This imperialism is a particular cause of the conceptual *impasse* in which meteorology now finds itself.

15.5.1 The shortcomings of the models

Now models, when applied to atmospheric phenomena (where success is doubtless more likely than in other domains), exhibit serious shortcomings. In the course of this book, we have several times mentioned the most serious.

- The use of statistical relationships leads to an erroneous confusion of co-variations and actual physical relationships. This is a particularly serious fault, leading to simplistic reasoning.
- Such statistical correlations are seen as cause-and-effect relationships, but the direction of the relationship is not determined, and the effect is often considered to precede the cause: more evidence of lack of understanding of the reality of phenomena.
- 'Teleconnections' may be established, often merely because of the availability of data files, with no climatological, synoptic, or meteorological analysis having been undertaken to determine the direction and protocol of the research.
- Modelling shies away from the use of concepts, but in fact, in climatic matters, it resembles the 'climatological' school of the early 20th century in its use and abuse of means, and, meteorologically, it bases itself on the 'dynamical' school (1939), giving primacy to phenomena at altitude.
- With its basic cells, infinitely multipliable, modelling presents the atmosphere as a homogeneous *block* within which interactions may take place without restriction. In the real atmosphere, with its units of circulation, there are discontinuities, aerological stratification, and breaks, which can prevent direct propagation from one cell to other.
- The cellular, '*pointilliste*' approach contains within it neither a general perception of phenomena, nor (in particular) the notion of the general circulation of the atmosphere. The general organisation is even supposed to emerge by itself from the particular elements. Therefore, models do not provide the context within which all phenomena can be included in their proper places, and they do not specify the direction of the cause-and-effect relationships between phenomena.
- The scale of the cells is insufficient to include all phenomena, particularly local or even regional meteorological phenomena. Moreover, climatology works on a regional basis, which models do not encompass.

Therefore, models, disconnected from reality, are *sui generis* fundamentally incapable of providing a global image of climate but this is, paradoxically, exactly what they claim to do best. What is more, they have largely contributed to the cessation of research on concepts which might give insights into the *ensemble* of atmospheric phenomena.

15.5.2 Statistical and deterministic modelling

So models symbolise nearly all the faults of today's climatology, while claiming to establish scenarios for the climate of the future. Can they do this? Is this claim serious? Since the best judges are the specialists within the discipline itself, let us hear from modellers Gidaglia and Rittaud (2004):

- '*In order to be able to work with a predictive model from which a simulation may be realised, one needs to ensure that one has a very good understanding of the fundamental phenomena involved*'. It must be said that this is not what is happening with climatological models, which instead, as the late lamented meteorologist Giraud put it, '*short-circuit reasoning*' rather than employing it: there is still a wide lack of understanding of meteorological phenomena.
- '*A deterministic model is one within which the workings of all the phenomena involved are completely described by a known set of equations* (*which may or may not resolve them explicitly*)'. Such a model is, with some exceptions, as yet unachievable, not only because of the complex nature of phenomena, but also because there exists no unanimously accepted definition of '*the workings of all phenomena*'.
- '*Since these (deterministic) models are not always easy to set up, one usually has to settle for a statistical model* (*i.e., one which adds a dash of randomness to the equations*) *... unfortunately, as all statistical models contain by definition a "black box"* (*i.e., an 'area of residual ignorance' channelled by probabilities*)*, the predictive ability of such a model is low*'. Climatic models are statistical, and their 'black boxes' contain parameterisations which are really admissions of ignorance.

There is also an 'invisible box' containing everything that has been brushed aside, for the sake of convenience, and/or deliberately, because nobody knows how to include it, or perhaps there is no desire to include it. In this box we might find, for example, the true roles of water vapour, the ME, the Trade Inversion (TI), units of circulation, satellite imaging ... and I would also, in a one-sided way, put in this box the contributions of Mobile Polar Highs (MPHs) and anticyclonic agglutinations (AAs), general circulation and its modes of working ... The 'invisible box' is most likely a very big one! So, the principal shortcoming of today's statistical models is a result of '*the fact that certain phenomena are poorly understood, and not easily put into equations, which ... are as yet the only instruments of dialogue with the computer*' (Gidaglia and Rittaud, 2004). The 'boxes', both black and invisible, therefore severely limit the 'predictive' power of models.

Consequently, we are fooling ourselves if we overestimate the value of predictions by models. Their calculations are of necessity correct, but the bases for their scenarios are often erroneous: the mathematics and the information processing may be 'spot on', but they do not guarantee anything but fiction if they are built upon a defective view of meteorology–climatology, especially if the latter is of poor quality to start with.

The predictions can therefore be reliable only if the modelling has moved on to a deterministic level.

15.6 THE RECENT PAST AND THE FAR FUTURE: THE EXAMPLE OF THE ARCTIC

As chance would have it, at the time of writing, the Arctic Climate Impact Assessment (ACIA) is available online (`www. acia.uaf.edu`, since 8 November 2004). The world's press got to work on this without delay, and their coverage gives another insight into the way the media deal with such things. To choose but one example from the many: The *Financial Times* of 2 November 2004 was echoed by the *Tribune de Genève* of 6–7 November, with the headline: '*The Arctic is melting ... polar bears homeless by 2070*', followed by '*the melting ice sheet gushes forth, a result* (*the report confirms*) *of climatic warming due to greenhouse gases*'. It's all there: the scientific reference, the doomed polar bears, the greenhouse effect, and the exaggeration bordering on falsehood! And where does it appear in the newspaper? Under the respectable heading of 'Science'!

15.6.1 The lessons of observations

It is worth looking at this report, given the important role of the Arctic in circulation in the northern hemisphere (cf. Chapters 8–13). In the general conclusion, two pages on *Changes in climate in the ACIA sub-regions* (there are four sub-regions around the pole) are devoted to the evolution of temperature over the last 50 years, from 1954 to 2003. Now, maximum temperatures in the Arctic date from the 1940s (cf. Figure 55, Chapter 10). On the chart for ACIA sub-region I, which covers east Greenland, Iceland, Norway, Sweden, Finland, north-west Russia, and adjacent seas, there is no indication from mean annual temperatures that the Arctic is getting warmer. Some parts have warmed up by about 1°C in 50 years (northern parts of east Greenland, Scandinavia, and north-west Russia), but some, southern Greenland, Iceland, and the North Atlantic, and also northern Scandinavia, have cooled by 1°C (Figure 141, left).

The chart for winter temperatures shows the same schema, exaggerating the area in which cooling has occurred, and revealing that temperatures '*over the Arctic and the North Atlantic Oceans have remained very cold in winter*', with a fall of the order of 2°C, including the south, west, and north of Greenland, and extending into eastern Canada (cf. sub-region IV). However, in Scandinavia (with the exception of the north where the cold zone has increased in size), and in north-west Russia, temperatures have risen by 2–3°C (Figure 141, right).

Each series of charts of recent temperature evolution for the four sub-regions is followed by two charts (annual/winter) for '*projected surface air temperature change*', as far ahead as 2090. All these charts are 'red' (i.e., they specify generalised warming). Not unexpectedly, given that the predictions are based upon scenarios for

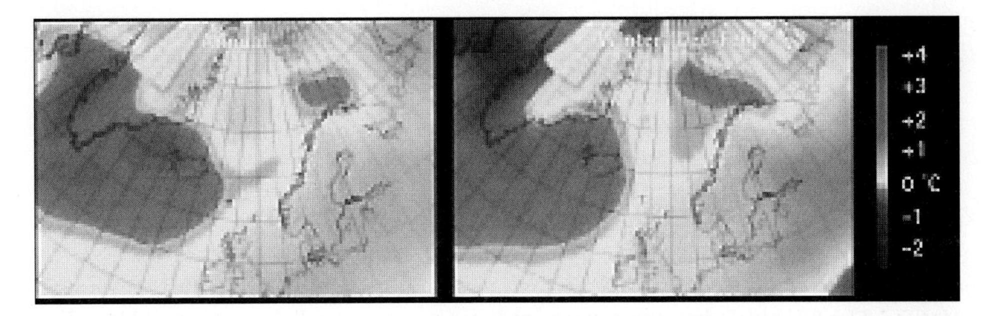

Figure 141. Observed surface air temperature change: 1954–2003, over Arctic sub-region I. (*left*) Annual; (*right*) winter. Cooling trend appears as dark areas (for more detail see: acia.udf.edu).
From *Arctic Climate Impact Assessment* (2004).

emissive gases by the IPCC. But, it does not matter in the case of these ultra-hypothetical charts.

One essential question should have been asked before these projections for the climate of the future were made: a question which does not appear in the ACIA report. It is: 'To what can we attribute this distribution of the evolution of temperatures over the last few decades, given that such a distribution cannot be the result of mere chance; and in particular, **what is it that causes the perfect organisation of areas which exhibit cooling and warming**?'

We can immediately make the following statements:

- First, the Arctic is not undergoing a homogeneous evolution.
- Second, the thermal fields are identical on the annual and winter scales, so permanent mechanisms are at work.
- Third, winter sees an increase in intensity (as happens elsewhere), and it is observed that the most marked rises in temperature occur at the same time as the most marked falls. Also, the values for these rises and falls are about the same (i.e., of the order of 2°C).
- Fourth, in the North Pacific (sub-region III), eastern Siberia has experienced marked cooling, while Alaska and the Canadian Yukon have experienced marked warming (cf. Chapter 13), again, configurations comparable (allowing for wide gaps in purely Arctic data) to those of the adjacent North Atlantic aerological space.

So, the lack of any uniform thermal trend, and/or the alternation of warming and cooling trends, makes it **impossible mechanically to apply the same rates of increase (the same rule of three) to all sectors of the Arctic**, using the argument that all the curves of the IPCC scenarios are rising.

The ACIA certainly tells us that '*some sub-regional variations are likely to result from shifts in atmospheric circulation patterns ... Region I is particularly susceptible to changes in the North Atlantic Oscillation ...*'. But we are not discussing only a few 'sub-regional variations', but all variations (increases and decreases) across the whole of the Arctic, whose climatic evolution is indissociable from that of the northern hemisphere; moreover, the very fact that the NAO is thought to be a '*possible response to increasing greenhouse gas concentrations*' (note the reference to the greenhouse effect, by now a 'Pavlovian reflex'), clearly shows that the real mechanism of Arctic climatic evolution is still misunderstood (cf. Chapters 12 and 13).

15.6.2 The dynamic of temperature in the Arctic

The statements above indicate that it is absolutely necessary at this stage to **take account of the true dynamic of the meridional exchanges** responsible for the Arctic thermal field (and the wind and pressure fields). **The thermal signal can be only a modest one**, as alternating fluxes from north and south cancel out as far as mean values are concerned. However, in spite of the modest nature of the signal, the resulting mean thermal field seems particularly well organised, and quite revealing of the aerological dynamic, with its origins and destination in the Arctic. Regions exhibiting cooling are associated with the favoured trajectories of cold air leaving the Arctic (MPHs), and those exhibiting warming are associated with cyclonic warm air flowing into the Arctic from the south (to the fore of MPHs). The co-variation of cooling and warming, eloquently confirmed by the winter charts (Figure 141, right), represents a real physical relationship as defined by the MPH concept (Chapters 12 and 13).

The facts of radiation (upon which models are based) mean that it is necessarily cold at the poles. Overall variations in the thermal deficit, ever present though not always marked, result mainly from external causes (cf. Chapter 6). With the exception of these (non-dynamical) factors, warmth can only come from advection, its distribution determined by the geographical factor (channelling both warm and cold advections). This means that any warming in the Arctic not associated with radiation is necessarily of dynamical origin: for example warming in winter, which shows the greatest increase, but is certainly not associated with (absent) insolation. The intensification of meridional exchanges increases advections of warm air (and of warm water, drawn in as part of the same movement), and corresponds therefore to an increase in the vigour of MPHs, the engines of circulation.

Let us remind ourselves of the rules governing these dynamical mechanisms, and of their thermal outcomes (Chapters 8 and 9).

- When there is an attenuated polar thermal deficit, weaker MPHs will cause only slight localised warming, and the (mathematical) resultant of the mean temperature remains close to 'normal'.
- When there is an accentuated polar thermal deficit, stronger MPHs will cause greater localised warming, and the (mathematical) resultant of the mean temperature is above 'normal'.

Also, variations in the respective areas of MPHs and lows are also capable of causing variations in the extent of thermal fields, with cold air occupying much less space than warm air, and also moving out more rapidly from polar regions at times of rapid circulation (Pommier, 2004).

One cannot therefore make projections about the thermal future of the Arctic, colouring the whole map red! **It is an absolute physical impossibility**. For some well defined regions of the Arctic to become warmer, it is necessary that other regions, equally well defined, become colder (as shown on the winter charts). All the ACIA charts showing thermal 'projections' for the Arctic should be in both blue and red (cold and warm), and the deeper the blue in the cold areas, the deeper the red should be in the warm ones.

CONSEQUENTLY, THE *ACIA* CHARTS FOR 'PROJECTED SURFACE TEMPERATURES' FOR THE PERIOD UNTIL 2090 IN THE ARCTIC *ARE ABSOLUTELY WRONG.*

The supposed significance of these imaginative projections is of no real interest, neither for the Arctic, nor for the whole of northern hemisphere circulation. *With the exception of the charts of temperature trends over the last 50 years, which are very useful,* **the lengthy ACIA report is of no scientific interest at all**. In fact, the 'key impacts' on the environment, the economy, and people's lives are (at least half of them) but fictions, imaginings, and stories which bring in the inevitable threatened polar bears (there they are, prominently, on page 3: '*sea ice retreat will decrease habitats for polar bears*', ACIA, 2004). An image leapt upon by the media!

15.7 A GULF OF MISUNDERSTANDING: THE 'MISSING LINK'

The example above highlights the reasons why there is a gulf of misunderstanding between the defenders of the IPCC scenario and those who question it. It also points out clearly where the gulf lies: it is a question of the breakdown in the comprehension of phenomena. The climatological analysis of 50 years of observations in the Arctic has been very traditional, and based upon means, but it is regionalised, and is therefore of great value. It would be even more precious if it were used as the basis of a logical argument. But this reflection is absent. There is, though, a missing link in the chain, a gap of thought, of reflection on the true representativeness of data and on their dynamical significance. The gulf is now deepened by (most of) the ACIA report, and whatever follows it will be of absolutely no use, and wrong.

There are many examples which can be used to illustrate this severe lack of understanding, or indeed the animosity which arises from this impassioned debate. Let us look at the website of the Laboratoire de Météorologie Dynamique–Ecole Normale Supérieure LMD–ENS (Rocca, 2002): a title reads '*Understanding climate change through the theory of radiative–convective equilibrium*'. After an introductory presentation of the physics of climate change, with emphasis on the principle of the greenhouse effect, reference is made to observed facts: the Vostok-based palaeoclimatic curve, the 'hockey stick', the IPCC's official curve of mean global temperature, and thermal outcomes from models. The conclusion supports the idea of global warming. Then, global radiative–convective equilibrium is discussed, according to the principles of convection, thermodynamics, hydrostatics, the conservation of energy, and the empirical parameterisation of the convective flux. Then follow equations for radiative transfer and radiative forcing by clouds, the respective contributions of absorbants, with particular attention to the question of the doubling of the concentration of CO_2, as well as the relationship between saturating vapour pressure and temperature. The work is crowned by the presentation of the results of global modelling, according to the IPCC emission scenarios, notably showing '*enormous warming in the Arctic*'.

Here we see a knowledgeable exposition and careful mathematics at work, with references to physical principles based upon fundamental equations, and a rigorous demonstration. So such an account is a priori not likely to be challenged, and its authors may well feel that they have correctly established the principle of 'climate change'. A sceptical reception to it would inevitably be much frowned upon by the institution in question. However, the references presented are brought in 'as found' (e.g., the 'hockey stick'!), without analysis, taken as gospel, and, in the final recommendations, it is suggested merely that we might '*better understand the physics of clouds*' (cf. Chapter 5)! The primal role of water vapour as a greenhouse gas is not taken into account, the observational data are only considered in their first degree, and absolute faith in the model (often for convenience reduced to a vertical column) is apparent. *Ipso facto*, the shortcomings already mentioned will be incorporated. With no reference, again, to the aerological dynamic, climate change as presented is, although mathematically sound, still essentially theoretical, and has no solid links to the climatology of what is real.

This kind of 'missing link' approach is seen everywhere: the 'reality gap' leads to the juxtaposition of two sets of ideas, which may be loosely strung together, or even completely divorced from one another. For example, Beltrando (1990) collected together pluviometric data from East Africa, but when it came to processing them, he called upon computer specialists for support, as many others do. He thanks them for their patience in the light of his '*numerous requests*'. One often sees this dichotomy, within research institutions, between climatologists and computer specialists, the result of which is that the last or next-to-last parts of a climatological analysis are reduced to a mere computerised exercise, unconnected to meteorological reality (as previously seen in the case of the ACIA).

The participants may be justifiably satisfied with their own contributions: it is only the juxtaposition that makes the final product incoherent. This difference in perceptions is recognised by Lindzen (pers. commun., 12 January 2004): '*I think that differences in scientific culture make communication difficult*'. This 'cultural' difference is normally contributed to in equal, or nearly equal amounts, by co-authors from two different disciplines.

Each draws on the other for support, without necessarily being able to assess the validity of conclusions. For example, I was very surprised by Bard's assertion that a slight difference in density might produce the dramatic climate outcomes which he puts forward. His reply was revealing: '*it is in fact necessary to take into account the opposing effects of a decline in salinity and cooling ... Obviously, this is what the various modellers interested in this aspect do*' (pers. commun., 29 April 2004). A confident statement but quite misplaced, since the proposed scenario of his article (Bard, 2004) is physically unfounded (cf. Chapter 14)!

Differences in perception are worsened by shortcomings in modelling practice, since only phenomena which it can easily (or wants to) deal with are retained. Another aggravating factor is that of the inclusion of incompletely described or unsolved phenomena from the discipline of climatology, which gives the model the task of finding out what it has not itself provided. In this way, meteorological facts are misrepresented, mutilated, or transformed as they pass from one discipline to the other, and this can lead only to highly debatable or even worthless results. We might well ascribe this, if we are feeling charitable, to 'teething troubles' in modelling, but the problem will persist unless the discipline becomes a totally deterministic one. Then, it will have achieved '*a very good understanding of the fundamental phenomena involved*' (Gidaglia and Rittaud, 2004), and models will be able faithfully to reproduce these phenomena. This can only happen when climatology overcomes its own failings and finds its way out of the conceptual *impasse* in which it sits. All the hopes raised by soaring levels of computer capability are only a diversion, and will count for nothing until the missing 'link' between meteorology and modelling is put in place.

15.8 WITH AND WITHOUT MPHs

At this point in our conclusion, we need to pause, to ask about the appropriateness of the constant references in the book to the MPH concept. There is a simple and practical reason for this. There are many people who feel qualified to swiftly dismiss criticisms of both the limitations of models and the erring ways of climatology. As an old English proverb has it, '*He who has a mind to beat his dog, will easily find a stick*'! And the stick has already been wielded: rather than confront arguments to which they most often have no answers, these critics prefer to discredit some particular aspect of a work. Take for example the ironic words of Rochas (2002), president of the editorial committee of the journal *La Météorologie*, commenting on Lenoir's book *Climat de Panique*, 2001: '*it must be said that Yves Lenoir has a joker up his sleeve: he proposes to explain everything ... with MPHs*'. Then, of MPHs themselves: '*they are little beasts born in the polar regions, in the shape of thin, cold lenses ...*', and later '*I have never understood how MPHs and AAs can explain things, nor the mechanisms which create them or cause them to engender depressions*' (Rochas, 2002). In a similar vein, Jancovici (2002), who styles himself an 'engineer–climatologist' but has no qualifications in this field, considers that Lenoir's reference to MPHs shows his lack of knowledge of the basics of climatology (cf. `www.lcre`). It is quite easy, with this mindset, to reject any argument contradicting the greenhouse effect scenario by discrediting MPHs! So, in the interests of objectivity, we must read the arguments of this book twice: once, omitting MPHs, and again, including them.

15.8.1 First reading: without MPHs

Chapters 2 to 7 do not deal with the MPH concept. It appears as an important item in Chapter 8, which discusses general circulation. Critics of those who do not take account of the role of water vapour, critics of the unrepresentativeness of past greenhouse gas levels, which show that the IPCC scenario is not a long-term phenomenon, do not need to refer to MPHs. The same applies to critics of the 'hockey stick' curve, of the official curve of supposed mean global temperature, and of the way in which *TOPEX/Poseidon* data have been used to announce a non-existent rise in sea levels. There is no need to refer to MPHs to bring to light obvious facts revealed by direct observation of phenomena such as cooling, warming, higher rainfall values, or aridity. Nor are they vital in the presentation of a thoughtful review of the 'state of the art' in meteorology, or in describing the conceptual *impasse* in which it finds itself.

Briefly, **the MPH concept is not needed in a demonstration of the vacuity of the IPCC's 'greenhouse effect' scenario**. So it is not possible to reject the preceding arguments out of hand, by discrediting MPHs. To still the critics, a detailed point-by-point argument will be required.

15.8.2 Second reading: with MPHs

If we take MPHs into account, the shortcomings and inconsistencies of current concepts are apparent. As Lenoir (2001) says: if the MPH concept is allowed, then '*one can no longer talk nonsense about extreme events*'. The concept offers a coherent if restrictive framework, applicable to all scales of intensity, space, and time, and that framework is general circulation, within which any phenomenon may be, and indeed must be, located in its correct place (which is particularly useful to guard against reversing the chain of phenomena). It also offers multiple explanations for observed phenomena: the dynamic of glaciation and deglaciation; the specific behaviours of the Antarctic or the Arctic, both of them interacting with the dynamic of the circulation of adjacent aerological units; the dynamic of precipitation and extreme weather, or the lack of precipitation resulting from the formation of AAs; climatic peculiarities dictated by local geography (as in Mediterranean regions), etc ...

In brief, the MPH concept throws light upon aspects ignored by the scenarios of models (i.e., the dynamic of meteorological phenomena), offering an alternative solution, the only one at present able to unblock the conceptual *impasse*. This does not *ipso facto* mean that the solution is 'definitive', but it is currently the only concept able to clear a path forward for the discipline, and to put an end to climatological inconsistencies. It should therefore be taken up, if only temporarily. This would spell a revolution for the modellers, and would give us the opportunity to see whether models are really able to make the transition, and to represent the real facts.

The MPH concept represents, though, only one among many other possible objections: in tropical meteorology, there are other shortcomings which might be attended to. In particular, account should be taken of: the notion of the AA within the dynamic of circulation; the double structure of the inclined meteorological equator (IME) and VME; the unproductive nature of the TI; the dynamic of disturbances; and not least, the trajectories of tropical cyclones. It seems that the subject needs to be revisited several times.

15.9 THE REAL STATE OF CLIMATOLOGY

In the introduction to this book, I asked a fundamental question: is climatology in a position to be able, or even want, to encompass today's world, through the intermediary of the greenhouse effect scenario which imposes draconian economic measures – and the world of tomorrow, through the intermediary of its predictions? The answer is exact: NO, assuredly NO.

The erring ways of climatology have become too numerous, and the models too imperfect. There is no point in repeating all the shortcomings, gaps, and errors previously mentioned. Let it suffice to remind ourselves that **the current state of climatology does not justify its pretensions**.

After the appearance of the ACIA report analysed above, another example emerged in early December 2004, eloquently illustrating the real state of climatology in both France and Europe.

The *Centre National de Recherches Météorologiques* (CNRM) announced that '*the thermometer could be registering an extra 4–7 degrees on average in summer in France during the period 2070–2100 ... it would make the heatwave of 2003 look like a 'cool summer*'.' (communiqué from the AFP, 6 December 2004, taken up by *Le Monde* that same day). Dequé held that '*in certain regions the thermometer will show 9, or even 10 extra degrees in summer in 2070–2100 in comparison with today*'. This warming is attributed without any prevarication to '*the concentration of gases emitted as a result of human activities*'. This rise will be '*2 to 4 degrees on winter afternoons*'. These results represent the distillation of data from 10 European numerical models, employing the IPCC's A2 scenario which '*foresees a tripling in CO_2 concentrations compared with the pre-industrial era*'.

This announcement from the CNRM followed the publication in *Nature* on 2 December of the article by Stott *et al.* (2004), of the Hadley Centre (HC), entitled '*Human contribution to the European heatwave of 2003*'. This article began with: '*evidence is mounting that most of the warming observed in recent decades has been caused by increasing atmospheric concentrations of greenhouse gases*'. Its conclusions predict that '*European mean summer temperatures exceeding those of 2003 (will) increase rapidly under the SRES A2 scenario*' and, by the end of the century, '*2003 would be classed as an anomalously cold summer*' (Stott *et al.*, 2004).

So the HC and the CNRM have the same view (and overbid) on the cause of the heatwave and the nature of future summers. Some points to consider:

- Greenhouse gases are held to be the cause of the heatwave, in the usual 'Pavlovian' way, even though this has not been demonstrated by any means, and can only be speculation.

- **The cause of the heatwave is totally unknown**. We need only refer to Chapter 11, to the section dealing with this episode of heat, and especially look again at Figures 74 and 75 (Chapter 11).
- The rise in temperature is associated with an increase in surface pressure. As Allen and Lord (2004) remind us: '*The immediate cause of the heatwave was a persistent anticyclone over north-west Europe*'. In spite of this, neither the HC nor the CNRM makes any mention of atmospheric pressure.
- The approach is only a statistical one, without reference to meteorology, and it is particularly simplistic. It boils down to an elementary classroom problem: 'If we turn up the thermostat on the heating from 1 to 3, will the temperature rise?' So banal (a simple schoolchild's 'rule-of-three') that we must wonder about the pertinence of such statements, especially if they refer to 2070!
- **The approach in question is not a climatological one**: first, because it is based upon an (erroneous) postulate, hastily introduced, concerning the role attributed to humans. **I challenge the HC and the CNRM to clearly show how the greenhouse effect could physically be responsible for the heatwave**. What is more, their approach (like that of the ACIA), takes into account neither the real dynamic of the establishment of the thermal field, on the synoptic scale (Figures 73, 74, and 75, Chapter 11), nor its recent evolution (Figures 76 and 77, Chapter 11).

A period of hot weather has absolutely no connection with the greenhouse effect, since it requires high pressures, great anticyclonic stability, an absence of surface winds, and strong insolation. It must be considered that such circumstances (above a heterogeneous surface, both oceanic and continental), can only be supplied by the aerological dynamic. However, **no reference is made to the dynamic of meridional exchanges**, even though the formation of a marked (if temporary) AA requires powerful and active MPHs. Now, if temperature is rising at this point, it will also be rising at the Pole, and the MPHs will therefore be weaker: lessening the likelihood of a heatwave. A rise in temperature would lower pressures, and reduce the possibility of raised pressures being maintained. Also, the slowing of circulation (warm scenario) would diminish the chances of a summer AA over Europe (cf. Figure 73, Chapter 11). If the temperature goes up by '*9 or 10°C*' in Europe during the summer (cf. Dequé, above), what will then be happening to the temperature in the other regions of the North Atlantic unit of circulation? We ought not to forget that, in August 2003, there was no heatwave in North America, or in Scandinavia; neither should we forget (like the ACIA), the real synoptic dynamic ...

In summary – and to dispense with any further unnecessary listing of inconsistencies – **this scenario involving warming is absurd**, not least because, once again, Europe's climate does not evolve in isolation ... as in a model, where phenomena are imagined *in situ*, immobile and disconnected from meteorological reality.

To sum up what has already been said in Chapter 11, and about these latest predictions:

- The heatwave of August 2003 was not foreseen, and remains unexplained.
- In summer 2004, with no prediction having been made, the weather was cool and

damp, and Météo-France laid itself open to ridicule by announcing completely redundant '*heatwave alerts*'. Large numbers of French holidaymakers flocked to Atlantic beaches, to avoid the hypothetical heat, and they were not disappointed by the cold, wet weather: although Météo-France had not predicted this either!

- What will summer 2005 be like? Will we, by April/May, or as late as June, be getting the 'official' forecast from Météo-France for the coming summer? Now *there* is a challenge for them!
- However ... we do know that summer 2070 will be a hot one! Good news: well done, HC and CRNM! That's one holiday we can organise with full knowledge of the facts.

We may smile, but remember the communiqué from the AFP: '*The heatwave has claimed 15,000 lives in mainland France*'. Perhaps it is not so funny. It would be better if we looked seriously into the causes of phenomena, and concentrated on showing (if it is possible) if the heatwave is really linked with the alleged greenhouse effect, as well as predicting the weather for next summer (and even this winter!) – instead of deploying ranks of computers to play superficial and derisory electronic forecasting games. There are more important things.

15.10 THE PRIORITIES FOR CLIMATOLOGY

As Kondratyev (2004) stated (see above), for climatology to be focused on supposed global warming is prejudicial to the discipline. The greenhouse effect is not what climatology is all about, but it has by now 'cornered the market'. In other words, '*the focus on an imaginary anthropogenic greenhouse effect performed by the (political) scientific community has effectively prevented good research being done and demoralized many scientists for decades in the field of climate science*' (Jelbring, 2004, `www.climatesceptics`). **Climatology is wasting its time, in fact, by trying to demonstrate, at all costs, non-existent warming**. It is also sometimes risking its soul: for example, there are those who try to read into observations of glaciers, and satellite measurements (e.g., from *TOPEX/Poseidon*), what is not there. Climatology has better, urgent things to do.

15.10.1 Regional variations

An additional unwelcome effect of the IPCC's scenario and of its digital context is the suggestion that the 'climate' behaves in the same way everywhere on the Earth's surface, and especially that it is getting warmer everywhere: the *leitmotiv* being '*climatic warming*'. It is obvious to anybody, however, that there is no real 'global' climate, but that there is a great variety of climates, depending on latitude, the geographical environment, and aerological dynamics.

If the Earth is not experiencing a general warming, this is not to say, however, that the climate is not evolving. 'Climate change', which is definitely not synonymous with 'global warming', is a permanent fact: **constant evolution is a property of climate**.

To ask '*has climate change already begun*?' is as ridiculous as to wonder '*is the climate going to topple*?'. The climate has been changing since the dawn of time, and in a highly organised fashion! Climatic evolution also works on the regional scale. Some regions are becoming warmer, and others are cooling. Some are getting wetter, and others drier. Instead of stirring everything into the cooking pot of computer-averaged data processing, we should be highlighting these particular facets of climate. We should be discovering and publicising the specific dynamics which lie at their origin.

There is no point in colouring charts for a hypothetical remote future in red, assuming warming to be automatic, and then constructing scenarios which can never happen. Where is the sense in predicting warming for an area which is currently experiencing cooling, for example, the eastern part of North America (cf. Figure 141, right)? This region, in the '*last winter season, January–March 2003, experienced one of the longest and coldest winters in many years*' (Khandekar, 2003); the season here is characterised by waves of extreme cold and record snowfalls. As long as we lack understanding of the actual causes of these low temperatures, and consequently of the hypothetical reason why this area should miraculously become warmer, making such automatic predictions is little short of buffoonery! The priority in this domain should therefore be the establishment of an honest appraisal of regional climatic behaviours, explaining them in depth and integrating them into general circulation. Most importantly, we should be studying observed realities if we really intend to 'programme' the future. This future is necessarily the near future, because it is no use trying to look as far ahead as 2100, unless one is trying to achieve the serenity of the forecaster who knows that (s)he is unlikely ever to be contradicted!

How are we to handle natural climatic risks, and how are we to take, in the short and the medium terms, realistic precautions and preventive measures, if we do not know the answers to these fundamental questions?

15.10.2 Predicting extreme phenomena

Understanding and predicting intense or extreme meteorological phenomena, and the efficiency of early warning systems, are also matters of high priority. Weather-alert procedures must necessarily allow time for those involved to anticipate events and their possible consequences, and an accurate assessment of the intensity of the phenomena in question is needed if correct precautionary action is to be taken. Any forecast of 'rain' or 'snow' which gives no indication of their degree of severity is really no forecast at all. Every 'weather drama' to hit France, for example, evokes the usual comments and pious reflections, to no positive end. Events at Grand-Bornand, Nîmes, and Vaison-la-Romaine, and in the *départements* of the Aude and the Pyrénées–Orientales; the storms which lashed western Europe in December 1999: none of these catastrophes was properly foreseen, in the true sense of the word. After each of these tragedies, the same, eternal question: '*Are we doing everything we can, in all directions, to stop these disasters happening again*?'.

A similar situation occurred in the Gard *département*, at Nîmes, where torrential rain fell on 7–8 September 2002: rain had been forecast 24 hours earlier, though nothing in any way out of the ordinary. The amount of rain, however, far exceeded the predictions. The '*orange alert*' (level 3) was announced only 12 hours before the event, and the '*red alert*' (the highest, at level 4) after the deluge had begun! Really, the prediction of 'bad weather' 24 hours in advance, particularly if a 7-day forecast is the official claim, is no great task, with the wealth of modern (and especially satellite) technology in support. Four days later, on 12 September 2002, the model-based '*current possibilities*' measure was again estimated, and Météo-France broadcast a new alert; but to no useful purpose, since all that transpired was rain or bad weather! It goes without saying that 'crying wolf' simply to make amends for a 'miss' does not constitute prediction.

And what are we to make of the heavy snow of the weekend of 4–5 January, 2003 in north-eastern France? The alert was not sounded until Saturday the 4th, at 12.30 p.m., and the agencies involved heard nothing until 2.00 p.m., when drivers were already trapped on blocked roads, and passengers were stranded at airports! The media carried various 'explanations' (excuses) from Météo-France officials for this new *débâcle*, one of which was that '*our three numerical prediction models were in disagreement*': so the culprit was named, and by the director of Météo-France himself. The model was to blame! But nothing was said about the actual capability of the numerical models, whose inefficiency is plain to see, as their users themselves admit: far from 'disagreeing', the models were in harmony, in that they were all equally wrong! As usual – and after every extreme weather event (and there have been many in the last decade, just as there have been many victims) – it is concluded that nothing else could be done, but to persist in error, awaiting the arrival of the new magic super-machine which will finally perform the desired miracle!

It seems that the understanding of phenomena, and meteorological prediction, are not inseparable. According to Joly, of the CNRM, '*prediction is not explanation*' (in *La Recherche*, 276, vol. 26, p. 480, 1995). This view (like the lessons which should have been learned from the FASTEX non-experiment) firstly underlines the fact that, lacking 'knowledge' of the real processes which initiate disturbances, particularly in middle latitudes, models are quite unable to predict the evolution of the weather. It also raises the problem of which approach to adopt: one based on statistics and probabilities, or a deterministic one. Meteorological prediction models use the former approach, and observation and 'comprehension' of phenomena are in this case a priori not indispensable. However, a statistical, probability-oriented approach does not bring 'good' results, unless the development of processes remains 'normal' (i.e., does not deviate from the 'mean'): in other words, when there is nothing unusual to predict. The '*limits of predictability*' (cf. Météo-France website) are thus soon reached.

What the weather *does* depend upon in our latitudes is not calculated abstractions, but clearly identified agents: MPHs. Each MPH creates particular kinds of weather at each stage of its evolution, as a function of its initial and acquired potentialities, and the differing circumstances encountered *en route*. Because the deterministic approach attributes a specific weather type and its evolution to an identified agent, it is the most appropriate choice if reliable predictions are to be made. This however presupposes a deep understanding of phenomena, the identification of the factors responsible for them, and their efficient integration into the models. It is a considerable challenge for the meteorological community, conscious of its obligation to provide results, as befits its mission to serve the common good.

In climatology, there are many subjects still awaiting clarification. Other priorities, not all of which we mention here, remain to be fulfilled. For example, palaeoclimatology must examine the representativeness of its findings about past greenhouse gas levels; it should ask itself about the real links between CO_2/CH_4 and temperature, and should make a particular point of trying to determine the exact role of the ocean as a reservoir of CO_2, in the past as in the present. It should also define the real mechanisms initiating glaciation, and deglaciation, and the more or less prolonged returns to cold conditions during the deglaciations (cf. Chapters 9 and 14). The 'mystery' enshrouding El Niño events should also be dispelled, through the definition of its correct place in the chain of processes. Careful use of satellite data can make a useful contribution to revealing more about the involvement of the VME, at the same time establishing the truth about what is really happening with sea levels. We can again recall the actual causes of heatwaves or long-term droughts, or the dynamics of rain-making disturbances ...

In summary, climatologists have much better things to do with their time than to waste it, and their resources, in looking for (or even inventing) proofs of the improbable IPCC scenario.

15.11 HUMANKIND: RESPONSIBLE OR IRRESPONSIBLE?

15.11.1 Human responsibility

Are humans responsible for the climate and the way it changes? Definitely not. We are of course entirely responsible for our despoliation of nature. We are responsible for the pollution which our activities cause. But any effect we have upon the climate is only on a local scale, the best example being the urban heat domes over our cities. However, our influence goes no further, even if we would like to think so. **The human race cannot possibly affect the climate on a global scale** (or, at least, not yet).

It is not always easy to decide whether humanity is responsible, or not, for the propagation of the greenhouse effect scenario.

- Some people feel the need for self-flagellation: as Stott (BBC, Open2.NET, 2004) put it, '*people want to explain it in terms of human action and human faults. In other words, we always need a Noah myth, and a Noah myth that says we have sinned*'.
- Other people do not know where to draw the line between the fine sentiments of militant ecologists and actual climate science, and they determinedly salve their consciences by accusing humanity, even if they lack good scientific arguments.
- And there are some, and too many usually called 'scientists' belong to this group, who show unquestioning faith in the IPCC reports, press communiqués, and newspaper articles, with little thought for searching out the truth. How

should we categorise this passive acceptance? It is really difficult to understand '*how anyone in his right mind could possibly believe in the claims of those who say mankind's enriching of the air with CO_2 is a greater threat to the biosphere, to humanity, to civilization, to whatever than either world terror or potential nuclear warfare. Clearly, it just ain't so.*' (Idso, 2004).

15.11.2 Human irresponsibility

Lack of responsibility is (in the main) an involuntary thing, it might be said, but *irresponsibility* is a matter of choice. To practise medicine without a licence is against the law, and the doctor is held to be responsible for the diagnosis. There is a 'precautionary principle' involved, and the doctor has to take care not to pursue paths where mistakes might be made. Unfortunately, this does not seem to apply in the case of **the illicit paths of climatology**! And it is a great pity! Anyone can say or write anything, with impunity, even if it implicates a whole scientific discipline, or indeed the future of a nation, or nations. Ought we to be making this illegal too? Perhaps so, since voluntary codes of ethics seem no longer to be sufficient.

This irresponsibility is found among some who occupy, or lay claim to, positions in science; most of the media; and politicians whose mandate is to 'spread the right message' (Spencer, 2004, www.climatesceptics).

A prime example of this lack of responsibility: the scientists who disseminate erroneous 'news', to match the flavour of the times, such as the staggering rise in temperature (see above), or the aforementioned (supposed) rise in sea levels. And there are those self-appointed climatologists whose busy photocopiers churn out the IPCC 'hymn sheets', and who contribute to the outpourings of the 'popular science' publications, a notable example already mentioned being that of Le Treut and Jancovici (2001). We might also cite those who fritter away precious resources trying to predict the climate for 2100, and yet pay little mind to averting the dramatic results of real weather happening today. Also, and it has to be admitted, there are some who see 'science' as just another rung on the ladder of their own ambition. Spencer (2004) deplores them: '*It wasn't long after I became a research scientist that I learned that scientists aren't the unbiased, impartial seekers of truth I always thought they were. Scientists have their own agendas, philosophies, preconceived notions, and pet theories ... I wish all those global warming extremists would simply confess their faith and stop giving science a bad name*'. When irresponsibility passes from deliberate fallacy and mere censure, to become totalitarianism and a refusal to engage in debate, at conferences or in print, the very spirit of science fades into nothingness.

The media, too, share the (ir)responsibility: newspapers and magazines always claim that articles represent 'scientific fact', and the popular 'science' sections of such publications cannot resist making a 'scoop' out of some banal news item, in the constant search for sensation at all costs. Often, a hard-pressed journalist, knowing little about the subject in hand (nobody knows everything!), and with the impatient editor hovering, just copies the press communiqué, or skim-reads the item, or a résumé of it, and jots down or highlights whatever might make a catchy headline or effect. It is not unknown for 'creative' proofs to be invented: remember those drowned Dutch windmills in the previous chapter? Whatever code of practice the media might follow when verifying their sources does not seem to extend to differentiating between qualified and unqualified informants and their 'authoritative' statements on various subjects. There are newspapers and television channels with tame 'in-house' experts, always ready to give (only) the 'correct' version of events!

These faults are not exclusive to the popular press. Respectable scientific publications also suffer from them. Editorial committees may often show bias, rather than encourage debate about ideas. In a recent article *What is Happening to Science?*, Michaels (2004) reminds us that '*some things are sacred to scientists: facts, data, quantitative analysis, and the journal Nature, long recognized as the world's most prestigious science periodical*'. He has his doubts, though, in the light of some 'dubious' articles, about the way that *Nature* is going: '*This is nothing but tragic, junk science, published by what is (formerly?) the most prestigious science periodical in the world*'. So 'pure' science too is tainted, like the press and the television, by the desire for a 'scoop'.

Why does the press in general present this hypothetical warming as an 'apocalypse'? Such hand-wringing, at which the media excel, may well get the anti-pollution message across, but scientifically speaking, the message is based on a total untruth. However, when we read tearful copy about the homeless braving a cold spell, or drivers stranded in snowdrifts, we may discern the real motivation behind such presentations, 'news' devoid of real information!

Journalists like Grenier and Kohler, who genuinely seek to inform, are rare indeed. The inconsistencies, contradictions, and exaggerations have become so ingrained that we can but agree with author Michael Crichton, who said on ITV on 9 December 2004, on the subject of climate change: '*No-one really understands what's going on ...*' . It is amazing that the great majority of journalists have not realised this themselves, especially in the French media, where ignorance is bliss! Another question frequently recurs: what does the word 'scientific' actually mean when applied to a journalist, and what does the heading 'Science', which includes the subheading 'climate', signify? Would it not be better included under the heading 'News', even 'Economics'? Moreover, when tragedy does strike, as it did in the recent heatwave, even fewer commentators seem surprised or even indignant that we can spend so much public money to no purpose, organising needless conferences and symposia, and setting our sights on the year 2100, instead of turning the efforts of our public services towards the improvement of our knowledge and ability to forecast the weather right now.

Irresponsibility among politicians is self-evident. 'Climatology' in France, and in some other countries, illustrates this fact well. Since signatures were put to the Climate Change Convention of 1992, the greenhouse effect scenario has become a state 'religion':

- For Météo-France, which is now wedded to global warming and receives generous funding to pursue the phantasms of its supercomputers.
- For the CNRS, now the Centre for Foregone Conclusions (or imposed ones),

where research is offered no alternative the conclusion (as in the CNRM) is written before the introduction!

- for the *Mission Interministérielle de l'Effet de Serre* (Interministerial Task Force on Greenhouse), charged with the control of pollution, which labelled Lomborg (2002) a dangerous '*negationist*' (in *Le Nouvel Observateur*, 2 October 2002, p. 22), because he dared to question the 'established facts'!
- for the *Commission Française du Développement Durable* (French Sustainable Development Commission), which organised a 'Citizens' Conference' in February 2002, to '*inform the lay public ... in the most objective and complete manner*', but neglected to invite any detractors to put *their* points of view.
- For the French delegation to the IPCC, which, to date, has not included any recognised climatologist as an attentive observer of meteorological reality.
- For the *Office Parlementaire d'Evaluation des Choix Scientifiques et Technologiques* (Parliamentary Office for the Evaluation of Scientific and Technological Choices) of the *Assemblée Nationale*, whose representative (unknown in the bibliography of climatology) interviewed 89 people, with not a single 'sceptic' among them.

So this is how a 'national priority' is handled, and why it can be declared during an international conference that '*the house is burning*'! Disbelievers in the official religion are declared 'heretics', and gagged. What is more, the 'solution', at every level of state administration, is nothing but a vast operation in **shrugging off responsibility**, since any real reflection on the subject will be pointless. The decision makers, as faithful servants of the state, may feel the need to hide behind the official dogma; or they may have no desire or ability to stand out from the crowd.

The same atmosphere pervades IPCC meetings, to which no 'heretics' are ever invited, as officially demonstrated at the symposium held at the Moscow Academy of Sciences in July 2004. Products of the collusion of science with politics are the *Summaries for Policymakers*, which are a harmonious catalogue of '**Instructions for Parroting Climatical Correctness**' (IPCC?)!

15.12 THE REHABILITATION OF CLIMATOLOGY

This book and many other publications strongly maintain that **the supposed global warming is a fable**. Or, as Hämeranta (moderator, climatesceptics, 2004) put it: '*Well, the war is over, but the battle continues*' ...

The IPCC's climatology, or, to be more precise, the climatology of the *Summary for Policymakers*, is not climatology. It visits upon the whole of the discipline a distorted, deplorable image which is extremely schematic and simplistic.

Climatology itself gains no merit from these 'greenhouse effect days', which in the final analysis will be seen as just another stage in its history. Of course, climatology feels honoured (without mentioning any names) to have been raised into the limelight by the advent of the IPCC. It is, though, particularly surprising to see how certain scientists, elevated to the ranks of the IPCC, will so easily allow themselves and their discipline to be overshadowed by politicians with no scientific agenda of their own.

Climatology, a science of relative things, felt suddenly able to deal in dubious certainties, and soon began wasting its time and losing its soul. This unfortunately continues: the runaway train is rolling, with the inertia inherent in any enterprise, and also because of the dogged persistence of purposely created institutions, both national and international, which will not easily give up their identities – what Lenoir (2001) described as the 'climatocracy'. There are too many well-paid posts to lose, too many all-expenses-paid overseas trips to sacrifice, too many individual and collective interests at stake. We can only hope that minds will focus upon freeing climatology from its political and economic thrall, and that the climatologists themselves will realise that their discipline has lost its true way (more often because of themselves); it should not become an excuse or a justification.

15.12.1 Climatology is now of no use in the framework of the IPCC

There has been no scientific debate on this subject for 10 years now. The IPCC seems able to move this *affaire* (and its own interests) forward only with media *coups* which at once discredit and disrupt the scientific community. The climate change 'machine' is more and more a politico-commercial one, and no longer needs its false climatic justification. After the vigorous debate at the Moscow Academy of Sciences, Russia's recent support for the Kyoto Protocol is clearly shown to be without 'climatic' motivation.

... From the outset, I made the decision not to discuss this economic aspect, and I shall not dwell upon it: the non-involvement of climatology will make no difference whatsoever to the construction of a new economic world order. I have discussed the irresponsibility of scientists and of the media, but I have not touched upon where the real responsibility lies, and the politico-financial stake in this well orchestrated operation. **Leave climatology to work in peace**.

Climatology has been tarnished enough; it must get back to the pursuit of its proper ends.

It must first recover from the misconceptions, affectations, and wounds which this episode in its history has visited upon it, i.e.:

- shake off the serious delusion that it is a completely mature science;
- cure itself of the disorder which causes it to overuse 'averaged' data;
- overcome the confusion, caused by these data, between the apparent and real climates;
- correct the short-sightedness which leads it to mix up co-variations and physical links;
- cure itself of the disorientation which causes it to treat relationships back-to-front;
- stop backing away from its responsibilities and hiding behind models;

- replace the paralysing view of phenomena always seen *in situ* and motionless, by a dynamical view, since all is permanently in motion;
- fill in the 'missing link' between meteorology and the computer processing to reach the level of deterministic forecasting;
- do something about its frequent lack of knowledge about meteorology and the physical reality of phenomena;
- get over its disdain for the techniques of direct observation and analysis of documents;
- lose its taste for the 'scoop', media jargon, the sensational, and even 'star status', all of which have led to piles of opportunistic publications and fewer in-depth analyses;
- contain its fond imaginings of horizontal and vertical circulations in air and water, going in all directions, and replace them with something real;
- avoid, in the most serious cases, opting for the immaterial, the magical and the mysterious;
- renounce the screen of global warming, an answer of ignorance and laziness, behind which it conceals its erring ways; and
- especially give its attention, with absolute urgency, to (too many) subjects not yet clarified.

The best remedy for all these ills will doubtless be serene, hard, objective, and detailed work.

Not the least of climatology's further tasks will be **to make a calm analysis of what climate change is really all about, for weather and climate do evolve, by their very nature**. Instead of spending its energies on vain conjectures about the very hypothetical climate of the year 2100, or of the next millennium, climatology should be addressing the real questions, which are more immediate and more urgent. It should avoid going down false tracks, although and perhaps because they are the easiest!

15.12.2 Climatology has so many better things to do

The priority of meteorology should be to draw back from its conceptual *impasse*, and seek a better understanding of normal and extreme phenomena. For a long time now, meteorology has turned its back on conditions at the surface, and on reality as revealed in particular by satellites, and has spurned research into the concepts which will point the way towards a coherent appreciation of phenomena. The advent of computers has speeded this divorce from reality, and has caused the firm revival of the spirit of the climatological school of the early 20th century. Real forecasting will progress only through a progressive *rapprochement* of the virtual and the concrete, using proper, deterministic models. It is a radical, and a fundamental proposition, and it may be delayed for many reasons, good and bad, but it will come. The considerable progress already made in computer technology can serve only to maintain an 'illusion of progress', but cannot fill in the missing link, or links, in the chain of processes.

When (and not before) models finally provide answers to the biggest question in climatology, which is **to reconstruct, understand, and explain unambiguously the (verifiable) evolution of the weather of the last century**, climatic modelling will be succeeding in its task.

One thing that needs to be explained is how the **climatic turning point of the 1970s** came about, principally, but not entirely, in the northern hemisphere. The real causes of recent climatic evolution need to be found. This will be the key to foreseeing how current trends will develop, without recourse to the crystal ball. In particular, will the recent trend (Figure 141) initiated by cooling in the western Arctic, and whose causes are still not well understood, continue at the same intensity, or will it instead slow down, or even go into reverse? Some avenues of approach have been suggested (e.g., the influence of volcanism on the density of dust in the Arctic). An alternative hypothesis from Radionov and Marshunova (1992) was that atmospheric turbidity over Russia has been a factor in the extinction of solar radiation in the Arctic for the past 30 years. The greater turbidity in the Arctic is also associated with air flowing in from the south, with growing intensity (cf. Chapters 12 and 13). The actual role of solar activity also gives rise to many questions. To answer, or even begin to answer, these questions (cf. Figure 20, Chapter 6) is of capital importance in the study of the northern hemisphere's climatic destiny.

Instead of always seeking justifications and excuses, such as that the 'chaotic' nature of weather precludes prediction, we ought to be admitting to ourselves that **meteorological phenomena are strictly organised, and respond to clearly identified causes**. The fact that we have not (yet!) isolated (or acknowledged) these causes does not mean that we ought to resort to magic, or chaos, or almighty El Niño, or the greenhouse effect, or whatever . . .!

Those who assess weather risks and plan our strategies have pressing concerns, and they need ready answers about the weather as a matter of priority: sooner, rather than in 100 years' time. In France, for example, where our official meteorologists have a duty of public service, we may ask these questions:

- Can we hope for regular and clement weather ('warm' scenario, slow circulation), or must we brace ourselves for more irregular and more violent weather ('cool' scenario, rapid circulation), with cold and heat waves, drought and floods, storms?
- Will the kind of rainfall pattern that has repeatedly caused winter floods in Brittany, autumn floods along the southern Rhône Valley, and long dry spells in both winter and summer, persist?
- In the mountains, will the current *régime* whereby there is often a shortage of snow in mid-winter, but plenty of snow and rain in spring and autumn, become the rule? Will regional development in the mountains have to adapt?

There are numerous such questions, especially those put by the country people, too many to list here. They are more serious, more demanding, and much more urgent and immediate than the question of what castles we are to build in the air of 2100! They are wide-ranging questions about the details of risk and land management, not just in France but elsewhere too. Simple and practical examples: should we be attaching our outdoor TV aerials more firmly, and cutting trees shorter? Should we be installing snow cannons in the mountains, or promoting seaside resorts and stocking up on swimsuits? Should we be making better preparations for cold snaps, and spending more on heating – or on air conditioning units? The list goes on.

To find answers to such questions, and to get things done, we have to shrug off the numbing influence of simplistic equations, ready-made formulas, and routine and outmoded concepts; we must do without the easy reassurance of the false certainties served up by models. And **we must break with the IPCC as soon as possible** ... it will no doubt be able to find a new 'terror' to put in the place of the alleged 'climate change'!

15.12.3 Climatology for tomorrow

The 'greenhouse effect' or 'global warming' experience has not been an entirely negative one. One of its positive aspects is that some of its excesses, and especially those of the *Summary for Policymakers*, have caused an outcry, and intense reflection, among individuals and some particularly active groups: articles and books have denounced its improbabilities, extravagances, and falsehoods. There has also been a beneficial *triage* among scientists.

The experience has also allowed us to show that **climatology is a rigorous and complex scientific discipline**, and that there is no such thing as 'popular climatology', and less still a 'tabloid climatology'.

What climatology is *not*, is an excuse for planet-wide manipulations; neither is it a scare story, a source of sensational copy for journalists short of lines, or a provider of disaster movie scenarios. **The protection of Nature and the fight against pollution need no false climatic premisses to underpin them: their necessity is self-evident**.

Let those who lay claim to knowledge of the 'climate' make sure that they know what they are talking about, and take time to learn the facts – and this includes some scientists, it must be said. The destiny of climatology is not to be a recreation, a supplier of news items and catastrophes: **its destiny, as a science of Nature, is to resolve, objectively, disinterestedly, and responsibly, some of humanity's fundamental problems**.

Bibliography and references

(With such a subject, bibliography can only be indicative)

Abley, M. (1998). *Le Grand Verglas*. Récit en images de la tempête de jenvier 1998. The Gazette, Montréal, 192 p.

Académie des Sciences – USA (1992). C.S Silver, R.S. DeFries. Traduction française de *'One Earth. One Future. Our Changing Global Environment'. Une planète, un avenir*. Préface de J. Grinevald. Sang de la Terre. 192 p.

Aceituno, P. (1992). El Niño, the Southern Oscillation, and ENSO: Confusing names for a complex ocean–atmosphere interaction. *Bull. Am. Met. Soc.*, vol. 73, n° 4, 483–485.

ACIA (2004). Arctic Climate Impact Assessment. (amap.no/acia/index.html)

Agee, E.M. (1991). Trends in cyclone and anticyclone frequency and comparison with periods of warming and cooling over the Northern Hemisphere. *J. of Climate*, vol. 4, n° 2, 263–267.

AGU special report (1995). Chapman conference probes water vapour in the climate system. *EOS*, Feb. 14, p. 67.

AGU: American Geophysical Union (1999). Position statement adopted on climate change and greenhouse gases. *EOS*, vol. 80, n° 5, p. 49.

Alexandre, P. (1987). *Le climat en Europe au Moyen Age*. Ed. Ec. Hautes Etudes en Sciences Sociales, Paris.

Allan, J. and Komar, P. (2000). Are ocean waves heights increasing in the Eastern North Pacific? *EOS*, vol. 81, n° 47, 561, 566–567.

Allègre, C. (2001). *Histoires de Terre*. Fayard, Paris.

Allen, M.R. and Lord, R. (2004) The blame game. *Nature*, vol. 432, 551–552.

Alley, R.B., Meese, D.A., Shuman, C.A., Gow, A.J., Taylor, K.C., Grootes, P.M., White, J.W.C., Ram, M., Waddington, E.D., Mayewski, P.A. and Zielinski, G.A. (1993). Abrupt increase in Greenland snow accumulation at the end of the Younger Dryas event. *Nature*, vol. 362, 527–529.

Alley, R.B. *et al.* (1997). Holocene climatic instability: A prominent, widespread event 8,200 years ago. *Geology*, 25, 483–486.

Am. Geophys. Union (1995). Climate Change. Ann. Geophys. Union, *EOS*, 185–187.

Amraoui, L. (2000). *Les précipitations au Maroc atlantique: Quelques aspects de la variabilité spatiale et de l'évolution temporelle*. Mémoire LCRE, Univ. Lyon, 90 p.

Amraoui, L. (2004). Les pluies diluviennes de novembre 2002 au Maroc. *Revue de Géographie de Lyon, RGL* (à paraître).

Amraoui, L., Leroux, M. and Atillah, A. (2004). Suivi spatio-temporel des températures de surface marine et de l'upwelling maroco-mauritanien à l'aide de la télédétection. *Bull. Ass. Inter. Clim.*

Amraoui, L. (2005). *La dynamique aérienne et maritime sur les régions atlantiques du Maroc et de la Mauritanie*. Th. Univ. Lyon 3, LCRE.

Anderson, D.E. (1997). Younger Dryas research and its implications for understanding abrupt climatic change. *Progress in Physical Geography*, 21(2), 230–249.

Andreae, M.O. (1996). Raising dust in the greenhouse. *Nature*, vol. 380, 389–390.

Angell, J.K. (1989). On the relation between atmospheric ozone and sunspot number. *J. of Climate*, 2(11), 1404–1416.

Angel, J.R. and Isard, S.A. (1998). The frequency and intensity of Great Lakes cyclones. *Journal of Climate*, 11, 61–71.

Arbogast, Ph. and Joly, A. (1997). Identification des précurseurs d'une cyclogenèse. *C.R. Académie des Sciences*, Paris, Géophysique externe, Météorologie, Elsevier, 1998, 326, 227–230.

Arking, A. (1991). The radiative effects of clouds and their impact on climate. *Bull. Am. Met. Soc.*, vol. 71, n° 6, 795–813.

Arrhenius, S. (1896). On the influence of carbonic acid in the air upon the temperature of the ground. *Philosophical Magazine*, vol. 41, 237–276.

Assel, R.A., Janowiak, J.E., Young, S. and Boyce, D. (1996). Winter 1994 weather and ice conditions for the Laurentian Great Lakes. *Bull. Am. Met. Soc.*, vol. 77, n° 1, 72–88.

Aubert, S. (1994). L'évolution récente de la pression atmosphérique en Roumanie. *Publ. Ass. Int. de Clim.*, vol. 7, 331–336.

Aubert, S. (2002). Evolution de la température et de la pression en Roumanie. Comm. Colloque: Evolution récente du climat, Lab. de Clim. (LCRE), Lyon.

Aubert, S. (2005). *Le climat de la Roumanie et son évolution récente*. Th. Univ. Lyon 3, Labor. Clim., Risq., Env., LCRE.

Audran, E. (1998). Variations et dynamique des vents forts sur le littoral de la Bretagne de 1949 à 1996. Th. en cours, Géolittomer, Brest, comm. pers.

Avery, S.K., Try, P.D., Anthes, R.A. and Hallgren, R.E. (1996). Open letter to Ben Santer. *Bulletin of the American Meteorological Society*, vol. 77, n° 9, 1961–1962.

Baliunas, S. (1996). Are human activities causing global warming? (`www.mitosyfraudes.com`)

Balling, R.C. and Idso, S.B. (1989). Historical temperature trends in the United States and the effect of urban population growth. *Journ. of Geophys.* Res., vol. 94, n° D3, 3359–3363.

Bard, E. (2002). L'effet de serre. *La Recherche*, 356, 50–53.

Bard, E. (2004). Le climat peut-il basculer? *La Recherche*, n° 373, 30–36.

Barber, D.C, Dyke, A., Hillaire-Marcel, C., Jennings, A.E., Andrews, J.T., Kerwin, M.W., Bilodeau, G., McNeely, R., Southon, J., Morehead, M.D. and Gagnon J.M. (1999). Forcing of the cold event of 8,200 years ago by catastrophic drainage of Laurentide lakes. *Nature*, vol. 400, 344–348.

Barbier, E. (1999). Le cyclone Mitch. *Comm. Laboratoire de Climatologie*, LCRE, Lyon.

Barbier, E. (2004). *La dynamique du temps et du climat en Amérique centrale*. Thèse de Climatologie. Univ. J. Moulin, Lab. de Clim., Risq., Env., LCRE, Lyon, 545 p.

Barnola, J.M., Raynaud, D., Korotkevich, Y.S. and Lorius, C. (1987). Vostok ice core provides 160,000-year record of atmosheric CO_2. *Nature*, vol. 329, 408–413.

Barnola, J.M. (1994). Les glaces: Archives des gaz à effet de serre. *Lettre n° 5 PIGB-PMRC*, CNRS. (website `pigb`)

Barsugli,, J.J., Whitaker J.S., Loughe, A.F., Sardeshmukh, P.D. and Toth, Z. (1999). The effect of the 1997/98 El Niño on individual large-scale events. *Bull. Am. Met. Soc.*, vol. 80, n° 7, 1399–1411.

Bartlein, P.J., Anderson, K.H., Anderson, P.M., Edwards, M.E., Mock, C.J., Thompson, R.S., Webb, R.S., Webb III, T. and Whitlock, C. (1998). Paleoclimate simulations for North America over the past 21,00 years: Features of the simulated climate and comparisons with paleoenvironmental data. *Quaternary Science Reviews*, vol. 17, 549–585.

Baudin, F. and Koutchmy, S. (1997). Les faux airs tranquilles du Soleil. *La Recherche*, 303, 32–36.

Becker, F. (1992). Peut-on mesurer la température terrestre? Les mesures satellitaires. *La Recherche*, 243, vol. 2, 588–591.

Beltrando, G. (1990). *Variabilité interannuelle des précipitations en Afrique orientale (Kenya, Ouganda, Tanzanie) et relations avec la dynamique de l'atmosphère*. Th. Un. Aix-M. II, 223 p.

Beltrando, G. and Chémery, L. (1995). *Dictionnaire du Climat*. Larousse, Références, 344 p.

Benestad, R. (2002). *Solar activity and Earth's climate*. Springer–Praxis, New York, 288 p.

Beniston, M. *et al.* (1997). Climate modelers meet in Switzerland (Worshop on high resolution climate modeling), Wengen, CH. *EOS*, 78(43), 484.

Benoist, J.P., Jouzel, J., Lorius, C., Merlivat, L. and Pourchet, M. (1982). Isotope climatic record over the last 2,5 ka from Dome C, Antarctica, ice cores. *Annals of Glaciology*, 3, 17–22.

Beran, M. (2004). Cyclonic storms over western Europe. In J. Daly, *Still Waiting For Global Warming*.

Berger, A. (1986). L'hiver nucléaire. *La Recherche*, n° 179, 880–890.

Berger, A., Dickinson, R.E. and Kidson, J.W. (eds) (1989). *Understanding Climate Change*. Geophysical Monograph 52, IUGG vol. 7, AGU.

Berger A. (1992). *Le Climat de la Terre*. De Boeck Université, Bruxelles, 479 p.

Bergeron T. (1928). Uber die dreidimensional verknupfende wetteranalyse. *Geofysiske Publikasjoner*. Vol. V, No 6, 1–11.

Betts, A. (1990). Greenhouse warming and the tropical water budget. *Bull. Am. Met. Soc.*, vol. 71, n° 10, 1464–1465.

Bezinge, A. (1999). Die gletscher des Schweizer Alpen. *Wasser-Energie-Luft*, heft1/2, Baden.

Bindschadler, R.A., Alley, R.B., Anderson, J., Shipp, S., Borns, H., Fastook, J., Jacobs, S., Raymond, C.F. and Shuman, C.A. (1998). What is happening to the West Antarctic Ice Sheet? *EOS*, vol. 79, n° 22, 257, 264–265.

Bjerknes, V. (1904). Das Problem der Wettervorhersage, betrachtet vom Standpunkte der Mechanik und der Physik. *Meteorologische Zeitschrift*, 21, 1–7. Trad. française in *La Météorologie*, 1995, n° 9, 55–62.

Bjerknes, J. (1923). L'évolution des cyclones et la circulation atmosphérique, d'après la théorie du front polaire. Trad. *Off. Nat.* Mét. (1), n° 6, Paris.

Bjerknes, J. and Solberg, H. (1921). Life cycles of cyclones and the polar front theory of atmospheric circulation. *Publikajoner Geeophysic. Institut*, vol. III, n° 1.

Bjerknes, J. (1969). Atmospheric teleconnections from the equatorial Pacific. *Mon. Weather Review*, 97, 163–172.

Blair, T.A. (1942). *Climatology, General and Regional*. Prentice Hall, New York, p. 101.

Bluth, G.J.S., Doiron, S.D., Schnetzler, C.C., Kruger, A.J. and Walter, S.S. (1992). Global tracking of the SO_2 clouds from the June 1991 Mount Pinatubo eruptions. *Geophys. Res. Letters*, 19, 151–154.

Bolin, B., Doos, B.R., Jäger, J. and Warrick, R.A. edit. (1986). *The Greenhouse Effect, Climatic Change, and Ecosystems*, Scope 29, Wiley, 541 p.

Bolin, B., Jäger, J. and Doos, B.R. (1986). The greenhouse effect, climatic change, and ecosystems. A synthesis of present knowledge. In: *The Greenhouse Effect, Climatic Change, and Ecosystems,* Scope 29, Wiley, 1–32.

Bolin, B. (1986). How much CO_2 will remain in the atmosphere? The carbon cycle and projections for the future. In: *The Greenhouse Effect, Climatic Change, and Ecosystems,* Scope 29, Wiley, 93–155.

Bolin, B., Houghton, J. and Filho, L.G.M., (1996). Attachment 3, in Open letter to B. Santer. *Bulletin of the American Meteorological Society*, vol. 77, n° 9, 1965–1966.

Bonneau, M., Bessemoulin, P. and Buffet, M. (1987). La nouvelle maladie des forêts: Description sommaire et causes possibles. *La Météorologie*, n° 16, 35–43.

Bonnissent, J. (1992). Cours de météorologie tropicale. *Ecole Nationale de la Météorologie*, Météo-France, Toulouse. Tome 2: Circulation générale, 41 p. (cf., figures: 5.19, 5.20, 5.21, 5.23).

Bopp, L., Legendre, L. and Monfray, P. (2002). La pompe à carbone va-t-elle se gripper? *La Recherche*, 355, 48–51.

Borisenkov, Ye. P., Tsvetkov, A.V. and Agapanov, S.V. (1983). On some characteristics of insolation changes in the past and the future. *Climate Change*, 5, 237–244.

Bouligand, R. (2000). Les surcotes à Brest depuis un siècle: Analyse des paramètres météorologiques influents à l'échelle locale. *Norois*, n° 186, 201–217.

Bouws, E., Janninck, D. and Komen, G.J. (1996). The increasing wave height in the North Atlantic Ocean. *Bull. Am. Met. Soc.*, 77(10), 2275–2277.

Brad Adams, J., Mann, M.M. and Ammann, C.M. (2003). Proxy evidence for an El Niño-like response to volcanic forcing. *Nature*, vol. 426, 274–278.

Bradley, R.L. (2003). *Climate alarmism reconsidered.* IEA, Institute of Economic Affairs, London, 176 p. (online)

Braithwaite, R.J. (2002). Glacier mass balance: the first 50 years of international monitoring. *Progress in Physical Geography*, 26, I, 76–95.

Braithwaite, R.J. and Zhang, Y. (1999). Relationships between interannual variability of glacier mass balance and climate. *Journal of Glaciology*, 45(151), 456–462.

Broecker, W.S. (1995). Chaotic climate. *Scientific American*, 44–50.

Broecker, W.S (1997). Thermohaline circulation, the Achilles heel of our climate system: Will man-made CO_2 upset the current balance? *Science*, 278, 1582–1588.

Broecker, W.S. (2000). Abrupt climatic changes: causal constraints provided by the paleoclimate record. *Earth Science Reviews*, 51, 137–154.

Broecker, W.S. (2001). Was the medieval warm period global? *Science*, 291, 1497–1499.

Bunyard, P (2000). Comment la crise climatique pourrait échapper à tout contrôle. *L'Ecologiste*, vol. 1, n° 2, 27–32.

Cabanes, C., Cazenave, A. and Le Provost, C. (2001). Sea level rise during past 40 years determined from satellite and in situ observations. *Science*, vol. 294, 840–842.

Cabanes, C., Cazenave, A. and Remy, F. (2002). 2,5 millimètres par an. In La Mer, *La Recherche*, 355, 64–66.

Caillon, N., Severinghaus, J.P., Jouzel, J., Barnola, J.M., Kang, J. and Lipenkov, V.Y. (2003). Timing of atmospheric CO_2 and Antarctic temperature changes across Termination III. *Science*, vol. 299, 1728–1731.

Callendar, G.S. (1938) The artificial production of CO_2 and its influence on temperature. *Quat. Jour. of The R. Met. Soc.*, 64, 223–237.

Callendar, G.S. (1949). Can CO_2 influence climate? *Weather*, 4, 310–314.

Callendar, G.S. (1958). On the amount of CO_2 in the atmosphere. *Tellus*, 10, 243–248.

Carlowitz, M. (1996). Did water vapor drive climate cooling ? *EOS*, vol. 77, n° 33, 321–322.

Carslaw, K.S., Harrison, R.G. and Kirkby, J. (2002). Cosmic rays, clouds, and climate. *Science*, vol. 298, 1732–1736.

Cassou, C. (2004). Du changement climatique aux régimes de temps: L'oscillation nord-atlantique. *La Recherche*, n° 45, 21–32.

Catchpole, A.J.W. and Hanuta, I. (1989). Severe summer ice in Hudson Strait and Hudson Bay following major volcanic eruptions, 1751 to 1889. *Climatic Change* 14, 61–79.

Cazenave, A. (2003). Les variations actuelles du niveau de la mer. *La Météorologie*, n° 45, 38–42.

Cazenave, A., *et al.* (2004). 2,8 millimètres par an. Le risque climatique. N° spécial. *Dossiers de la Recherche*, 46–51.

CCSM (2003). Community Climate System Model Strategic Business Plan, 2004–2008. (www. ccsm. ucar.edu)

Cess, R.D., Potter, G.L., Blanchet, J.P., Boer, G.J., Genio, A.D.D., Deque, M., Dymnikov, V., Galin, V., Gates, W.L., Ghan, S.J., *et al.* (1990). Intercomparison and interpretation of climate feedback processes in 19 atmospheric general circulation models. *J. Geophys. Res.*, 95, 16601–16615.

Cess, R.D. (1991). Positive about water feedback. *Nature*, vol. 349, 462–463.

Chalon, J.P. and Joly, A. (1996). FASTEX: un programme d'étude des tempêtes atlantiques et des systèmes nuageux associés. *La Météorologie*, n° 16, 41–47.

Chalon, J.P. (2002). *Combien pèse un nuage*? Coll. Bulles de Sciences. EDP Sciences, Paris, 188 p.

Changnon, S.A. (1992) Inadvertent weather modification in urban areas: Lessons for global climate change. *Bull. Am. Met. Soc.* Vol. 73, n° 5, 619–627.

Changnon, S.T., Kunkel, K.E., Reinke, B.C. (1996). Impacts and responses to the 1995 heat wave: A call to action. *Bull. Am. Met. Soc.*, vol. 77, n° 7, 1487–1506.

Changnon, S.A. (1999). Impacts of 1997–1998 El-Niño-generated weather in the United States. *Bull. Am. Met. Soc.*, vol. 80, n° 9, 1819–1827.

Chappell, A. and Agnew, C.T. (2004). Modelling climate change in West African Sahel rainfall (1931–1990) as an artifact of changing station locations. *Int. Journal of Climatology*, vol. 24, n° 5, 547–554.

Charney, J.G. (1975). Dynamics of deserts and drought in the Sahel. *Quarterly J. Royal Meteorological Society*, 101, 193–202.

Chen, P., Hoerling, M.P. and Dole, R.M. (2001). The origin of the subtropical anticyclones. *Journal of the Atmosph. Sciences*, vol. 58, 1827–1835.

Chenoweth, M. (1996). Ship's logbooks and 'the year without a summer'. *Bull. Am. Met. Soc.* vol. 77, n° 9, 2077–2093.

Chevassus-au-Louis, N. (2003). Enquête sur les experts du climat. *La Recherche*, n° 370, 59–63.

Chenoweth, M. (1996). Ship's logbooks and the 'year without a summer'. *Bull. Am. Met. Soc.*, vol. 77, n° 9, 2077–2093.

Christy, J.R., *et al.* (1997). How Accurate Are Satellite Thermometers? *Nature*, 389, 342–343.

Christy, J.R. and Spencer, R.W. (2000). MSU tropospheric temperatures: Data set construction and radiosonde comparisons. *J. Atmos. Oceanic. Tech.*, 17, 1153–1170.

Christy, J.R., Spencer, R.W. (2003). Reliability of Satellite Data Sets. *Science*, 301, 1046–1047.

Christy, J.R. and Norris W.B. (2004). What we may conclude about global tropospheric temperature trends? *Geophys. Res. Letters*, 31, L06211.

Church, J.A. (2001) How fast are sea levels rising? *Science*, vol 294, 802–804.

Chylek, P., Box, J.E. and Lesins, G. (2004). Global Warming and the Greenland Ice Sheet. *Climatic Change*, vol. 63, n° 1–2, 201–221.

Clarke, G., Leverington, D., Teller, J. and Dyke, A. (2003). Superlakes, megafloods, and abrupt climate change. *Science*, vol. 301, 922–923.

CLIMAP project members (1981). *Seasonal reconstruction of the Earth's surface at the Last Glacial Maximum.* Boulder, CO, Geological Society of America Map and Chart Series MC-36.

Coiffier, J. (1997). *Eléments de prévision numérique du temps.* Cours et Manuels, n° 12. Ecole Nationale de la Météorologie. Météo-France, Toulouse, 84 p.

Coiffier, J. (2000). Un demi-siècle de prévision numérique du temps. *La Météorologie*, n° 30, 11–31.

COHMAP members (1988). Climatic changes of the last 18,000 years: Observations and model simulations. *Science*, 24, 1043–1052.

Comby, J. (1990). La catastrophe du Grand-Bornand, composantes météorologiques. *Revue de Géographie de Lyon*, vol. 65, n°2, 118–122.

Comby, J. (1998). *Les Paroxysmes Pluviométriques dans le Couloir Rhodanien et ses Marges.* Th. Univ. Lyon3, Laboratoire de Climatologie, Risques, Environnement (LCRE), 668 p.

Comby, J. (2000). Eléments d'évaluation des conséquences socio-économiques des intempéries des 12 et 13 novembre 1999. *Revue de Géographie de Lyon*, vol. 75, n° 3, 227–244.

Comiso, J.C. (2003). Warming trends in the Arctic from clear sky satellite observations. *Journal of Climate*, vol. 16, n° 21, 3498–3510.

Comiso, J.C. and Parkinson, C.L. (2004). Satellite-observed changes in the Arctic. *Physics Today*, August 2004, 38–44.

Conte, M. and Palmieri, S. (1990). Tendenze evolutive del clima d'Italia. *Proc. of Giornata Ambiente-Atmosfera.* Acc. Naz. Lincei, Roma.

Corcoran, T. (2004). Comment about 'hockey stick', in the Financial Post, 13 July 2004.

Cortijo, E., Duplessy, J.C., Labeyrie, L., Leclaire, H., Duprat, J. and van Weering, T.C.E. (1994). Eemian cooling in the Norwegian Sea and North Atlantic ocean preceding continental ice-sheet growth. *Nature*, vol. 372, 446–449.

Courtin, R., Mckay, C.P. and Pollack, J. (1992). L'effet de serre dans le système solaire. *La Recherche*, 243, vol. 23, 543–549.

Courtney, R.S. (1999). Global warming: How it all began. (www.john-daly.com)

Courtney, R.S. (2004). Significant differences in determinations of mean global surface temperature trends for recent decades. RSC Environmental Services, *Comm. to 'Climate sceptics'*, 10 p.

Croll, J. (1875). *Climate and Time in Their Geological Relations. A Theory of Secular Changes of the Earth's Climate.* New York, Appleton.

Curry, R., Dickson, B. and Yashayaev, I. (2003). A change in the freshwater balance of the Atlantic Ocean over the past four decades. *Nature*, 426, 826–829.

Dady, G. (1979). Plaidoyer pour l'atmosphère. Tribune libre, *Bull. d'Inform.*, Dir. Mét., 43, 18–20.

Dady, G. (1995). La méthode réductionniste. *Le Monde*, 24 February 1995.

Daly, J.L. (1997). The 'hockey stick': A new low in climate science. (http://www.john-daly.com/hockey)

Daly, J.L. (2000). Report to the Greening Earth Society 'The surface record: global mean temperature and how it is determinded at surface level'. (http://www.john-daly.com)

Daly, J.L. (2001) Falsification de l'histoire climatique pour 'prouver' le réchauffement global. *Fusion*, n° 87, 32–46.

Daly, J.L. (2003, 2004). Still waiting for greenhouse. What the stations say. (http://www.john-daly.com)

Damon, P.E. and Kunen, S.M. (1976) Global cooling? No, southern hemisphere warming trends may indicate the onset of the CO_2 greenhouse effect. *Science*, vol. 193, n° 4252, 447–453.

Dansgaard, W., White, J.W.C. and Johnsen, S.J. (1989). The abrupt termination of the Younger Dryas climate event. *Nature*, vol. 339, 532–533.

Davis, O.K. (1988). The effect of latitudinal variations of insolation maxima on the desertification during the late Quaternary. *Proc. IGCP 252*, Unesco, 41–58.

De Félice, P. (1999). Circulation générale de l'atmosphère. *Conf. Soc. Météo. de France*, Paris, website SMF. (http:www.smf.asso.fr/con99_1.html)

Defant, A. (1921). Die zirkulation der atmosphäre in den gemässigsten breiten der Erde. *Geografiska Annaler* 3, 209–266.

De Laat, A.T.J. and Maurellis, A.N. (2004). Industrial CO_2 emissions as a proxy for anthropogenic influence on lower tropospheric temperature trends. *Geophys. Res. Letters*, vol. 31, L05204.

Deléage, J.P. (1993). Les étapes de la prise de conscience (pp. 35–39). L'écologie politique ou la conscience planétaire (40–44). In: *L'Etat de l'Environnement dans le Monde*. La Découverte/FPH, Paris, 1993.

Delécluse, P. (1998). Heurs et malheurs de la prévision d'El Niño. *La Recherche*, n° 307, 72–77.

Delécluse, P. (2002). Quand le tapis roulant a des ratés. *La Recherche*, n° 311, 7–8.

Del Genio, A.D. (2002). The dust settles on water vapor feedback. *Science*, vol. 296, 665–666.

DeMenocal, P. and Bond, G. (1997). Holocene climate less stable than previously thought. *EOS*, n° 78(41), 453–454.

De Moor, G. and Veyre, P. (1991). *Les Bases de la Météorologie Dynamique*. Cours et Manuels n° 6, Direction de la Météorologie Nationale, Toulouse, 312 p.

Deneux, M. (2001). Rapport de l'Office Parlementaire d'Evaluation des Choix Scientifiques et Technologiques, sur *'Les changements climatiques en 2025, 2050 et 2100'*. Paris, Assemblée Nationale-Sénat.

Denton, G.H. and Hugues, T.J. (eds) (1981). *The Last Great Ice Sheets*. Wiley Interscience, New York, 484 p.

Deser, C. and Wallace, J.M. (1987). El Niño events and their relation to the Southern Oscillation: 1925–1986. *J. Geophys. Res.*, 92, 14189–14196.

Deser, C. and Blackmon, M.L. (1993). Surface climate variations over the North Atlantic Ocean during winter 1900–1989. *Journal of Climate*, vol. 6, n° 9, 1743–1753.

De Silva, Sh. (2003). Eruptions linked to El Niño. *Nature*, vol. 426, 239–241.

Dettinger, M. (1998). Workshop assesses 1997–1998 El Niño from Pacific climate perspectives. *EOS*, vol. 79, n° 36, 430–431.

Diaz, H.F. and McCabe, G.J. (1998). A possible connection between the 1878 yellow fever epidemic in the Southern United States and the 1877–1878 El Niño episode. *Bull. Am. Met. Soc.*, vol. 80, n° 1, 21–27.

Dickinson, R.E. (1986). How will climate change? The climate system and modelling of future climate. In: *The Greenhouse Effect, Climatic Change, and Ecosystems*. Scope 29, Wiley, 207–270.

Dicks, L. (2003). Coral bleaching caused by 'malaria of the oceans'. Climate Change (www.Newscientist.com)

Dickson, B., Yashayacv, I., Meincke, J., Turrel, B., Dye, S. and Holfort, J. (2002). Rapid freshening of the deep North Atlantic Ocean over the past four decades. *Nature*, 416, 832–837.

Dietzer, P. (2000). IPCC's most essentiel model errors. *Still waiting for greenhouse* (www.john-daly.com)

Ding, D. *et al.* (1998). Research about the development history of permafrost in Qinshai-Xisang plateau. *Geogr. Soc. of China.*

Dione, O. (1996). *Climatologie et hydrologie des hauts bassins du Sénégal et de la Gambie.* Th. Géogr. Phys. Univ. Lyon3, LCRE, 405 p. In: *Mémoires et Documents, Publ.* ORSTOM/IRD.

Djellouli, Y. and Daget, Ph. (1993). Conséquence de la sécheresse des deux dernières décennies sur les écosystèmes naturels algériens. *Publ. AIC*, vol. 6, 104–113.

Documentation Française (La) (1994). *La Planète Terre entre nos mains.* Guide pour la mise en uvre des engagements du Sommet planète Terre, Paris, 442 p.

Doran, P.T., Priscu, J.C., Lyons, W.B., Walsh, J.E., Fountain, A.G., McKnight, D.M., Moorhead, D.L., Virginia, R.A., Wall, D.H., Clow, G.D., *et al.* (2002). Antarctic climate cooling and terrestrial ecosystem response. *Nature*, 710, 1–3.

Douglass, D.H., Pearson, B.D., Singer, F., Knappenberger, P.C. and Michaels, P.J. (2004). Disparity of tropospheric and surface trends: New evidence. *Geophys. Res. Letters*, vol. 31, n° 13, L13208.

Duplessy, J.C. and Ruddiman, W.F. (1984). La fonte des glaces polaires. *La Recherche*, v. 15, 806–818.

Duplessy, J.C. (1989). Le Programme Géosphère-Biosphère. *La Météorologie*, n° 26, 14–23.

Duplessy, J.C. (1992) Les certitudes des paléoclimatologues. *La Recherche*, 243, vol. 2, 558–564.

Duplessy, J.C. and Morel, P. (1990). *Gros temps sur la planète.* Editions O. Jacob, Paris, 297 p.

Duplessy, J.C. (1997). Vers un refroidissement de l'Europe? *La Recherche*, n° 295, 52–56.

Dutton, E.G. and Christy, J.R. (1992). Solar radiative forcing at selected locations and evidence for global lower troposheric cooling following the eruptions of El Chichon and Pinatubo. *Geophys. Res. Letters*, 19, 2313–2316.

Dyurgerov, M. (2002). Glacier mass balance and regime: Data of measurements and analysis. *Occas.* Paper n° 55, Inst. of Arctic and Alpine Research, Univ. of Colorado.

Easterling, D.R., Meehl, G.A., Parmesan, C., Changnon, S.A., Karl, T.R. and Mearns, L.O. (2000). Climate extremes: observations, modeling, and impacts. *Science*, vol. 289, 2068–2073.

EEA, European Environment Agency (2004). Impacts of climate change in Europe, an indicator-based assessment. (`www. eea.eu.int/climate`)

Elliot, W. and Gaffen, D. (1995). Chapman conference probes water vapor in the climate system. *EOS*, AGU, February 14, p. 67.

Erhardt, R.D. (1990). Reconstructed annual minimum temperatures for the Gulf States, 1799–1988. *Journal of Climate*, vol. 3(6), 678–684.

Eschenbach, W. (2002). Same old Arctic warming. In: J. Daly, *Still Waiting for Global Warming.*

Eschenbach, W. (2003). What the satellite record reveals. In: J. Daly, *Still Waiting for Global Warming.*

Essex, C. and McKitrick, R. (2002). *Taken by Storm: The Troubled Science, Policy and Politics of Global Warming.* Toronto, Key Porter.

Fabre, C. (2001). Analyse météorologique des pluies diluviennes du 21 au 25 octobre 2000 sur le Levante espagnol. *Mémoire LCRE*, Univ. Lyon3, 148 p.

Farrell, B.F. (1989). Cyclogenesis as an initial value problem: Theory and implications. *Palmen Memorial Symposium, Am. Met. Soc.*, 80–82.

Farrell, B.F. (1994). Evolution and revolution in cyclogenesis theory. *The life cycles of extra-tropical cyclones.* An International Symposium. Gronas and Shapiro (eds), Bergen, vol. 1, 101–110.

Favre, A. (2002). L'évolution récente de la dynamique aérologique dans le Pacifique Nord. *Ecole Normale Supérieure, Fête de la Science*, Lyon.

Favre, A. and Pommier, A. (2003). Le changement climatique dans les espaces aérologiques du Pacifique Nord-Est et de l'Atlantique Nord-Est, *Printemps de la géographie du CNFG*, ENS Lyon.

Favre, A, Pommier, A. and Gershunov, A. (2004). Trajectoires des Anticyclones et Dépressions dans le Pacifique Nord et l'Atlantique Nord de 1950 à 2000. *Journées de Climatologie, Commission 'climat et société'*, CNFG, Nancy.

Favre, A. and Gershunov, A. (2004). Evolution de l'activité anticyclonique polaire hivernale et de la circulation de la basse troposphère dans le Pacifique Nord de 1950 à 2000. *Coll. Ass. Int. de Climatologie, AIC*, (F) Caen.

Fellous, J.L. (2003). *Avis de tempêtes. La nouvelle donne climatique*. Odile Jacob, Sciences. Paris, 337 p.

Ferrel, W. (1859). The motions of fluids and solids relative to the earth's surface. *Math. Mon.*, vol. 1, 140–406.

Flaherty, T.H. éd. (1985). *La Planète Terre: Prairies et Toundras*. Editions Time-Life, Amsterdam.

Fléaux, R., Clause, L. and Jubelin, F. (2000). Pourquoi il y aura de nouvelles tempêtes. *Sciences et Avenir*, 645, nov., 80–93.

Flohn, H., Kapala, A., Knoche, H.R. and Machel, H. (1990). Recent changes of the tropical water and energy budget and of midlatitude circulations. *Climate Dynamics*, 4, 237–252.

Flückiger, *et al.* (2002). cf. website LGGE (lgge, 2004).

Folland, C.K., Karl, T.R. and Vinnikov, K. Ya. (1990). Observed climate variations and change. Chap. 7, *The IPCC Scientific Assessment*, Cambridge Univ. Press, 201–238.

Fong Chao, B. (1996). 'Concrete' testimony to Milankovitch cycle in earth's changing obliquity. *EOS*, vol. 77, n° 44.

Fontaine, B. (1990). *Etude comparée des moussons indienne et ouest africaine*. Th. Univ. Dijon.

Forbes, G.S., Blackall, R.M. and Taylor, P.L. (1993). 'Blizzard of the century', the storm of 12–14 March 1993 over the eastern United States. *The Met. Magazine*, vol. 122, 1452, 153–162.

Foukal, P. (1994). Study of solar irradiance variations holds key to climate questions. *EOS*, vol. 75, n° 33, 377–381.

Foukal, P. (2003). Can slow variations in solar luminosity provide missing link between the Sun and climate? *EOS*, vol. 84, n° 22, 205–208.

Fourier, J.B. (1827). Mémoire sur les températures du globe terrestre et des espaces planétaires. *Mém. Acad. R. Sc. Inst. de France*, 7, 569–604.

Francou, B., Vuille, M., Wagnon, P., Mendoza, J. and Sicart, J.E. (2003). Tropical climate change recorded by a glacier in the central Andes during the last decades of the 20th century: Chacaltaya, Bolivia, 16°S. *Journal of Geophysical Research*, 108.

Francou, B. and Wagnon, P. (2004). Recul des glaciers dans les Andes tropicales sur les dernières décennies. *Lettre n° 16, PIGB-PMRC*, CNRS.

Friis-Christensen, E. and Lassen, K. (1991). Length of the solar cycle: An indicator of solar activity closely associated with climate. *Science*, 254, 698–700.

Furrer, G., Burga, C., Holzhauser, H.P. and Maisch, M. (1987). Zur Gletcher-Vegetations-und-Klima geschichte des Schweiz seit der späteiszeit. *Geogr. Helvet.*, n° 2, Bern.

Gachon, Ph. (1994). Evolution des pressions et des températures au Canada de 1960 à 1990. *Publ. AIC*, vol. 7, 256–267.

Gaffen, D.J. (1998). Falling satellites, rising temperatures? *Nature*, vol. 394, 615–616.

Ganopolski, A. and Rahmstorf, S. (2001). Rapid changes of glacial climate simulated in a coupled climate model. *Nature*, 409, 153–158.

Garcia, I.P. (1996). Major cold air outbreaks affecting coffee and citrus plantations in the eastern and northeastern Mexico. *Atmosfera*, 9, 47–68.

Gates, W.L. *et al.* (1999). An overview of the results of the atmospheric model intercomparison project (AMIP I). *Bull. Am. Met. Soc.*, vol. 80 n° 1, 29–55.

Ge, Q., Zheng, J., Fang, X., Man, Z., Zhang, X., Zhang, P. and Wang, W.C. (2003). Winter half-year temperature reconstruction for the middle and lower reaches of the Yellow River and Yangtze River, China, during the past 2000 years. *The Holocene*, 13, 933–940.

Gergye, A. (1994). Mettre en perspective les rigueurs de cet hiver. PCC Info, *Delta*, vol. 5, n° 1, *Environnement Canada*, Progr. Can. des Chang. à l'échelle du Globe, 16.

Gerlach, T.M. (1991). Present day CO_2 emissions from volcanoes. *EOS*, vol. 72, n° 23, 249, 254.

Gershunov, A. and Barnett, T.P. (1998). Interdecadal modulation of ENSO teleconnections. *Bull. Am. Met. Soc.*, vol. 79, n° 12, 2715–2726.

Gershunov, A., Barnett, T.P. and Cayan, D.R. (1999). North Pacific interdecadal oscillation seen as factor in ENSO-related North American climate anomalies. *EOS*, vol. 80 (n° 3), 25, 29–30.

Gianini, A., *et al.* (2003). Warming Indian ocean wringing moisture from the Sahel, 1930–2000. Int. Res. Inst. for Clim. Pred. *Science*, vol. 302. (www.sciencemag.org/cgi/1089357)

Gidaglia, J.M. and Rittaud, B. (2004). La simulation numérique. Back to Basics. *La Recherche*, n° 380, 73–76.

Gillett, N.P. and Thompson, D.W.J. (2003). Simulation of recent southern hemisphere climate change. *Science*, vol. 302, 273–275.

Gil Olcina, A., *et al.* (1983). Sequias e inondacionas en El Campo. *Bol. de Inf. Agr.* n° 103, Bilbao, 11–13.

Gil Olcina, A. and Olcina Cantos, J. (1997). *Climatologia general.* Ariel Geografia, Barcelona, 577 p.

Gil Olcina, A. and Morales Gil, A. (2001) *Causas y consecuencias de las sequias en Espana.* Univ. Alicante, 574 p.

Gloersen, P. (1995). Modulation of hemispheric sea-ice cover by ENSO events. *Nature*, vol. 373, 503–506.

Godard, A. (2001). Changement climatique et effet de serre additionnel d'origine anthropique: Un débat parfois obscur. *Ann. de Géogr.*, n° 617, A. Colin, 78–89.

Goldsmith, T. (2000). Pour un programme d'urgence. *Ecologiste*, édition française de *The Ecologist*, vol. 1, n° 2, p. 3.

Golitsyn, G.S. and MacCracken, M.C. (1987). Possible consequences of a major nuclear war. *WMO/TD-n° 201, WCP-142*, Geneva.

Goni, M.F.S. and d'Errico, F. (2004). Les hommes face aux soubresauts du climat. *La Recherche*, n° 373, 34–36.

Goodridge, J.D. (1996). Comments on 'Regional simulations of greenhouse warming including natural variability'. *Bull. Am. Met. Soc.*, vol. 77, n° 7, 1588–1589.

Goody, R., Anderson, J. and North, G. (1998). Testing climate models: an approach. *Bull. Am. Met. Soc.*, Vol. 79, n° 11, 2541–2549.

Gordon, A.L., Zebiak, S.E. and Bryan, K. (1992). Climate variability and the Atlantic Ocean. *EOS*, vol. 73, n° 15, 161, 164–165.

Gornitz, V., *et al.* (1982). Global sea level trend in the past century. *Science*, 215, 1611–1614.

Goudriaan, J. (1992) Où va le gaz carbonique? Le rôle de la végétation. *La Recherche*, 243, vol. 2, 588–605.

Graham, N.E. and Diaz, H.F. (2001). Evidence for intensification of North Pacific winter cyclones since 1948. *Bull. of the AMS*, vol. 82, n° 9, 1869–1893.

Gray, V. (2003). The cause of global warming. (http://vision.net.au/daly, 16 p.)

Gray, V. (2003). Regional temperature change. (http://vision.net.au/daly/, 17 p.)

Grenier, E. (1992). Editorial. *Industrie et Environnement*, n° 56, Paris.

Grenier, E. (1998). Réchauffement climatique: N'appelez plus cela de la science. *Industrie et Environnement*, n° 208, Paris.

Grenier, E. (2002) La fin du Lyssenko du climat. *Industrie et Environnement*, n° 263, Paris.

Grenier, E. (2003). La courbe en crosse de hockey retyrouve une allure plus normale. *Industrie et Environnement*, n° 279, Paris.

Gribbin, J. (1989). *Le ciel déchiré. Pouvons-nous sauver la couche d'ozone?* Editions Sang de la Terre, Paris, 235 p.

Grinevald, J. (1992) De Carnot à Gaÿa: Histoire de l'effet de serre. *La Recherche*, 243, vol. 2, 532–540.

Grinewald, J. (1993). Préface à l'édition française. *Une Planète, un Avenir*. Sang de la Terre, Paris.

Grootes, P.M., Stuiver, M., White, J.W.C., Johnsen, S. and Jouzel, J. (1993). Comparison of oxygen isotope records from GISP2 and GRIP Greenland ice cores. *Nature*, vol. 366, 552–554.

Grousset, F. (2001). Les changements abrupts du climat depuis 60.000 ans. *Quaternaire*, 12(4), 203–211.

Grove, J.M. (1988). *The Little Ice Age*. Methuen, London and New York, 498 p.

Gulev, S.K. and Hasse, L. (1999). Changes of wind waves in the North Atlantic over the last 30 years. *Int. Journ. of Climat.*, 19, 1091–1117.

Gullett, D.W. and Skinner, W.R. (1992). L'état du climat au Canada: Les variations de la température au Canada 1895–1991. *Environnement Canada, Rapport EDE*, n° 92-2, 36 p.

Hadley, G. (1735). Concerning the cause of the general trade-winds. *Philo. Trans.*, London, vol. 39, 58–62.

Hadley Center, Met Office (2003). Climate Change, observations and predictions. *Defra*, Dec.

Haig, J.D. (1996). The impact of solar variability on climate. *Science*, 272, 981–984.

Hall, A. and Manabe, S. (1997). Can local linear stochastic theory explain sea surface temperature and salinity variability? *Clim. Dyn.*, 13, 167–180.

Hall, A. and Manabe, S. (1999). The role of water vapor feedback in unperturbed climate variability and global warming. *Journal of Climate*, 12, 2327–2346.

Halley, E. (1686). An historical account of the trade-winds and monsoons observable in the seas between and near the tropics with an attempt to assign the physical cause of said winds. *Philo. Trans*, London, 26, 153–168.

Halpert, M.S., Ropelewski, C.F., Karl, T.R., Angell, J.K., Stowe, L.L., Heim, R.R., Miller, A.J. and Rodenhuis, D.R. (1993). 1992 brings return to moderate global temperatures. *EOS*, vol. 74 , n° 38, 437–439.

Handler, P. and Andsager, K. (1990). Volcanic aerosols, El Niño and the Southern Oscillation. *International Journal of Climatology*, 10, 413–424.

Hannah, J. (2004). An updated analysis of long-term sea level change in New Zealand, *Geophys. Res. Lett.*, 31, L03307.

Hansen, J., Johnson, D., Lacis, A., Lebedeff, S., Lee, P., Rind, D. and Russell, G (1981). Climate impact of increasing carbon dioxide. *Science*, 213, 957–966.

Hansen, J., Fung, I., Lacis, A., Lebedeff, S., Rind, D., Ruedy, R. and Russell, G. (1988). Global climate changes as forecast by GISS three-dimensional model. *J. of Geoph. Research*, 93, 9341–9364.

Hansen, J. (2003). Can we defuse the global warming time bomb? (www.naturalscience.com)

Hare, S.R. (1996). *Low-frequency climate variability and salmon production*. Ph.D. Th., Univ. Washington, Seattle, 306 pp.

Hastenrath, S. (1991). *Climate Dynamics of the Tropics*. Kluwer Acad. Publ., 448 p.

Hebbeln, D., Dokken, T., Andersen, E.S., Hald, M. and Elverhol, A. (1994). Moisture supply for northern ice-sheet growth during the Last Glacial Maximum. *Nature*, vol. 370, 357–359.

Hecht, L. (2004). Glaciers are growing around the world, included the United States. *The next ice age – now*. (www.iceagenow.com)

Held, I.M. and Soden, B.J. (2000). Water vapor feedback and global warming. *Ann. Rev. Energy Env.*, 25, 441–475.

Henderson-Sellers, A. and Robinson, P.J. (1986). *Contemporary Climatology*. Longman, Harlow, England.

Henderson-Sellers, A., Zhang, H., Berz, G., Emanuel, K., Gray, W., Landsea, C., Holland, G., Lighthill, J., Shieh, S-L., Webster, P. and McGuggie, K. (1998). Tropical cyclones and global climate change: A post IPCC assessment. *Bull. Am. Met. Soc.*, vol. 79, n° 1, 19–37.

Herman, B.M., Zeng, X., Chase, T. and Pielke, R. Sr. (2002). More heat over Greenhouse gases. *Physics Today*, May 2002, 15. See Physics Tod., June 2001, 19 and Physics Tod., December 2001, 12.

Hieb, M. (2004) . Water vapor rules the greenhouse system. (http://www.clearlight.com/~mhieb/WVFossils/greenhouse_data.html)

Houghton, J.T., Callender, B.A., Varney, S.K. (1992). *Climate Change 1992. The supplementary report to the IPCC Scientific Assessment*. Published for the IPCC, Cambridge University Press.

Houssais, M.N., Gascard, J.C. (2001). L'Océan Glacial Arctique aux avant-postes du changement climatique global. *Lettre PIGB-PMRC*, n° 12, 24–31.

Hoyt, D. (2004). A critical examination of climate change. (www.warwickhughes)

Hulme, M. (1998). Global warming. *Progress in Physical Geography*, 22(3), 398–406.

Humboldt, A. (de). (1846). *Cosmos. Essai d'une description physique du monde*. Traduction H. Faye, Gide et Cie, Paris, p. 4.

Hunt, A.G., Baduini, C.L., Brodeur, R.D., Coyle, K.O., Kachel, N.B., Napp, J.M., Salo, S.A., Schumacher, J.D., Stanebo, P.J., Stockwell, D.A., Whiledge, T.E. and Zeeman, S.I. (1999). The Bering Sea in 1998: The second consecutive year of extreme weather-forced anomalies. *EOS*, 80(47), 565–567.

Hunt, A.G. and Tsonis, A.A. (2000). The Pacific Decadal Oscillation and long-term climate prediction. *EOS*, Nov. 28, 581.

Hurrell, J.W. and Trenberth, K.E. (1999). Global sea-surface temperature analyses: Multiple problems and their implications for climate analysis, modeling, and reanalysis. *Bull. Am. Met. Soc.*, vol. 80, n° 12, 2661–2678.

Hurrell, J.W., Kushnir, Y. and Visbeck, M. (2001). The North Atlantic Oscillation. *Science*, vol. 291, 603–604.

Hurrell, J.W., Kushnir, Y., Ottersen, G. and Visbeck, M. (2003). An overview of the North Atlantic Oscillation. The North Atlantic Oscillation, *Geophys. Monog*. 134, AGU, 1–35.

Idso, S. B. (1986). Nuclear Winter and the Greenhouse Effect. *Nature*, 321, 122.

Idso, S. B. (1987). A clarification of my Position on the CO_2/Climate Connection. *Climatic Change* 10, 81–86.

Idso, C.D. and Idso, K.E. (2000). The global surface air temperature record must be wrong. *Center for the Study of carbon dioxide and global change*, vol. 3, n° 12.

Indermühle, A., Stocker, T.F., Joos, F., Fischer, H., Smith, H.J., Wahlen, M., Deck, B. Mastroianni, D., Tschumi, J., Blunier, T., Meyer, R. and Stauffer, B. (1999). Holocene carbon-cycle dynamics based on CO_2 trapped in ice at Taylor Dome, Antarctica. *Nature*, vol. 398, 121–126.

IPCC: Climate Change – Reports of IPCC Working Group I:

- **Climate Change 1990**: The IPCC Scientific Assessment. Edited by J.T. Houghton, G.J Jenkins and J.J. Ephraums. Cambridge University Press, Published for the IPCC–WMO–UNEP.
- **Climate Change 1992**: The Supplementary Report to the IPCC Scientific Assessment. Edited by J.T. Houghton, B.A. Callander and S.K. Varney. Cambridge University Press, Published for the IPCC–WMO–UNEP.
- **Climate Change 1995**: The Science of Climate Change. IPCC Working group I. Edited by J.T. Houghton, L.G. Meira-Filho, J. Bruce, Hoesung Lee, B.A. Callender, E. Haites, N. Harris and K. Maskell. Cambridge University Press, Published for the IPCC–WMO–UNEP.
- **Climate Change 2001**: The IPCC Scientific Assessment. Edited by J.T. Houghton, Y. Ding, D.J. Griggs, M. Noguer, P.J. van der Linden, X. Dai, K. Maskell and C.A. Johnson. Cambridge University Press, Published for the IPCC–WMO–UNEP.

Janicot, S. and Fontaine, B. (1993). L'évolution des idées sur la variabilité interannuelle récente des précipitations en Afrique de l'Ouest. *La Météorologie*, n°1, 28–49.

Janowiak, J.E. (1990). The global climate of December 1989–February 1990: Extreme temperature variations in North America, persistant warmth in Europe and Asia, and the return to ENSO-like conditions in the Western Pacific. *J. of Climate*, vol. 3, n° 6, 685–709.

Jaworowski, Z. (1996). Greenhouse Gases in Polar Ice – Artifacts or Atmospheric Reality? Umwelttagung 1996, *Umwelt und Chemie*, Gesellschaft Deutscher Chemiker, Ulm, 7–10.

Jaworowski, Z. (2004). Solar cycles, not CO_2, determine climate. *21th Century Science and Technology*, winter 2003–2004, 52–65.

Jaworowski, Z. (2004). Climate Change: Incorrect information on pre-industrail CO_2. Statement for the US Senate Committee on Commerce, Science and Transportation. (www.john-daly.com/zjiceco2.htm)

Jelbring, H. (1998). *Wind controlled climate*. Doctoral Thesis. Paleogeophysics and Geodynamics. Stockholm University, Sweden, P 304–312.

Jones, P.D., Groisman, P. Ya., Coughlan, M., Plummer, N., Wang, W.C. and Karl, T.R. (1990). Assessment of urbanization effects in time series of surface air temperature over land. *Nature*, vol. 347, 169–172.

Jones, P.D. and Briffa, K.R. (1992). Global surface air temperature variations during the twentieth century: part 1, spatial, temporal and seasonal details. *The Holocene*, 2.2, 165–179.

Jones, P.D., Osborn, T.J., Briffa, K.R., Folland, C.K., Horton, E.B., Alexander, L.V., Parker, D.E. and Rayner, N.A. (2001). Adjusting for sampling density in grid box land and ocean surface temperature time series. *J. Geophys. Res.*, 106, D4, 3371–3380.

Jones, P.D. and Mann, M.E. (2004). Climate over past millenia. *Rev. Geophys.* n° 42, 2–42.

Joselyn, J.A., Anderson, J.B., Coffey, H., Harvey, K., Hathaway, D., Heckman, G., Hildner, E., Mende, W., Schatten, K., Thompson, R., *et al.* (1997) Panel achieves concensus prediction of solar cycle 23. *EOS*, vol. 78, n° 20, 205, 211–212.

Kahl, J.D., Charlevoix, D.J., Zaltseva, N.A., Schnell, R.C. and Serreze, M.C. (1993). Absence of evidence for greenhouse warming over the Arctic Ocean in the past 40 years. *Nature*, 131, 335–337.

Kahl, J.D., Jansen, M., Pulrang, M.A. (2000). Fifty-year record of North Polar temperatures shows warming. *EOS*, 1, 5.

Kalkstein, L.S., Jamason, P.F., Greene, J.S., Libby, J. and Robinson, L. (1996). The Philadelphia hot weather-health watch/warning system: Development and application, summer 1995. *Bull. Am. Met. Soc.*, vol. 77, n° 7, 1519–1528.

Kalkstein, L.S. and Greene, J.S. (1999). An evaluation of climate mortality relationships in large US cities and the possible impacts of climate change. *Env. Health Perspect.*, 105, 84–93.

Kanamitsu, M.W. and Ebisuzaki, *et al.* (2002). NCEP-DOE AMIP-II Reanalysis. *Bull. Am. Met Soc.*, n° 83, 1631–1643.

Kandel, R. and Fouquart, Y. (1992). Le bilan radiatif de la Terre. *La Recherche*, 241, vol. 23, 316–324.

Kaplan, A., Kushnir, Y., Cane, M., Clement, A. and Blumenthal, M.B. (1997). Reduced spaces optimal analysis for historical data sets: 136 years of Atlantic sea surface temperatures. *Geophys. Res. Letters*, 27, 27,835–27,860.

Karl, T.R., Diaz, H.F. and Kukla, G. (1988). Urbanization: Its detection and effect in the United States climate record. *Journal of Climate*, vol. 1, n° 11, 1099–1123.

Karl, T. and Jones, P.D. (1989). Urban bias in area-averaged surface air temperature trends. *Bull. Am. Met. Soc.*, vol. 70, n° 3, 265–270.

Karl, T., Jones, P.D., Knight, R.W., Kukla, G., Plummer, N., Razuvayev, V., Gallo, K.P., Lindseay, J., Charlson, R.J. and Peterson, T.C. (1993). Asymmetric trends of daily maximum temperature. *Bull. Am. Met. Soc.*, vol. 74, n° 6, 1007–1023.

Karl, T. *et al.* (1995). Critical issues for long-term climate monitoring. *Climate Change*, 31, 185–221.

Karl, T.R., Knight, R.W., Easterling, D.R. and Quayle, R.G. (1996). Indices of climate change for the United States. *Bull. Am. Met. Soc.*, vol. 77, n° 2, 279–292.

Karl, T.R. and Knight, R.W. (1997). The 1995 Chicago heat wave: How likely is a recurrence? *Bull. of Am. Met. Soc.*, vol. 78, n° 6, 1107–1119.

Karl, T.R. and Knight, R.W. (1998). Secular trends of precipitation amount, frequency and intensity in the United States. *Bull. Am. Met. Soc.*, vol. 79, n° 2, 231–240.

Kaser, G., Douglas, R.H., Mölg, T., Bradley, R.S. and Tharsis, M.H. (2004). Modern glacier retreat on Kilimanjaro as evidence of climate change: Observations and facts. *International Journal of Climatology*, vol. 24, n° 3, 329–339.

Kauffman, J.M. (2004). Water in the atmosphere. *Journal of Chemical Education*, vol. 81, n° 8, 1229–1230.

Keeling, C. D. (1960). The concentration and isotopic abundances of carbon dioxide in the atmosphere. *Tellus* 12, 200–203.

Keigwin, L. (1996). The Little Ice Age and Medieval Warm Period in the Sargasso Sea. *Science*, 269, 676-7–679.

Keller, C.F. (1999). Comment: Human contribution to climate change increasingly clear. *EOS*, 80(33), 368, 371–372.

Kellog, W.W. (1987). Mankind's impact on climate: The evolution of an awareness. *Climate Change*, n° 10, 113–136.

Khandekar, M.L. (2003). Comment on WMO statement on extreme weather events. *EOS*, vol. 84, n° 41, 428

Khandekar, M.L. (2004). Are Computer Model Projections reliable enough for climate policy? *Energy and Environment*, 15, 2004, 521–525.

Khodri, M., Leclainche, Y., Ramstein, G., Braconnot, P., Marti, O. and Cortijo, E. (2001). Simulating the amplification of orbital forcing by ocean feedbacks in the last glaciation. *Nature*, vol. 410, 570–573.

Kiehl, J. T. and Trenberth, K. E. (1997). Earth's annual global mean energy budget. *Bull. Amer. Met. Soc.*, 78, 197–208.

Kieffer, *et al.* (2000), In: S.R. Courtney (ed.), On believing melting tropical glaciers. (`www.climatesceptics`, Feb. 2004.)

Kingsnorth, P. (2001) Noyade dans un océan vert. *L'Ecologiste*, version française de *The Ecologist*, vol. 1, n° 2, p. 9.

Kininmonth, W. (2004). *Climate Change, A Natural Hazard*. Multi-Science Co, London, UK.

Kirkby, J., Mangini, A. and Muller, R.A. (2004). The glacial cycles and cosmic rays. *Earth and Planetary Science Letters*. July 2004.

Klitgaard-Kristensen, D., Sejrup, H.P., Haflidason, H., Johnsen, S. and Spurk, M. (1998). A regional 8200 cal. yr BP cooling event in northwest Europe, induced by final stages of the Laurentide ice-sheet deglaciation? *J. Quat. Sc.*, 13, 165–169.

Kocin, P.J. (1988). Meteorological analyses of the March 1888 'Blizzard of 88'. *EOS*, vol. 69(10), 137, 146–147.

Kocin, P.J. and Uccellini, L.W. (1990). Snowstorms along the northeastern coast of the United States: 1955 to 1985. *Met. Monographs*, vol. 22, n° 44, Am. Met Soc., 280 p.

Kocin, P.J., Schumacher, P.N., Morales, R.F. and Uccellini, L.W. (1995). Overview of the 12–14 March 1993 superstorm. *Bull. Am. Met. Soc.*, vol. 76, n° 2, 165–182.

Kohler, P. (2001). Quand le mercure s'affole. *Le Spectacle du Monde*, n° 469, 104–110.

Kohler, P. (2002). *L'imposture Verte*. A. Michel, Paris, 393.

Kohler, P. (2003). *Global Warming et Médias: Un Exemple de Désinformation*. Texte d'une conférence, DEA, Univ. Lyon 3.

Kondratyev, K.Ya. (1997). Comments on 'open letter to Ben Santer'. *Bulletin of the American Meteorological Society*, vol. 78 n°4, 689–691.

Kondratyev, K.Ya (1999). *Climatic Effects of Aerosols and Clouds*. Springer–Praxis, 264 p.

Kondratyev, K.Ya (2002). Global climate change and the Kyoto protocol. *Idojaras*, vol. 106, 1–37.

Kondratyev, K. Ya., Grigoryev, Al. A. and Varotsos, C. (2002). *Environmental Disasters: Anthropogenic and Natural*. Springer–Praxis, 484 pp.

Kondratyev, K. Ya. (2003). Uncertainties of global climate change observations and simulation modelling. *World Climate Change Conference*, Moscow.

Kondratyev, K.Y., Krapivin, W.F. and Varotsos, C.A. (2003). *Global Carbon Cycle and Climate Change*. Springer–Praxis, 368 p.

Kondratyev, K. Ya (2004). Key aspects of global climate change. *World Climate Change Conference*, Moscow, 23 p.

Koslov, M.V. and Berlina, N.G. (2002). Decline in length of the summer season on the Kola Peninsula. Russia. *Climate Change*, 54, 387–398.

Koutchmy, S. and Vial, J.C. (1990). Le Soleil 24 heures sur 24. *La Recherche*, n° 217, vol. 21, 10–19.

Krishnamurti, T.N., Kanamitsu, N., Koss, J.W. and Lee, J.D. (1973). Tropical east–west circulation during the northern winter. *Journal of Atmosph. Sc.*, 30, 780–787.

Kukla, G.J., Angel, J.K., Korshover, J., Dronia, H., Hoshiai, M., Namias, J., Rodewald, M., Yamamoto, R. and Iwashima, T. (1977). New data on climatic trends. *Nature*, vol. 270, n° 5638, 573–580.

Kukla, G. (1989). Recent climate change in the United States. *Carbon dioxide res. div.*, US Dep. of Energy, Res. Proj. 4 p.

Kukla, G. (1990). Present, past and future precipitation: Can we trust the models? In: R. Paepe *et al.* (eds), *Greenhouse Effect, Sea-level and Drought*. NATO, C235, 109–114.

Kukla, G. ed. (1991). Interglacial-glacial transitions. *Mallorca Group, report*, April 1991.

Kukla, G., Knight, R., Gavin, J. and Karl, T. (1992). Recent temperature trends are they reinforced by insolation shifts? *NATO ASI Series. Vol. 13. Start of a glacial.* Springer-Verlag.

Kukla, G. and Went, E. (eds) (1992) *Start of a glacial.* Mallorca Group, NATO ASI Series, vol. 13. Springer-Verlag, Berlin.

Kukla, G. and Gavin, J. (2003). Milankovitch climate reinforcements. *Glob. and Planet. Change.* (online)

Kukla, G. (2004). Central Arctic: Battleground of natural and man-made climate forcing. *EOS*, vol. 85, n° 20, 200.

Kunkel, K.E., Changnon, S.A., Reinke, B.C. and Arritt, R.W. (1996). The July 1995 heat wave in the Midwest: A climatic perspective and critical weather factors. *Bull. Am. Met. Soc.* Vol. 77, n° 7, 1507–1518.

Kunkel, K.E., Pielker, A. and Changnon, S.A. (1999). Temporal fluctuations in weather and climate extremes that cause economic and human health impacts: A review. *Bull. Am. Met. Soc.*, 80(6), 1077–1098.

Kushnir, Y., Cardone, V.J., Greenwood, J.G. and Cane, M.A. (1997). The recent increase in North Atlantic wave heights. *Journal of Climat.*, 10, 2107–2113.

Kushnir, Y. (1999). Europe's winter prospects. *Nature*, 398, 289–290.

Kutiel, H. and Paz, S. (2000). Variations temporaires et spatiales de la température de surface de la mer en Méditerranée. *Publ. Ass. Int. Climat.*, vol. 12, Dakar.

Labasse, B. and Foechterlé, V. (1999). La force dévastatrice des anticyclones. *Science et Vie*, 979, 69–73.

Labeyrie, J. (1985). *L'homme et le Climat.* Denoël, Paris,

Labeyrie, L.D., Pichon, J.J., Labracherie, M., Ippolito, P., Duprat, J. and Duplessy, J.C. (1986). Melting history of Antarctica during the past 60,000 years. *Nature*, vol 322, 701–706.

Labeyrie, L. and Jouzel, J. (1999). Les soubresauts millénaires du climat. *La Recherche*, n° 308.

Lamb, H. H. (1970). Volcanic dust in the atmosphere, with a chronology and assessment of its meteorological significance. *Phil. Tr. R. Soc.*, London A, 266, 425–533.

Lamb, H.H. (1965). *The Changing Climate.* Methuen, London, 236 p.

Lamb, H.H. (1972–1977). *Climate: Past, Present and Future* (Vol. 1 and 2). Methuen, London.

Lamb, H.H. (1984). Climate in the last thousand years: Natural climatic fluctuations and change (Chap. 2). In: H. Flohn and R. Fantechi (eds), *The Climate of Europe: Past, Present, and Future.* D. Reidel, Dordrecht, pp. 25–64.

Lamb, H. H. (1995). *Climate, History, and the Modern World* (2nd ed.). Routledge, London, 433 p.

Lambeck, K. and Johnston, P. (1995). Land subsidence and sea-level change: Contribution from the melting of the last great ice sheets and the isostatic adjustment of the Earth. In: F.B.J. Barends *et al.* (eds), *Land subsidence.* Balkema, Rotterdam, pp. 3–18.

Lambert, G. (1987). Le gaz carbonique dans l'atmosphère. *La Recherche*, 189, vol. 18, 778–787.

Lambert, G. (1992) Les gaz à effet de serre. *La Recherche*, 243, vol. 2, 550–556.

Landscheidt, T. (1998). Solar activity – A dominant factor in climate dynamics. (www.john-daly)

Landscheidt, T. (2000). Solar forcing of El Niño and La Nia. In: Vázquez M. and Schmieder B. (eds), *The Solar Cycle and Terrestrial Climate.* Europ. Space Agency, Spec. Publ. 463, 135–140.

Landscheidt, T. (2000). New confirmation of strong solar forcing of climate. (www.john-daly.com)

Landscheidt, T. (2001). Trends in Pacific Decadal Oscillation subjected to solar forcing. (www.john-daly)

Landscheidt, T. (2003). New Little Ice Age instead of global warming. Myths and Frauds, FAEC.

Landsea, C.W., Nicholls, N., Gray, W.M. and Avila, L.A. (1999). Downwards trends in the frequency of intense Atlantic hurricanes during the past five decades. *Geophys. Res. Letters*, 23, 1697–1700.

Langley, S.P. (1884). Reserches on solar heat. *Prof. Papers of the signal service*, 15, 123.

Lanzerotti, L.J. (1999). Position statement adopted on climate change and greenhouse gases. *EOS*, vol. 80 n° 5, p. 49.

Lean, J. and Rind, D. (1994). Solar variability: Implications for global change. *EOS*, vol. 75, n° 1, 1, 5–7.

Lean, J., Beer, J. and Bradley, R. (1995). Reconstruction of solar irradiance since 1610: Implications for climate change. *Geophys. Res. Letters*, 22, 3195–3198.

Ledley, T.S., Sundquist, E.T., Schwartz, S.E., Hall, D.K., Fellows, J.D. and Killen, T.L (1999). Climate change and greenhouse gases. *EOS*, vol. 80, n° 39, 453, 454, 457–458.

Lefauconnier, B. (2002). L'amaigrissement des glaciers. *Sciences et Avenir*, h.s., 74–79.

Lehman, S.J. and Keigwin, L.D. (1992). Sudden changes in North Atlantic circulation during the last deglaciation. *Nature*, vol. 356, 757–762.

Lemasson, L. and Regnauld, D.H. (1997). Evolution trentenaire des vents littoraux sur le Grand Ouest français. *Norois*, 44 n° 175, 417–431 .

Lenoir, Y. (1992). La vérité sur l'effet de serre. Le dossier d'une manipulation planétaire. Sciences et société, *Editions La Découverte*, Paris, 172 p.

Lenoir, Y. (2001). *Climat de Panique*. Favre Ed., Lausanne, 217 p.

Leroux, M. (1970). *La dynamique des précipitations en Afrique Occidentale*. Th. Univ. Dakar. Publ. Dir. Exploit. Météor., n° 23, ASECNA, Dakar, 282 p.

Leroux, M. (1983). *Le climat de l'Afrique tropicale*. t1: 636 p., t2: notice et atlas de 250 cartes. Ed. Champion-Slatkine, Paris-Genève.

Leroux, M. (1986). L'Anticyclone Mobile Polaire: Facteur premier de la climatologie tempérée. *Bull. Assoc. Géogr. Fr.*, 4, Paris, 311–328.

Leroux, M. (1992). Perception 'statistique' et réalité dynamique. Transport méridien en masse extra-tropical, agglutination anticyclonique et circulation linéaire tropicale. *Publ. Ass. Intern. de Climatologie*, vol. 5, 157–167.

Leroux, M. (1993a).The Mobile Polar High: A new concept explaining present mechanisms of meridional airmass and energy exchanges and global propagation of palaeoclimatic changes. *Global and Planet. Change*, 7, 69–93.

Leroux, M. (1993b). La dynamique des situations météorologiques des 21–22 et 26–27 septembre 1992 dans le sud du couloir rhodanien. *Revue de Géographie de Lyon*, vol. 68, n° 2–3, 139–152.

Leroux, M. (1994a). La circulation zonale dite 'de Walker' en Afrique: Mythe ou réalité? *Publ. Ass. Intern. de Climatologie*, vol. 6, 487–496.

Leroux M. (1994b). Sécheresse et dynamique de la circulation dans l'Hémisphère Nord. *Publ. Assoc. Intern. de Climatologie*, vol. 6, 69–82.

Leroux, M. (1994c). Interprétation météorologique des paléoenvironnements observés en Afrique depuis 20,000 ans. *Géo-Eco-Trop*. 16(1-4), Bruxelles, 207–258.

Leroux, M. (1995). La dynamique de la Grande Sécheresse sahélienne. Numéro 'Sahel' de la *Revue de Géographie de Lyon*, vol. 70, M. Leroux éditeur, n° 3-4, 223–232.

Leroux, M. (1996a). *La dynamique du temps et du climat* (1ère édition). Masson, Paris, 310 p. (2ème édition). Masson-Sciences, Dunod (2000), 368 p.

Leroux, M. (1996b). Commentaire sur 'Débat sur le Front Polaire', de Thillet/Joly. *La Météorologie*, 16, 49–52.

Leroux, M. (1997). Climat local, climat global. *Revue de Géographie de Lyon*, vol. 72: Le climat urbain, M. Leroux éditeur, n° 4, 339–345.

Leroux, M. (1998). *Dynamic analysis of weather and climate*. Wiley–Praxis series in Atmospheric Physics, 365 p.

Leroux, M. (1999). Volcanisme et climat. Comm. Colloque: L'homme et le volcanisme. *Bull. Assoc. Géographes Français* (BAGF, 1999-4), Paris, 348–359.

Leroux, M. (2000).

- Analyse météorologique des pluies torrentielles des 12 et 13 novembre 1999 dans le Languedoc-Roussillon. 179–188.
- Les phénomènes extrêmes récents s'inscrivent-ils dans une évolution perceptible du temps? 261–270. in Sur la séquence orageuse de novembre 1999 (J. Béthemont – M. Leroux éds.) Numéro spécial de la *Revue de Géographie de Lyon* (RGL), vol. 75, n° 3.

Leroux, M. (2000a). L'évolution récente du temps ... et si on se trompait? Catastrophes Naturelles, aléas extrêmes et niveaux de protection de référence. Recueil des textes, sessions 2000, *Société Hydrotechnique de France* (SHF), 19 p, 14 fig.

Leroux, M. (2000b). Analyse météorologique des pluies torrentielles des 12 et 13 novembre 1999 dans le Languedoc-Roussillon. *Revue de Géographie de Lyon*, vol. 75, n° 3, 179–188.

Leroux, M. (2001). *The Meteorology and Climate of Tropical Africa*. Springer–Praxis, 550 p.

Leroux, M. (2002). 'Global warming': Mythe ou réalité? *Annales de Géographie*, 624, A. Colin, 115–137. (2003). Global warming: myth or reality? *Energy and Environment*, vol. 14, n° 6.

Leroux, M. (2003). Réchauffement global: une imposture scientifique!. *Fusion* n° 95, 36–58.

Le Roy Ladurie, E. (1967). *Histoire du climat depuis l'an Mil*. Flammarion, Paris.

Le Treut, H. and Kandel, R. (1992). Que nous apprennent les modèles du climat? *La Recherche*, 243, vol. 2, 572–582.

Le Treut, H. (1997). Climat: Pourquoi les modèles n'ont pas tort. *La Recherche*, 98, 68–73.

Le Treut, H. and Jancovici, J.M. (2001). *L'effet de serre. Allons-nous changer le climat?* Dominos, 233, Flammarion.

Levitus, S., Antonov, J.I., Wang, J., Delworth, T.L., Dixon, K.W. and Broccoli, A.J. (2001). Anthropogenic warming of Earth's climate system. *Science*, vol. 292, 267–271.

Le Vourc'h, J.Y., Fons, C. and Le Stum, M. (2001). *Météorologie générale et maritime*. Cours et Manuels, n° 14, Ecole Nationale de la Météorologie, Météo-France, Toulouse, 277 p.

Lewis, J.M. (1998). Clarifying the dynamics of the general circulation: Phillips's 1956 experiment. *Bull. Am. Met. Soc.*, vol. 79, n° 1, 39–60.

L'Hôte, Y. and Mahé G. (1996). Afrique de l'Ouest et centrale, précipitations moyennes annuelles (période 1951–1989). Orstom, Paris. (diffusion@bondy.ird.fr.)

Lindzen, R. (1990). Some coolness concerning global warming. *Bull. American Meteorological Society*, 71 n° 3, 288–299.

Lindzen, R.S. and Giannitsis, C. (2002). Reconciling observations of global temperature change. *Geophys. Res. Letters*, 29, 10.1029.

Litynski, J. (1994). Changements climatiques au Canada et l'évolution générale du climat. Publ. AIC, vol. 7, 287–293.

Litynski, J. (2000). Changements de température de la surface terrestre pendant la période 1931–1990. *Publ. Assoc. Intern. de Climatologie*, vol. 12, 289–297.

Litynski, J. and Genest, C. (2002). Fluctuations récentes de la température au Québec et dans le Grand Nord Canadien. *Publ. Ass. Int. Clim.*, vol. 14, 111–119.

Lomborg, B. (2001). *The Skeptical Environmentalist. Measuring the Real State of the World.* Cambridge University Press, 515 p.

Long, D.G., Ballantyne, J. and Bertoia, C. (2002). Is the number of Antarctic icebergs really increasing? *EOS*, vol. 83, n° 42, 469, 474.

Long, A. (2003). The coastal strip: Sea-level change, coastal evolution and land-ocean correlation. *Progress in Physical Geography*, n° 27-3, 423–434.

Lorenz, E.N. (1993). *The Essence of Chaos.* University of Washington Press, Seattle, USA, 227 p.

Lorius, C. (1973). Les calottes glaciaires, témoins de l'environnement. *La Recherche*, n° 34.

Lorius, C. (1983). Les données des carottes de glace de l'Antarctique: Évolution du climat et de l'environnement atmosphérique depuis le Dernier Maximum Glaciaire. *Bull. Inst. Géol. Bas. d'Aquitaine*, n° 33, CNRS, Cahiers du Quaternaire, 37–49.

Lotka, A.J. (1924). *Elements of Physical Biology.* Williams & Wilkins, Baltimore.

Lovelock, J.E. (2000). *Homage to Gaia. The Life of an Independent Scientist.* Oxford University Press, UK.

Lurçat F. (2003). *De la Science à L'ignorance.* Ed. du Rocher. Paris, 231 p.

Luterbacher J., Dietrich D., Xoplaki E., Grosjean M., Wanner H. (2004). European seasonal and annual temperature variability, trends, and extremes since 1500. *Science*, vol. 303, 1499–1503.

Lutzenberger J. (2000). La fièvre de Gaïa. *L'Ecologiste*, vol. 1, n° 2, p. 22.

Maduro R., Schauerhammer R. (1992). *Ozone, Un Trou pour Nien.* Editions Alcuin, 279 p.

Mahé, G. and L'Hôte, Y. (2004). Sahel: Une sécheresse persistante et un environnement profondément modifié. *La Météorologie*, n° 44, 2–3.

Maheras, P. (1989). Principal component analysis of western Mediterranean temperature variations 1866–1986. *Theor. Appl. Climat.*, 39, 137–145.

Maheras, P., Kutiel, H. and Kolyva-Machera, F. (1996). Variations spatiales et temporelles des températures hivernales au-dessus de la Méditerranée durant la dernière période séculaire. *Publ. Assoc. Intern. de Clim.*, AIC, vol. 9, 454–462.

Maheras, P., Kutiel, H., Patrikas, J., Floca, E. and Agnagnostopolou, C.H. (2000). Climatologie objective des dépressions cypriotes. *Publ. Ass. Int. Climat.*, vol. 12, Dakar.

Mahfouf, J.F. and Borrel, L. (1995). L'impact climatique des éruptions volcaniques. *La Météorologie*, 8e sér. (10), 10–27.

Makrogiannis, T.J. and Sashamanoglou, C.S. (1990). Time variation of the mean sea level pressure over the major Mediterranean area. *Theor. Appl. Climat.*, 41, 149–156.

Malardel, S. (2004). Front contre front. *Atmosphériques*, Grand Angle, Météo-France, avril, 8–9.

Manabe, S. and Wetherald, R.T. (1967). Thermal equilibrium of the atmosphere with a given distribution of relative, humidity. *Journal of Atmospheric Science*, 21, 361–385.

Manabe, S. (1971). Estimate of future changes of climate due to increase of CO_2 in the air. In: Matthews *et al.* eds), *Man's Impact on Climate.* Cambridge, MIT Press, MA, pp. 249–264.

Manabe, S. and Wetherald, R.T. (1975). The effects of doubling the CO_2 concentration on the climate of a GCM. *Journal of Atmospheric Science*, 32, 3–15.

Mangerud, J. (1991). The last interglacial/glacial cycle in northern Europe. *Quaternary Landscapes*, Univ. Minnesota Press, 38–75.

Mann, M.E., Bradley, R.S. and Hughes, M.K. (1998). Global-scale temperature patterns and climate forcing over the past six centuries. *Nature*, n° 392, 779–787.

Mann, M.E., Bradley, R.S. and Hughes, M.K. (1999). Northern hemisphere temperatures in the past millenium: inferences, uncertainties and limitations. *Geoph. Res. Letters*, 30, 759–762.

Mann, M., Amman, C., Bradley, R., Briffa, K., Jones, Ph., Osborn, T., Crowley, T., Hughes, M., Oppenheimer, M., Overpeck, J., Rutherford, S., Trenberth, K. and Wigley, T. (2003). On past temperatures and anomalous late-20th century warmth. *EOS*, vol. 84, n° 27, 256–257. Response to W. Soon *et al.* (2003), *EOS*, 84, 273, 276.

Mann, M.E., Bradley, R.S. and Hughes, M.K. (2004). Corrigendum: Global-scale temperature patterns and climate forcing over the past six centuries. *Nature*, n° 6995, vol. 430, p. 105.

Mantua, N.J., Hare, S.R., Zhang, Y., Wallace, J.M. and Francis, R.C. (1997). A Pacific Interdecadal Climate Oscillation with Impacts on Salmon Production, *Bull. of the American Meteorological Society*, vol. 78, n°. 6, 1069–1079.

Mantua, N.J. (2004). The Pacific Decadal Oscillation. (www.atmos.washington.edu)

Martin-Ferrari, D. (1992). *L'Ecologie*. Alphabétique Retz, n° 19, 144 p.

Martin, F. (1999). La tempête de verglas de janvier 1998 au Canada. *Mém. LCRE,* Lyon.

Masato, S. and Yoshimura, J. (2004). A mechanism of tropical precipitation change due to CO_2 increase. *Journal of Climate*, vol. 17, n° 1, 238–243.

Maslanick, J.A., Serreze, M.C. and Barry, R.G. (1996). Recent decreases in Arctic summer ice cover and linkages to atmospheric circulation anomalies. *Geophys. Res. Letters*, 23, 1677–1680.

Mason, B.J. (1979). Some results of climate experiments with numerical models. World Climate Conference, *WCC/Overview paper 9, WMO*, Geneva, 32 p.

Mason, S.J. and Goddard, L. (2001). Probabilistic precipitations anomalies associated with ENSO. *Bull. Am. Met. Soc.*, vol. 82, n°4, 619–638.

Mass, C.F. and Portman, D.A. (1989). Major volcanic eruptions and climate: A critical evaluation. *Journal of Climate*, vol. 2(6), 566–593.

Masson-Delmotte, V. and Chappellaz, J. (2002). Au coeur de la glace, les secrets du climat. *La Météorologie*, n° 37, 18–25.

Masood, E. (1996). Climate Report subject to scientific cleansing. *Nature*, 381, 546.

Mauny, R. (1961). *Tableau géographique de l'Ouest Africain au Moyen Age*. Mém. Institut Français d'Afrique Noire (IFAN), Dakar, 587 p.

Mayr, F. (1964). Untersuchhungen über Ausmass und folgen der Klima- und Gletcherschwankungen seit dem Beginn der postglacialen Wärmezeit. *Zeitchrift für Geomorph.* 8, 257–285.

McCabe, G. J., Palecki, M.A. and Betancourt, J.L. (2004). Pacific and Atlantic Ocean influences on multidecadal drought frequency in theUnited States. *Proc. Natl. Acad. Sci.* USA.

McCarthy, M. (2003). The four degrees: How Europe's hottest summer shows global warming is transforming our world. *The Independent*. (news.independent.co.uk/world/environment)

McClain, E.P. (1978). The giant iceberg Trolltunga. NOAA. *Mariner's Weather Log.*, in Bull. Inform. 44, Dir. Mét., Paris, 1979.

McIntyre, S. and McKitrick, R. (2003). Corrections to the Mann *et al.* (1998) proxy data base and northern Hemisphere average temperature series. *Energy and Environment*, vol. 14, n° 6, 751–771.

McPhaden, M.J. (1999). The Child Prodigy of 1997–1998. *Nature*, vol. 398, 559–561.

Mears, C.A., Schabel, M.C. and Wentz, F.J. (2003). A reanalysis of the MSU channel 2 tropospheric temperature record. *Journal of Climate*, vol. 16, n° 22, 3650–3664.

Menard, Y. (2003). L'océan physique révélé par l'altimétrie satellitale. *La Lettre, Medias* n°12, 9–18.

Merilees, P.E. (ed) (2003). Community Climate System Model Science Plan (2004–2008), NCAR. (www.ccsm/ncar/edu)

Météo-France (1992). Le point sur l'évaluation scientifique de l'évolution du climat. Rapport du GIEC, *Phénomènes remarquables*, num. sp. n° 7.
Metzl, N. (2002). Puits ou source de CO_2? *Sciences et Avenir*, h.s.: L'aventure polaire, 88–92.
Michaels, P.J. and Stoocksbury, D.E. (1992). Global warming: A reduced threat? *Bull. Am. Met. Soc.*, 73, n° 10, 1563–1576.
Michaels, P.J. (1992). *Sound and fury. The science and politics of global warming.* Cato Institute, Washington DC, 196 p.
Michaels, P.M. and Balling, Jr. R.C. (2000). *The Satanic Gases: Clearing the Air About Global Warming.* Cato Institute, Washington, DC.
Michaels, P.J., Singer, F.S. and Douglass, D.H. (2004). Settling global warming science. (www.techcentralstation.com/081204D)
Milankovitch, M. (1924). *Théorie Mathématique des Phénomènes Thermiques Produits par la Radiation Solaire.* Gauthier-Villars, Paris.
Miner, T., Sousounis, P.J., Wallman, J. and Mann, G. (2000). Hurricane Huron. *Bull. Am. Met . Soc.*, vol. 81, n° 2, 223–236.
Minnich, R.A., Vizcaino, E.F. and Dezzani, R.J. (2000). The El Niño/Southern Oscillation and precipitation variability in Baja California, Mexico. *Atmosfera*, n° 13, 1–20.
Minnis, P., Harrison, E.F., Stowe, L.L., Gibson, G.G., Denn, F.M., Doelling, D.R. and Smith, W.L (1993). Radiative climate forcing by the Mount Pinatubo eruption. *Science*, 259, 1411–1415.
Minster, J.F. and Merlivat, L. (1992) Où va le gaz carbonique? Le rôle des océans. *La Recherche*, 243, vol. 2, 592–588.
Mitchell, J.F.B. (1989). The 'greenhouse' effect and climate change. *Rev. in Geophysics*, 29, 30–60.
Mohnen, V. (1988). Le danger des pluies acides. *Pour la Science*, n° 132, 54–62.
Moisselin, J.M., Schneider, M., Canellas, C. and Mestre, O. (2002). Les changements climatiques en France au XXe siècle. *La Météorologie*, n° 38, 45–56.
Möller, F. (1963) On the influence of changes in the CO_2 concentration in air on the radiation balance of the earth's surface and on the climate. *Journal of Geophysical Research*, 68, 3877–3886.
Moore, G.W.K., Holdsworth, G. and Alverson, K. (2002). Climate change in the North Pacific region over the past three centuries. *Nature*, vol. 420, 401–403.
Morel, R. (1995). La sécheresse en Afrique de l'Ouest. In: 'Sahel: la Grande Sécheresse', *Rev. de Géogr. de Lyon*, M. Leroux éd., n° 3–4, 215–222.
Morgan, M.R., Drinkwater, K.F. and Pocklington, R. (1993). Temperature trends at coastal stations in Eastern Canada. *Clim. Bull.*, 27(3), Envir. Canada, 135–153 .
Morgan, M.R. and Pocklington, R. (1995). Northern hemispheric temperature trends from instrumental surface air records. *SCMO Bulletin*, vol. 23, n° 4-5, 35.
Mörner, N-A. (2003). Estimating future sea level changes from past records. *Paleogeophysics and Geodynamics*, Stockholm Univ., S-10691.
Mörner, N-A., Tooley, M. and Possnert, G. (2004). New perspectives for the future of the Maldives. *Global and Planetary Change*, vol. 40, 1-2, 177–182.
Mott, R.J., Grant, D.R., Stea, R. and Occhietti, S. (1986). Late-glacial climate oscillation in Atlantic Canada equivalent to the Allerod/Younger Dryas event. *Nature*, vol. 323, 247–250.
Moum, J.N. and Caldwell, D.R. (1994). Experiment explores the dynamics of ocean mixing. *EOS*, vol. 75, n° 42, 489, 494–495.
Muller, R.A. and MacDonald, G.J. (2000). *Ice ages and astronomical causes.* Springer-Praxis Books in Environmmental Sciences, London, Berlin, N.Y., 318 p.
Mysak, L.A. (2001). Patterns of Arctic circulation. *Science*, vol. 293, 1269–1270.

Nakada, M. and Lambeck, K. (1988). The melting history of the late Pleistocene Antarctic ice sheet. *Nature*, vol. 333, 36–40.

Nakamura, H., *et al.* (2002). Interannual and decadal modulations recently observed in the Pacific storm track activity and east asian winter monsoon, *Journal of Climate*, vol. 15, n° 14, 1855–1874.

Nalbantis, I., Mamassis, N. and Koutsoyiannis, D. (1993). Le phénomène récent de sécheresse persistante et l'alimentation en eau de la cité d'Athènes. *Publ. AIC*, vol. 6, 123–132.

Nastos, P. (1993). Changements de la pluviosité en région hellénique pendant la période 1858–1992. *Publ. AIC*, vol. 6, 183–190.

Ndong, J.B. and Dione, O. (1994). Dynamique de la sécheresse en Afrique sahélienne: Cas du Sénégal. *Publ. Ass. Int. de Clim.*, vol. 7, 415–420.

Ndong, J.B. (1996). *L'évolution récente du climat et de l'environnement au Sénégal*. Th. Géogr. Phys. Univ. Lyon3, LCRE, 485 p.

Newhall, C.G. and Self, S. (1982). The volcanic explosivity index (VEI): An estimate of explosive magnitude of historical volcanism. *Journal Geophys. Res.*, 87, 1231–1238.

Niebauer, H.J. (1988). Effects of ENSO and North Pacific weather patterns on interannual variability in the subarctic Bearing Sea. *EOS*, 23, 119–120.

Nouaceur, Z. (1999). *L'évolution du climat et des lithométéores en Mauritanie*. Th. Univ. Lyon 3, Labor. de Climatologie (LCRE), 485 p.

O'Hirok, W. and Gautier, C. (2003). Absorption of shortwave radiation in a cloudy atmosphere: Observed and theoretical estimates during the second Atmospheric Radiation Measurement Enhanced Shortwave Experiment (ARESE), *J. Geophys. Res.*, 108(D14).

Olausson, E. (1985). The Glacial oceans. *Palaeogeography, -climatology, -ecology*, n° 50, 291–301.

Omar Haroun, S. (1997). *L'évolution récente du climat et hydrologie du Nil au Soudan*. Th. Univ. Lyon 3, Labor. de Clim., Risq., Env., LCRE, 418 p.

Oppenheimer, M. (1998). Global Warming and the stability of the West Antarctic Ice Sheet. *Nature*, 393, 325–32.

Otterman, J., *et al.* (2002). North-Atlantic surface winds examined as the source of winter warming in Europe. *Geophysical Research Letters*, 29, 181–184.

Overland, J.E. and Stabeno, P.J. (2004). Is the climate of the Bering Sea warming and affecting the ecosystem? *EOS*, vol. 85, n° 33, 309–310, 312.

Pagney, P. (1994). *Les climats de la Terre*. Masson, Paris, 166 p.

Paillard, D. and Parrenin, F. (2004). Le paradoxe de la fonte des glaces. *La Recherche*, h-s. n° 15, 28–31.

Palecki, M.A., Changnon, S.A. and Kunkel, K.E. (2001). The nature and impacts of the July heay wave in the Midwestern United States: learning from the lessons of 1995. *Bull. Am. Met. Soc.*, vol. 82, n° 7, 1353–1367.

Palmen, E. (1951). The aerology of extratropical distrurbances. *Compendium of Meteorology, Boston*, 599–620.

Palmen, E. and Newton, W. (1969). Atmospheric circulation systems, their structure and physical interpretation. *Geophys. Res.*, 13, 603 p.

Pang, K.D. and Yau, K.K. (2002). Ancient observations link changes in Sun's brightness and Earth's climate. *EOS*, vol. 83, n° 43, 481, 489–490.

Parker, D.E. and Folland, C.K. (1992). Peut-on mesurer la température terrestre? Les mesures traditionnelles. *La Recherche*, 243, vol. 2 , 584–588.

Parker, D.E., Gordon, M., Cullum, D.P.N., Sexton, D.M.H., Folland, C.K. and Rayner, M. (1997). A new global gridded radiosonde temperature data base and recent temperature trends. *Geophys. Res. Letters*, 24, 1499–1502.

Parkinson, C.L. (1992). Spatial patterns of increases and decreases in the length of the sea ice season in the North Polar region, 1979, 1986. *J. Geophys. Res.*, 97, 14377–14388.

Parkinson, C.L., Cavalieri, D.J., Gloersen, P., Zwally, H.J. and Comiso, J.C. (2000). Arctic sea-ice extents, areas, and trends, 1978–1996. *J. Geophys. Res.*, 104, 20,837–20,856.

Parkinson, C.L. (2002). *Annals of Glaciology*, 34, 435. (www.gsfc.nasa.gov./topstory/20020820)

Paskoff, R. (2001). *L'élévation du niveau de la mer et les espaces ctiers.* Institut Océanographique, Collection Propos, Paris.

Peixoto, J.P. and Oort, A.H. (1983). The atmospheric branch of the hydrological cycle and climate. Variations in the global water budget. A. Street-Perrott *et al.* (eds). D. Reidel Publ., 5–65.

Petersen, A.C. (2000). Philosophy of climate science. *Bull. AM. Met. Soc.*, vol. 81, n° 2, 265–271.

Peterson, T.C. and Vose, R.S. (1997). An overview of the Global Historical Climatology Network temperature data base. *Bull. Am. Met. Soc.*, vol. 78, n° 12, 2837–2849.

Petit, J.R., Jouzel, J., Raynaud, D., Barkov, N.I., Barnola, J.M., Basile, I., Benders, M., Chappellaz, J., Davis, M., Delaygue, G., *et al.* (1999). Climate and atmospheric history of the past 420,000 years from the Vostok ice core, Antarctica. *Nature*, vol. 399, 429–435.

Pfeffer, W.T., Cohn, J., Meier, M. and Krimmel, R.M. (2000). Alaskan glacier beats a dramatic retreat. *EOS*, vol. 81, 48, 577–578.

Pielke, R.A. and Landsea, C.N. (1999). La Niña, El Niño, and Atlantic hurricane damages in the United States. *Bull. Am. Met. Soc.*, vol. 80, n° 10, 2027–2034.

Pielke, R.A. (2004) Limitations of models and observations. *COMET COMAP Symposium* on planetary boundary.processes. (online: ppt 15.pdf)

Pielke, R.A. and Chase, T.N. (2004). NCEP Reanalysis Summary of the summer 2003 European heat wave (20 Feb.). (website)

Pierce, F. (2001). We are all guilty! It's official, people are to blame for global warming. *New Scientist*, 169. (http://.newscientist.com/archive)

Pirazzoli, P. (1996). Etat de la mer et niveaux marins. *Bull. Ass. Géogr. Fr.*, n° 4, 283–290.

Pittock, A.B. (1983). Solar variability, weather and climate: An update. *Quart. J. R. Met. Soc.*, 109, 23–55.

Planton, S. (2000). Tempêtes et changement climatique. *Aménagement et Nature*, n° 137, 67–72.

Planton, S. and Bessemoulin P. (2000). Le climat s'emballe-t-il? *La Recherche*, 335, 46–49.

Plass, G.N. (1956). The CO_2 theory of climate change. *Tellus*, 8, 140–153.

Polyakov, I., Akasofu, S.-I., Bhatt, U., Colony, R., Ikeda, M., Makshats, A., Swingley, C., Walsh, D. and Walsh, J. (2002). Trends and variations in Arctic climate system. *EOS*, 19 Nov. 547–548.

Polyakov, I.V., Alekseev, G.V., Bekryaev, R.V., Bhatt, U., Colony, R.L., Johnson, M.A., Karklin, V., Makshtas, A.P., Walsh, D. and Yulin, A.V. (2002). Observationally based assessment of polar amplification of global warming. *Geophysical Research Letters*, vol. 29, n° 18.

Polyakov, I.V., Bekryaev, R.V., Alekseev, G.V, Bhatt, U., Colony, R.L. Johnson, M.A., Makshtas, A.P. and Walsh, D. (2003). Variability and trends of air temperature and pressure in the maritime Arctic, 1875–2000. Vol. *Journal of Climate*, 16, n° 12, 2067–2077.

Pommier, A. (2002). L'évolution récente de la dynamique aérologique dans l'Atlantique nord. *Ecole Normale Supérieure, Lyon, Fête de la Science* (Th. Lab. Clim. Risq. Envir., LCRE, Lyon).

Pommier, A. (2004). Relationships between the Features Variations of Highs and Lows in the North Atlantic Region and North Atlantic Oscillation from 1950 to 2000, 1st *International CLIVAR Science Conference*, Baltimore.

Pommier, A. (2004). Relationships between frequency and surface of Mobile Polar Highs (MPHs) and Lows in the North Atlantic Aerological space during winter (JFM) from 1950 to 2000, *4th annual meeting of the European Meteorological Society.*

Pommier, A., Favre, A., Gershunov, A. and Leroux, M. (2004). Trajectoires des anticyclones et des dépressions dans le Pacifique Nord et dans l'Atlantique Nord de 1950 à 2001. Comm. *Colloque Commission de Climatologie, CNFG*, Nancy.

Pommier, A. and Leroux, M. (2004). L'intensité de la dynamique aérologique et les dépressions profondes dans l'espace aérologique Nord-Atlantique de 1950 à 2000. Coll. *Ass. Intern. Clim., AIC, Caen.*

Porter, S.C. (1986). Pattern and forcing of Northern Hemisphere glacier variations during the last millenium. *Quaternary Research*, 26, 27–48.

Postel-Vinay, O. (2002). Les pôles fondent-ils? *La Recherche*, n° 358, 34–43.

Radionov, V.F. and Marshunova, M.S. (1992). Long-term variations in the turbidity of the Arctic atmosphere in Russia. *Atmosphere–Ocean*, 30(4), 531–549.

Rahmstorf, S. (1997). On the freshwater forcing and transport of the Atlantic thermohaline circulation. *Nature*, 388, 825–826.

Rahmstorf, S. (1999). Shifting seas in the greenhouse? *Nature*, vol. 399, 523–524.

Rahmstorf, S. (2000). The thermohaline ocean circulation: A system with dangerous thresholds? *Climatic Change*, 46, 247–256.

Rahmstorf, S. and Alley, R. (2002). Stochastic resonance in glacial climate. *EOS*, n° 12, 129, 135.

Ramade, F. (1987). *Eléments d'Ecologie. Ecologie fondamentale.* McGraw-Hill, Paris, 403 p.

Ramanathan, V., Pitcher, E.J., Malone, R.C. and Blackmon, M.L. (1983). The response of a spectral GCM to refinements in radiative processes. *Journal of Atmospheric Science*, 40, 605–630.

Ramanathan, V. (1988). The greenhouse theory of climate change: a test by an inadvertent global experiment. *Science*, 240, 293–299.

Ramanathan, V., Barkstrom, B.R. and Harrison, E.F. (1989). Climate and the Earth's radiation budget. *Physics Today*, 5, 22–32.

Ramanathan, V., Cess, R.D., Harrison, E.F., Minnis, P., Barkstrom, B.R., Ahmad, E. and Hartmann, D. (1989). Cloud-radiative forcing and climate: Results from the Earth Radiation Budget Experiment. *Science*, vol. 243, 57–63.

Ramanathan, V., Crutzen, P.J., Kiehl, J.T. and Rosenfeld, D. (2001). Aerosols, climate and the hydrological cycle. *Science*, vol. 294, 2119–2124 (www.sciencemag.org)

Rampino, M.R. (1989). Distant effects of the Tambora eruption of April 1815. *EOS*, 1559.

Randall, D., Khairoutdinov, M., Arakawa, A. and Grabowski, W. (2003). Breaking the Cloud Parameterization Deadlock. *Bulletin of the American Meteorological Society*: Vol. 84, No. 11, 1547–1564.

Rasool, I. (1993). *Système Terre.* Dominos n° 12, Flammarion.

Raspopov, OM., Dergachev, V.A. and Kolstrom, T. (2004). Periodicity of climate conditions and solar variability derived from dendrochronological and other paleoclimatic data in high latitudes. *Palaeogeography, -climatology, -ecology*, 209, 127–139.

Raval, A. and Ramanathan, V. (1989). Observational determination of the greenhouse effect. *Nature*, 342, 758–761.

Redmond, K.T. (2002). The depiction of drought. *Bull. Am. Met. Soc.*, BAMS, 1143–1147.

Reynaud, J. (1994). Evolution récente de la pression en surface et des températures dans l'espace Atlantique Nord, du Groenland à la Scandinavie. *Publ. AIC*, vol. 7, 268–278.

Reynaud, L. (2003). Influence de l'Oscillation Nord-Atlantique sur les glaciers alpins et scandinaves. *Lettre n° 15, PIGB-PMRC*. (CNRS website)

Richard, Y. (1993). *Relations entre la variabilité pluviométrique en Afrique australe tropicale et la circulation océano-atmosphérique*. Th. Univ. Aix-Marseille I, 252 p + fig.

Richardson, L. F. (1922). *Weather Prediction by Numerical Process*. Cambridge University Press, UK (reprinted, Dover, NY 1965).

Riehl, H. (1969). Mechanisms of the general circulation of the troposphere. General Clim., (Chap. 1). In: *World Survey of Climatology*, Elsevier, vol. 2, 1–37.

Rignot, E. and Thomas, R.H. (2002). Mass balance of polar ice sheets. *Science*, vol. 297, 1502–1506.

Rignot, E. *et al.* (2004). Accelerated ice discharge from the Antarctic Peninsula following the collapse of Larsen B Ice Shelf. *Geophysical Research Letters*, 31, L18401.

Rigor, I.G., Colony, R.L. and Martin, S. (2000). Variations in surface air temperature observations in the Arctic, 1979–1997. *J. of Climate*, AMS, vol. 13, n° 5, 896–914.

Robinson, W.A., Reudy, R. and Hansen, J.E. (2002). General circulation model simulations of recent cooling in the east-central United States. *Geophys. Research Letter*, vol. 107, n°. D24, 4748.

Robock, A., Free, M.P. (1995). Ice cores as an index of global volcanism from 1850 to the present. *Journ. of Geophys. Res*, vol. 100, 11549–11567.

Robock, A. (2002) Blowin' in the wind: Research priorities for climate effets of volcanic eruptions; *EOS*, 15 Oct., 472.

Rocca, R. (2002). Comprendre le changement climatique à l'aide de la théorie de l'équilibre radiatif-convectif, site www: *Lab. de Météorologie Dynamique-Ecole Normale Supérieure*, Paris.

Rochas, M. and Javelle, J.P. (1993). *La Météorologie. La prévision numérique du temps et du climat*. Syros, coll. 'Comprendre', Paris, 264 p.

Rochas, M. (2002). Lu pour vous: Climat de panique, d'Yves Lenoir. *La Météorologie*, 38, 68–69.

Rodbell, D.T. (2000). The Younger Dryas: cold, cold everywhere? *Science*, vol. 290, 285–286.

Rogers, J.C. (1989). Seasonal temperature variability over the North Atlantic Arctic. *Proc. 13th Annual Climate Diagnostics Workshop*, 1988, NOAA-NWS, 170–178.

Rogers, J.C. and Rohli, R.V. (1991). Florida citrus freeze and polar anticyclones in the Great Plains. *Journal of Climate*, vol. 4, n° 11, 1103–1113.

Rogers, J.C. (1997). North Atlantic storm track variability and its association to the North Atlantic Oscillation and climate variability of northern Europe. *Journal of Climate*, vol. 10, n° 7, 1635–1647

Ropelewski, C.F. and Halpert, M.S. (1987). Global and regional scale precipitations patterns associated with the El Niño/Southern Oscillation (ENSO). *Mon. Weather Rev*. 115, 1606–1626.

Roqueplo, Ph. (1993). *Climats sous surveillance*. Limites et conditions de l'expertise scientifique. Economica, 401 p.

Rossby, C.G. and Weightman, R.H. (1939). Relations between variations in the intensity of the zonal circulation of the atmosphere and the displacement of the semi-permanent centers of action. *Journ. Mar. Res*., 2(1), 38–55.

Rossby, C.G. (1941). The scientific basis of modern meteorology. *Climate and Man. Yearbook of Agric*. New York, 599–655.

Rossby, C.G. (1949). On the nature of the general circulation of the lower atmosphere. *The Atmosphere of the Earth and Planets*, Kuiper, Chicago.

Rothrock, D.A., Yu, Y., Maykut, G.A. (1999). Thinning of the Arctic sea-ice cover. *Geophys. Res. Letters*, 26, 3469–3472.

Rotty, R.M. and Marland, G. (1980). Constraints on fossil fuel use. In: *Interactions of Energy and Climate*. D. Reidel, pp. 191–212.

Rougerie, F., Salvat, B. and Tatarata-Couraud, M. (1992). La mort blanche des coraux. *La Recherche*, n° 245, 827–834.

Rowland, F.S. and Molina, M. (1975). Chlorofluoromethanes in the environment. *Review of Geophysics and Space Studies*, vol. 13, n° 1.

Roy, R.W. and Christy, J.R. (2003). Global temperature Report 1978–2003. (http://www.uah.edu/News/climate/25years.pdf)

Ruddiman, W.F. and McIntyre, A. (1979). Warmth of the subpolar North Atlantic Ocean during northern hemisphere ice-sheet growth. *Science,* 204, 173–175.

Rudels (1994). Surface circulation in the Arctic Ocean and the Greenland Sea. In Houssais, M.N. and Gascard, J.C., *PIGB-MRC, Lettre* n° 12, 2001.

Rupa Kumar, K. and Hingane, L.S. (1988). Long-term variations of surface air temperature at major industrial cities of India. *Climatic Change*, 13, 287–307.

Sabatier, F. and Provansal, M. (2002). La Camargue sera-t-elle submergée? *La Recherche*, 355, 72–73.

Sadourny, R. (1992). L'homme modifie-t-il le climat? *La Recherche*, 243, vol. 2, 522–527.

Sadourny, R. (1994). *Le climat de la Terre*, Dominos n° 28, Flammarion, 126 p.

Sagna, P. (1988). *L'importance pluviométrique des lignes de grains en Afrique Occidentale*. Th. Clim. Univ. de Dakar, LCTA.

Sagna, P. (1994). L'évolution de la mousson et des précipitations au Sénégal de 1974 à 1993. *Publ. Ass. Int. de Clim.*, vol. 7, 311–317.

Sagna, P. (2001). Contribution des images satellitaires Meteosat à l'analyse des saisons pluvieuses au Sénégal de 1980 à 1999. *Publ. Ass. Int. de Clim.*, vol. 13, 335–343.

Sagna, P. (2004). *Le climat du littoral et des iles de l'ouest de l'Afrique occidentale*. Th. Univ. Ch. A. Diop, LCTA, Dakar.

Sala, J.Q. and Chiva, E.M. (1996). L'élévation de la température en Espagne méditerranéenne: tendance naturelle ou effet de l'urbanisation? *Publ. Assoc. Intern. de Clim.,* vol. 9, 487–495.

Sarmiento, J.L. and Gruber, N. (2002). Sinks for anthropogenic carbon. *Physics Today*, AIP, 30–36.

Sato, M., Hansen, J.E., McCormick, M.P. and Pollack, J.B. (1993). Stratospheric optical depth. *Journ. of Geophys. Res.*, 98, 22,987–22,994.

Schär, C., Vidale, P.L., Lüthi, D., Frei, C., Häberli, C., Liniger, M.A. and Appenzeller, C. (2004). The role of increasing temperature variability in European summer heat waves. *Nature*, vol. 427, 332–335.

Schär, C. and Jendritzky, G. (2004). Hot news from summer 2003. *Nature*, vol . 432, 559–560.

Schiermeier, Q. (2004). Greenland's climate: a rising tide. *Nature*, vol. 428, 114–115.

Schimel, D.S., House, J.I., Hibbard, K.A., Bousquet, P., Ciais, P., Peylin, P., Braswell, B.H., Apps, M.J., Baker, D., Bondeau, A., *et al.* (2001). Recent patterns and mechanisms of carbon exchange by terrestrial ecosystems. *Nature*, vol. 414, 169–172.

Schmidt, M.W., Howard, J.S. and David, W.L. (2004). Links between salinity variation in the Caribbean and North Atlantic thermohaline circulation. *Nature*, vol. 428, 160–163.

Schneider, D. J. (1994). Unnammed aerial sampling of a volcanic ash cloud. *EOS*, 75, n° 12, 137–138.

Schneider, E.K., Kirtman, B.P. and Lindzen, R.S. (1999). Tropospheric water vapor and climate sensitivity. *J. Atm. Sci.*, 36, 1649–1658.

Schneider, S.H. (1989). The greenhouse effect: Science and policy. *Science*, 243, 771–781.

Schneider, S.H. (1990) The global warming debate heats up: an analysis and perspective. *Bull. Am. Met. Soc.*, vol. 71, n° 9, 1292–1303.

Schneider, E.K., Kirtman, B.P. and Lindzen, R.S. (1999). Tropospheric water vapor and climate sensitivity. *J. Atm. Sci.*, 36, 1649–1658.

Schonher, T. and Nicholson, S.E. (1989). The relationship between California rainfall and ENSO events. *J. of Climate*, vol. 2(11), 1258–1269.

Schonwiese, C. and Rapp, J. (1997). *Climate trend atlas of Europe based on observations 1891–1990*. Kluwer Acad. Publ., 228 p.

Schubert, S.D., Suarez, M.J., Pegion, P.J., Koster, R.D. and Bacmeister, J.T. (2004). On the cause of the 1930s Dust Bowl. *Science*, vol. 303, n° 5665, 1855–1859.

Schwing, F., Moore, C. (2000). A year without summer for California, or a harbinger of a climate shift? *EOS*, vol. 81, n° 27, 301, 304–305.

Seager, R., Battisti, D.S., Yin, J., Gordon, N., Naik, N., Clement, A.C. and Cane, M.A. (2002). Is the Gulf Stream responsible for Europe mild winters? *Quat. J. of R. Met. Society*, vol. 128, n° 586, 2563–2586.

Seager, R. (2003). Gulf Stream, la fin d'un mythe. *La Recherche*, n° 361, 40–45.

Seitz, F. (1996). A major deception on 'global warming'. reprinted of the Wall Street Journal, 1996. *Bulletin of the American Meteorological Society*, vol. 77, n°9, 1962–1963.

Self, S. and Rampino, M.R. (1988). The relationship between volcanic eruptions and climate change: still a conundrum? *EOS*, vol. 69, n° 6, 74-75, 85–86.

Serreze, M.C., Box, J.E., Barry, R.G. and Walsh, J.E. (1993). Characteristics of Arctic synoptic activity, 1952–1989. *Meteorol. Atmos. Phys.*, 51, 147–164 .

Serreze, M.C., Carse, F., Barry, R.G. and Rogers, J.C. (1997). Icelandic low cyclone activity: climatological features, linkages with the NAO, and relationships with recent changes in the Northern Hemisphere circulation. *Journ. of Climate*, 10, 453–464.

Servain, J. and Seva, M. (1987). On relationships between tropical Atlantic sea surface temperature, wind stress and regional precipitaation indices: 1964–1984. *Ocean-Air Interactions*, 1, 183–190.

Shaviv, N.J. and Veizer, J. (2003) Geological Society of America, *GSA Today*, 4.

Shaw, G.E. (1995). The Arctic haze phenomenon. Bull. Amer. Met. Soc., vol. 76, n° 12, 2403–2413.

Sheperd, J.M. and Jin, M. (2004). Linkages between the built urban environment and Earth's climate system. *EOS*, vol. 85, n° 23, 227–228.

Siegert, M.J. (1997). Quantitative reconstructions of the last glaciation of the Barents Sea: a review of ice-sheet modelling problems. *Progress in Physical Geography*, 21(2), 200–229.

Singer, S.F. (1970). *Global Effects of Environmental Pollution*. Springer-Verlag, NY.

Singer, F., Boe, B.A., Decker, F.W., Frank, N., Gold, T., Gray, W., Linden, H.R., Lindzen, R., Michaels, P.J., Nierenberg, W.A., Porch, W. and Stevenson, R. (1997). Comments on 'open letter to Ben Santer'. *Bulletin of the American Meteorological Society*, vol. 78 n°1, 81–82.

Singer, F. (1999). Human contribution to climate change remains questionable. Forum, *EOS*, 20 April, 185–187.

Singer, F. (2002). Le cas scientifique contre la convention sur les changements climatiques. *The Science & Environmental Policy Project*, SEPP. (`http//www.sepp.org`)

Six, D., Reynaud, L. and Letréguilly, Y. (2001). Comparaison des bilans glaciaires, 1967–1997, in L. Reynaud, *PIGB-PMRC*, Lettre n° 15, 2003.

SMIC: Study of Man's Impact on Climate (1971). *Inadvertent climate modification*. Cambridge, MA, MIT Press.

Smith, E. (1999). Atlantic and East Coast hurricanes 1900–1998: A frequency and intensity study for the twenty-first century. *Bull. Amer. Met. Soc.*, vol. 80, n° 12, 2717–2720.

Smith, S.R. and O'Brien, J.J. (2001). Regional snowfall distributions associated with ENSO: Implications for seasonal forecasting. *Bull. Am. Met. Soc.*, vol. 82, n° 6, 1179–1191.

Soon, W., Baliunas, S., Idso, S.B., Kondratyev, K. Ya and Posmentier, E.S. (2001) . Modeling climatic effects of anthropogenic carbon dioxide emissions: Unknowns and uncertainties. *Myths and Frauds, Clim Res.*, vol. 18, 259–275.

Soon, W. and Baliunas, S. (2003). Global warming. *Progress in Physical Geography*, 27, 3, 448–455.

Soon, W., Baliunas, S. and Legates, D. (2003). Comment on 'On past temperatures and anomalous late-20th century warmth'. *EOS*, 84, 473.

Soon, W. and Yaskell, S.H. (2003). Year without a summer. *Myths and Frauds, FAEC.* (website)

Soon, W., Baliunas, C., Idso, C., Idso, S. and Legates, D.R. (2003). Reconstructing climatic and environmental changes of the past 1000 years: A reappraisal. *Energy and Environment*, vol. 14, n° 6, 233–296.

Soon, W. (2004). Winter weather wonder. (`www.techcentralstation.com`)

Soon, W. (2004). What is the earth's 20th century temperature trend? (online) *TCS*. Soon W., Legates D., Baliunas S. (Feb. 2004) in *Geophys. Res. Letters*, vol. 31.

Spencer, R. (2004). When is global warming really a cooling? (`www.techcentralstation.com`)

Spencer, R. (2004). Let them confess their faith. Comm. 6 Feb. 2004. (in `www.climate sceptics` and `www.techcentralstation`)

Stanhill, G. (1999). Climate change science is now big science. Forum, *EOS*, vol. 80, n° 35, 396–397.

Steig, E.J. (2001). No two latitudes alike. *Science*, vol. 293, 2015–2016.

Stephens, B.B. and Keeling, R.F. (2000). The influence of Antarctic sea ice on glacial-interglacial CO_2 variations. *Nature*, vol. 404, 171–174.

Stephenson, D.B., Wanner, H., Brönniman, S. and Luterbacher, J. (2003) The history of scientific research on the North Atlantic Oscillation. The North Atlantic Oscillation, *Geophys. Monog.* 134, AGU, 37–50.

Stevenson, R. (2001). L'océan se réchauffe mais ce n'est pas le réchauffement global! *Fusion*, n° 84, 37–43. from 21st Century Science et Technology, 2000.

Stott, P.A., Tett, S.F.B., Jones, G.S., Allen, M.R., Ingram, W.J. and Mitchell, J.F.B. (2001). Attribution of twentieth century temperature change to natural and anthropogenic causes. *Climate Dynamics*, 17, 1–21.

Stott, P.A. (2004). Alternative Arguments. *BBC, The open University*, Global warming.

Stott, P.A., Stone, D.A. and Allen, M.R. (2004). Human contribution to the European heatwave of 2003. *Nature*, vol. 432, 610–613.

Suess, H. E. (1957). Radiocarbon Concentration in Modern Wood. *Science* 122, 415–417.

Svensmark, H. and Friis-Christensen, E. (1997). Variation of cosmic ray flux and global cloud coverage – A missing link in solar-climate relationships. *Journal of Atmospheric and Solar Terrestrial Physics*, 59, 1225–1232.

Talagrand, O. (1988). La dynamique des atmosphères planétaires. *La Recherche*, n° 202, vol. 19, 1011–1021.

Tanner, W.F. (1992). 3000 years of sea level change. *Bull. Am. Met. Soc.*, vol. 73, n° 3, 297–303.

Tegen, I, Lacis, A.A. and Fung, I. (1996). The influence on climate forcing of mineral aerosols from disturbed soils. *Nature*, vol 380, 419–422.

Tett, S.F.B., Stott, P.A., Allen, M.R., Ingram, J.W. and Mitchell, J.F.B. (1999). Causes of twentieth-century temperature change near the Earth's surface. *Nature*, vol. 399, 569–572.

Thiede, J. and Mangerud, J. (1999). New map revises extent of Last Ice sheet over Barents and Kara seas. *EOS*, vol. 80, n° 42, 493–494.

Thieme, H. (2002). Greenhouse gaz hypothesis violates fundamentals of Physics. (http://people.freenet.de/klima/indexe.htm)
Thieme, H. (2002). On the phenomenon of atmospheric backradiation. (http://people.freenet.de/klima/indexe.htm)
Thieme, H. (2003). Does man really affect weather and climate? Are the interactions relly understood? (http://people.freenet.de/ klima/indexe.htm)
Thillet, J.J. and Joly, A. (1995). Débat sur le front polaire. *La Météorologie*, 12, 58–67.
Thomas, R.H. (2001). Remote sensing reveals shrinking Greenland ice sheet. *EOS*, vol. 82, n° 34, 369, 372–373.
Thompson, D.W.J. and Wallace, J.M. (1998). The Arctic oscillation signature in the winter time geopotential height and temperature fields. *Geophys. Res. Lett.*, 25, 1297–1300.
Thompson, D.W.J., Lee, S. and Baldwin, M.P. (2003). Atmospheric processes governing the Northern Hemisphere Annular Mode/North Atlantic Oscillation. The North Atlantic Oscillation, *Geophys. Monog.* 134, AGU, 81–112.
Thuillier, P. (1992). L'humanité saisie par l'effet de serre. *La Recherche*, 243, vol. 2 , 515–517.
Time (1972). 'Another Ice Age?' *Time*, 13 Nov., p. 81.
Time (1974). 'Another Ice Age?' *Time*, 26 June, p. 86.
Toupet, Ch. (1992). *Le Sahel.* Géographie, Nathan-Université, Paris, 192 p.
Trenberth, K.E. (1990). Recent observed interdecadal climate changes in the Northern Hemisphere. *Bull. Am. Met. Soc.*, vol. 71, n° 7, 988–993.
Trenberth, K.E. and Hurrell, J.W. (1994). Decadal atmosphere–ocean variations in the Pacific, *Climate Dynamics*, 9, 303–319.
Trenberth, K.E. (1997). The definitions of El Niño. *Bull. Am. Met. Soc.*, vol. 78, n° 12, 2771–2777.
Trenberth, K., Overpeck, J. and Salomon, S. (2004). Exploring drought and its implications for the future. *EOS*, vol. 85, n° 3, 27.
Triplet, J.P. and Roche, G. (1988, 1996). *Météorologie générale.* Ecole Nat. de la Météorologie, 317 p.
Trzaska, S. (2003). El Niño et la pluie en Afrique du Sud. *La Lettre de Medias*, n° 12.
Tyndall, J. (1861) On the absorption and radiation of heat by gases and vapours, and the physical connexion of radiation, absorption and conduction? *Philos. Magazine*, 22, 169–194, 273–285.
UNEP-WMO (2002). Climate Change. Information. *Rapport du GIEC (IPCC)*, Genève.
Uriarte Cantolla, A. (2003). *Historia del Clima de la Tierra.* Servicio Central de Publicaciones del Gobierno Vasco, Dir. de la Meteorologia y Climatologia, San-Sebastian, 306 p.
Usoskin, I.G., Solanki, S.M., Schüssler, M., Mursula, K. and Alanko K. (2003). A millenium scale sunspot number reconstruction: Evidence of an unusually active sun since the 1940's. arXiv: astro-ph/0310823v1, *APS/123-QED*, 1–4.
Van Loon, H. and Rogers, J.C (1978). The seasaw in winter temperatures between Greenland and northern Europe (Part I: general description). *Monthly Weather Review*, 22(9), 1949–1970.
Vaughan, D.G. and Doake, C.S.M. (1996). Recent atmospheric warming and retreat of ice shelves on the Antarctic Peninsula. *Nature*, vol. 379, 328–330.
Vaughan, D.G., Marshall, G.J., Connolley, W.M., King, J.C. and Mulvaney, R. (2001). Devil in the detail. *Science*, vol. 293, 1777–1779.
Vernadsky, V.I. (1945). The Biosphere and the Noösphere. *American Scientist*, Jan., pp. 1–12.
Vernin, Ch. (2003). Prévoir le temps prendra du temps. *Arts et Métiers Magazine*, 30–31.
Veum, T., Jansen, E., Arnold, M., Beyer, I. and Duplessy, J.C. (1992). Water mass exchange between the North Atlantic and the Norwegian Sea during the past 28,000 years. *Nature*, vol. 356, 783–785.

Veyre, C. (2000). L'évolution de la température en France de 1950 à 1995. *Mém. Lab. Clim., Risq., Env.*, LCRE, Univ. Lyon.

Vigneau, J.P. (2000). *Géoclimatologie*. Ellipses, Paris.

Vigouroux, G. (2004). Les hauts et les bas de la neige dans les Alpes françaises. *Revue de Géographie de Lyon, Géocarrefour*, 'Evénements extrêmes', J. Béthemont, M. Leroux (éd).

Vinnikov, K.Y. and Grody, N.C. (2003). Global warming trend of mean tropospheric temperature observed by satellites. *Science*, vol. 302, 269–272.

Visbeck, M. (2002). The ocean's role in Atlantic climate variability. *Science*, vol. 297, 2223–2224.

Visbeck, M., Chassignet, E.P., Curry, R.G., Delworth, T.L., Dickson, R.R. and Krahmann, G. (2003). The ocean's response to North Atlantic Oscillation variability. The North Atlantic Oscillation. *Geophys. Monograph 134, AGU*, 113–145.

Vivian, R. (2002). Glaciers et climats: ne faites pas dire aux glaciers ce qu'ils ne disent pas! (http://virtedit.online.fr/article1.html)

Von Grafenstein, U., Erlenkeuser, H., Muller, J., Jouzel, J. and Johnsen, S. (1998). The cold event 8,200 years ago documented in oxygen isotope records of precipitation in Europe and Greenland. *Climate. Dynamics*, 14, 73–81.

Von Storch, H., Güss, S. and Heimann, M. (1999). *Das Klimasystem und seine Modellierung*. Springer-Verlag, Berlin, p. 83.

Wagnon, P. and Vincent, Ch. (2004). Le changement climatique enregistré par les glaciers. L'observatoire GLACIOCLIM. *Lettre n° 16, PIGB-PMRC*, CNRS.

Walker, G.T. and Bliss, E.W. (1932). World Weather V, *Memoirs of the R. Met. Soc.*, 4, n° 36, 53–84.

Wallace, J.M. and Gutzler, D.S. (1981). Teleconnections in the geopotential height field during the Northern Hemisphere winter. *Monthly Weather Review*, 109, 784–812.

Wallace, J.M. and David Thompson, W.J. (2002). Annular Modes and Climate Prediction. *Physics Today*, Feb., pp. 28–33.

Wallen, C.C. (1984). *Present Century Climate Fluctuations in the Northern Hemisphere and Examples of Their Impact*. WCP-87, WMO/TD, No. 6, WMO, 85 p.

Walsh, J.E., Chapman, W.L. and Shy, T.L. (1996). Recent decrease of sea level pressure in the central Artic. *Journ. of Climate*, n° 9, 480–485

Wanner, H. (1999). Le balancier de l'Atlantique Nord. *La Recherche*, 321, 72–73.

Waple, A.M. (1999). The sun-climate relationship in recent centuries: A review. *Progress in Physical Geography*, 23, 3, 309–328.

Ward and Dubos (1972). *Only One Earth: the care and maintenance of a small planet*. Norton.

WASA Group (1998). Changing waves and storms in the Northeast Atlantic? *Bull. of the Am. Met. Soc.*, vol. 795, 741–760.

WCED (1987). World Commission on Environment and Development. *Our common future*. Oxford University Press.

Weart, S.R. (1997). The discovery of the risk of Global Warming. *Physics Today*, January, 34–40.

Weart, S.R. (2003). *The discovery of Global Warming*. Harvard University Press. Simple models of climate. General Circulation Models of Climate.]www.aip.org/history/climate). In *Physics Today*, n° 8, 30–32.

Wellington, G.M., Glynn, P.W., Strong, A.E., Navarrete, S.A., Wieters, E. and Hubbard, D. (2001). Crisis on coral reefs linked to climate change. *EOS*, vol. 82, n° 1, 1–3.

White, J.W.C. and Steig, J.C. (1998). Timing is everything in a game of two hemispheres. *Nature*, vol. 394, 717–718.

Whitfield, J. (2003). News feature: Too hot to handle. *Nature*, vol. 425, 338–339.

Whitten, R.C. (2001). More heat over greenhouse gases. *Physics Today*, 19.
Wielicki, B.A., Cess, R.D., King, M.D., Randall, D.A. and Harrison, E.F. (1995). Mission to planet Earth: Role of clouds and radiation in climate. *Bull. Am. Met. Soc.*, vol. 76, n° 11, 2125–2153.
Wigley, T.M.L. and Raper, S.C.B. (1987). Thermal expansion of sea water associated with global warming. *Nature*, vol. 330, 127–131.
Wild, M. and Ohmura, A. (1999). The role of clouds and the cloud-free atmosphere in the problem of underestimated absorption of solar radiation in GCM atmospheres. *Physics and Chemistry of the Earth*, 24B: 261–268,
Wijngaard, J.B., Klein Tank, A.M.G. and Können, G.P. (2003). Homogeneity of 20th century European daily temperature and precipitation series. *Intern. Journal of Climatology*, vol. 23(6), 679–692.
White, R.W. (1989). Greenhouse policy and climate uncertainty. *Bull. Am. Met. Soc.*, 70, n° 9, 1123–1128.
WMO (1979). World climate conference. A conference of experts on climate and mankind. Declaration and supporting documents, Geneva.
WCP-WMO (1981). On the assessment of the role of CO_2 on climate variations and their impact. Report of meeting of experts (WMO/UNEP/ICSU) in Villach, Austria, 1980, World Climate Programme, Geneva.
WMO-WCP (1981). On the assessment of the role of CO_2 on climate variations and their impact (Villach, Austria, November 1980), *WMO/ICSU/UNEP*, Geneva.
WMO-WCP (1983). WMO project on research and monitoring of atmospheric CO_2. Report of the JSC/CAS meeting of experts on detection of possible climate change. *WCP-29*, ICSU-WMO, Geneva.
WMO-WCP (1986). Report of the International Conference on the assessment of the role of carbon dioxide and of other greenhouse gases in climate variations and associated impacts. Villach, Austria, 9–15 October 1985, WCP, WMO, n° 661.
WMO-WCP (1988). Developing policies for responding to climate change. A summary of the discussions and recommandations of the workshops held in Villach (Sept.–Oct. 1987) and Bellagio (Nov. 1987) under the auspices of the Beijer Institute, Stockholm. WCIP-1, WMO/TD-N 225, *WMO-UNEP*, Geneva.
WMO. Region VI (each year). *Annual Bulletin of the Climate, Europe and Middle east.*
WMO. (1993 to 2004). *WMO statement on the status of the global climate in...* (yearly, since 1993).
WMO (2004). The 1997–1998 El Niño event in brief. (`www.wmo.ch`)
Woodhouse, C.A. and Overpeck, J.T. (1998). 2000 years of drought variability in the central United States. *Bull. Am. Met. Soc.*, vol. 79, n° 12, 2693–2714.
Woodworth, P.L., Aarup, Th., Merrifield, M., Mitchum, G.T. and Le Provost, C. (2003). Measuring progress of the global sea level observing system. *EOS*, vol. 84, n° 50, 565.
Wu, P., Wood, R. and Stott, P. (2004). Does the recent freshening trend in the North Atlantic indicate a weakening thermohaline circulation? *Geophysical Research Letters*, 31.
Wyrtki, K. (1979). El Niño. *La Recherche*, 10, 1212–1220.
Zhisheng, A., Porter, S.C., Weijian, Z., Yanchou, L., Donahue, D.J., Head, M.J., Xihuo, W., Jianzhang, R. and Hongbo, Z. (1993). Episode of strengthened summer monsoon climate of Younger Dryas age on the loess plateau of central China. *Quaternary Research*, 39, 45–54.

Index

Printing: Mercedes-Druck, Berlin
Binding: Stein + Lehmann, Berlin